THE BALTIC SEA

FURTHER TITLES IN THIS SERIES

1 J.L. MERO
THE MINERAL RESOURCES OF THE SEA
2 L.M. FOMIN
THE DYNAMIC METHOD IN OCEANOGRAPHY
3 E.J.F. WOOD
MICROBIOLOGY OF OCEANS AND ESTUARIES
4 G. NEUMANN
OCEAN CURRENTS
5 N.G. JERLOV
OPTICAL OCEANOGRAPHY
6 V. VACQUIER
GEOMAGNETISM IN MARINE GEOLOGY
7 W.J. WALLACE
THE DEVELOPMENT OF THE CHLORINITY/SALINITY CONCEPT IN OCEANOGRAPHY
8 E. LISITZIN
SEA-LEVEL CHANGES
9 R.H. PARKER
THE STUDY OF BENTHIC COMMUNITIES
10 J.C.J. NIHOUL (Editor)
MODELLING OF MARINE SYSTEMS
11 O.I. MAMAYEV
TEMPERATURE-SALINITY ANALYSIS OF WORLD OCEAN WATERS
12 E.J. FERGUSON WOOD and R.E. JOHANNES (Editors)
TROPICAL MARINE POLLUTION
13 E. STEEMANN NIELSEN
MARINE PHOTOSYNTHESIS
14 N.G. JERLOV
MARINE OPTICS
15 G.P. GLASBY (Editor)
MARINE MANGANESE DEPOSITS
16 V.M. KAMENKOVICH
FUNDAMENTAL OF OCEAN DYNAMICS
17 R.A. GEYER (Editor)
SUBMERSIBLES AND THEIR USE IN OCEANOGRAPHY AND OCEAN ENGINEERING
18 J.W. CARUTHERS
FUNDAMENTALS OF MARINE ACOUSTICS
19 J.C.J. NIHOUL (Editor)
BOTTOM TURBULENCE
20 P.H. LEBLOND and L.A. MYSAK
WAVES IN THE OCEAN
21 C.C. VON DER BORCH (Editor)
SYNTHESIS OF DEEP-SEA DRILLING RESULTS IN THE INDIAN OCEAN
22 P. DEHLINGER
MARINE GRAVITY
23 J.C.J. NIHOUL (Editor)
HYDRODYNAMICS OF ESTUARIES AND FJORDS
24 F.T. BANNER, M.B. COLLINS and K.S. MASSIE (Editors)
THE NORTH-WEST EUROPEAN SHELF SEAS: THE SEA BED AND THE SEA IN MOTION
25 J.C.J. NIHOUL (Editor)
MARINE FORECASTING
26 H.-G. RAMMING and Z. KOWALIK
NUMERICAL MODELLING OF MARINE HYDRODYNAMICS
27 R.A. GEYER (Editor)
MARINE ENVIRONMENTAL POLLUTION
28 J.C.J. NIHOUL (Editor)
MARINE TURBULENCE
29 M. WALDICHUK, G.B. KULLENBERG and M.J. ORREN (Editors)
MARINE POLLUTANT TRANSFER PROCESSES

Elsevier Oceanography Series, 30

THE BALTIC SEA

edited by

AARNO VOIPIO

Institute of Marine Research, Helsinki, Finland

ELSEVIER SCIENTIFIC PUBLISHING COMPANY

Amsterdam — Oxford — New York 1981

ELSEVIER SCIENTIFIC PUBLISHING COMPANY
1, Molenwerf 1014 AG Amsterdam
P.O. Box 211, 1000 AE Amsterdam, The Netherlands

Distributors for the United States and Canada:

ELSEVIER/NORTH-HOLLAND INC.
52, Vanderbilt Avenue
New York, N.Y. 10017

Library of Congress Cataloging in Publication Data

Main entry under title:

The Baltic Sea.

(Elsevier oceanography series ; 30)
Includes bibliographies and index.
1. Oceanography--Baltic Sea. 2. Marine biology--Baltic Sea. 3. Fisheries--Baltic Sea. 4. Marine pollution--Baltic Sea. I. Voipio, Aarno, 1928-
GC571.B24 551.46'134 80-17385
ISBN 0-444-41884-9

ISBN 0-444-41884-9 (Vol. 30)
ISBN 0-444-41623-4 (Series)

Printed in The Netherlands

PREFACE

The Baltic Sea is neither an ocean nor a lake, but a large brackish-water basin with very pronounced density stratification prevailing the whole year round. In addition to this, its recent geological history has been very complicated, resulting in profound changes in the hydrographic conditions and subsequently also in the biological features of this sea during the last ten thousand years.

The above special features make it rather difficult for a deep-sea oceanographer to find the marked differences in physical properties between the shallow Baltic Sea and, for instance, the shelf seas with a similar mean depth. A limnologist, on the other hand, seems to have equal problems in remembering that in the Baltic Sea the thermal convection never extends to the bottom in basins whose depth is greather than the mean.

The surprisingly great difficulties encountered in discussing the Baltic Sea conditions which colleagues representing either deep-sea oceanography or limnology gave me an incentive to accept the kind invitation of the Elsevier Scientific Publishing Company to edit a volume on the Baltic Sea for inclusion in their Oceanography Series.

The publication of the present volume has also made it possible to present a large body of unpublished material available in Finland and Sweden, using it to expand the summary of the rather fragmentary earlier studies on the geology of this sea. I was also encouraged by the fact that most of the persons from whom I requested contributions on the other and perhaps better known areas of Baltic marine sciences agreed without hesitation to participate in this project.

The only major disadvantage in having several authors has been the difficulty of synchronizing the delivery of the manuscripts. As the editor, I wish to express my sympathy with those contributors who were able to send in their manuscripts within the agreed time. I also hope that the reader will understand that it has been impossible to update those articles which were completed before the original deadlines.

It is my pleasant duty to record my gratitude to the late Professor Ilmo Hela and to Professor Kalervo Rankama, who persuaded me to accept the task of editor. Professor Rankama has always been ready to advise me on the work and has checked the English language of the manuscripts. Ms. Mirja

Ristola and Ms. Terttu Someroja have given me indispensable help in the editorial work.

Finally, I wish to extend my warm thanks to my wife, Raija, for her constant understanding and encouragement during this work.

January 1980
AARNO VOIPIO

CONTENTS

Preface V
List of contributors XIII

Chapter 1. GEOLOGY OF THE BALTIC SEA
B. Winterhalter, T. Flodén, H. Ignatius, S. Axberg and L. Niemistö

A. Pre-Quaternary geology of the Baltic Sea (T. Flodén and B. Winterhalter)
Introduction 1
Bothnian Bay 23
Bothnian Sea 25
Åland Sea 28
Gulf of Finland 31
Baltic Proper 33
B. Quaternary geology of the Baltic Sea (H. Ignatius, S. Axberg, L. Niemistö and B. Winterhalter)
Introduction 54
Evolution of the Baltic Sea 58
Stratigraphy of the clay sediments 63
Geomorphology of the Baltic Sea floor 69
Quaternary sediments of the Baltic Sea 86
C. Natural resources (B. Winterhalter)
Hydrocarbons 105
Ferromanganese concretions 107
Amber, phosphorite and glauconite 114
Sand and gravel 110
Placer deposits 116
References 117

Chapter 2. HYDROLOGY OF THE BALTIC SEA
U. Ehlin

Hydromorphology 123
River inflow 125
Precipitation and evaporation 126
Water transport through the Danish Sounds 129

Water storage and water exchange 133
References 133

Chapter 3. PHYSICAL OCEANOGRAPHY
G. Kullenberg

Introduction 135
Salinity and temperature distribution 135
Salinity. 135
Temperature 139
Long-term variations 143
Causes of long-term variations 147
Density stratification and its variability 149
Circulation 150
Mean circulation 150
Time-dependent motion 151
Coastal boundary layer 155
Theoretical considerations 157
Mixing conditions . . . .
Small-scale motion 160
Vertical and horizontal motion 162
Optical properties, heat balance and ice conditions 167
Optical properties . . . .
Heat balance 170
Ice conditions 174
References 175

Chapter 4. CHEMICAL OCEANOGRAPHY
K. Grasshoff and A. Voipio

Anomalies in the composition of the Baltic Sea water 183
Distribution of dissolved oxygen 188
Nutrients 194
Trace metals. 203
Sediment—water interactions 205
Dissolved organic matter 211
References 213

Chapter 5. BIOLOGICAL OCEANOGRAPHY
G. Hällfors, Å. Niemi, H. Ackefors, J. Lassig and E. Leppäkoski

A. Introduction (G. Hällfors and Å. Niemi) 219
B. Vegetation and primary production (G. Hällfors and Å. Niemi). . . 220

Phytoplankton, general 220
Phytoplankton production, succession and regulating factors . . 221
Phytoplankton in near-shore areas 229
Benthic vegetation, general aspects 229
Zonation of the benthic vegetation 232
Littoral primary production 236
Trophic status of the Baltic Sea 237
C. Zooplankton (H. Ackefors) 238
Introduction 238
Sampling technique 239
Composition of the fauna 242
Environment and fauna 245
Vertical distribution and diel migration 249
Food web 251
Dynamics of the plankton community 251
D. Benthic fauna of the Baltic Sea (J. Lassig and E. Lappäkoski) . . . 254
Origin of the benthic fauna 254
Littoral zone 256
Sublittoral zone 257
Production and utilization of zoobenthos 262
Benthos research 264
References 265

Chapter 6. FISHES AND FISHERIES
E. Ojaveer, A. Lindroth, O. Bagge, H. Lehtonen and J. Toivonen

A. Fish fauna of the Baltic Sea (E. Ojaveer) 275
B. Marine pelagic fishes (E. Ojaveer) 276
Geographical distributions and groups 276
Seasonal distribution patterns and migrations 279
Spawning, larval and adolescent phase 281
Fecundity 284
Feeding 285
Growth and age 287
Year-class abundance 288
Catches and mortality 290
Other fishes 291
C. Anadromous and catadromous fishes (A. Lindroth) 292
Salmon 293
Sea-running brown trout 303
Grayling 306
Whitefish 306
Vimba 307

Lamprey 308
European eel 309
D. Demersal fishes (O. Bagge) 311
Cod 312
Flounder 320
Plaice 323
Turbot 325
Brill 327
Dab. 327
Sandeel 328
Greater sandeel 328
Snake blenny 329
Four-bearded rockling 329
Father lasher 330
Sea scorpion 330
Four-horned cottus 330
Eelpout 331
Black goby 331
Sand goby 331
Lumpsucker 332
Sea snail 332
Gold sinny 333
Butterfish 333
E. Fresh-water fishes (H. Lehtonen and J. Toivonen) 333
Species composition and distribution. 333
Migrations 338
Spawning grounds and reproduction 338
Growth 339
Fishery and catches 340
References 341

Chapter 7. POLLUTION
B.I. Dybern and S.H. Fonselius

Introduction. 351
Global aspects of pollution 351
Spreading and accumulation of pollutants in the Baltic Sea . . . 352
Sensitivity of the Baltic ecosystems 353
Eutrophication. 353
Waste discharge from communities and industries 353
Oxygen utilization 354
Effects on the ecosystems 358
Toxic matter. 362
Organochlorine compounds 362

Metals. 365
Oil pollution. 369
Radioactive pollution 371
Physical pollution. 372
Warm-water effects. 372
Solid waste. 373
Extraction of sand and gravel. 374
Other kinds of physical pollution 375
References 376

Chapter 8. INTERNATIONAL MANAGEMENT AND COOPERATION
V. Sjöblom and A. Voipio

A. International management of the Baltic Sea fisheries (V. Sjöblom) . 383
B. International cooperation as the basis of the protection of the marine environment of the Baltic Sea area (A. Voipio) 386

Author index 391
Subject index 405

LIST OF CONTRIBUTORS

H. ACKEFORS	University of Stockholm, Department of Zoology, Box 6801, S-113 86 Stockholm, Sweden
S. AXBERG	University of Stockholm, Department of Geology, Box 6801, S-113 86 Stockholm, Sweden
O. BAGGE	Danmarks Fiskeri- og Havundersøgelser, Charlottenlund Slot, DK-2920 Charlottenlund, Denmark
B.I. DYBERN	National Board of Fisheries, Institute of Marine Research, Biological Department, S-453 00 Lysekil, Sweden
U. EHLIN	Swedish Meteorological and Hydrological Institute, Fack, S-601 01 Norrköping, Sweden
T. FLODÉN	University of Stockholm, Department of Geology, Box 6801, S-113 86 Stockholm, Sweden
S.H. FONSELIUS	National Board of Fisheries, Institute of Marine Research, Hydrographic Department, Box 2566 S-403 17 Göteborg, Sweden
K. GRASSHOFF	Institut für Meereskunde an der Universität Kiel, 23 Kiel, Düsternbrooker Weg 20, Federal Republic of Germany
G. HÄLLFORS	University of Helsinki, Tvärminne Zoological Station, SF-10850 Tvärminne, Finland
H. IGNATIUS	Geological Survey of Finland, SF-02150 Espoo 15, Finland
G. KULLENBERG	University of Copenhagen, Institute of Physical Oceanography, Haraldsgade 6, DK-2200 Copenhagen N, Denmark
J. LASSIG	Institute of Marine Research, P.O. Box 166, SF-00141 Helsinki 14, Finland
H. LEHTONEN	Institute of Finnish Game and Fisheries Research, Fisheries Division, Box 193, SF-00131 Helsinki 13, Finland
E. LEPPÄKOSKI	Åbo Akademi, Porthansgatan 3-5, SF-20500 Åbo 50, Finland
A. LINDROTH	Rättviksvägen 29, S-161 42 Bromma, Sweden (formerly: University of Umeå, Institute of Ecological Zoology, Umeå, Sweden)

Å. NIEMI	University of Helsinki, Department of Botany, Ecological laboratory, Apollonkatu 5 B 45, SF-00100 Helsinki 10, Finland
L. NIEMISTÖ	Institute of Marine Research, P.O. Box 166, SF-00141 Helsinki 14, Finland
E. OJAVEER	Tallinn Department of the Baltic Fishery Research Institute, Apteegi 1-2, 200001 Tallinn, Esthonian SSR, USSR
V. SJÖBLOM	University of Helsinki, Department of Limnology, Viikki, SF-00710 Helsinki 71, Finland
J. TOIVONEN	Institute of Finnish Game and Fisheries Research, Fisheries Division, P.O. Box 193, SF-00131 Helsinki 13, Finland
A. VOIPIO	Institute of Marine Research, P.O. Box 166, SF-00141 Helsinki 14, Finland
B. WINTERHALTER	Geological Survey of Finland, SF-02150 Espoo 15, Finland

Chapter 1

GEOLOGY OF THE BALTIC SEA

BORIS WINTERHALTER, TOM FLODÉN, HEIKKI IGNATIUS, STEFAN AXBERG and LAURI NIEMISTÖ

A. PRE-QUATERNARY GEOLOGY OF THE BALTIC SEA*

Introduction

The Baltic Sea including its adjoining gulfs (Fig. 1.1), as we know it today, fills a complex depression within the East European platform and its southwestern border zone. The Precambrian crystalline basement of the Baltic Shield is exposed along the major part of the western, northern and eastern coasts of the Baltic Sea (Fig. 1.2). Only the southeastern and southern part of the present-day marine area exhibits a coastline consisting of sedimentary rocks being part of the East European sedimentary complex (Fig. 1.3a, b).

The crystalline basement of the Baltic region is represented by various metamorphic and igneous rocks referable to Svecokarelian, and Gothian orogenies. Anorogenic rapakivi—granite intrusions are common in several localities both as supramarine (Vorma, 1976) and submarine outcrops (e.g., Winterhalter, 1967). They were formed during Middle Proterozoic (1.65 Ga) postorogenic activities. For a comprehensive presentation of the Precambrian the reader is referred to Rankama (1963).

Prior to the evolution of the sub-Cambrian peneplane (cf. pp. 31 and 36) a more or less uniform deposition of sandstones, generally reddish arkose and siltstones occurred in Late Proterozoic. Some of the sandstones are known to exhibit dyke intrusions of diabase. The sedimentary deposits are often referred to as Jotnian. Their age is approximately 1.3 Ga (Middle Riphean) according to Simonen (1971). Today the erosional remnants of these unmetamorphosed sedimentary rocks are found exposed both on the sea floor and on the adjacent land areas (Winterhalter, 1972; Flodén, 1973).

The intense postdepositionary erosion leading to the formation of the sub-Cambrian peneplane explains the rather thin and patchy distribution of the Proterozoic sedimentary rocks. These rocks have only been preserved in tectonic depressions, where they often attain considerable thicknesses, e.g. the Satakunta and Gävle sandstones on land, and the submarine deposits in the Åland Sea (p. 28) and the Landsort Deep (p. 36).

* By Tom Flodén and Boris Winterhalter.

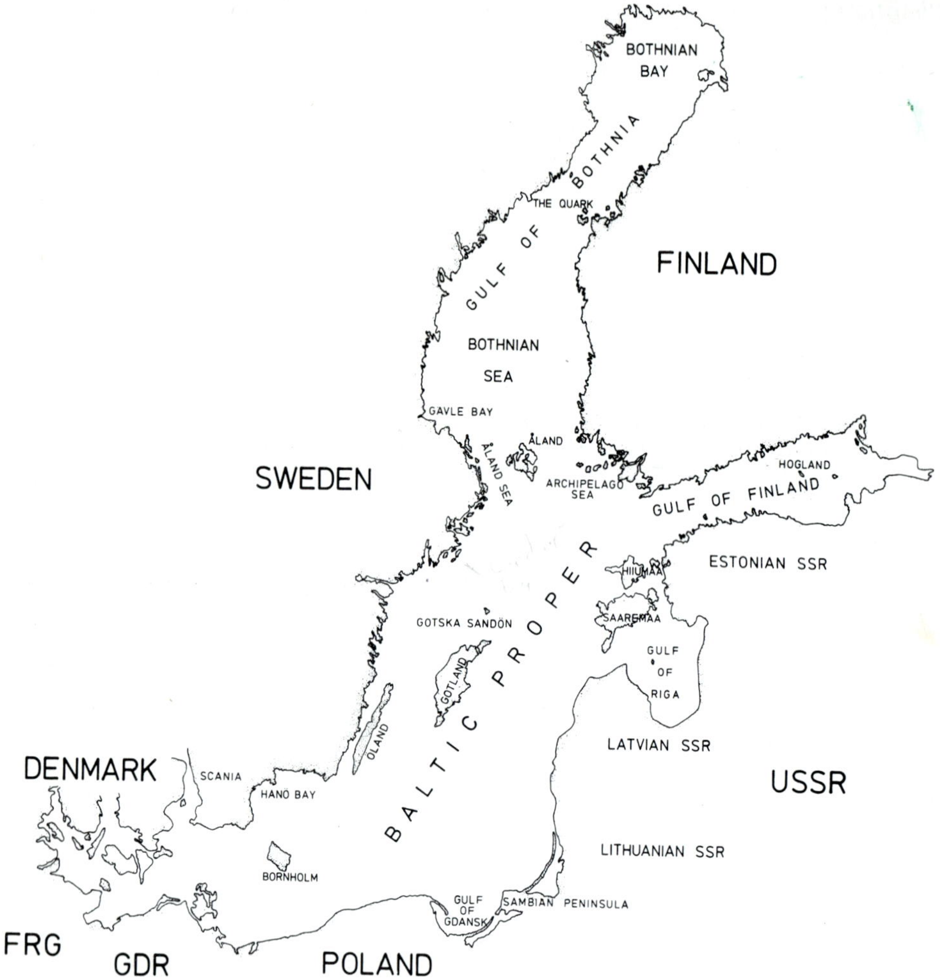

Fig. 1.1. Index map of the Baltic Sea region.

A "blue clay" sequence found outcropping NW of Leningrad in the eastern part of the Gulf of Finland was later observed in several drillings in, e.g., Esthonia and Latvia. This sequence was formerly confused with the Lower Cambrian Blue Clay, but now it is known that it is of Late Precambrian age constituting part of the arenaceous—argillaceous Vendian sedimentary rocks. The Muhos-Formation in Finland and its submarine extension in the Bothnian Bay was considered by Veltheim (1969) to be of Jotnian (Middle Riphean) age. Current investigations of the sediments cored on the Hailuoto Island indicate an upper Riphean or even Vendian age (R. Tynni, pers. commun., 1979.) possibly comparable with the Vendian of Estonia.

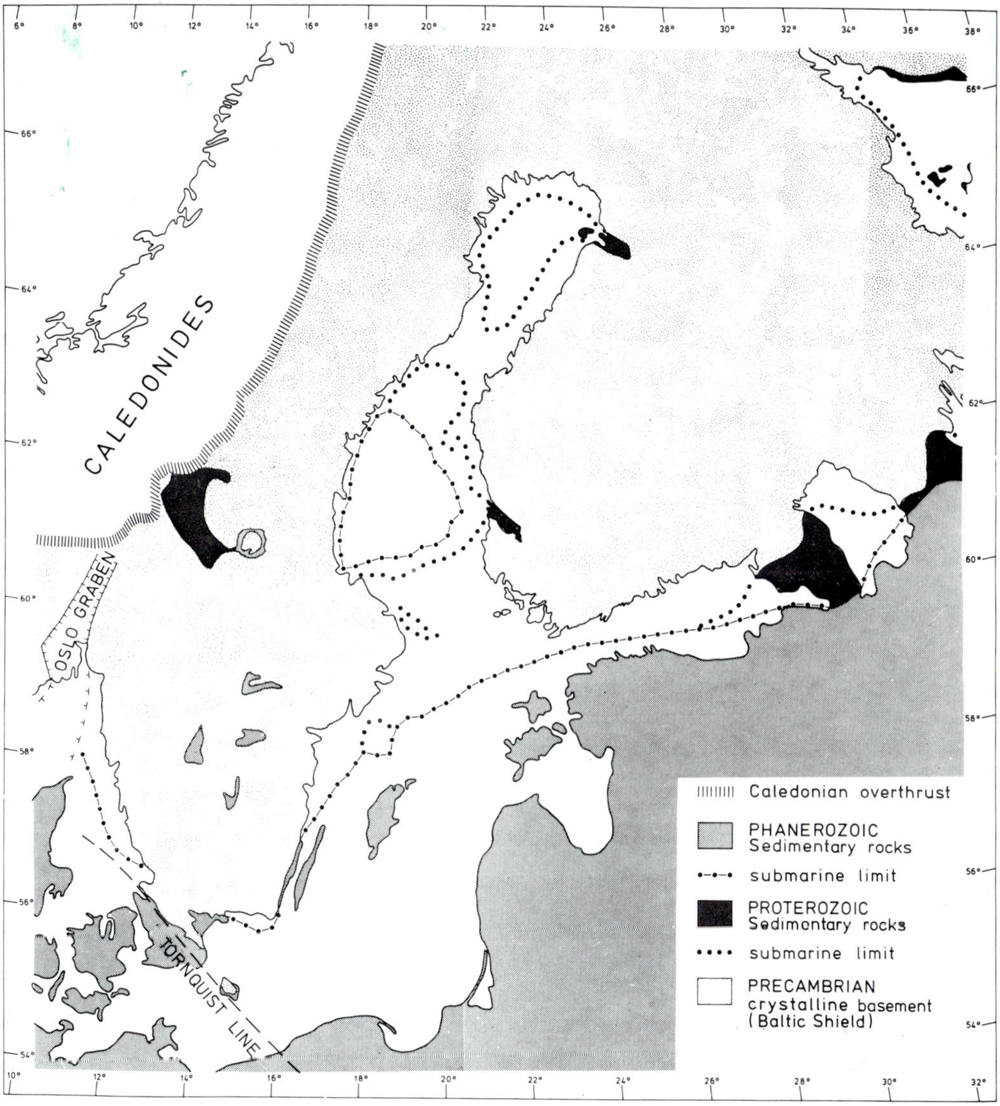

Fig. 1.2. Baltic Shield and the northwestern part of the East European Platform bounded in the west by the Caledonides and in the southwest by the Tornquist Line.

The Early Paleozoic seas exhibiting a multitude of depositional and erosional phases covered the major part of the present-day Baltic Sea region. The exact maximum extent of Paleozoic sedimentation in the north is unknown. However, Cambrian and Ordovician sedimentary rocks occur in situ in the Bothnian Sea, and Cambrian sandstone erratics have been found along the coast of the Bothnian Bay. The sedimentary rock sequence in the Bothnian Bay may in addition to the Muhos-Formation contain in the north central part of the Bay beds of Lower Cambrian sedimentary rocks.

A new occurrence of Lower Cambrian sedimentary rocks has been found in the coastal area of the NE Bothnian Sea, just south of the town of Vaasa (Laurén et al., 1978). The total thickness of the deposit has been estimated at 300—400 m. Slightly over 100 m was penetrated by drilling. The recovered sedimentary sequence is comparable with the Lontova and Lükati beds of Esthonia.

In the Baltic Sea, the total thickness of the Paleozoic sequence increases rapidly towards the southeast measuring over 3 km off the Polish coast. Mesozoic and Tertiary sedimentary rocks are known only from the southern part of the Baltic Sea.

Within a large part of the Baltic Sea the sedimentary strata are more or less horizontal suggesting very little postdepositional deformation. Numerous shallow block faults and fractures have, however, been revealed by continuous seismic profiling. In contrast, strong block faulting has occurred most dramatically along the "Tornquist Line" (Tornquist, 1913), the major fracture zone that denotes the southwest border of the East European Platform (see Fig. 1.2). Along the Tornquist Line, from Scania in the NW to the Polish coast in the SE, the maximum vertical displacement has been estimated to exceed 7000 m.

The major tectonic features: faults, fractures, and lineaments are shown in the maps, Fig. 1.4a, b. The figures are based on data gathered both from literature (e.g., Härme, 1961; Tuominen et al., 1973), bathymetric maps and available seismic reflection and refraction profiles. The detection of faults and fractures in seismic and echosounding profiles is seldom difficult (Fig. 1.12). However, the rather limited number of available profiles and the fact that the profiles often transect possible lineaments at small angles makes it obvious that only the most persistent features can be reliably evaluated and have been included in the maps. Whenever possible the downthrown side is also noted.

No attempt has been made to interpret the ages of the various fractures and faults noted on the map. It is clear, however, that many of the major lineaments denote tectonic zones, that have been activated during several geological events since the Archean. Some of the tectonic events can be related to the various orogenies of northern and central Europe. Epeirogenic movements have also obviously been very active, not forgetting the Late Pleistocene and Holocene crustal uplift still active within the Baltic region (Fig. 1.5). It has a maximum annual rate of almost 10 mm in the central part of the Bothnian Bay. The distribution of earthquake epicentres is shown in Fig. 1.6. Large-scale neotectonic faulting seems, however, to be very scarce. Vertical displacements of some meters have been observed in acoustic profiles only rarely (Fig. 1.7).

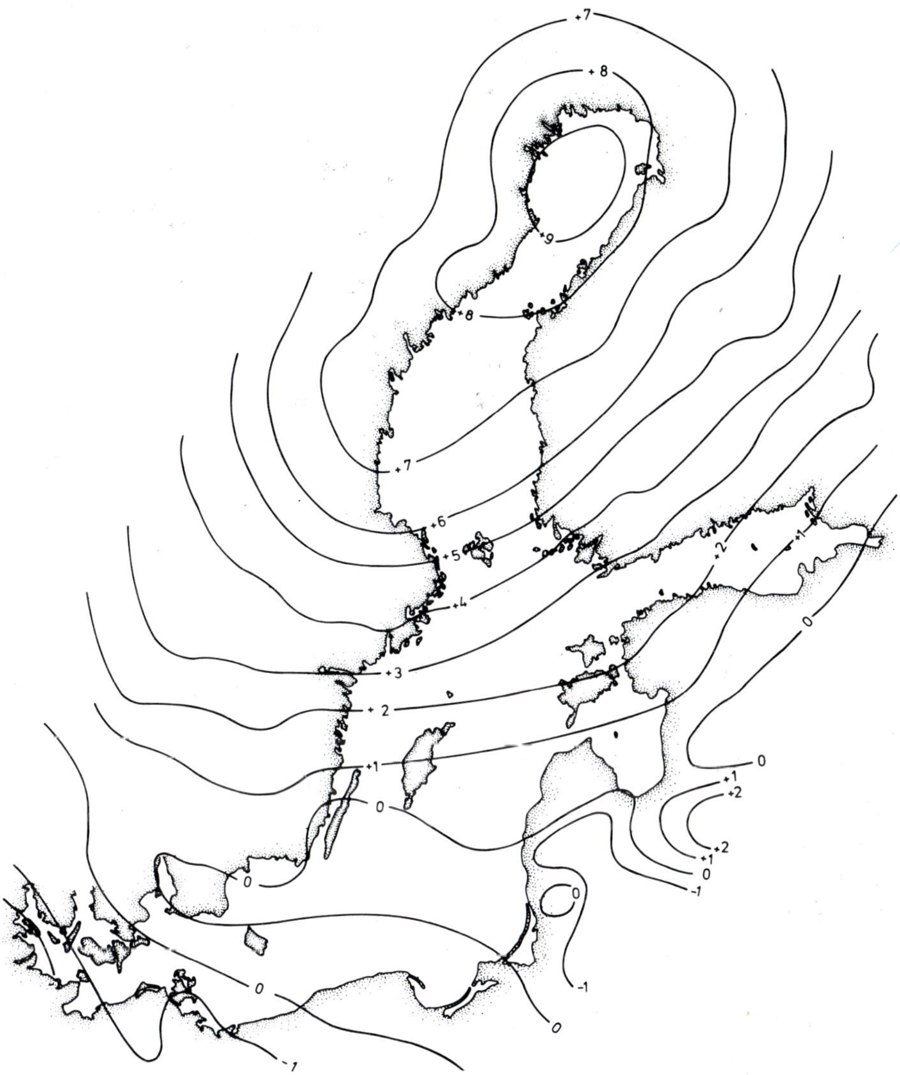

Fig. 1.5. Present-day relative crustal uplift (and submergence) in the Baltic Sea region. The isobases (mm a^{-1}) are based on precise levellings, and tide-gauge data from a number of papers (e.g., Boulanger et al., 1975; Kukkamäki, 1975; Lillienberg et al., 1975; Liszkowski, 1975; Bergqvist, 1977; Mörner, 1977).

The following, more detailed, description of the pre-Quaternary geology of the Baltic Sea and its adjoining gulfs is based on the results of marine geological research conducted by the Marine Geological Department of the University of Stockholm and the Geological Survey of Finland. Further information has been acquired from published papers and unpublished

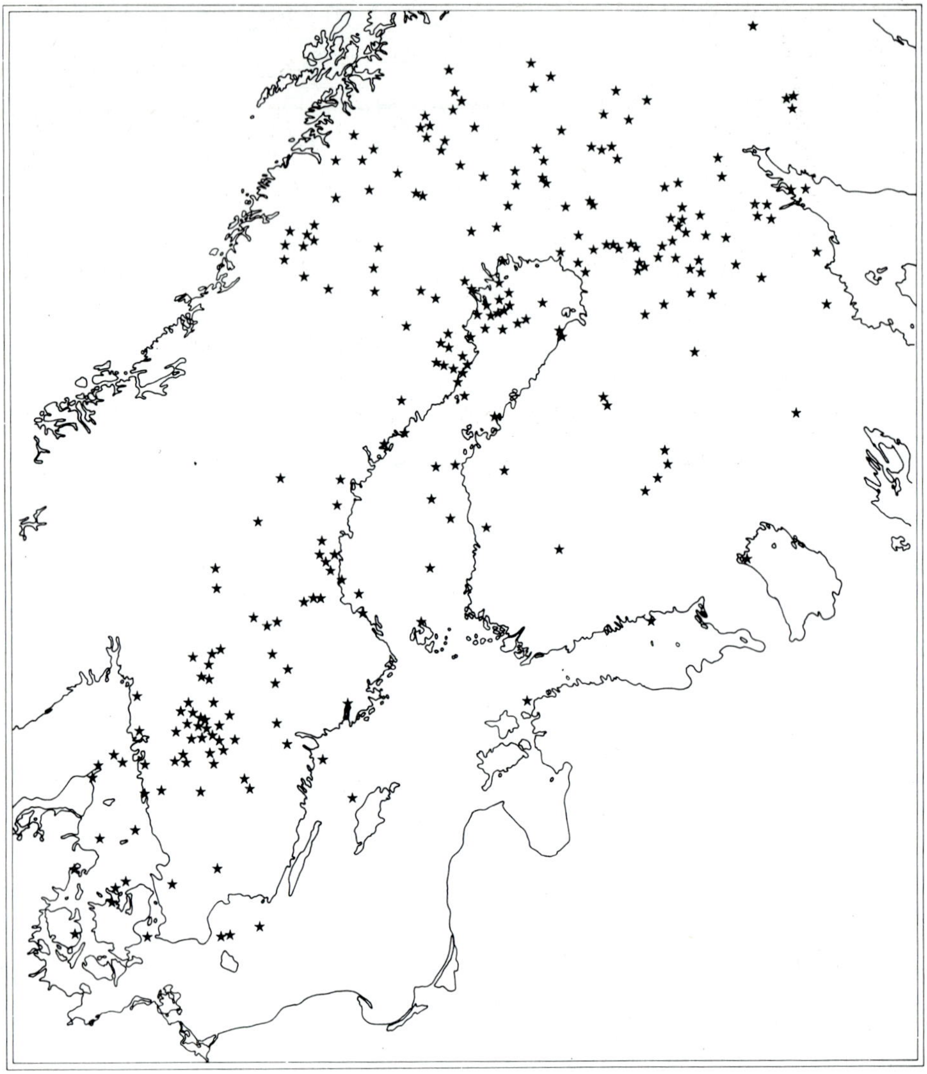

Fig. 1.6. Seismicity of the Baltic Sea and its adjoining land areas based on earthquake epicenter data from Panasenko (1977) and Penttilä (1978). Norwegian and offshore Norwegian earthquakes are omitted.

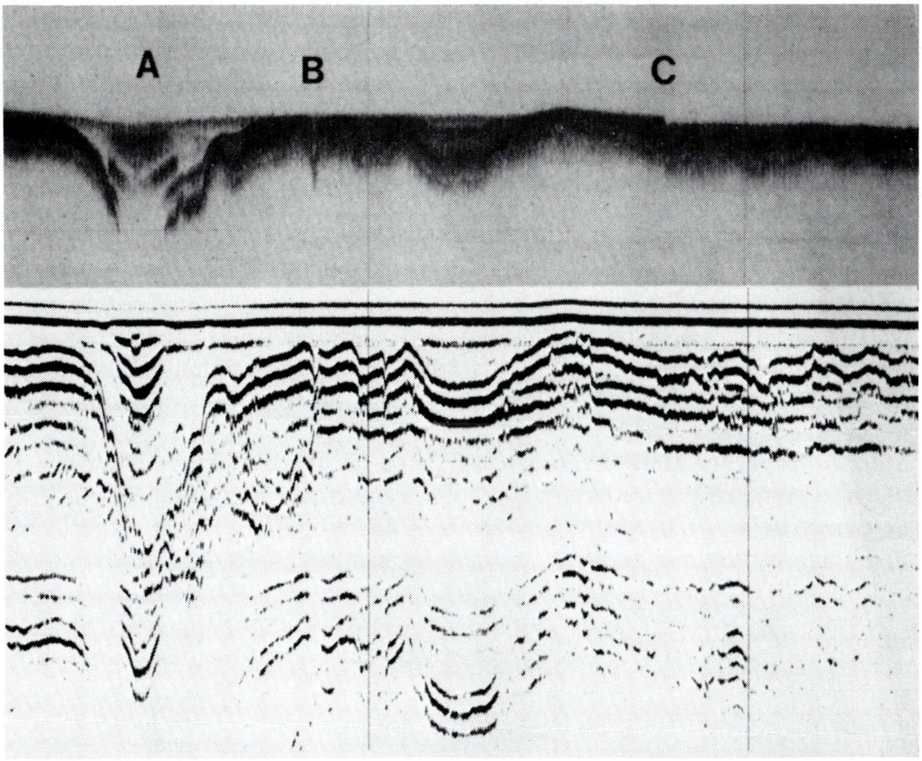

Fig. 1.7. Tectonic elements in an area south of Öland. The upper figure is an echo sounding profile and the lower one is a reflection profile. *A* = lineament in the sedimentary bedrock; *B* and *C* = features in Quaternary strata that may be associated with neotonic movements.

manuscripts updated by personal communications with many colleagues working within the Baltic Sea.

Bothnian Bay

The crystalline complex forming the outcropping bedrock in the adjacent land areas does not exhibit any fundamental differences between the two sides of the Bothnian Bay. Svecokarelian granites and gneisses dominate, except for the northern shore of the Bay, where the bedrock is characterized by Karelian schists. The crystalline complex can be assumed to continue under the Bothnian Bay without any significant variations.

The existence of erosional remnants of unmetamorphosed sedimentary rocks on the bottom of the Bothnian Bay, placewise covering the crystalline basement, has been anticipated for a long time due to the numerous finds of erratics along the shores of the Bay. Likewise, the Proterozoic Muhos-Formation (Tynni, 1978) found both on the Finnish mainland and on the Hailuoto Island in the NE part of the Bay was assumed to have a submarine continuation (Veltheim, 1969). Subsequent seismic reflection profiling and refraction shooting have verified this assumption but, in addition, they have in fact shown that a major part of the bottom of the Bothnian Bay is covered by sedimentary rocks. Sound-velocity measurements indicate considerable lithological variations. These are comparable both with the transitions between sandy and silty beds observed in the drill cores from the Muhos-Formation on the Hailuoto Island and those found in the Bothnian Sea. The total thickness of the sedimentary rock strata seems to exhibit considerable variations in different parts of the marine area, varying from depressions with several hundred meters of sediments to thin erosional remains as outliers smoothening out the otherwise rather irregular relief of the crystalline basement.

Due to the rather limited geological data so far available from the Bothnian Bay, the extent of the sedimentary bedrock denoted on the map in Fig. 1.3a should not be taken literally but more as an indication of the relation between the sedimentary bedrock and the crystalline basement. The complexity of the Pleistocene deposits covering the sedimentary rocks makes interpretation of the available reflection profiles very difficult (Fig. 1.8).

Except for some erratics of evidently Cambrian Sandstone discovered along

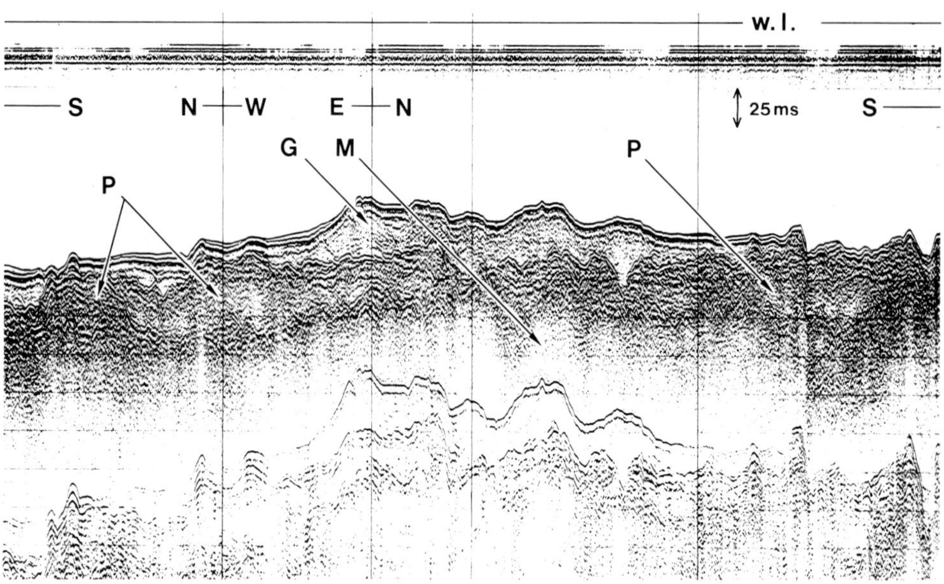

Fig. 1.8. Part of a reflection profile showing the complicated nature of the sedimentary deposits (and bedrock) in the Bothnian Bay. *G* = glacial drift; *P* = pre-Weichselian sediments; *M* = Middle or Late Proterozoic sedimentary rock (see text); *w.l.* = water line, i.e., sea surface. Length of profile is approximately 20 km.

the eastern shore of the Bothnian Bay indicating the probable existence of Lower Cambrian deposits within the submarine area, no younger sedimentary rocks have as yet been ascertained. The available reflection profiles do point towards the existence of Cambrian sedimentary rocks in the north-central part of the Bay.

Bothnian Sea

In contrast to the Bothnian Bay, the geology of the Bothnian Sea has been studied in considerable detail (Veltheim, 1962; Winterhalter, 1972; Thorslund and Axberg, 1979). Thus the various bedrock boundaries given in the map (Fig. 1.3a) are based on rather reliable data. In fact, the only difficulties encountered in the evaluation of the available marine seismic data were to distinguish between the crystalline basement and the Late Proterozoic arkose (Jotnian sandstone), especially when the latter occurs as thin erosional remnants within a rather rugged basement topography.

Late Proterozoic sedimentary rocks, commonly referred to as Jotnian (Riphean) sandstones are known from several localities in Finland and Sweden. Within the coastal part of the Bothnian Sea such Late Proterozoic sedimentary rocks have been described from two major areas, the Satakunta-Formation in Finland (Simonen and Kouvo, 1955) and the Gävle-Formation in Sweden (Gorbatschev, 1967). In both cases, the age of the rocks has been estimated to be Middle Riphean. The maximum thickness of the sequence has in both localities been estimated to be approximately 1000 m. The strata have been preserved through downfaulting.

The extremely high seismic velocity of the sandstone (close to 5000 ms^{-1}) makes it difficult to detect the depth of the basement since the basement exhibits a similar or only slightly higher velocity. Sonobuoy refraction data indicate that the sandstone may be present below at least part of the Paleozoic rocks in the central parts of the Bothnian Sea. One verified exception is the Finngrunden Shoals in the Gävle Bay, where drillings have revealed a weathered crystalline basement directly below the Paleozoic (Thorslund, 1970). The shoal areas have been studied in detail by Thorslund and Axberg (1979). They noted that the shoals coincide with an uplifted part of the crystalline basement, and that the Paleozoic sequence is locally reduced in thickness within the shoal areas.

As demonstrated in the map in Fig. 1.3 a substantial part of the bottom of the Bothnian Sea consists of Cambrian and Ordovician sedimentary rocks. A cross-section of the sea is given in Fig. 1.9. In the Gävle Bay, the Paleozoic strata consist of about 40 m of Cambrian clays with subordinate layers of sandstone followed by some 50 m of Ordovician limestone. Fig. 1.10 shows that both the Cambrian and the Ordovician sequences increase generally in thickness towards the NW-central part of the formation (Winterhalter, 1972). In the northern part of the Paleozoic area, the Cambrian strata have attained

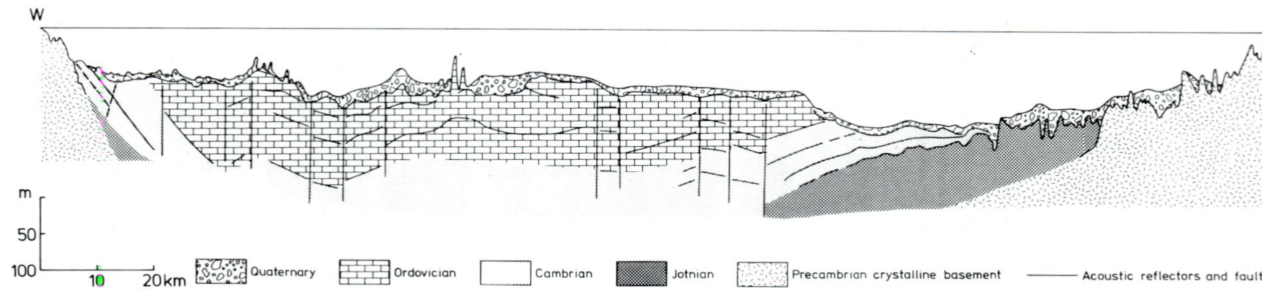

Fig. 1.9. Schematic geological profile (W—E) across the southern part of the Bothnian Sea from Söderhamn in Sweden to Rauma in Finland. The faulting off the Swedish coast is in reality considerably more complex than shown in the figure.

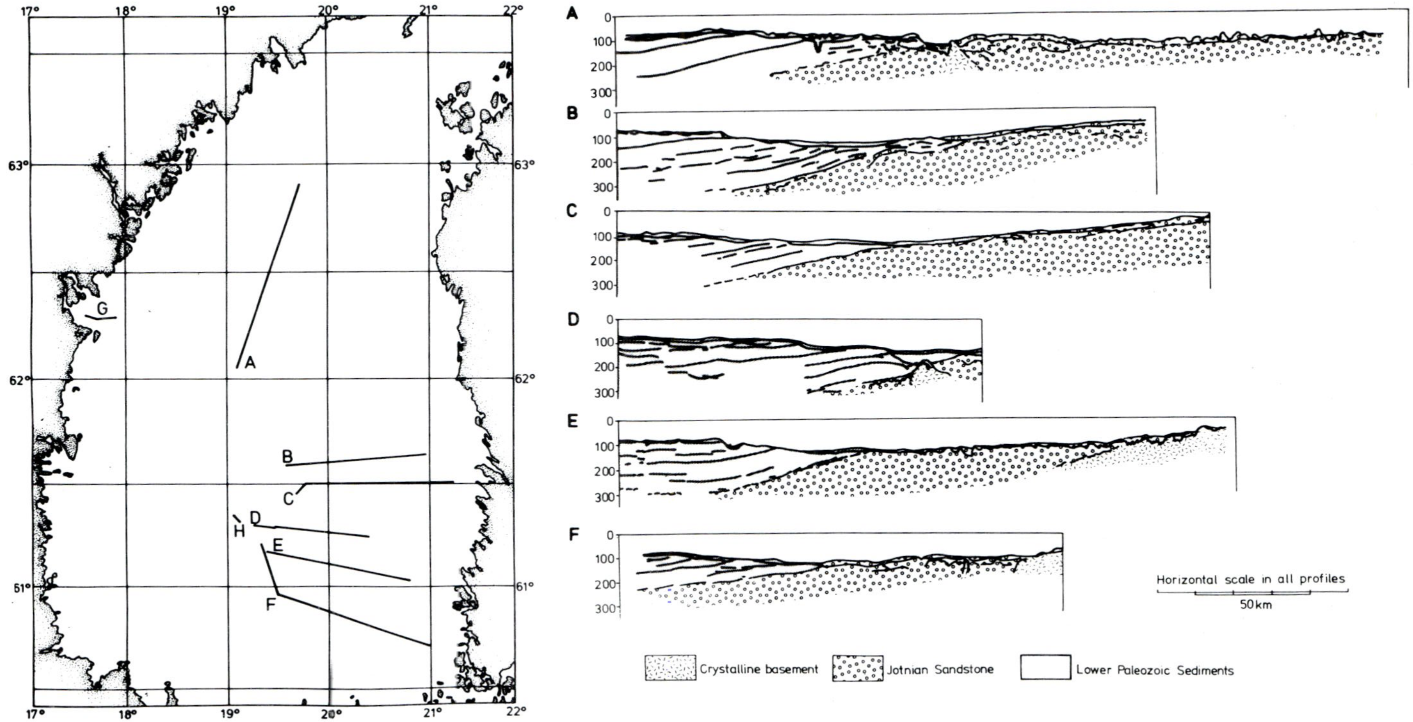

Fig. 1.10. Interpreted reflection profiles from the central and eastern parts of the Bothnian Sea showing the gentle sloping of the sub-Cambrian peneplane towards the west. Profiles *G* and *H* are shown in Figs. 1.11 and 1.12, respectively.

a thickness of about 200 m. Although the Ordovician sedimentary sequence also exhibits a general increase in thickness towards the north, the maximum total thickness of these beds (200—350 m) is to be found in the central part of the outcrop area. This is probably caused by more intense postdepositional erosion of the northern part of the formation.

The youngest stratigraphic level so far dated within the Bothnian Sea is the *Tretaspis* - stage of the lowest part of the Upper Ordovician represented by a hard calcilutitic limestone. This reddish or greyish limestone is commonly known as 'Baltic limestone' (Thorslund, 1960). The well-developed peneplane surface separating the Cambrian sedimentary rocks and the Jotnian sandstone is evident in the profiles in Fig. 1.10. This well-preserved peneplane surface is typical of the eastern part of the formation. The western extent is governed by a set of strong faults along the Swedish coastline (see Figs. 1.4a and 1.11). Minor faulting exists in the central parts of the Paleozoic formation (Fig. 1.12). For a more comprehensive description of the geology of the Bothnian Sea the reader is referred to Winterhalter (1972).

Åland Sea

The Åland Sea encompasses a local tectonic depression of considerable vertical dimensions evidently genetically closely related to the previously mentioned grabenlike occurrences of Jotnian sandstones in the vicinity of Gävle, in the Satakunta area, and the Muhos-Formation in the Bothnian Bay. The main fault line runs south from the Bothnian Sea between the islands, Märket and Understen, close to the Swedish coast, first south and southeast

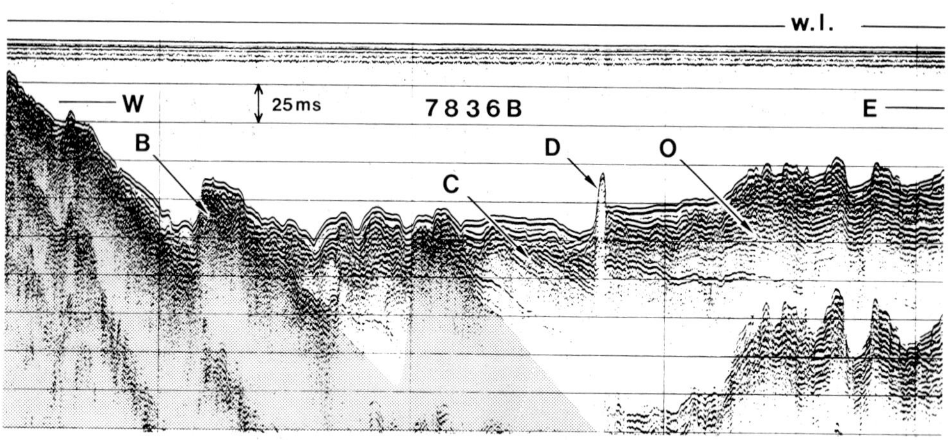

Fig. 1.11. Detail of a reflection profile recorded near Sundsvall in Sweden showing the complicated (faulted) contact between the Precambrian basement (*B*) and the sedimentary rocks off the Swedish coast: *C* = Cambrian; *O* = Ordovician; *D* = diabase intrusion. For location of profile see Fig. 1.10, line *G*.

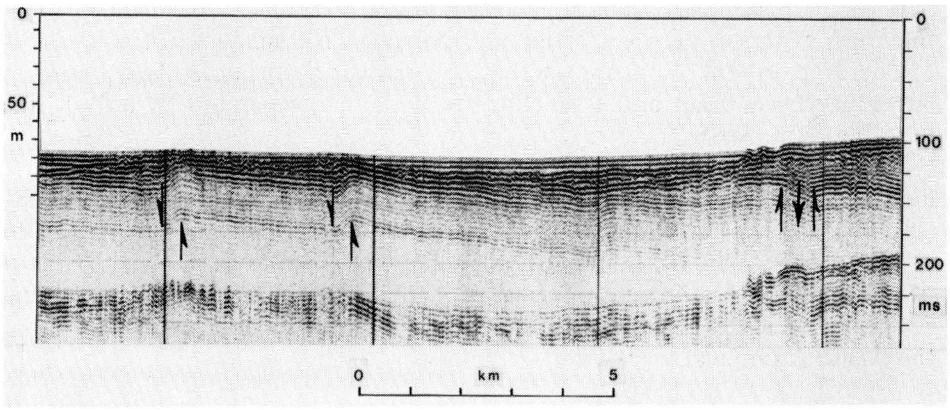

Fig. 1.12. Block faulting in the southern part of the main Paleozoic formation in the Bothnian Sea. For location of profile see line *H* in Fig. 1.10.

and then turning due east (Fig. 1.4a). This fault forms the southwestern and southern limit of the Jotnian sandstone formation. In the north and northeast, the sandstone is bounded by the Middle Proterozoic rapakivi massif of Åland (Winterhalter, 1967). The boundary itself is located within the erosional trough forming the Deep of the Åland Sea (Fig. 1.13). A substantial

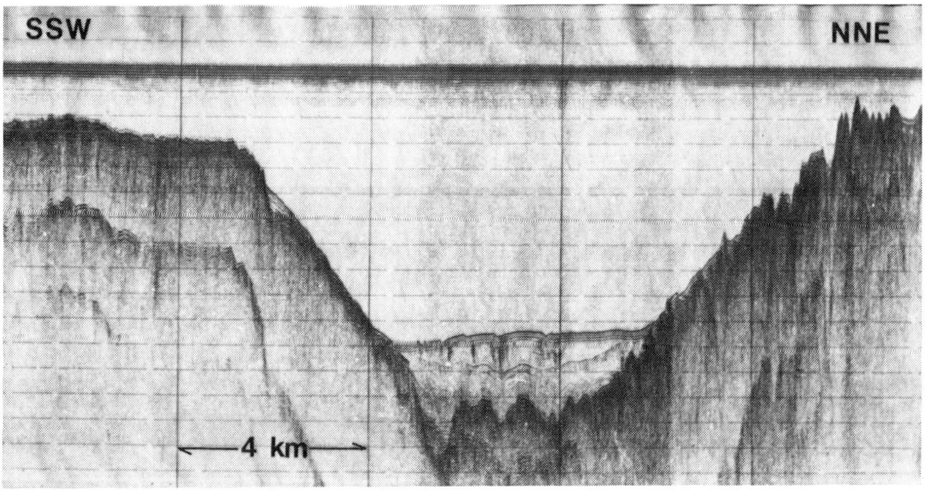

Fig. 1.13. Reflection profile across the southern part of the trough-like depression, the Åland Deep, trending NW—SE in the Åland Sea (Fig. 1.14). Although acoustic penetration in the Jotnian sandstone, in the left part of the profile, is very limited, the difference in the bedrock characteristics on both sides of the trough is obvious. The rugged relief in the right part of the profile is typical of the Middle Proterozoic Åland-massif. Vertical scale lines are 25 ms apart.

part of the Åland Sea probably consists of arkosic sandstone. The maximum thickness of this sequence is estimated to exceed 700 m.

The map in Fig. 1.14 shows that the sandstone occupies two basins, a larger northern basin and a smaller southern one, separated by a fault line trending ENE—WSW.

West of Mariehamn on Åland the sub-Jotnian peneplane, in part detectable in seismic reflection profiles, evidently dips approximately 2.5° towards the southwest. If the dip and strike of the Jotnian sandstone beds are assumed to coincide with those of the underlying peneplane, it follows that the NE limit of the sedimentary formation is erosionally induced and that the Jotnian sandstone of much the same thickness, viz., 700 m, has once covered most of the Åland Archipelago.

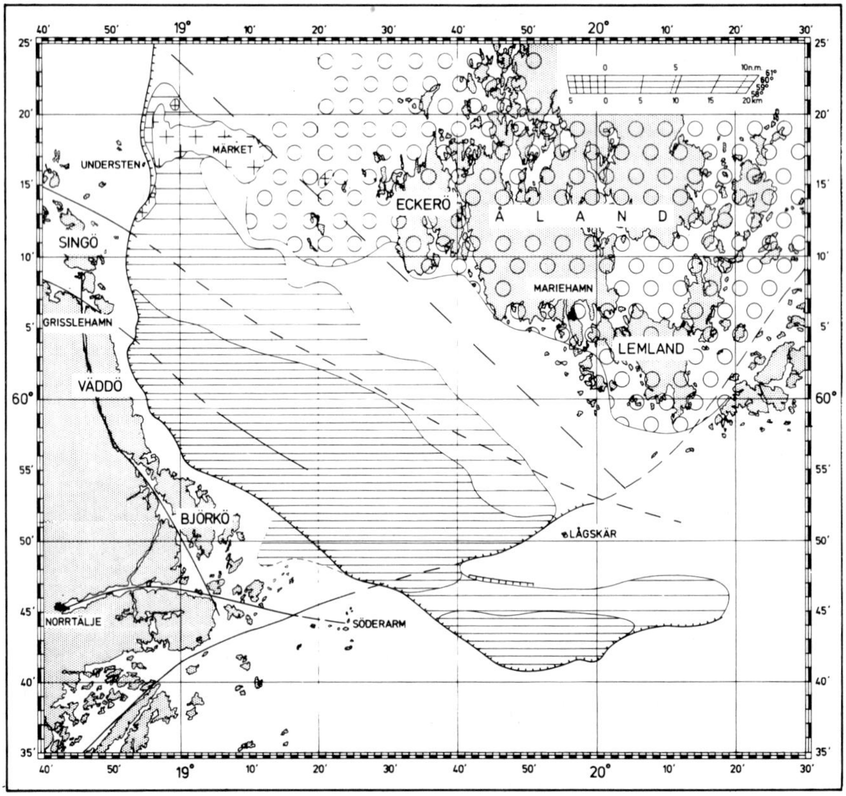

Fig. 1.14. Bedrock of the Åland Sea area. Most of the Åland islands consist of anorogenic rapakivi granite (Proterozoic) marked with circles on the map. Crosses denote a diabase intrusion in the Märket area. The horizontal striation shows the extent of the Jotnian (Riphean) sandstone. The double striation indicates the assumed extent of the sub-Cambrian peneplane related to the clastic dykes of Cambrian age found in the Åland islands (e.g., Martinsson, 1956).

No traces of Paleozoic or younger sedimentary rocks have so far been detected in the Åland Sea. Middle Ordovician limestone and underlying Cambrian sandstones and siltstones are, however, known to fill a substantial part of the bottom of the Bay of Lumparn in the southeastern part of the main island of Åland. A large number of clastic dykes of Cambrian age (Martinsson, 1974) are known at various localities within the Åland Archipelago.

Prior to the deposition of Lower Paleozoic sediments, the Jotnian sandstone and even some of the crystalline basement must have been eroded, since the deposition of the Cambrian clastic dykes found in many places on the Åland Islands occur in virtually unweathered bedrock. The Lower Paleozoic deposits in the Bay of Lumparn have been preserved from erosion through downfaulting. Similar deposits of Paleozoic sedimentary rocks probably occur also within sheltered localities in the Åland Sea, in the Archipelago Sea east of the Åland Islands and in the Swedish archipelago further southwest.

It is evident that the Åland Islands and the adjacent areas have formerly been covered by an unknown thickness of sedimentary rocks subsequently removed by erosion except for the few tectonically depressed areas where these rocks have been more or less preserved.

Gulf of Finland

The Gulf of Finland separates a crystalline bedrock area in the north (Finland) from an area of Lower Paleozoic sedimentary rocks in the south (Esthonian SSR). Using the principle of topconstancy it is possible to establish the present-day position of the sub-Cambrian peneplane on the bottom of the Gulf (Fig. 1.15). Thus one may conclude that the present-day land surface of southern Finland coincides well with the sub-Cambrian peneplane, or, in this area, rather with the sub-Vendian peneplane. This assumption is substantiated by the finds of allegedly Cambrian sedimentary rocks as clastic dykes in the crystalline bedrock of southwestern Finland (see Martinsson, 1974). Thus it seems obvious that, at least the Lower Paleozoic sedimentary rocks forming the north coast of Esthonia once covered the entire Gulf of Finland and, possibly, also a substantial part of southern and western Finland.

The stratigraphy and the former extent of the Vendian sediments of the Esthonian SSR and especially the eastern part of the Gulf of Finland pose many unanswered questions. Due to the scarcity of data on the Vendian, the northern boundary delineated on the map (Fig. 1.3a) must be considered tentative, except for the shore of the Karelian Isthmus, north of Leningrad, where the Vendian "Blue Clay" is found in outcrops and in many drillholes.

The delineation of the present-day northern extent of the Paleozoic sedimentary rocks in the Gulf of Finland has encountered considerable difficulties due to the scarcity of available continuous seismic profiles and other

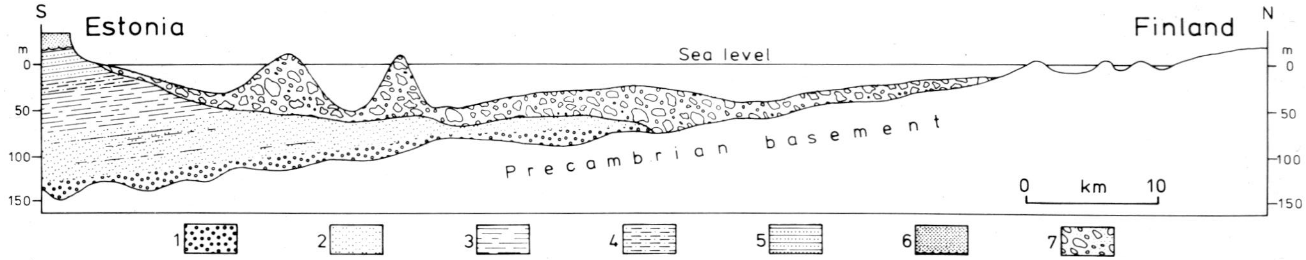

Fig. 1.15. Section (hypothetical) across the Gulf of Finland according to Öpik (1956). *1—6*, represent the Lower Cambrian (probably also Vendian). *1* = the basal conglomerate; *2* = the lower sandstone with *Platysolenites* and clay interbeds; *3* and *4* = Lontova beds or Blue Clay (proper), with sandstone interbeds on top (*4*); *5* = Lükati beds; *6* = Kakumägi beds and lower part of Tiskri sandstone; *7* = Quaternary deposits.

seismic data. The boundary has been drawn on the map (Fig. 1.3a) by correlating land data from the Esthonian SSR with bathymetric data from nautical charts and a few echo-sounding profiles across the Gulf. This has brought the boundary, at least in the western part of the Gulf, rather close to the Esthonian coast. It should be emphasized, that outliers in the form of erosional remnants of Cambrian and evidently also of Vendian sedimentary rocks probably extend a lot farther north; in fact there are indications of sedimentary rocks quite near the Finnish coast southeast of Hanko and possibly also west of the Hogland Island, which itself consists of Proterozoic crystalline bedrock.

Baltic Proper

General outline

The bulk of the Baltic Proper lies within the Baltic Shield, part of the East European Platform. Only a small area in the southwest lies outside, separated by the "Tornquist Line", forming the Fennoscandian Border Zone. The Tornquist Line extends from central Jutland in Denmark in the NW through Scania and Bornholm to Poland in the SE. The southeastern part of the Baltic Proper forms a subsided area in the East European Platform; the Baltic Syneclise (Depression), that contains Paleozoic and Mesozoic sedimentary rocks of a considerable thickness.

Southeast of Gotland the sedimentary rocks reach a thickness of 2000 m (Fig. 1.16), and further southeast off the Lithuanian coast, more than 3000 m. SW of the Tornquist Line the crystalline basement occurs at a depth of 5000—7000 m.

The northern part of the Baltic Proper occupies an area that consists predominantly of the exposed crystalline basement rocks of Early and Middle Proterozoic age. The term Baltic Shield is generally used for the crystalline complex that forms most of cratonic Fennoscandia.

Crystalline basement

The Precambrian crystalline basement is exposed within the northern and northwestern part of the Baltic Proper and within a limited area in the northernmost part of the Hanö Bay on the Blekinge coast. No attempt will be made to give a petrological description of the basement due to the scarcity of reliable data. Probably an interpolation of land data might give an acceptable picture of the crystalline basement on the bottom of the Baltic Proper, but it lies outside the scope of the present treatment. A comprehensive description of the relevant Precambrian basement is given in Rankama (1963).

Although the present surface of the crystalline basement is rugged in detail, it still exhibits a general flatness referable to the sub-Cambrian peneplane. North of Öland, this peneplane coincides well with the general trend

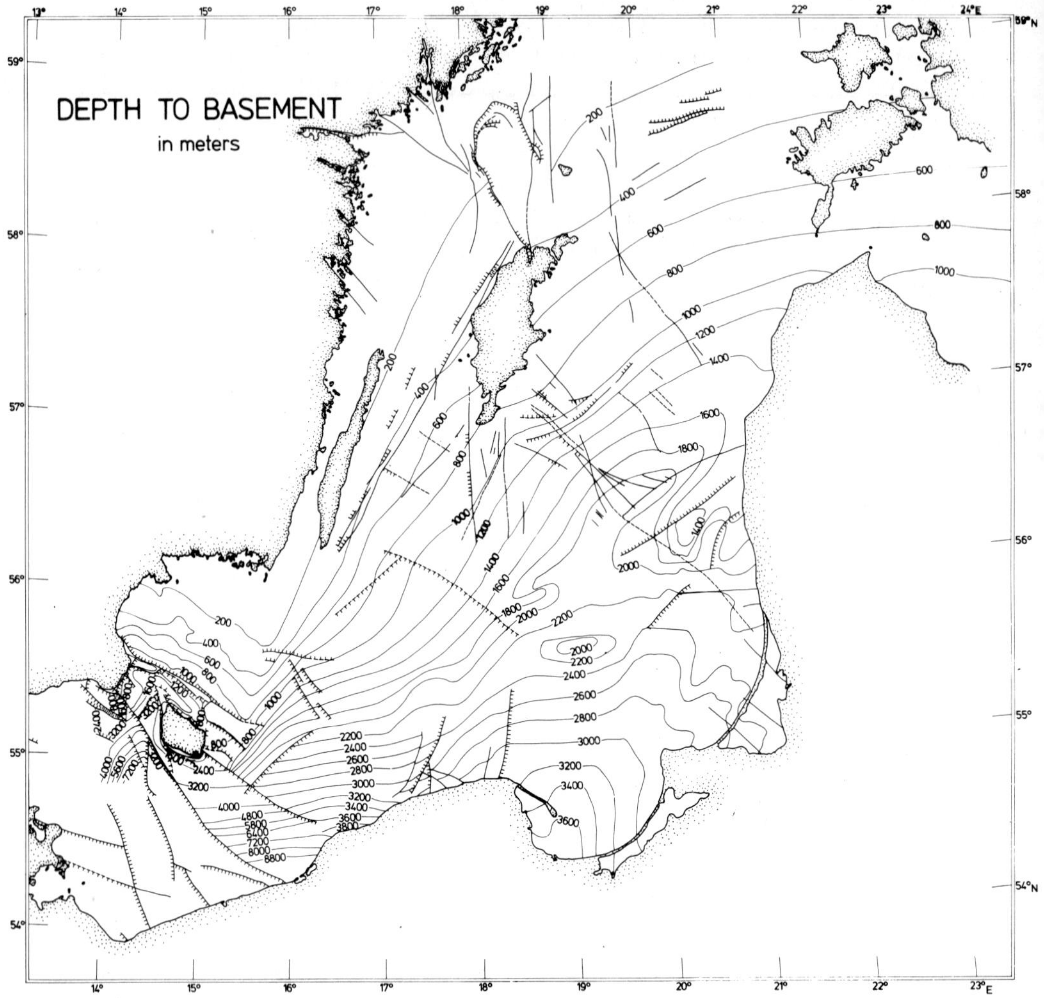

Fig. 1.16. Map showing the depth to the basement, in meters, in the central and southern parts of the Baltic Proper.

of the bedrock on land having a gentle dip towards ESE with a relief of only about 20 m in the boundary zone of the Cambrian—Silurian sedimentary rocks. A completely different relief is encountered further north (NW of Gotska Sandön) in an area of intense faulting partly manifested in the Landsort Deep (Fig. 1.17). Further north and northeast, the crystalline basement once more attains a gentle relief (Fromm, 1943), dipping slightly towards SSE. The two-fold trending of the sub-Cambrian peneplane, i.e., ESE north of Öland, and SSE east of Gotska Sandön, indicates the considerable influence

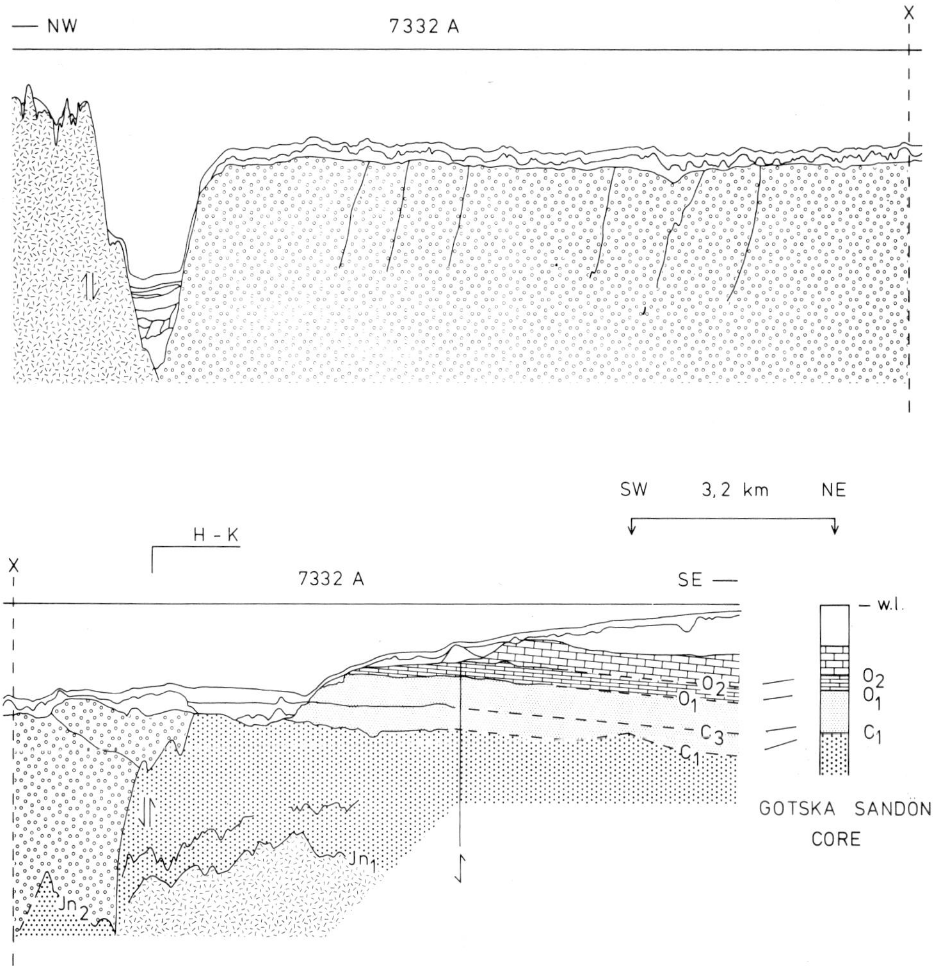

Fig. 1.17. Geological section across a Jotnian fault basin in the northern Baltic Proper. The NW—SE profile starts from the Landsort Trench in the NW and ends south of Gotska Sandön. The vertical scale is exaggerated 25 times. The Landsort Trench is 5 km wide in this section, and the depth to bedrock at the base of the trench is 580 m. The basic difference in erosional form between areas of crystalline bedrock (NW of the trench) and sedimentary rocks (small circles) is obvious. Six reflectors dipping about 10° NW exist in the Jotnian sandstone SE of the trench. These have been interpreted as volcanic dykes. The drill core from Gotska Sandön (Thorslund, 1938) has been used as reference when interpreting the right hand part of the profile. Jn_1 = bottom of the lower Jotnian sedimentary unit; Jn_2 = bottom of the upper Jotnian sandstone; C_1 = Precambrian/Cambrian boundary; O_1 = Cambrian/Ordovician boundary; O_2 = approximate level of the Middle/Upper Ordovician boundary.

that the fracture system in the area of the Landsort Deep has had on the present configuration of the Baltic Depression.

The peneplane off the Blekinge coast (SW of Öland) dips gently towards the south. The relief of the crystalline bedrock is 50—60 m. The relief of the sea-floor is, however, much less, due to the infilling of sedimentary rocks as erosional remains. The primary peneplane is evidently contemporaneous with the sub-Cambrian peneplane in the northern part of the Baltic Proper, but a second peneplanation must have occurred during the Late Paleozoic (Kumpas, 1978). This erosional stage seems to have caused the levelling of both the crystalline basement and the sedimentary bedrock.

The crystalline basement is intensely block-faulted along the Tornquist Line. In southern Scania and around Bornholm, vertical displacements of the order of 1000 m and more are common (Fig. 1.18). The geology of the area seems to be very complicated making the available seismic data rather insufficient for a reliable interpretation.

Proterozoic rocks

Jotnian sandstone of the same type as described from the Gulf of Bothnia and the Åland Sea forms the submarine bedrock southeast of the Landsort Deep (Fig. 1.17). The sediments were downfaulted soon after deposition and thus sheltered from being eroded away during the formation of the sub-Cambrian peneplane and the subsequent erosional stages in the evolution of the Baltic Sea area. Thus, the situation is analogous to the other Jotnian sandstone deposits mentioned above. The maximum vertical displacement, around 1000 m, occurs along the line from the southern end of the Landsort Deep to Gotland. The northeastern boundary exhibits vertical displacements

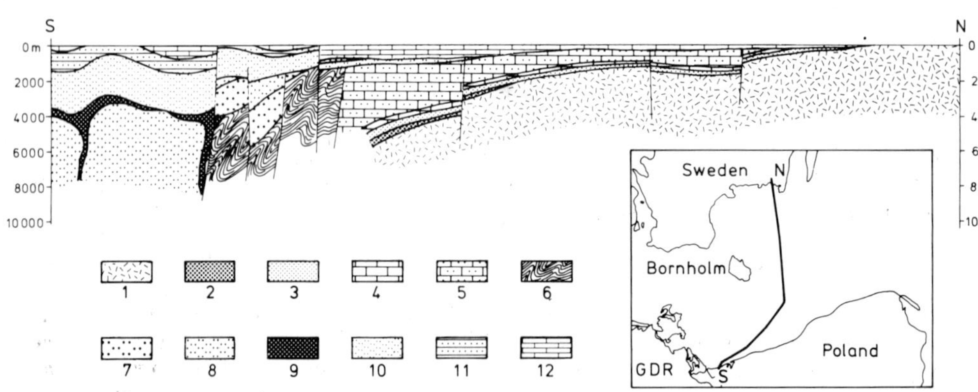

Fig. 1.18. Geological profile across the southwestern Baltic Sea according to Dadlez (1976). *1* = Precambrian crystalline basement; *2* = Proterozoic (Vendian?) sedimentary rocks; *3* = Cambrian; *4* = Ordovician; *5* = Silurian; *6* = folded Ordovician and Silurian; *7* = Devonian; *8* = Carboniferous; *9* = lower Permian volcanics; *10* = Permian and Triassic; *11* = Jurassic; *12* = Cretaceous.

7609 / 11

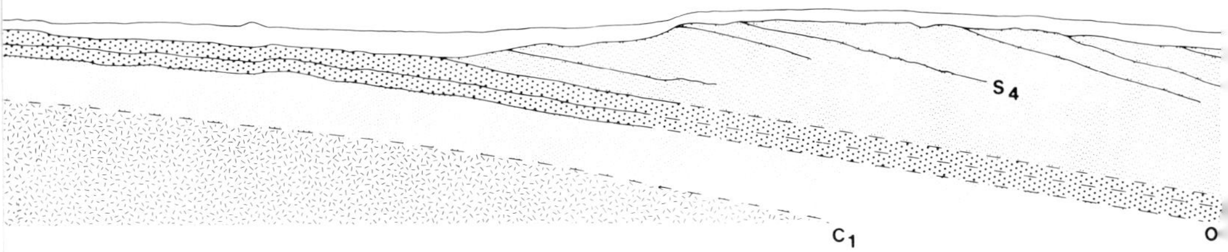

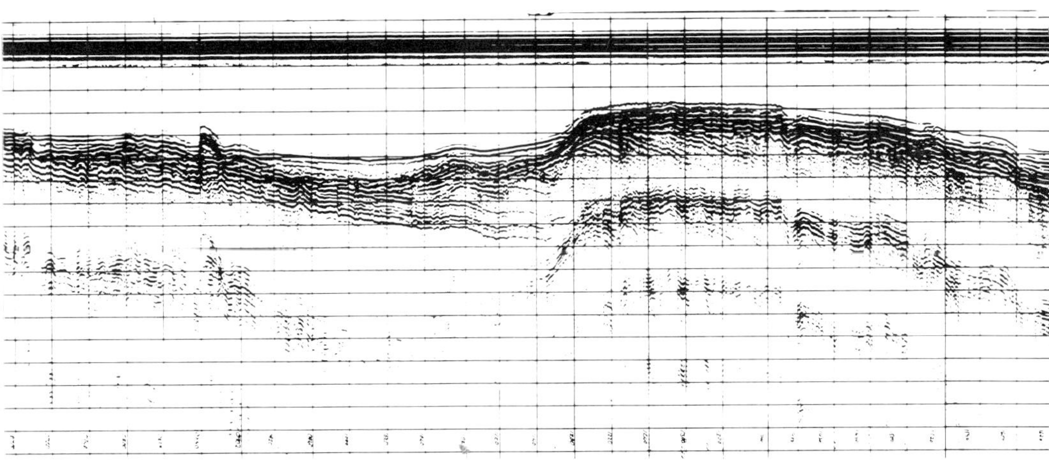

N—S line
t a point
wn in the

pp. 40-42

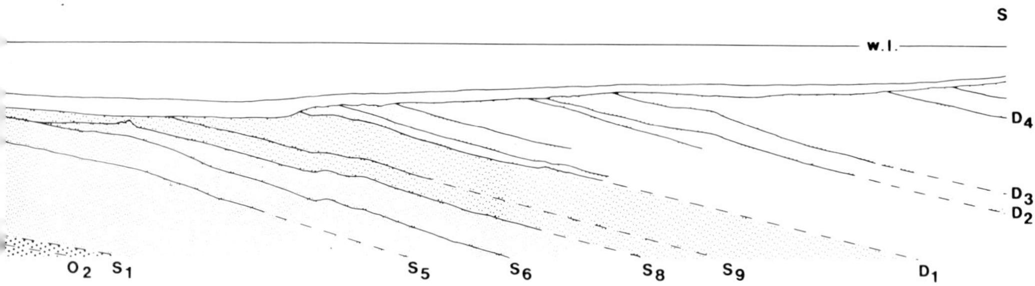

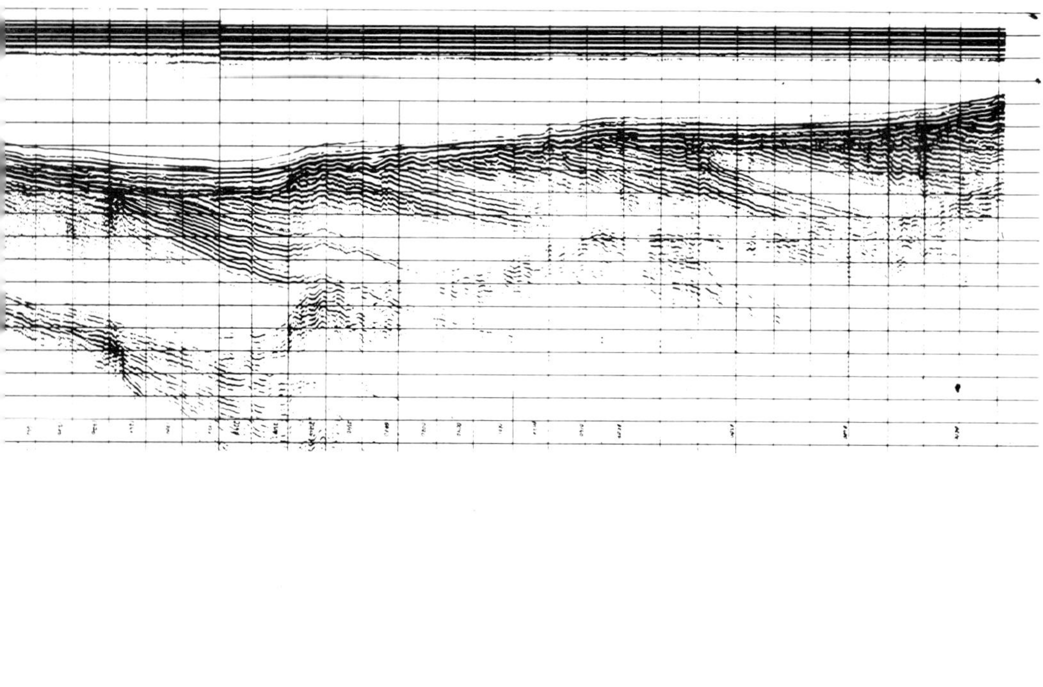

in the range of 100—200 m. The Landsort Deep forms the northwestern limit of the sandstone. Its southeastern extension has not yet been verified due to the considerable thickness of the overlying younger rocks, hampering the interpretation of the seismic data. It does, however, seem to reach Fårö and northern Gotland.

The Jotnian sedimentation was followed by intense denudation, leading to the formation of the peneplane. Towards the end of the Proterozoic, the Vendian sea reached the Baltic Sea area from the east and arenaceous to argillaceous sediments were deposited on the peneplane. The sedimentation was more or less continuous into the Cambrian.

At the beginning of the Lontova stage, which is generally accepted as constituting the lowermost Cambrian strata in Esthonia (Fig. 1.19), the sea had reached as far west as the mouth of the Gulf of Finland. By the end of the Lontova stage the main part of the Baltic Proper was inundated. The Lontova—Lükati boundary was followed by a distinct break in sedimentation that terminated the westward transgression.

The Lükati sea spread from the southwestern margin of the East European Platform across the Baltic Sea to the Moscow Basin. This transgression introduced the Lower Cambrian trilobite fauna into the Baltic Sea area. Thus, the stratigraphic position of the Precambrian/Cambrian boundary in the Baltic Sea is rather controversial. In the map Fig. 1.3b, the boundary is drawn at the base of the Lontova stage in Esthonia (e.g., Brangulis et al., 1974). A somewhat higher level of the boundary, at the base of the Lükati stage, has been proposed by, e.g., Plissov et al. (1975).

Paleozoic rocks

Paleozoic sediments once covered the Baltic Depression, the Bothnian Sea and substantial parts of the Swedish mainland. With the end of the Ordovician the extent of the Paleozoic Sea decreased rapidly, and in the Middle Silurian only a narrow gulf extended from the southwest between Gotland and the coastal area of the U.S.S.R. At the end of the Paleozoic, sedimentation occurred only in the southwestern part of the Baltic Proper off the Lithuanian coast. The total thickness of the Paleozoic strata increases towards SSE (Fig. 1.20) and exceeds 3000 m in the Gulf of Gdansk (see Fig. 1.16).

Cambrian sedimentary rocks are exposed along the erosional escarpment (Fig. 1.21) stretching from northern Esthonia to the Kalmar Strait (between Öland and the Swedish mainland) and further south in the direction of Bornholm. Within the Hanö Bay, only erosional remnants of Lower Cambrian quartzites of a limited thickness have been observed under the Mesozoic sedimentary rocks. In the central part of the sedimentary basin, southeast of Gotland, the Cambrian sedimentary rocks attain a thickness of more than 300 m under a cover of Ordovician limestones.

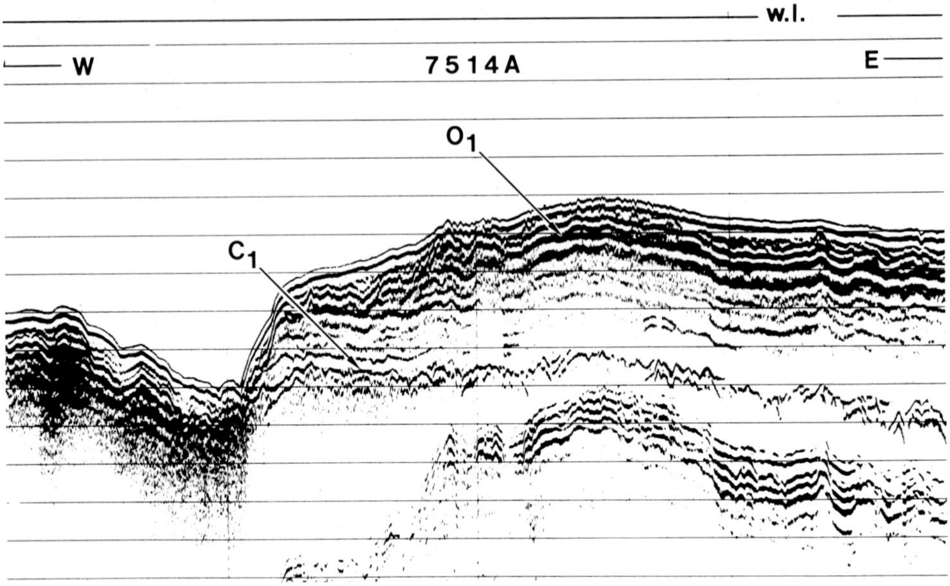

Fig. 1.21. Cambrian Klint west of Visby (Gotland). The profile is 20 km long. The vertical scale lines are at 25 ms intervals, equivalent of a water depth of approximately 18 m. C_1 = Precambrian/Cambrian boundary; and O_1 = Cambrian/Ordovician boundary.

The Paleozoic Sea had its greatest extent during the Lower Cambrian, covering the entire Baltic Depression, the Åland Islands, and reaching as far north as the Bothnian Bay. Contrary to the argillaceous Lower Cambrian sedimentary rocks of the Bothnian Sea, the sedimentary rocks of the Baltic Proper consist of loosely consolidated arenaceous sediments with only minor intercalations of clay. Locally the sandstones may be quartzitic as, e.g., in the northern Kalmar Strait and in the western Hanö Bay.

In the central parts of the bassin, SE of Gotland, the thickness of the Lower Cambrian sandstone is approximately 200 m and decreases successively towards N and NW. Thicknesses of 78 m and 72.5 m were recorded in drillings at Böda Hamn on northern Öland and on Gotska Sandön, respectively.

The Middle Cambrian Sea had receded south to a line between Gotska Sandön and Hiiumaa west of the mainland in the Esthonian SSR. The thickness increases towards the south, being over 200 m off the coast of Poland.

The Upper Cambrian sedimentary rocks are restricted to the southern Baltic Proper. The northern limit extends from northern Öland across central Gotland to the coast of the Lithuanian SSR in the southeasternmost part of the Baltic Proper. The thickness of the beds is very modest being only 2 m in southern Gotland and 13 m in southern Öland.

The Middle and Upper Cambrian sedimentary rocks consist of argillaceous schists in the western part of the basin with sandy intercalations increasing towards the east. Arenaceous sedimentary rocks dominate in the eastern Baltic Proper. Alum shales have been encountered in drill cores from Öland.

Ordovician limestones form the bedrock in a wide zone of the bottom of the Baltic Proper from the northern Esthonian SSR to the eastern part of the Hanö Bay (Fig. 1.3b). The northern and northwestern margins of this zone exist as a steep escarpment (Fig. 1.22). The explanation for the formation of this well-developed Ordovician klint is that the poorly consolidated Cambrian sediments were overlain by the erosionally more resistant Ordovician limestones. The age of the klint is obviously pre-Quaternary, although the Pleistocene glacial erosion has somewhat changed and probably also accentuated the relief.

Erosional remnants of Ordovician limestones are found also beyond the klint, e.g., in the Bay of Lumparn (cf. p. 31) thus indicating the existence of a connection in the Ordovician Sea between the Baltic Proper and the Bothnian Sea.

The Ordovician of the Baltic Proper can on seismic grounds be divided into two sequences; a lower sequence representing the Lower and Middle Ordovician and an Upper sequence of Upper Ordovician sedimentary rocks. The latter possibly includes the *Macrourus* sandstone of uppermost Middle Ordovician. Especially the Upper Ordovician strata exhibit a multitude of structures that have been interpreted as algal reefs. These reefs are generally roundish with a height of 20—30 m. Sometimes several reef generations of both Middle and Upper Ordovician age have been observed to rest on top of one another. Reefs have been discovered virtually in all parts of the Baltic Sea either exposed as in the area north and west of Gotland where they form topographic elevations like the Hall Banks or covered by Silurian deposits. Due to the reef structures the boundary between the Ordovician and the Silurian is quite irregular. In fact the upper parts of the reefs may in some

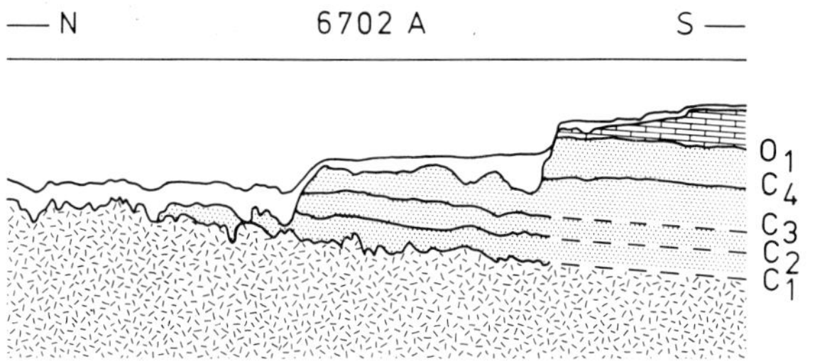

Fig. 1.22. Ordovician Klint west of Hiiumaa. C_1 = the sub-Cambrian peneplane; C_2, C_3, C_4 = reflectors in the Cambrian; O_1 = Cambrian/Ordovician boundary. The formation of the erosional klint has been enhanced by the resistant Ordovician limestone covering the softer Cambrian sedimentary rocks.

places protrude more than 20 m into the Silurian strata. This is the case, e.g., under Gotland and south and southeast of the island.

The Lower Ordovician strata consist mainly of quartzitic sandstones, of argillaceous limestones often with glauconitic layers, and of shales with *Dictyonema* in the NE and *Didymograptus* in the SW. The greatest thickness of the Lower Ordovician (80—110 m) is found in the Esthonian SSR while the beds in coastal Poland and on Bornholm are 24 m and 14 m thick, respectively. The Middle Ordovician strata consisting mainly of reddish limestones with marly intercalations attain a thickness of 35—55 m in the central and southern parts of the Baltic Proper. Further north the predominantly calcareous sediments attain a thickness of 30—40 m. The Upper Ordovician, exhibiting a thickness of about 60 m in the central parts of the Baltic Proper, consists chiefly of detrital limestones and marls, partly bituminous in the upper parts. The thickness of the Upper Ordovician strata increases slightly towards the east, ranging from 75 m to 95 m in the coastal area of the southern Latvian SSR.

Silurian sedimentary rocks are exposed over a vast area in the Baltic Sea Proper (Fig. 1.3b). The erosionally induced limit of the strata runs west from central Hiiumaa, the Esthonian SSR, and then curves towards the southwest passing just northwest of Gotland and southeast of Öland. Silurian deposits, from Upper Llandovery to Upper Ludlow, are exposed on Gotland. The Silurian klint, forming part of the western and northern coastline of Gotland also constitutes a marked submarine topographic feature along a considerable part of the sea floor from Gotland to Esthonia. It is similar to the Ordovician klint consisting of readily eroded sedimentary rocks below a resistant cover of reef limestones forming the Visby beds of Gotland.

During the early Silurian, the Baltic Basin consisted of a broad gulf open towards the border of the East European Platform in the south. This gulf diminished in size and by the end of the Silurian its northern shore barely reached up to the line between southern Gotland and the island of Saaremaa west of the Esthonian mainland. The coastal zone of the gulf was characterized by the deposition of calcareous sediments and the formation of banks with coral reefs. Further out, in deep water, successively more fine-grained calcareous sediments with an increasing proportion of argillaceous material were deposited. The regression seems to have taken place in steps, possibly related to the Caledonian folding. The orogenic processes led to a temporary increase in sediment transport into the gulf causing occasional breaks in the reef formation.

The geological evolution during the Silurian must have been considerably more complicated than during the early Paleozoic. The typical occurrence southwards of consistently younger calcareous banks with dispersed reef structures, resting on top of argillaceous open-sea deposits, is well observed in the cliffs of Gotland and on the adjacent sea floor. Thus, the Silurian deposits in the present Baltic Proper bear witness of the formerly widespread occurrence of reefs that have been later partly or wholly removed by erosion. The majority of the reefs seem to have been formed as bank reefs. Only the Burgsvik Reef of southern Gotland seems to bear the characteristic features of a true barrier reef. It can be followed on seismic reflection profiles from southern Gotland to the Latvian coast (Fig. 1.23). The Burgsvik Reef marks the end of large-scale reef formation in the Baltic Basin preserved in the existing sedimentary deposits.

Subsequent sedimentation of some 250 m of marly—arenaceous deposits took place during the Downtonian, marking the end of the Silurian Period. A minor transgression occurred during the Downtonian, but the extent of the basin is unknown since the near shore sediments are nowhere preserved. The

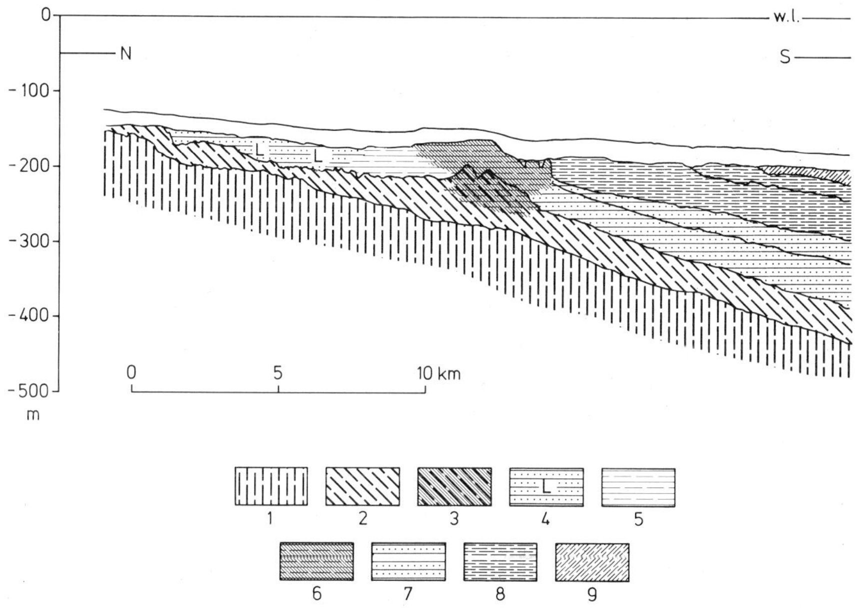

Fig. 1.23. Geological section showing the facial subdivision of the late Ludlowian (Upper Silurian) SE of Gotland. The unit names refer to their nearest equivalents in the Silurian of Gotland. On the open shelf-sea sediments of upper Hemse age (*1*) rest the probable equivalents to the Eke beds (*2*), ranging from lagoonal sediments in the N part of the section to open shelf-sea sediments in the S. The central part of the section shows a distinct bank, presumably rich in bioherms (*3*). The following unit exhibits an even more distinct subdivision between lagoonal sediments (*4*) equivalent to the Burgsvik beds on Gotland, bank sediments (*5—6*) equivalent to the Hamra (*5*) and Sundre (*6*) beds, and open shelf-sea sediments (*7*) that are unknown on Gotland. The subdivision between mainly bedded (*5*) and mainly biohermal (*6*) bank sediments is introduced for the sake of comparison with the deposits on Gotland. They cannot be referred to any distinct reflector in the acoustic profile. The onset of the Downtonian was transgressive in Gotland, and the sediments (*8—9*) in the section were evidently formed in an open shelf-sea environment.

present-day northern erosional boundary of the Downtonian off southern Gotland runs immediately southeast of the Burgsvik Reef while further east the Downtonian strata extend across the reef, reaching as far north as the southern part of Saaremaa.

The total thickness of the Silurian deposits is approximately 500 m in southern Gotland, increasing to nearly 2000 m in the southeastern part of the Baltic Proper.

Devonian sedimentary rocks are restricted to the eastern part of the Baltic Proper and probably never extended far beyond their presentday limit (Fig. 1.3b). The maximum sediment thickness of some 900 m is found along the Latvian and Lithuanian coasts (Fig. 1.24).

The Lower Devonian sedimentary rocks that are characterized by mainly terrigenic, continental, reddish arenites and argillites, exhibit a gradual transition from the underlying purely marine Downtonian sedimentary rocks. The intense erosion following the folding of the Scandinavian mountain range must have been the main source of sediment. The Lower Devonian strata, attaining a thickness of some 350 m are known to locally contain bitumen, e.g., in coastal Lithuania.

The Middle Devonian rocks, more widely distributed than the Lower Devonian strata, consist of 70—80 m of sandstones, overlain by dolomitic marls 60—170 m thick. The uppermost beds consist of reddish arenites.

During the Upper Devonian, the area of active sedimentation again decreased to what it was during the latter part of the Lower Devonian. The rocks consist mainly of dolomites and dolomitic marls with a maximum thickness of 500 m. Layers of gypsum and anhydrite are common in the middle parts of the Upper Devonian strata especially in the coastal parts of Latvia and Lithuania.

Carboniferous deposits, lying conformably on the Devonian, occur in a very limited part of the southeastern Baltic Sea approximately between the coastal cities of Liepaja (Latvian SSR) and Klaipeda (Lithuanian SSR). The thickness of the Carboniferous strata in the Baltic Proper ranges from 70 m to 130 m. They consist of a lower unit with marly and argillaceous layers and an upper unit of fine-grained sandstones with thin intercalations of clay. Both units belong to the Lower Carboniferous.

Permian sedimentary rocks consisting of terrigenic calciargillites and salt deposits are widely distributed in Latvia, Lithuania, and Poland. The original sediments were discordantly deposited on partly deeply eroded substratum of older sedimentary rocks. No Lower Permian rocks have been detected within the Baltic Sea. Upper Permian (Zechstein) strata are, however, known from the southern Baltic Sea off the Polish and Lithuanian coasts. Within the East European Platform part of the Baltic Sea a maximum thickness of 350 m of Upper Permian strata occur in the Gulf of Gdansk.

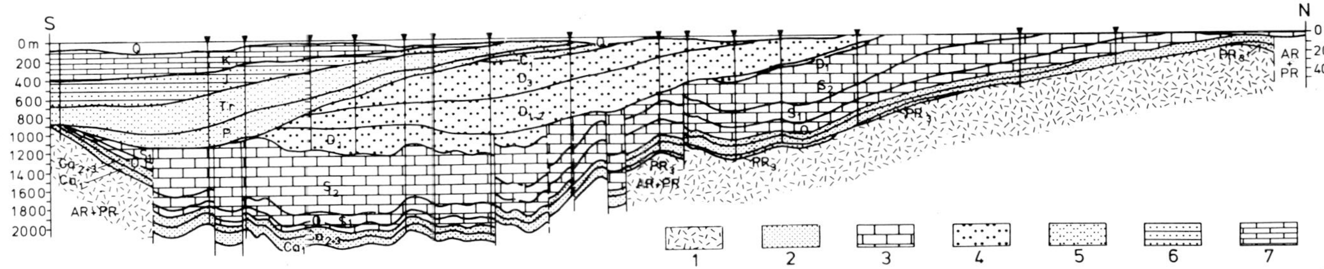

Fig. 1.24. Geological cross section along the eastern Baltic Sea coastline from the northern part of Hiiumaa (Esthonia) south to the Polish border according to Volkolakov et al. (1977). *1* = Precambrian basement; *2* = Cambrian; *3* = Ordovician and Silurian; *4* = Devonian; *5* = Permian and Triassic; *6* = Jurassic; *7* = Cretaceous; *Q* = Quaternary. The interpretation is based on drill-hole data.

Permian sedimentary rocks are encountered over large areas in the Danish—Polish depression in Denmark, GDR, and Poland. No exact data on the lithology of these rocks and their total thickness are as yet available. No Permian strata have been detected in Scania nor on Bornholm.

Mesozoic rocks

The Mesozoic seas were mainly restricted to the areas southwest of the East European Platform. The Danish—Polish depression, initiated by the Late Hercynian tectonic movements, formed an area of major deposition during the entire Mesozoic Era.

The bulk of the thick deposits (4000—5000 m) in the central parts of the depression are of Triassic and Late Cretaceous age, while rather modest sedimentation took place during the Jurassic and Early Cretaceous.

The maximum marine transgression during the Cretaceous extended far into the Baltic Sea region. How far north the sea actually extended cannot be ascertained, but the northern boundary of the preserved Mesozoic sedimentary rocks runs from Lithuania in the east to the Hanö Bay (Fig. 1.25) in the west.

The Mesozoic sedimentary rocks within the Fennoscandian Border Zone, across Scania, Bornholm and further southeast into Poland, are intensely block-faulted (see Fig. 1.18) along mainly NW—SE faultlines. The strata of this zone are also characterized by rapid lithological and stratigraphical changes.

Triassic sedimentation occurred in the Danish—Polish Depression from northwestern Poland to Scania. The sediments varied from chiefly marine in the actual depression to mainly continental along the marginal parts of the East European Platform. In northern Poland the continental sedimentation is represented by Lower Triassic (Buntsandstein), arenites, argillites, and limestones deposited in lagunal, fluvial and lacustrine environments. Middle Triassic marine sedimentary rocks (Muschelkalk) are absent in northernmost Poland and, consequently, most probably also in the southwestern Baltic Proper where Mesozoic deposits crop out. The sedimentary sequence is concluded with the deposition of continental (brackish to marine) sediments now present as sandstones and shales (Keuper and Rhaetian).

Triassic sedimentation in Scania is mainly continental and is in many aspects similar to the sedimentation in northern Poland. The *Muschelkalk* does, however, exhibit marine phases in the form of marly and calcareous intercalations. The Rhaetian deposits are represented in Scania by grayish, brownish, or blackish clays, sandstones, and shales containing coal layers. The Hanö Bay had during the Mesozoic evolved into a sedimentary basin confined in the south by the horst system of the Fennoscandian Border Zone. The oldest sediments were deposited in this basin during the Late Triassic (Rhaetian). The sediments are mainly continental, arenaceous and

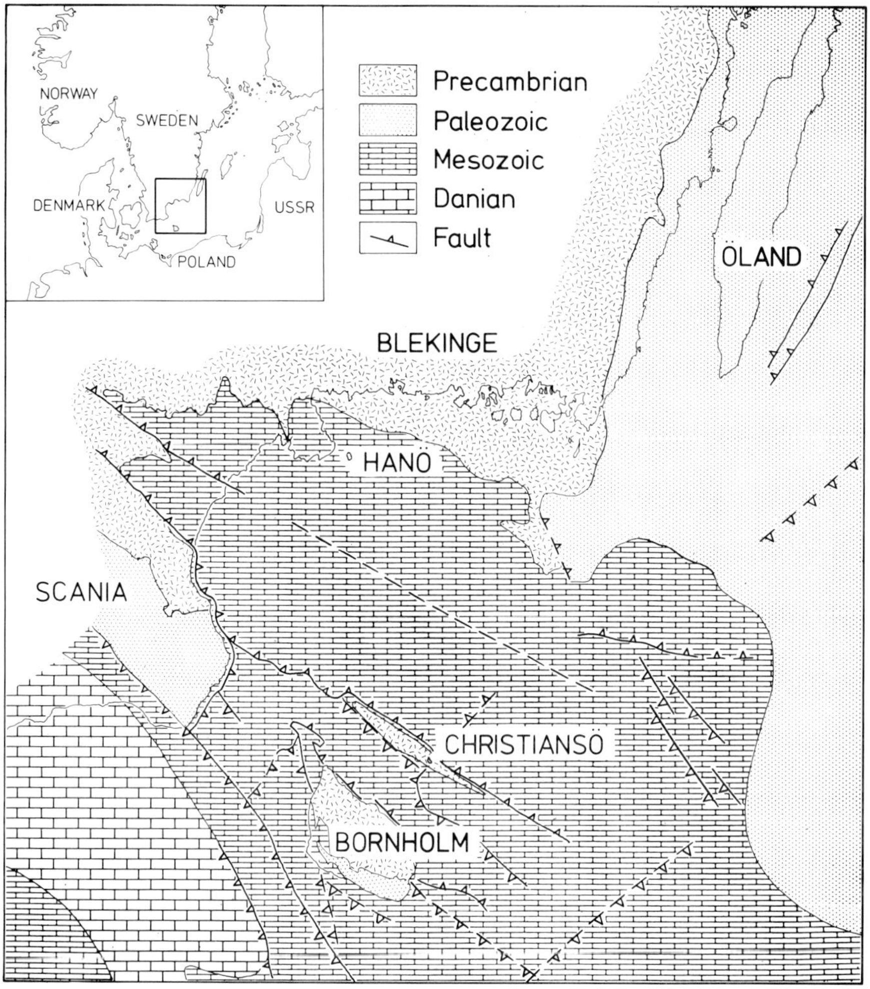

Fig. 1.25. Distribution of Mesozoic sedimentary rocks and major fault lines in the Hanö Bay, southwestern Baltic Proper, according to Kumpas (1978, and unpublished data).

argillaceous, but marine activity is also discernible, contrary to the Rhaetian sedimentation of Scania. The total thickness of the deposits in the Hanö Bay amounts to 50 m.

The Triassic within the Platform area of the southeastern Baltic Proper is also rather limited in thickness (100 m). Within the Danish—Polish Depression, the strata exhibit a tremendous increase in thickness (Fig. 1.18), being over 2000 m in Denmark.

Jurassic sedimentation occurred mainly within the Danish—Polish Depression and its northeastern margin including Scania, the Bornholm region, and the Hanö Bay. Limited marine transgressions have probably occurred in the southeastern Baltic Proper. The sediments deposited alternate from marine to lowland continental with deltaic marshland. The block-faulting of the Fennoscandian Border Zone caused considerable variations in the thicknesses of the Jurassic strata, in the horst regions of Scania and Bornholm.

Lower Jurassic (Lias) sedimentation commenced with a continental phase followed by several marine transgressions from the North Sea Basin in the southwest towards central Poland in the southeast. In the Danish—Polish Depression the Lower Jurassic sediments attain a thickness of 800 m in the coastal area of northwestern Poland. The Lower Jurassic of Bornholm is subdivided into 350 m of continental sediments overlain by 100 m of marine sediments and topped by 270 m of a second sequence of continental sedimentation. The sedimentary rocks are mainly sandy and clayey, with coal intercalations in the continentally deposited strata. In the Hanö Bay the Lower Jurassic is only 40 m thick. In southeastern Sjaelland the thickness of the mainly marine argillites amounts from 100 m to 200 m.

During the Middle Jurassic, sedimentation took place only in a narrow depression running from Scania in a southeasterly direction through northwestern Poland. Primarily marine deposition prevailed in northern Poland. In Scania, the sedimentation environment was mainly continental with only minor marine influence. The Middle Jurassic strata of northern Poland are less than 200 m thick and are completely absent on Bornholm. In the Hanö Bay they attain a thickness of 50 m, and consist of sandstones that are argillaceous in the lower parts with coal intercalations and minor calcareous layers.

The Jurassic Sea invaded even the southeastern part of the Baltic Proper for a short time during the transition into the Late Jurassic. The marine sediments deposited are, however, of a limited thickness. Thus the Upper Jurassic strata exposed in the southeastern Baltic Proper are less than 100 m thick. Off the northwest coast of Poland the thickness exceeds 100 m, but is only 20 m in the Hanö Bay.

The entire Jurassic sequence and part of the Triassic sedimentary sequence are exposed along the bottom off the Polish coast in the central parts of the Danish—Polish Depression. In the Bornholm region, Jurassic sediments exist only near the coast except for the northwestern area where the steeply dipping strata occur almost halfway across to the Scanian coast.

Cretaceous sedimentary rocks consisting predominantly of Upper Cretaceous marine limestones attain a considerable thickness (more than 100 m) in southwestern Scania and in the adjacent parts of the southwestern Baltic Sea within the Danish—Polish Depression (Fig. 1.18). Beyond the depression, in the coastal area of the southeastern Baltic Proper, especially off the coast of NE Poland, the strata consisting of Upper Cretaceous marine deposits are less than 100 m thick, their thickness decreases rapidly towards the north.

The transition from the Late Jurassic to the Early Cretaceous does not exhibit any marked changes in the predominantly continental deposition of sandy and clayey sediments. The absence of coal-bearing strata does, however, indicate climatic deterioration. The period is characterized by intense tectonic activity along the southwestern border of the East European Platform and within the Danish—Polish Depression. Movements along old fault lines and partly also new faulting caused a tilting of the older Mesozoic and Paleozoic deposits. The Lower Cretaceous sands and clays were therefore deposited discordantly on the older sedimentary rocks. The sedimentation also became intermittent, and the size of the sedimentary basin decreased.

A marked change occurred at the beginning of the Late Cretaceous, initiating a rather calm period of marine sedimentation. During its maximum extent the sea covered the Danish—Polish Depression and also the Hanö Bay and the adjacent land areas in northeastern Scania as well as the East European Platform off the northern and northeastern coast of Poland. It is possible that the Cretaceous sea reached originally as far north as southern Öland. Within the Depression the Late Cretaceous sedimentary rocks are mainly calcareous with sandy layers in the lower parts and thick chalk layers with flint in the upper parts. In the Hanö Bay area the sedimentary rocks are mainly sandy with calcareous intercalations.

The thickness of the Cretaceous, south of Scania, ranges from 1000 m to 2000 m. In the Hanö Bay thicknesses of 610 m and 670 m were recorded in two drill holes. More than 520 m of this sequence consisted of Upper Cretaceous sedimentary rocks.

Cainozoic rocks

Tertiary sedimentary rocks like the Mesozoic deposits are restricted to the southern parts of the Baltic Sea. In Scania, the Fennoscandian Border Zone forms the northeastern boundary for the Tertiary strata. The bedrock consists of Danian limestones with a thickness of 50—100 m and locally 170 m. Younger Tertiary (Paleocene and basal Eocene) sediments occur in Scania only sporadically but southwest, in Denmark, the Tertiary deposits become younger and thicker.

Glauconitic clays, 25 m thick, of Paleocene age have been detected in a drill hole in the Hanö Bay. Although very little is actually known of the extent of submarine deposits of Tertiary age in the southwestern Baltic Sea the large number of amber clumps found along the coast in Scania does indicate the existence of such deposits.

Paleogene and Neogene deposits have also been observed along the coasts of Poland and the Sambian Peninsula. The total thickness of the Paleogene has been estimated as 60—70 m and that of the Neogene as 6—45 m. The actual submarine extent of the Tertiary deposits today in the southeastern Baltic Proper is very poorly known.

It should finally be pointed out that Tertiary microfossils have been detected at several localities in Finland indicating the possible existence of minor Tertiary deposits within the northern parts of the Baltic Sea.

Quaternary deposits will be discussed in part B of this chapter.

B. QUATERNARY GEOLOGY OF THE BALTIC SEA*

Introduction

The entire Baltic Sea underwent multiple glaciation during the Pleistocene. Hence, the area has been repeatedly subjected to glacial erosion and accumulation. Information on possible interglacial deposits in the present marine area is very scarce and even the known deposits are restricted to the southernmost part of the Baltic Sea.

Marine interglacial deposits of Eemian age are according to V.K. Gudelis (1973, and pers. commun., 1979) known from the island of Suur-Prangli (NW Esthonian SSR), in the Frische Nehrung (spit) and in the delta of the Vistula River (at some places redeposited by glacier). According to Gudelis it is probable that marine deposits of Eemian age are present in the Ventspils area (Kurzene Peninsula) as well as on the coast of the Lithuanian SSR. It should be pointed out that the Latvian and Lithuanian interglacial marine deposits may be even older belonging to the Holsteinian Interglacial, since the radiocarbon dating of these deposits give an age beyond the limits of the method i.e., over 40 ka.

The movement of the continental ice sheet (Fig. 1.26) has caused both erosional and depositional features. On the basis of morphological data it seems that the effect of glacial erosion is not very pronounced. In fact, the general topographical outlines of the Baltic Sea area were established long before the beginning of the Pleistocene glaciation. Only when the glacial flow direction coincided with structurally weak zones in the bedrock, considerable deepening and widening of channels and valleys were caused by glacial gouging.

The material picked up and transported by the ice sheet was deposited as various types of glacial drift. All the common forms of glacial deposits, e.g., ground moraines, drumlins, and end moraines, occur on the bottom of the Baltic Sea. Not only the advance of the ice sheet but also the meltwater streams caused local erosion of the underlying bedrock (e.g., Gudelis and Litvin, 1976). Accumulation of the material transported by meltwater took place in the form of eskers and ice-marginal deltas. The finer material, silt

* By Heikki Ignatius, Stefan Axberg, Lauri Niemistö and Boris Winterhalter.

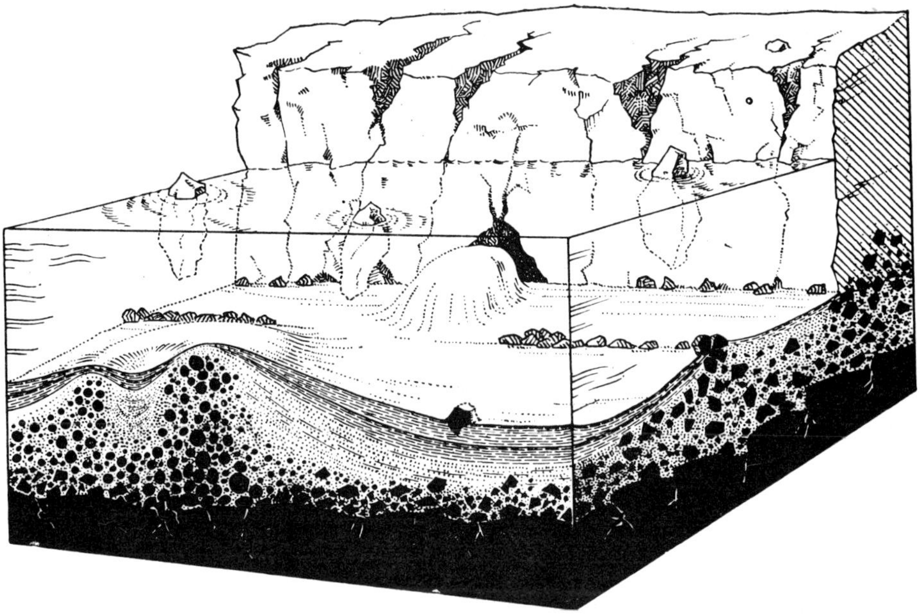

Fig. 1.26. Erosional and depositional effects of an advancing and receding ice sheet. From Magnusson et al,, 1957. The advancing ice has detached angular blocks and smaller rock fragments from the underlying bedrock (black) forming deposits of till. Melt water discharge from the ice-front induces accumulation of stratified (sorted) drift as e.g., kames and eskers. Two annual end moraines are shown in the sketch denoting the stepwise recession of the ice margin.

and clay, was transported to a greater distance and deposited as varved sediments.

The last deglaciation of the Baltic Sea basin began about 15 000 a B.P. when the Late Weichselian continental glacier had an ice marginal position just south of the present Baltic Sea (Fig. 1.27). The successive ice-marginal positions representing the gradual recession of the ice sheet have been presented by various authors (e.g., Sauramo, 1929, 1958; Hult De Geer, 1954, 1957; Aartolahti, 1972; Hyvärinen, 1975; Gudelis, 1976; and Mörner, 1977).

The ice-marginal positions for the present-day submarine area are mainly based on extrapolation from the adjacent land areas rather than on morphological and stratigraphic evidence from the sea floor and must therefore be accepted as hypothetical. This is especially the case when considering the controversiality of these positions even on land (see references in previous paragraph).

In the southern Baltic Sea, ice-marginal positions based on end moraines have been reported from the coastal shallow-water area (e.g., Kolp, 1965). Referring to Fig. 1.27, the deglaciation of the Gotland Deep area took place according to stratigraphic studies, 12 700—12 800 a B.P. (Ignatius and Niem-

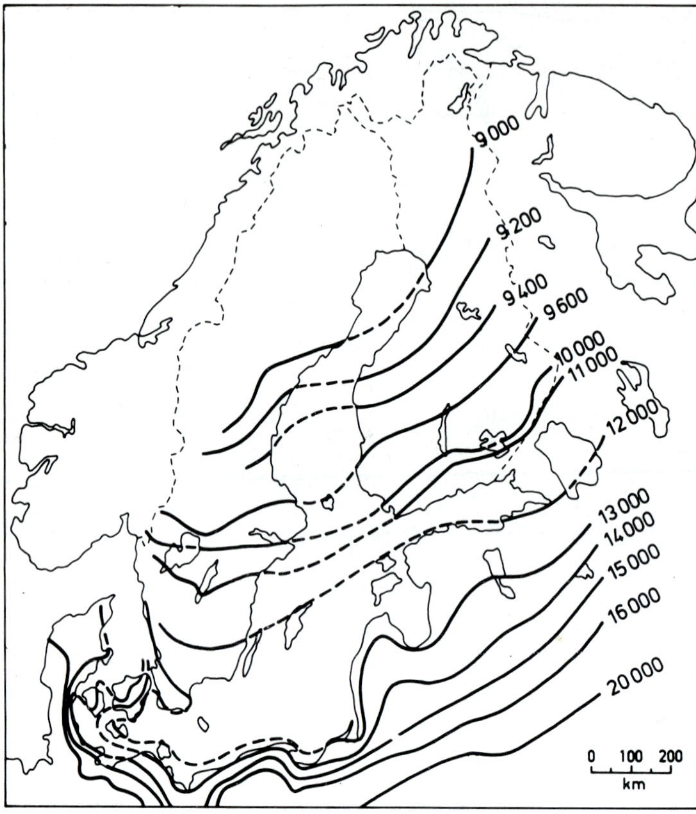

Fig. 1.27. Isochrons in years B.P. for the retreat of the Weichselian ice sheet across the Baltic Sea area. Compiled from various sources (see text).

istö, 1971). Using this date as a basis Mörner (1977) presented a rather hypothetical view of the ice-marginal positions in the central Baltic Sea.

Fig. 1.28 shows a recent concept of the extent of the continental ice sheet in late Younger Dryas time, when the ice sheet had retreated to the Second Salpausselkä (Hyvärinen, 1975). The three Salpausselkä end moraines in southern Finland have been correlated with the Moraines in Central Sweden (Donner, 1978).

In the submarine area, the continuation of the Third Salpausselkä can be traced from the coast of Finland westward. Thus an ice-marginal position with an age of about 10 000 a B.P. may rather reliably be established in the northern Baltic Proper. The submarine continuation of the First and Second Salpausselkä ridges beyond the coastal zone is uncertain. Nevertheless, a straighter line across the Northern Baltic Sea seems more likely than the southward bending line for the First Salpausselkä suggested by Mörner (1977).

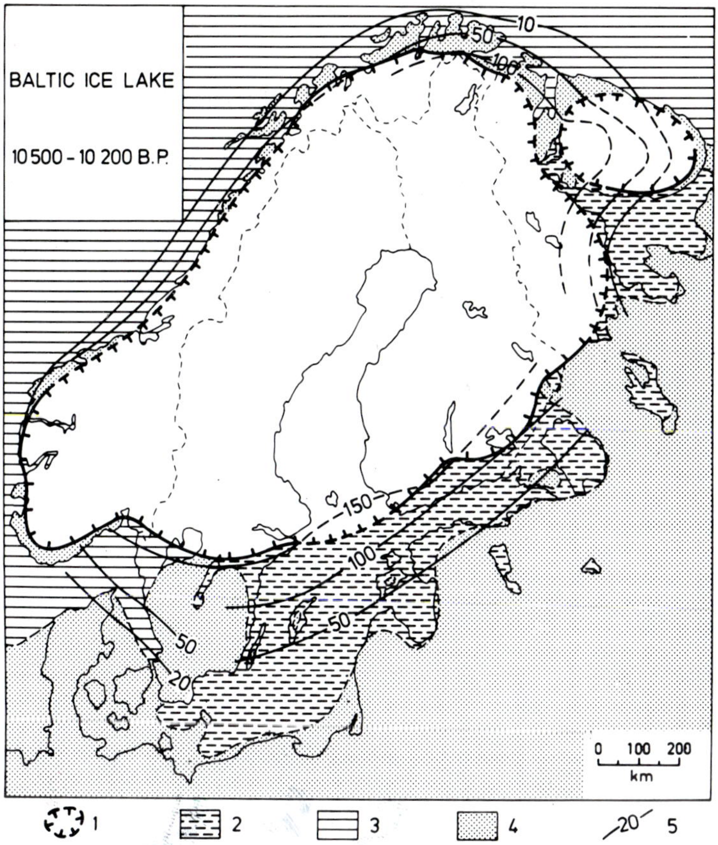

Fig. 1.28. Baltic Ice Lake and the extent of the ice sheet in late Younger Dryas time according to Hyvärinen (1975). The isobases in meters, modified from Sauramo (1958), show the height of the lake level with reference to the present-day sea level. *1* = ice margin; *2* = fresh water lake; *3* = marine; *4* = dry land; *5* = isobase with height in meters.

On the basis of Swedish (De Geer, 1940) and Finnish (Sauramo, 1929) varved-clay chronologies, recessional ice-marginal lines have been presented also for the submarine area in the Gulf of Bothnia. These lines have been shown to disagree with the trend of the ice-margin as indicated by end moraines in certain coastal areas (Aartolahti, 1972). According to varve chronology, the northwestern part of the Bothnian Bay was deglaciated by 8800 a B.P. (Lundqvist, 1961). Sediments believed to represent the late phase of the Preboreal Yoldia Sea, ending about 9000 a B.P. were found by Fromm (1965) on the Swedish coast of the Bothnian Bay, showing that the deglaciation must have taken place somewhat earlier. The view of an earlier deglaciation is in agreement with the ideas presented by Hyyppä (1966).

Evolution of the Baltic Sea

General aspects

The evolution of the Baltic Sea after the retreat of the last continental ice sheet (Weichselian glaciation) has been governed by several factors. The climatic change causing the melting of the ice had a two-fold effect. It freed the present Baltic Sea basin gradually with the recession of the ice margin, and caused an increase in the sea level of the world oceans. These together, with the isostatic rebound as an aftermath of the crustal downwarping caused by the weight of the continental ice sheet, are responsible for the various marine and lacustrine phases encountered in the evolution of the Baltic Sea. The question of glacial isostasy has recently been a source of heated debate in favour of plate tectonics. Mörner (1977) summed up the discussion around this question and presented the suggestion that the glacial isostatic rebound died out 2000—3000 a ago and that 'a tectonic factor of uncertain origin is responsible for the present uplift'.

The present-day knowledge of the various stages in the evolution of the Baltic Sea is based on geological, geomorphological, paleontological, and sedimentological data, and on various dating methods. The 'history' of the Baltic Sea as it is generally presented, especially in older literature, deals mainly with the history of shore-line displacement (Sauramo, 1958), and the hydrographic conditions have been deduced from sediments deposited in coastal lagoons and ponds instead of sediments from the deep basins of the Baltic Sea where the possibility to find a continuous sedimentary record is most obvious as shown by, e.g., Ignatius, 1958; Jerbo, 1965; Ignatius et al. 1968; Blažčišin, 1976a; and Kögler and Larsen, 1979.

The magnitude and duration of the connection of the Baltic Sea with the world ocean together with climatic fluctuations are recorded in the sedimentary strata as both lithostratigraphic and biostratigraphic units (diatom stratigraphy and pollen zonation). The following description on the evolution of the Baltic Sea and the description of its sediments are partly based on unpublished piston-coring data collected during the last two decades by the Geological Survey of Finland on marine geological cruises to various parts of the Baltic Sea.

The deglaciation of the Baltic Sea basin as delineated on a present-day map occurred between 15 000 a B.P. (southernmost Baltic Sea) and approximately 9000 a B.P. (Bothnian Bay). The history of the basin up to some 11 000 a B.P. is rather poorly and controversially known. Several authors postulate very early marine phases called either the Late Glacial Yoldia Sea (Sauramo, 1958), Karelian Ice Sea (Hyyppä, 1966), or the Baltic Ice Sea (Mörner et al., 1977), that had an open connection with the arctic White Sea in the east and possibly also a connection in the west with the Atlantic Ocean. The evidence of early salt-water phases is, however, very scarce and may have been based on the resedimentation of marine Eemian microfossils (see p. 54). In fact, Hyvärinen (1975) states categorically that the view of Sauramo (1958) of the existence of the Late Glacial Yoldia Sea must be abandoned. Considering all the facts presented in the previously cited literature and considering also the unpublished views of many workers in this field, if one insists on the existence of a connection with the world ocean it has most likely been in the west.

Baltic Ice-Lake

With the waning of the ice sheet the intensified crustal uplift together with the still rather slow eustatic sea-level rise must eventually have severed off any possible connections with the world ocean (Fig. 1.28). The sediments deposited during this stage contain very little biogenic matter, however, and the few macrofossils detected represent fresh-water species. To the north and northwest the Baltic Ice-Lake, as this stage is called, was bounded by the retreating ice margin. The meltwaters from the ice forced their way westward, possibly through the Danish Straits (Kolp, 1965). The subsequent retreat of the ice sheet opened a new outlet across the lowlands of central Sweden at Billingen causing the lake level to drop suddenly 26—29 m establishing a connection with the world ocean. This event, which according to Swedish varve chronology occurred at the time of 8213 a B.C. (Nilsson, 1968) marked the end of the Baltic Ice-Lake phase and the beginning of the Yoldia Sea phase (Fig. 1.29).

Yoldia Sea

Salt water intrusion through the widened channel across Central Sweden rapidly increased the salinity of the Baltic Sea water. This brackish cold-water stage in the evolution of the Baltic Sea, known as the (Preboreal) Yoldia Sea (Fig. 1.29) has derived its name from the arctic marine mollusc *Yoldia arctica*. This mollusc was introduced into the western Baltic Sea from the North Sea area together with other marine species. Due to the proximity of the ice margin, the northern part of the Yoldia Sea exhibited arctic conditions as witnessed by a very scarce arctic biota and the deposition of varved clays.

Synchronously with the deposition of varved clays in the north, homogenous clays, stained black by amorphous iron sulphides, were being deposited in the southern part of the sea (cf. Fig. 1.33). This is an indication of a somewhat higher production of biogenic matter and of more uniform conditions of sedimentation.

Towards the end of the Preboreal Yoldia Sea stage crustal uplift, being more rapid than the eustatic sea-level rise, restricted the inflow of saline water, thereby lowering the overall salinity of the water. This brackish-water phase has sometimes been called the Echeneis Sea, according to the diatom *Campylodiscus echeneis* found in littoral sediments of that time. This species has not been observed in Baltic deep-water sediments. In accordance with

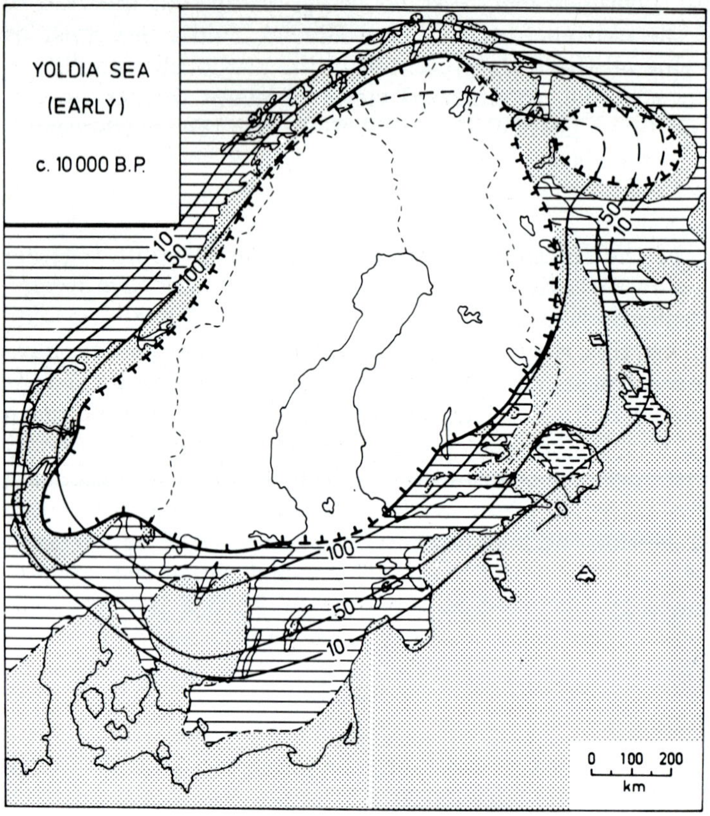

Fig. 1.29. Early Yoldia Sea and the extent of the ice sheet during the Preboreal time according to Hyvärinen (1975). The isobases in meters, modified from Sauramo (1958), show the height of the sea level with reference to the present-day sea level. For symbols see Fig. 1.28.

many authors (e.g., Alhonen, 1971) a separate phase between the Yoldia and Ancylus stages is not recognized here. In fact, Eronen (1974) even suggested that the importance of the ocean connection during the Yoldia Sea seems to be overemphasized possibly due to the redeposition of, e.g., Eemian sediments.

By the end of the Yoldia Sea phase the ice had vanished from Finland but was still present on the mainland in Sweden. The progressing crustal uplift finally superseded the eustatic sea-level rise cutting off the oceanic connection. As a result the Baltic Sea basin, isolated from the ocean, was changed into a fresh-water lake, the Ancylus Lake.

Ancylus Lake

The Ancylus Lake (Fig. 1.30) had its first outlet through the Svea River in southern Sweden (Munthe, 1927). The justification of the existence of the Svea River has been discussed in length by Fredén (1967). The fauna in the Ancylus Lake included, among others, the fresh-water snail, *Ancylus fluviatilis*, and the flora consisted of species typical of great lakes, e.g., the littoral diatom *Melosira arenaria*, *Melosira islandica ssp. helvetica* and *Stephanodiscus astreae* are typical in offshore sediments (Table 1.I). The Ancylus clays are homogeneous indicating the distant retreat of the remains of the continental ice sheet beyond the Baltic Sea basin.

Because of the differential land uplift, greater in the north than in the south, the water level of the Ancylus Lake tilted, transgressing in the south, and finally establishing a new contact with the rising ocean through the Danish Sounds. This initiated a new stage in the evolution of the Baltic Sea, the Litorina Sea at approximately 7500 a B.P.

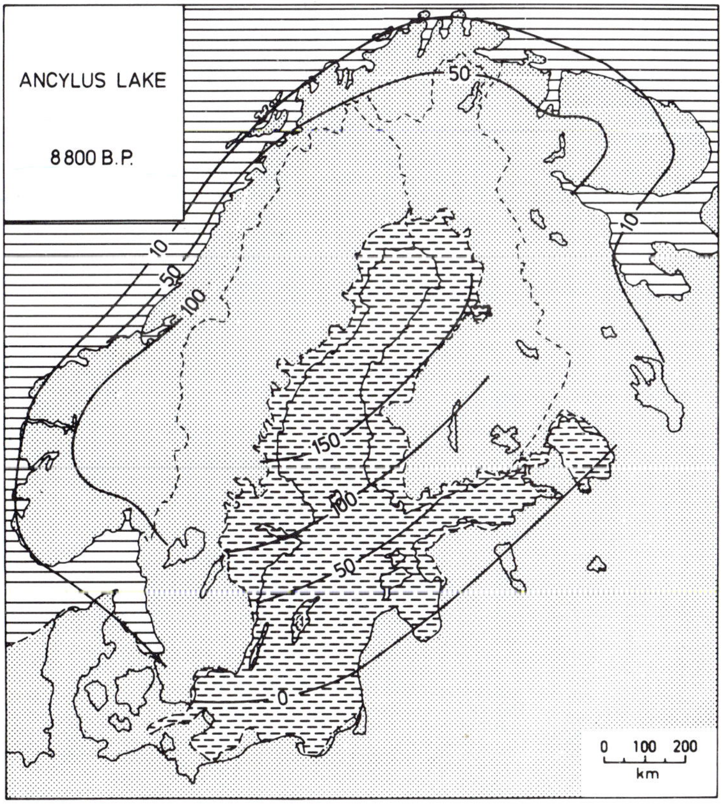

Fig. 1.30. Extent of the Ancylus Lake according to Sauramo (1958). The isobases in meters show the height of the lake level with reference to the present-day sea level. For symbols see Fig. 1.28.

TABLE 1.I

Diatom flora in Baltic Sea sediments (according to Ignatius and Tynni, 1978)

Stage of the Baltic Sea	Littoral	Deep water*
Present Baltic	*Melosira moniliformis, M. jurgensi, Cocconeis scutellum, Synedra tabulata, Mastogloia smithii, M. elliptica, Anomoeoneis sphaerophora, Navicula peregrina, N. elegans, Caloneis amphisbaena, Epithemia turgida*	Brackish water forms: *Coscinodiscus lacustris var. septentrionalis, Thalassiosira baltica* Marine forms, partly rather euryhaline: *Actinocyclus ehrenbergii, Thalassionema nitzschioides*
Litorina Sea	Benthos and ephiphyte species similar to those above, but the halofilous forms are more common: *Melosira moniliformis, Cocconeis scutellum, Synedra tabulata, S. crystallina, Hyalodiscus scoticus, Rhabdonema arcuatum, Diploneis didyma, D. interrupta, Mastogloia. elliptica, Navicula peregrina, N. digitoradiata, N. elegans, Amphora robusta, Nitzschia circumsuta, N. punctata, N. tryblionella* ...	Marine forms, partly rather euryhaline: *Actinocyclus ehrenbergii, Chaetoceros mitra, Ch. subsecundus, Rhizosolenia calcar avis, Thalassionema nitzschioides, Rhabdonema arcuatum* Brackish water forms: *Thalassiosira baltica, Diploneis didyma*
Ancylus Lake	Benthos and epiphyte forms: *Melosira arenaria, Cocconeis disculus, Diploneis mauleri, Caloneis latiuscula, Eunotia clevei, Epithemia hyndmanni. Cymbella prostrata, Cymatopleura elliptica* ...	Clear fresh water forms: *Stephanodiscus astraea, Melosira islandica ssp. helvetica, Cymatopleura elliptica, Cocconeis disculus, Diploneis domblittensis, Gyrosigma attenuatum*
Yoldia Sea	Benthos and epiphyte forms: *Synedra tabulata, Grammatophora oceanica, Rhabdonema arcuatum, Diploneis interrupta, D. smithii, Rhopaloidia musculus, Nitzschia navicularis* ...	Brackish water forms: *Thalassiosira baltica, Diploneis smithii, D. didyma* Clear fresh water forms: *Melosira islandica ssp. helvetica.* Diatom density often very low
Baltic Ice-Lake	Scarce diatoms, principally clear fresh water forms: *Melosira islandica ssp. helvetica*	

* pertains to the Gulf of Finland and the Gulf of Bothnia.

Some authors (e.g., Sauramo, 1958) suggest that a feeble marine influence began already about 8000 a B.P., basing their views on the occurrence of weakly halofilous diatoms, e.g., *Mastogloia* species. These species are, however, also known from the fresh-water littoral environment in proximity to carbonate sediments or limestone bedrock (R. Tynni, pers. commun., 1979). No evidence of a distinct brackish-water Mastogloia Sea phase has so far been observed in deep water sediments of the Baltic Sea. Thus, it is preferred to prolong the Ancylus Lake stage to approximately 7500 a B.P., in accordance with radiocarbon datings.

Litorina Sea

Sediment cores from various parts of the Baltic Sea indicate that the end of the Ancylus Lake stage and the beginning of the Litorina Sea stage is marked by an exceptionally sharp lithostratigraphic boundary (Jerbo, 1961; Ignatius et al, 1968; Blažčišin, 1976a). In fact, it is generally so sharp (Fig. 1.31) that a catastrophic event in the hydrographic conditions of the entire Baltic Sea would seem to be the only plausible explanation of the abrupt change, most probably synchronous, from the deposition of homogeneous gray clay to the deposition of a soft greenish mud rich in organic matter. The idea of a catastrophic drainage of the Ancylus Lake and the subsequent formation of the Litorina Sea (Mastogloia Sea) was proposed by Sauramo (1954). Kolp (1965) in his study of the southwestern Baltic Sea could not, however, find any proof of a catastrophic outflow of waters, causing the rapid regression of the Ancylus Lake.

According to diatom analyses, the initial phase of the Litorina Sea (Fig. 1.32) was definitely more saline than the present-day Baltic Sea. This is also evidenced by the molluscs *Mytilus edulis* and *Litorina littorea*, which inhabited more northerly coastal waters than at present. The slow decrease in salinity to reach the salinity of the present-day Baltic Sea is only feebly detectable in the sediments of the Baltic Proper. In the Gulf of Bothnia and especially in coastal lagoons this decrease in salinity is rather well established according to diatom analyses. Due to this recorded decrease, the last phase of the Litorina Sea at a time about 3000 a ago, is by some authors given the name Limnea Sea (see, e.g., Sauramo, 1958).

Stratigraphy of the clay sediments

The stratigraphic succession in the deep basins of the Baltic Sea is, on the whole, rather regular. In principle, the same stratigraphic units are recognizable in the entire area from the Gdansk Bay in the south to the Bothnian Bay in the north (Fig. 1.33). It should be emphasized that the lower part of the sedimentary sequence, i.e., the late-glacial sediment units are metachronous while the upper part, i.e., post-glacial sediments as major lithostratigraphic units are synchronous. This, of course, does not exclude regional and local variations in the sediment facies.

Fig. 1.31. Core section M1/69 from the Bothnian Sea showing the lower (dark) and upper (light gray) Ancylus sediments and the sharp boundary (X) separating the lowest part of the microlayered mud sediments of the Litorina Sea (uppermost).

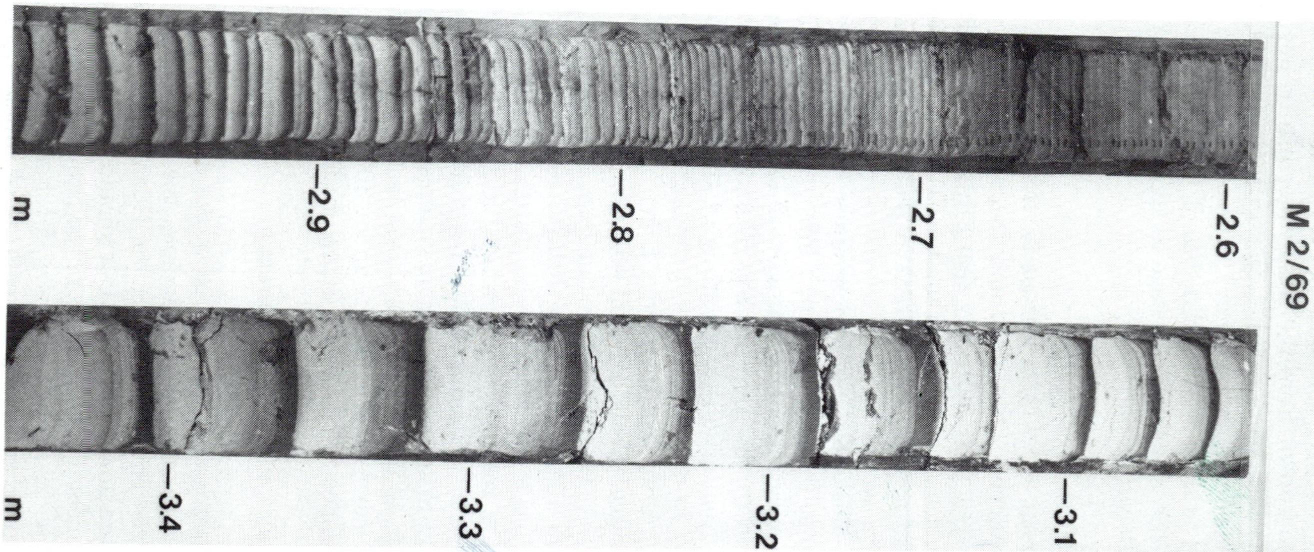

Fig. 1.32. Glacial varved clayey sediments from core M2/69, from the Bothnian Sea. The core section on the right shows the seasonal graded bedding of thick proximal varves deposited near the ice margin. The core section on the left also shows diatactic varves but of a decreasing thickness with increasing distance from the receding ice margin.

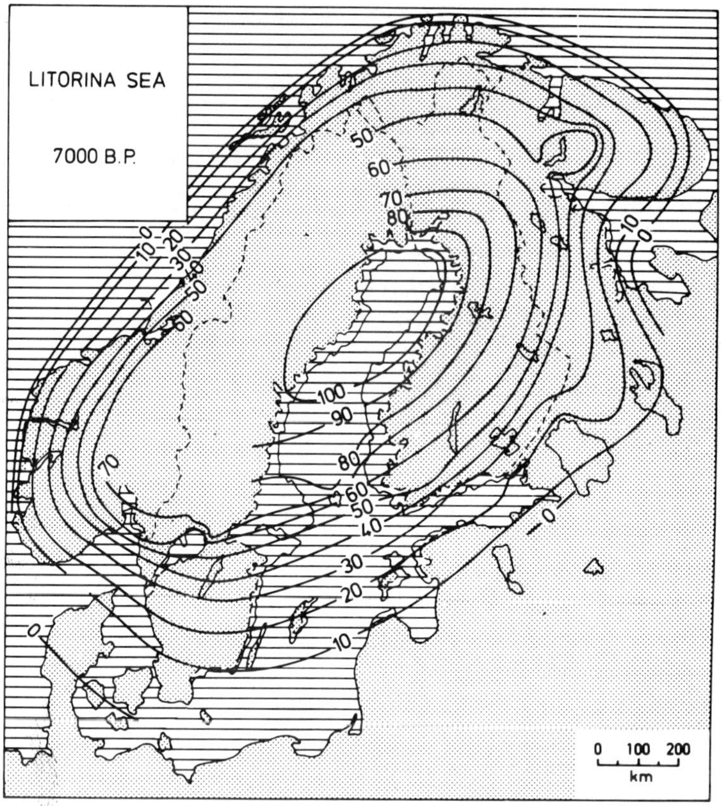

Fig. 1.32. Extent of the early Litorina Sea according to Sauramo (1958). The isobases in meters show the height of the sea level with reference to the present-day sea level. For symbols see Fig. 1.28.

The sedimentary sequence consists of three major lithostratigraphic units: glacial clay and silt, transition clay, and post-glacial mud. Referring to Gripenberg (1934) it is proposed to use the term mud instead of the term gyttja clay and clay and gyttja-banded clay used in Scandinavian literature to denote loose, water-laden, fine-grained post-glacial Baltic sediments, often rich in biogenic matter (see e.g., Jerbo, 1961; Ignatius et al., 1968).

Glacial clay. The basal sequence consists of glacial (late-glacial) clay and silts. Characteristic of these sediments is the more or less distinct varved texture (Fig. 1.34*). The term diatactic is used for sediments with a distinct graded bedding related to the annual rhythm of deposition of coarse and fine laminae (Sauramo, 1923). Sauramo also introduced the term symmict to denote varved sediments consisting of mixed material without a pronounced graded bedding. The glacial clays may be even rather homogeneous in the

* Figure 1.34 is shown on p. 64.

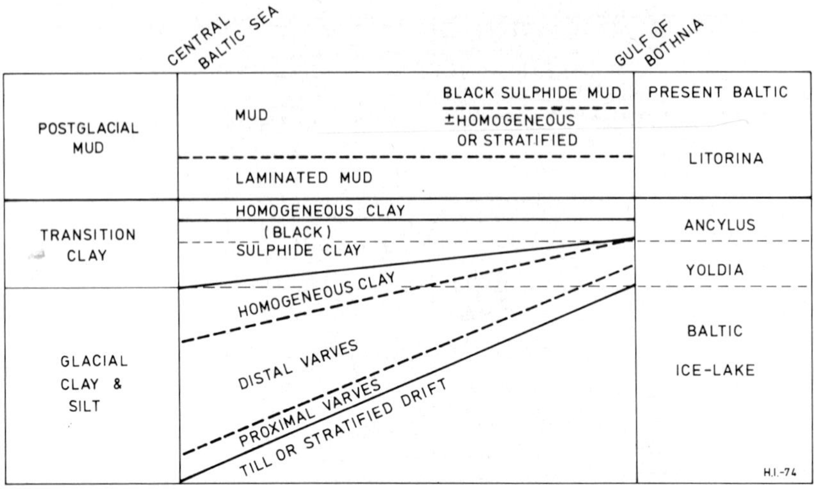

Fig. 1.33. Stratigraphy of the glacial and post-glacial sediments of the Baltic Sea. Note that the sedimentary facies of the Ancylus, Yoldia and Baltic Ice Lake deposits are time transgressive in a north-south direction.

upper part of the sequence. In the Gulf of Bothnia the varved glacial deposits grade without an intermediate homogeneous clay layer into the overlying sulphide clay in the transition clay sequence (Ignatius et al., 1968).

In addition to a great varve thickness, the basal strata or the so-called proximal varves contain normally also coarser-grained material, sand and even gravel. Higher up in the stratigraphic column the mean grain size decreases. This change in sediment facies is obviously the result of normal deglaciation, the retreating ice margin causing an ever diminishing supply of sediment material, thus resulting in an upward decreasing varve thickness (Fig. 1.35).

The glacial clays of the Baltic Sea are in general gray or brownish. Variations in color occur both regionally and stratigraphically. Regional variations in the basal sediments are mainly controlled by lithology. Thus the color portraying the mineralogy of the varved sediments in the Gulf of Bothnia is related to the occurrence of various sedimentary rocks on the sea floor (see part A of this chapter). Stratigraphic variations are obviously influenced also by the hydrographic conditions at the time of sedimentation. In the southern and central parts of the Baltic Sea the lower gray diatactic varved clay layer is generally overlain by a brown less distinctly diatactic varved clay sequence (Kögler and Larsen, 1979). This change could result from the inflow on saline water into the Baltic Sea basin.

Transition clay. In the deep-water facies of the Baltic Sea the glacial clay is covered by a lithostratigraphic unit corresponding to the change from a

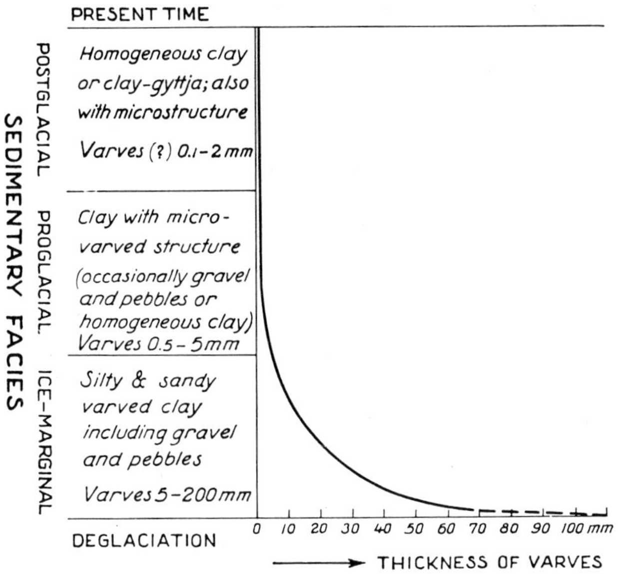

Fig. 1.35. Generalized curve showing the succession of changes in the rate of sedimentation in the deep basins of the Baltic Sea from the time of deglaciation until the present time. From Ignatius (1958).

late-glacial to a truly post-glacial sedimentation environment. The term, transition clay, will in the following be used to describe this unit, which, although Holocene or post-glacial in age, still exhibits characteristics of a glaciogene sediment, being, however, not as compact as the latter due to a higher content of water. It is typically low in biogenic matter, and sedimentologically it is a direct continuation of the conformably deposited glacial varved clays and silts. The transition clay is, however, also related with the overlying post-glacial muds as far as the amount of microfossils, pollen and diatoms, is concerned. This is especially true of the conditions in the southern and central parts of the Baltic Sea.

The transition clay consists of two units. The lower unit is characterized by the occurrence of black amorphous monosulphide-stained material (Papunen, 1968). In the basal part the sulphide material occurs in the form of black spots and streaks, sometimes as laminae. In the upper part the sulphide is often so dominant that the sediment gives the impression of a compact black layer. This is especially true of the sediments in the Gulf of Bothnia. The more or less compact black layer was deposited during the early Ancylus stage (Ignatius et al., 1968).

The upper part of the transition clay, with an age corresponding to the late Ancylus stage which by some authors is called the Mastogloia Sea (cf. Section p. 63), consists of a rather homogeneous gray, sometimes bluish gray clay, occasionally faintly laminated or streaky. Small pyrite and marcasite concretions, at most a few millimeters in diameter, are often observed in the middle part of the upper Ancylus sediments (Ignatius et al., 1968).

Post-glacial mud. The uppermost lithostratigraphic unit in a typical sediment from a sedimentary basin in the Baltic Sea consists of post-glacial mud sediments. Although, as stated on p. 67, the transition clay represents an intermediate type between the truly glacial clays and post-glacial muds, the boundary between the transition clay and the overlying mud is normally strikingly distinct (cf. p. 63). The abrupt change in sediment facies representing the Ancylus—Litorina boundary in the evolution of the Baltic Sea is generally recorded on sonic profiles as a good acoustic reflector (Fig. 1.36; Winterhalter, 1972; Blažčišin, 1976a).

The Litorina mud sediments of the Baltic Sea are characterized by a high content (10—15%) of organic matter. In the deep-water facies a very pronounced laminated structure is present (Ignatius, 1958). The laminated mud or "gyttja-banded clay" (Jerbo, 1961; Ignatius et al., 1968) has been observed in the entire Baltic Sea, although variations in facies do occur. In the central basins of the Baltic Sea the muds include laminae rich in carbonates

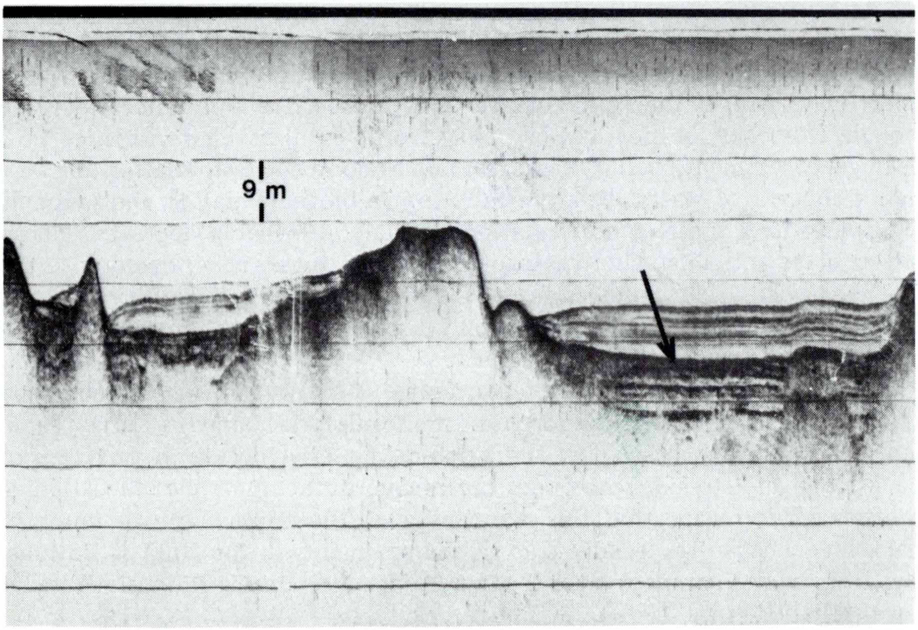

Fig. 1.36. Boundary between the Ancylus and Litorina sediments constitutes a good acoustic reflector (arrow) as the sonogram (4 kHz) from the Bothnian Sea indicates.

not present in the corresponding sediments in the Gulf of Bothnia. The laminated structure is not as well developed, or may even be absent in the upper part of the mud sequence, i.e., in the sediments of the Limnaea (post-Litorina) Sea. The upper part is characterized by the presence of black monosulphides. This is especially the case in the Bothnian Bay where the younger muds are entirely black with the exception of the topmost brown, oxidized layer, that is at most 2—3 cm thick (Ignatius, 1958; Tulkki, 1977).

Breaks in the sedimentary sequence may occur due to temporary changes in the current pattern even in deep-water sediments. In coastal areas such hiatuses are common, caused by sea-level variations during the evolution of the Baltic Sea. The sedimentation interrupted due to interaction of land uplift and "wave-base" erosion may resume in sheltered basins of the emerging coast line.

Geomorphology of the Baltic Sea floor

The present-day Baltic Sea is a shallow brackish sea with only a limited water exchange with the world ocean through the Danish Straits. Thus tidal sea level fluctuations are hardly noticeable. Long-term sea level changes have, however, been considerable mainly as a result of the crustal rebound (Fig. 1.5) following the diminishing weight of the waning continental ice sheet. Also eustatic changes in the world ocean have at times been effective in the Baltic Sea. Although it has not always been the case (see Section p. 58), today the southernmost coastline of the Baltic Sea is weakly transgressive, and the northern part is strongly regressive. The unstable sea level together with the geology of the coast are the main factors responsible for the great variety of coast line types in the various parts of the Baltic Sea (Fig. 1.37).

Although the Baltic Sea is a shallow sea, the morphology of the sea floor is as diverse as that of the coast lines. The main features are of pre-glacial origin. Glacial erosion and deposition together with later current- and wave-induced erosional and depositional processes have a rather limited role. The difference in the pre-glacial and the present-day sea floor relief ranges from a few meters to some tens of meters. Only in extreme cases have glacial and post-glacial processes caused more extensive changes.

The majority of the more outstanding morphological features in the Baltic Sea consist of various forms of deeps — depressions and troughs (trenches) in most cases partly filled with Quaternary sediments. The depths of the existing depressions are, however, rather modest: the depth of the Gotland Deep (east of Gotland) is 245 m. If the Quaternary deposits be removed the depth would be about 280 m. The Gdansk depression in the southeastern Baltic Proper is only 116 m deep. The Landsort Deep, north of Gotland, is the deepest place in the Baltic Proper with a depth of 459 m plus an estimated 150 m of Quaternary sediments. The trough forming the deepest part of the Åland Sea attains a maximum depth of a little less than 300 m. The Härnö-

Fig. 1.37. Distribution of morphogenetic types of present-day coasts of the Baltic Sea according to Gudelis (1967). I, non-altered coasts: *1* = archipelagic (skerries); *2* = fjords; *3* = fjards; *4* = bays and coves; *5* = faulted. II, coasts altered by processes other than wave induced: *6* = deltaic; *7* = marine alluvial accumulation. III, coasts altered by wave-induced processes: *8* = abraded and indented, *8a* = subtype with klints; *9* = abraded and accumulated, *9a* = bodden subtype; *10* = smoothened by abrasion; *11* = smoothened by abrasion and accumulation; *12* = smoothened by accumulation, *12a* = lagoonal subtype.

sand Deep in the northern part of the Bothnian Sea is 230 m deep (Winterhalter, 1972), while the deepest part in the Bothnian Bay is a mere 147 m.

Despite the several considerable deeps, the mean depths of the various parts of the Baltic Sea are rather modest: the Baltic Proper is 65 m, the Bothnian Sea 68 m, and the Bothnian Bay 43 m deep.

The general bathymetry of the Baltic Sea is rather well established. Detailed bathymetric data, a prerequisite for morphological analysis of the sea floor is, however, lacking in many areas, e.g., in parts of the Gulf of Finland and in the central part of the Baltic Proper. This insufficiency of data is most obvious in areas with rough topography where large depth fluctuations occur within short distances. It is most pronounced in areas where the bedrock consists of igneous and metamorphic rock predominantly of Precambrian age (see maps in Fig. 1.3a, b). In areas with a sedimentary bedrock the morphology is generally more gentle, and thus even a scarce network of depth data permits a rather reliable interpretation of the morphology of the sea floor.

The bathymetric maps in Fig. 1.38a, b have been prepared at the Geological Survey of Finland from a heterogeneous material, consisting of, e.g., unpublished marine survey data acquired from various research institutions in both Finland and countries surrounding the Baltic Sea, nautical charts and marine science publications containing bathymetric information in various forms. Physical dimensions of the Baltic Sea and its subareas are presented in Table 1.II.

The major factors that have affected the morphology of the present-day sea floor are summed up as follows:

(1) pre-glacial bedrock surface;
  (a) type of rock;
  (b) tectonism, fractures and faults;
(2) glacial erosion and deposition;
  (a) overdeepening of elongated depressions, scouring and gouging;
  (b) deposition of glacial drift;
(3) post-glacial sedimentary processes;
  (a) coastal erosion and accumulation;
  (b) "basin fill" type of sedimentation;
  (c) local erosion and deposition caused by bottom currents.

Considering the inhomogeneity and partial insufficiency of bathymetric data and the multitude of factors affecting the present-day morphology of the Baltic Sea, the following discussion will be a general description instead of a morphoanalytical treatment of the sea floor forms. For further details, the reader is referred to, e.g., Winterhalter (1972), Gudelis (1976), and Tulkki (1977).

The southern part of the Baltic Sea is rather shallow with depths rarely exceeding 50 m. The connection between the North Sea and the Baltic Sea has its deepest passage, 18 m, through the Danish Straits. The Sound, between Denmark and Sweden, forms a rather flat and shallow area. The sill depth of the Sound is only 8 m, although north of the sill the depth locally reaches 50 m. Depths exceeding 100 m in the southern Baltic Sea occur only in two areas: the Gdansk depression and the basin NE of Bornholm. The Gotland Deep and the Gdansk depression are bounded in the west by several shoals forming a broad ridge extending from Gotland to the coast of Poland in the

TABLE 1.II

Dimensions of the Baltic Sea and its subareas

Basin or Deep	Area (km^2)	Volume (km^3)	Max (m)	Mean (m)	Sill depth (m)
Baltic Proper	209,200	13,600	459	67	17
Arkona Basin			55		17
Bornholm Basin			105		45
Gotland Sea			245		60
Gdansk Basin			116		88
Gotland Deep			249		140
Central Basin			219		115
Landsort Deep			459		138
Western Gotland Deep			205		100
Gulf of Riga	18,100	410	51	28	20
Gulf of Finland	29,600	1130	123	38	a
Åland Sea	5200	410	301	77	b
Archipelago Sea	8300	200	104	23	a
Gulf of Bothnia	103,600	5830	294		70
Bothnian Sea	66,000	4340	294	68	70
Bothnian Bay	36,800	1490	147	43	25
Baltic Sea, total	374,000	21,580	459	60	17

[a] No clear sill.
[b] Several deep but narrow channels up to 150 m.

south. There are four main shallow areas, viz., Hoburg Bank with a minimum depth of 10 m, the Northern Midsjö Bank 9 m, the Southern Midsjö Bank 13 m, and the Lawica Slupska, 40 km N of Ustka in Poland with a minimum depth of 8 m. These shoals exhibit large areas with depths less than 20 m.

The area around the Hiiumaa and Saaremaa islands and also the adjacent Gulf of Riga exhibit a rather flat sea bottom rarely reaching deeper than 25 m. However, in the central part of the Gulf, the bottom slopes gently to a depth of 50 m.

North of Gotland to the shoal of Kopparstenarna the submarine steep sided ridge, of which the island of Gotska Sandön is a part, separates the northern part of the Gotland Deep, often called the Fårö Deep, from the Landsort Deep. The Landsort Deep, located along a major fault line, is a typical trough deepened by exaration (Flodén and Brännström, 1965).

The northern and northwestern boundary of the Paleozoic sedimentary rocks running in an arc from northern Öland, north of Gotland, into the Gulf of Finland constitutes a morphological scarp (Martinsson, 1958) of considerable dimension (see Section p. 43). North and northwest of this boundary, the morphology of the sea floor is governed by the rugged crystalline basement (Fromm, 1943) with narrow and rather deep valleys between sharp peaks of often exposed bedrock (Fig. 1.39).

The Gulf of Finland shows a general decrease in depth towards the east. The southeastern part of the Gulf exhibits characteristic large-scale forms, probably drumlinoid in origin, with a NNW—SSE direction rising some 25—50 m above the surrounding sea floor. Off the Finnish coast the bottom morphology is governed by the bedrock surface, rugged in the shallow zone and successively leveled off by late- and post-glacial sediments in the deeper parts.

The Baltic Proper is connected with the Gulf of Bothnia by a number of deep channels running both through the Archipelago Sea between the Åland Islands and the SW mainland of Finland and connecting the deep of the Åland Sea. The channels have formed at least partly by exaration along fracture zones that probably formed during the Precambrian. A set of very prominent channels link the Åland Deep with the Bothnian Sea. All the channels are characterized by very steep sides.

In the southern part of the Bothnian Sea the sea floor is very rugged, but further north the Paleozoic sedimentary rocks cover the crystalline rocks and change the bottom into gently undulating forms with thick post-glacial sediments filling the Eastern Basin. The northern boundary of this gentle topo-

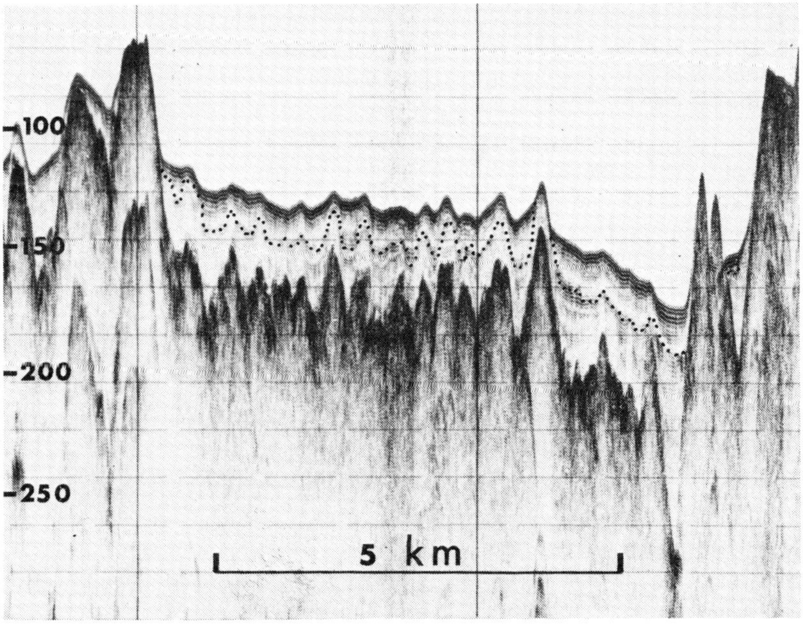

Fig. 1.39. Sonogram across rugged crystalline basement partially covered by Quaternary sediments north of the Paleozoic boundary in the northern Baltic Proper. The vertical scale is in milliseconds denoting two-way travel time.

graphy runs NW from Pori in Finland to Härnösand in Sweden along a major fracture zone (see Fig. 1.4). Further north the sea floor has a more rugged nature. The area around the Härnösand Deep is characterized by large-scale drift forms, sometimes exceeding 100 m in height (Winterhalter, 1972). There is a rather sharp contrast between the bottom morphology off the Finnish coast as compared to that off the Swedish coast. The Finnish side is largely rather flat whilst the Swedish side of the coastline is governed by a series of faults and fractures which make the bottom morphology highly irregular, even near the coast. Large shallow-water areas occur in the central and southern part of the Bothnian Sea.

In the north, the Bothnian Sea has a shallow connection, called the Quark, with the Bothnian Bay. The same general features, asymmetrical morphology, rugged on the Swedish side and gentle on the Finnish side, typical of the Bothnian Sea also hold true in the Bothnian Bay. The uneven sea floor in the central and northeastern part of the Bay is mainly caused by deposits of glacial drift of variable thickness and by channels running mainly NW-SE (Tulkki, 1977).

Quaternary sediments of the Baltic Sea

The geology of the bedrock is of utmost importance with respect to the character and distribution of unconsolidated sediments in the Baltic Sea. Besides being the major source of material deposited during and after glaciation, the old bedrock topography contributed much to the distribution and evolution of glacial deposits. A factor which has considerably affected the distribution of late-glacial and post-glacial sediments is the differential uplift of the Baltic Sea basin, together with the multiple transgressions and regressions that have occurred during the various phases of the evolution of the Baltic Sea.

The distribution of various sediments, e.g., coarse-grained vs. fine-grained ones, depends on a number of factors such as water depth, wind fetch, distribution of currents and the supply of material. Thus one can differentiate between at least five different zones of sedimentation (see Pratje, 1948; Gudelis, 1976):

(1) coastal sand accumulation zone, especially noteworthy in the southern and southeastern part of the Baltic Sea;

(2) relict clastic deposits, e.g., exposed glacial drift, with only minor evidence of reworking of the uppermost layer of sediments;

(3) zone of retarded sedimentation, also of non-deposition; increasing exposure to wave- and current-induced water motion following crustal uplift or eustatic sea level change;

(4) zone of erosion, not to be confused with local erosion in deep channels and on the flanks of topographic highs, the latter being a case of local increase in current velocities due to topographic flow restrictions; and

(5) zone of sedimentation (silts and muds) generally located well below the permanent halocline and the "wave-base" level including internal waves in the permanent halocline (50—80 m).

Due to the differential land uplift there is a marked difference in the sedimentation conditions between the southern Baltic Sea and the northern marine area. The sedimentation conditions in the southern part are rather stable, since sea-level fluctuations have been rather small for a considerable time span. In the northern part, however, the continuous regression has brought new sea floor areas into the regime of erosion processes. Former areas of active sedimentation pass first into a stage of non-sedimentation before being eroded (resuspended) and transported for deposition in tranquil basins. Especially in the northern part of the Baltic Sea basin, where a vast archipelago restricts free water circulation and decreases the depth of the "wave-base", even shallow basins can exhibit conditions of active sedimentation. This is true also of coastal lagoons and ponds common in various parts of the southern Baltic Sea.

A schematic map showing the distribution of Quaternary sediments in the Baltic Sea is given in Fig. 1.40a, b. The southern part is based mainly on a compilation of various published data together with some unpublished material from the Department of Geology at the University of Stockholm. The northern part is also a compilation from various sources. The Gulf of Bothnia has been redrawn from a map prepared by Ignatius et al. (1970a, b). The sediment distribution in the Gulf of Finland has been compiled at the Geological Survey of Finland from both new field data and from previously published material (e.g., Gudelis, 1976).

In the compilation of the sedimentary map it has been purposely departed from the conventionally adopted grain-size-dependent classification of the topmost sediment layer in favor of a genetic approach. The classification used is based on the acoustic characteristics of the sediment, in fact of the bulk of the sediment. Thus, for example, clays or muds covered by sands some decimetres thick are still classified as soft bottom sediments on the map. Analogously an area marked as sand and gravel is either a glaciofluvial deposit or a coastal accumulation of sand.

According to the principles mentioned above it is differentiated between three major bottom types: (1) hard bottom including till, and outcropping bedrock; (2) sand bottom including gravel; and (3) soft bottom to which mud, clay and silt have been assigned. When these sediment types occur in a patchy manner, not warranting a differentiation either due to small-scaleness or to a lack of sufficient data for a reliable interpretation, a combination of the types has been used. This gives a total of 12 separate composite groups with 8 groups common for both the northern and southern parts. The symbols used are identical in the two maps (Fig. 1.40a, b) and also in the following detail maps.

In the southwestern part of the Baltic Sea, in the region of the Danish Straits and the Sound, the main part of the bottom consists of sand (Fig. 1.41). Exceptions are only the deeper parts where soft bottoms dominate. In the Sound, several bottom areas consist of outcropping bedrock exposed to high-velocity currents. Within the southern and central parts of the Baltic Sea, the three main bottom types cover individually large areas. Soft bottoms are generally dominant in the deep parts, where the sediment thickness amounts to several tens of meters. Sand bottoms are present along the southern and eastern coastal zones. Hard bottoms occur, for example, north of Poland and also off the southeastern coast of Sweden. They are

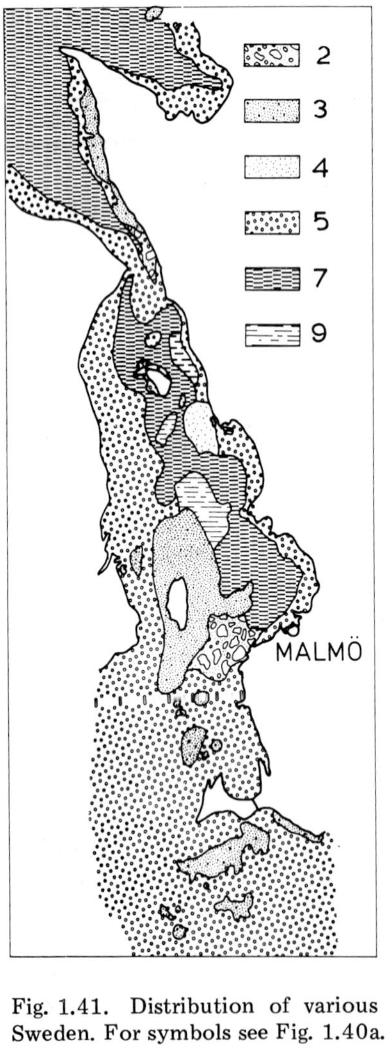

Fig. 1.41. Distribution of various bottom types in the Sound between Denmark and Sweden. For symbols see Fig. 1.40a.

separated by soft bottoms in deep basins and by sands in the central shoal areas.

The northern part of the Baltic Sea is characterized by an irregular topography with depressions, fairly flat areas and shoal regions. Around Gotland and Öland, in the Gulf of Riga, and also further north there exist sand and till deposits of considerable dimensions. Soft bottoms occur mainly where the water exceeds 80 m in depth. The northern part of the Baltic Proper is largely composed of small alternating hard and soft bottom areas too small to be distinguished on the map.

The area between the northern part of Gotland and the small island of Gotska Sandön is shown in Fig. 1.42. A conspicuous narrow ridge bounded by soft-bottoms runs in a N—S direction. It is a complex deposit of glacial drift with large amounts of redeposited sediments along the sides (see Mörner et al., 1977).

The Gulf of Finland is characterized by a greatly variable distribution of bottom types e.g., Logvinenko et al., 1978. The northern part consists of rocky areas alternating with clay sediments filling the deeper parts. Large deposits of glacial drift are typical for the southeastern part. There too, the deeps lying between are filled with soft sediments.

The Åland Sea is dominated by soft bottoms in the depression. The deepest and narrowest parts are eroded by bottom currents, hence the top surface consists of coarse-grained lag sediments. The area northeast of the depression is characterized by rough bedrock topography with consequently associated alternation of hard and soft bottoms.

In the Bothnian Sea, west of the basin area that links the Åland Deep with the Härnösand Deep, large areas are dominated by hard bottoms. Along the coast of Sweden, as well as that of Finland, hard bottoms separated by minor soft bottom areas are present. From the coast off the town of Pori in Finland, a broad esker extends northwestwards almost across the entire sea. The reason why it is not visible on more than a part of the map is that it is covered in the central parts by thick soft deposits. Near the shore, coastal processes have destroyed the original ridgelike form of the esker. A more typical esker system exists in the southwestern part of the Bothnian Sea where the Uppsala Esker (Hoppe, 1961) extends as an underwater ridge approximately 100 km towards the NNE. A detailed map of this Gävle Bay area is given in Fig. 1.43. The map is based on approximately 2000 km of continuous seismic reflection profiles and simultaneously obtained echo-sounding profiles. It illustrates well how complicated the sediment distribution is when detailed information is available.

Very extensive deposits of glacial drift occur in the northwestern part of the Bothnian Sea just off the coast of Sweden. Further to the east there is a large drumlin field. All of these drift forms are more or less exposed, with soft sediments in between and sometimes on the flanks. A more comprehensive description of the Quaternary sediments of the Bothnian Sea is given by Winterhalter (1972).

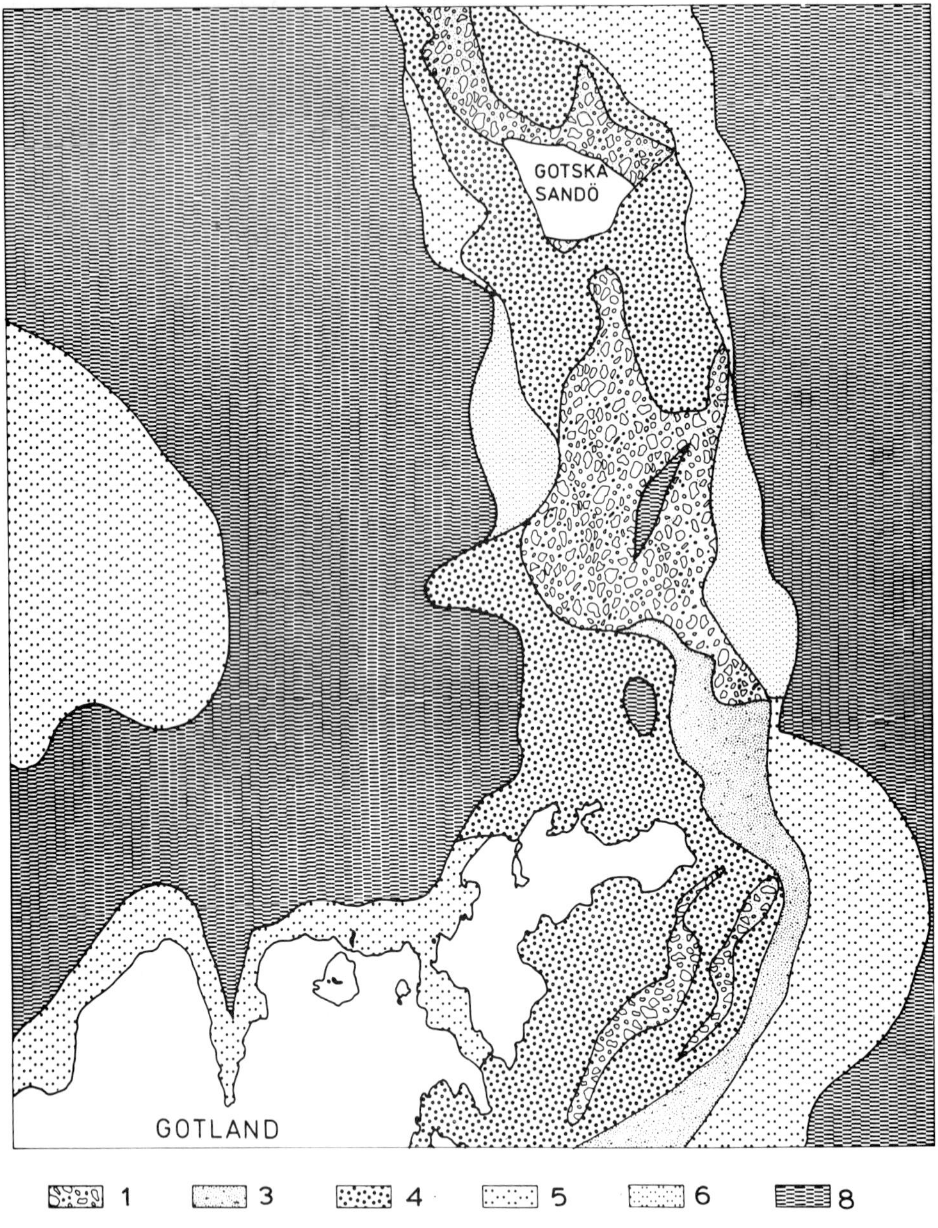

Fig. 1.42. Ridge of glacial drift north of Gotland. For symbols see Fig. 1.40b.

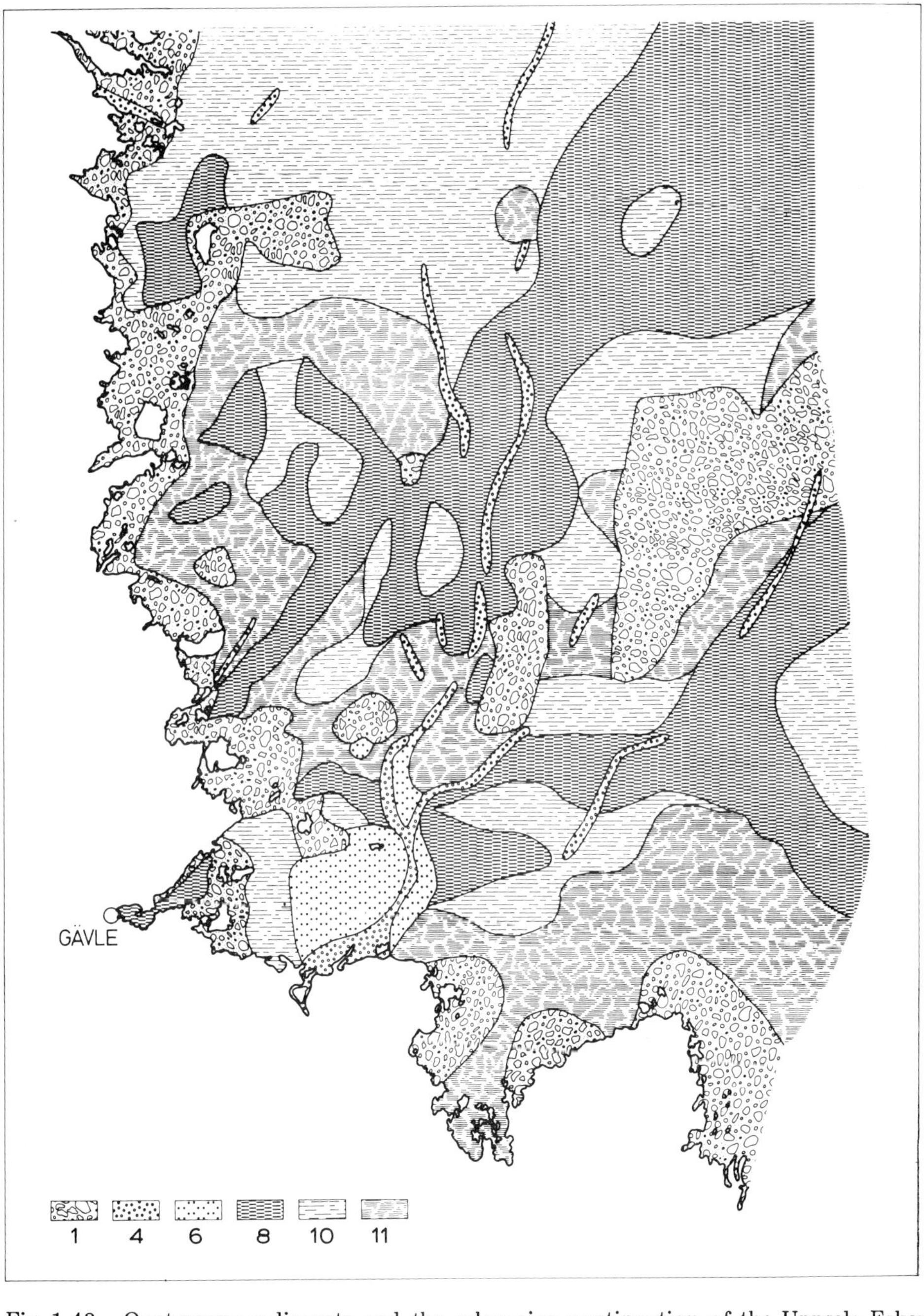

Fig. 1.43. Quaternary sediments and the submarine continuation of the Uppsala Esker (marked as sand and gravel) in the Gävle Bay, SW Bothnian Sea. For symbols see Fig. 1.40b.

The conditions in the Bothnian Bay, constituting the northernmost part of the Baltic Sea, are dominated by resedimentation of material derived from shallower areas. This dominance is due to the very rapid land uplift in the region and the vast shallow areas exposed to wave erosion. Sands are very common in the northeastern part of the Bay. Several of the eskers along the Bothnian Bay coast have submarine extensions. A review of the geomorphology and the sediments of the area has been prepared by Tulkki (1977).

Recent mud sediments

Continuity of sedimentation

The increased interest in the recent sediments of the Baltic Sea is caused by the need to learn more about the biogeochemical cycles of the elements and their compounds for which the recent sediments serve at least as a temporary and often also as an ultimate sink. In biogeochemical circulation, the time scales are of totally another order of magnitude than is usual in geological contexts.

However, before a sediment can be considered on a time scale, the continuity of sedimentation in the sampling area must be ascertained and an estimate of the rate of sedimentation must be made. It cannot be taken for granted that uninterrupted sedimentation always prevails in open basins located in a central part of a marine area and exhibiting sufficient water depth to exclude, e.g., wave-induced water motion which in turn might disturb sedimentation. If, however, the present annual accumulation of sediments agrees with the mean deposition of material during a longer stratigraphically determinable time span, e.g., 7000 a for the Litorina Sea stage and no hiatuses can be detected in the core samples, the measured rate of sedimentation can be considered reliable and indicative of continuous deposition. With the above principle in mind four different areas (basins) will be briefly discussed.

The Gotland Deep is a flat basin with gentle slopes. In the central part of the basin, the maximum total thickness of the Litorina and post-Litorina sediments amounts to 7 m. Considering the duration of 7000 a for the Litorina Sea to the present-day, this would imply an annual sedimentation rate of 1 mm a^{-1} (see Ignatius et al., 1971). By means of ^{210}Pb method, which is applicable for sediments younger than 100 a (see Niemistö and Voipio, 1974; Häsänen, 1977) a sample of the uppermost sediment from the Gotland Deep yielded annual sedimentation rates of 1.0—1.3 mm. This suggests that the rate of sedimentation has been surprisingly uniform for several thousand years. Kögler and Larsen (1979), among others, have pointed out that in the West Bornholm Basin the rate of the sedimentation varies in different parts, being greatest in the central parts and decreasing to nil at the fringes. They give for the West Bornholm Basin a sedimentation rate estimate of 0.5—1.5 mm a^{-1} depending on the state of consolidation of the sediments, the upper limit indicating the most recent deposition.

Since the various mass-balance calculations include estimates for deposition per time units, a knowledge of mean sedimentation rates would be highly desirable. However, the sedimentation rate varies even in a single basin, mainly governed by the local current pattern. Further more, several sediment zones with great variations in sedimentation rates may be distinguished (see p. 86). The results referred above, regarding the Baltic Proper, show that the rate in sedimentary basins hardly exceeds 1—2 mm a^{-1}. Therefore, the mean sedimentation rate for the entire Baltic Sea is probably lower by one order of magnitude. Some authors have used the value of 0.1 mm a^{-1} in their mass-balance calculations (e.g., Voipio and Niemistö, 1979). According to Pustelnikov (1977) the total rate of sedimentation for the entire Baltic Sea is 0.079 mm a^{-1} and 0.14 mm a^{-1} for the active sedimentation regions calculated according to the amount of material suspended in the water both of terrigenic and biogenic origin.

Winterhalter (1972) has estimated the mean thickness of post-glacial (Litorina and post-Litorina) sediments in the Bothnian Sea as 1.9 m as spread over an area of 53,000 km^2, the period corresponding to about 7000 a. This is equivalent to a mean rate of sedimentation of 0.27 mm a^{-1}. A core collected from an area with exceptionally favourable sedimentation conditions in the southern Bothnian Sea suggests a present-day sedimentation rate of 2.4 mm a^{-1}, according to ^{210}Pb dating (Viopio and Niemistö, 1979). The total thickness of the post-glacial sediment in the locality corresponds to a mean rate of sedimentation of 1 mm a^{-1}. The difference is probably mainly a result of sediment compaction.

A maximum sedimentation rate of 1.9 mm a^{-1} for the Bothnian Bay has been given by Tulkki (1977). This rate is representative for a very limited area of the Bay. No general estimates for the whole area have been given, but there are indications of sedimentation rates similar to or slightly lower than those found for the Bothnian Sea.

In closed and small basins sedimentation can be definitely higher than those mentioned above, in fact, rates up to 10 mm a^{-1} are observed near the coast. In cores taken from such basins the content of biogenic matter is rather high. Consequently, the decomposition of the biogenic organic substances causes a virtual decrease of the sedimentation rates deeper down in the sediment.

Geochemical comments

Biogeochemical processes in recent sediments are not adequately known. Ignatius et al. (1971) have studied the relationship between the known stagnation periods of the Gotland Deep and the vertical changes in redox potential in a sediment core taken from that basin. The analyzed core covered a time span from 1750 to 1970. Some of the redox-potential minima coincided with high water salinities at the beginning of the known stagnation periods in 1922—1932 and 1952—1969 and also 1894. Periodicity of different magnitudes was observed, two of these seem to have lasted about

60 years. The sporadic occurrence of hydrogen sulphide in the bottom-near waters of the Gotland Deep of today need not necessarily be attributed to human activity.

Although chemical information on recent muds from a pollution point of view is still very scanty, there are definite indications of an increasing contamination of bottom sediments of the Baltic Sea by various harmful substances. Thus, sediment cores from various parts of the Baltic Sea and especially from coastal areas influenced by, e.g., municipal waste discharge, do show an increasing concentration of some heavy metals, PCB, DDT, etc., towards the top of the sediment layer. Erlenkeuser et al. (1974) have discussed the sediments of the western Baltic Proper and the Kieler Bucht. They found a significant increase of metal contents towards the top of their cores; Cd, Pb, Zn, and Cu being enriched 7-, 4-, 3-, and 2-fold, respectively. Suess and Erlenkeuser (1975) found that the annual transport of anthropogenic zinc was of the order 100 mg m^{-2} in the Bornholm and Gotland Deeps and in the Kieler Bucht. They also stated, that the increase in zinc began in the middle of the 19th century in the Kieler Bucht area and clearly later in the Bornholm Basin and in the Gotland Deep. Niemistö and Tervo (1978) noted a similar increase of zinc beginning in about 1930—1940 in a core taken in the northern Baltic Proper. They found no significant increase of zinc in either the Bothnian Sea or the eastern Gulf of Finland. The observed trend seems to indicate that zinc dispersion is a slow process and that the northern parts of the Baltic Sea are so far uncontaminated. Near-shore studies have, however, revealed high Zn contents in the vicinity of some industrial establishments. Olausson et al. (1977) pointed out that the contents of heavy metals are higher in near-shore sediments than in the open-sea sediments, indicating a local influence rather than an overall contamination of the Baltic Sea.

Although the content of many elements and compounds in recent mud sediments has been determined at a number of localities in the Baltic Sea, very little is still known about the total flux of elements into the sediments. This is partly due to the insufficient knowledge of actual sedimentation conditions prevailing in the various parts of the Baltic Sea. Another factor causing uncertainty is the inadequacy of the available knowledge of processes taking place in the sediment/water interface (see Chapter 4, p. 205).

Niemistö et al. (1978) made an attempt to evaluate the net accumulation of two elements, iron and phosphorus, in the Bothnian Sea. The conditions for precipitation of iron are ideal in the Bothnian Sea. In the oxidizing environment iron precipitates readily as oxyhydrates and is trapped in sedimentary basins during normal deposition. Due to the disintegration of contemporaneously deposited matter a reducing environment is formed when buried under some 5—6 cm of new sediment. At least a part of the reduced and mobilized iron is immediately fixed in the form of monosulphides. The annual input of iron into the Bothnian Sea was estimated as about 100,000 tons while about 170,000 tons were stored annually in the sediments. Niemistö et al. (1978) assumed that the excess iron originates from material being eroded in the more shallow areas.

Niemistö et al. (1978) also found that the amount of phosphorus annually stored in the sediments of the Bothnian Sea seems to be greater than the annual input from rivers, rain and industry. A reasonable explanation would be that the excess of phosphorus originates in the Baltic Proper, where the conditions for phosphorus deposition are not favourable, due to the sporadic oxygen deficiency encountered in the deeps.

The variation of phosphorus content in the topmost centimeter of the sediment is very marked in the Gotland Deep, where values from 0.13% to 0.46% (of dry matter) (Fig. 1.44) have been observed during different years (e.g., Tarkiainen et al., 1974). The high phosphorus contents can be explained by the rapid co-precipitation with iron when anoxic conditions turn into oxidizing ones as a result of occasional inflow of new water into the deep.

The very high correlation coefficient of 0.943 for the iron/phosphorus ratio in recent muds in the Bothnian Sea, as compared with a correlation coefficient of —0.667 for the ratio in sediments in temporarily anoxic basin in the northern Baltic Proper (Niemistö et al. 1978), indicates a significant difference in entrapment of the elements under varying environmental conditions. In the Bothnian Sea virtually all phosphorus is co-precipitated with iron, while in the Baltic Proper the phosphorus in the sediments occurs in various forms; according to Blažčišin (1976b) as authigenic minerals, detrital particles, and organic complexes. The latter provide mobile phosphorus as a result of decomposition in an anoxic environment. Fig. 1.44 shows the distribution of iron and phosphorus in the Baltic Sea determined in the topmost 1 cm layer of recent mud. It is obvious that organic matter plays an important role in the transportation, deposition and remobilization of heavy metals, pollutants, etc. Thus a comparison of the C/N ratios of recent muds indicates that at least a part of the organic carbon in the Baltic Proper is of authigenic origin (C/N = 9.1). The C/N ratio of, e.g., 11.5 in the Bothnian Bay implies according to Gripenberg (1934), a strong influence of land humus (see also Chapter 4, p. 211).

Finally it is emphasized that although anthropogenic impact in the Baltic Sea, as deduced from recent muds, is evident in many instances, the assay data must be interpreted very critically until the processes influencing element transport and deposition are adequately understood. Since the major part of the Baltic Sea exhibits a regressive sea level and thus due to shoaling new bottom areas are being exposed to erosion, also the knowledge of the chemical and mineralogical composition of the older sediments is essential for comparison.

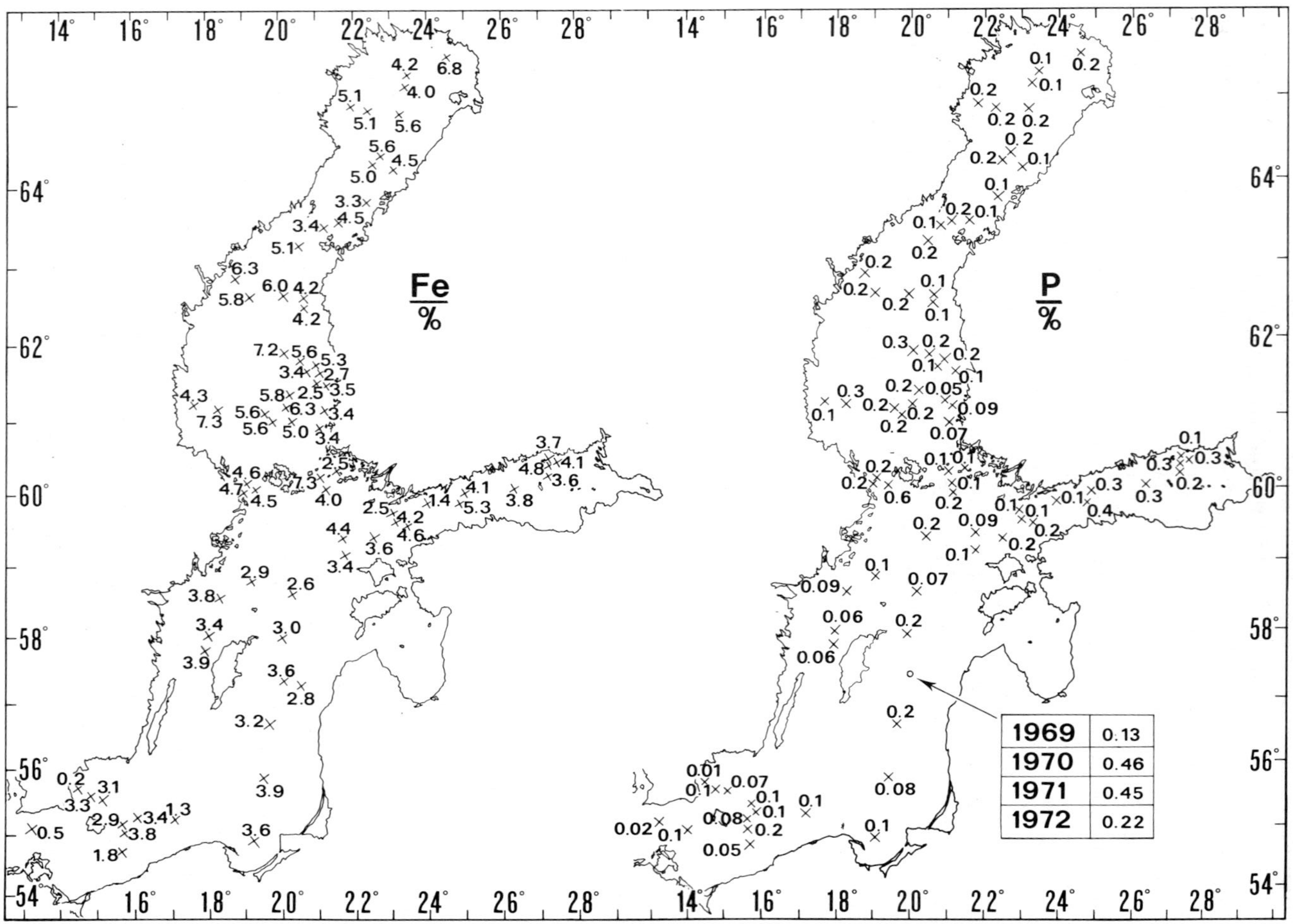

Fig. 1.44. Content of iron and phosphorus (per cent of dry matter) in the topmost centimeter of core samples from the

C. NATURAL RESOURCES*

Hydrocarbons

As stated in part A of this chapter, considerable parts of the marine area are characterized by non-deformed or slightly deformed sedimentary rocks. The fact that exploitable oil and gas deposits occur in more or less similar sedimentary sequences along the southern borderland of the Baltic Proper, especially in the Kaliningrad District of the USSR, makes it obvious that coastal states show increasing interest in the marine area. Except for the rather comprehensive study on the perspectives for oil and gas in the Soviet region of the Baltic Proper by Volkolakov et al. (1977) very little has, however, been published about actual exploration results.

In the coastal region of the southeastern Baltic Proper in the Latvian SSR, Lithuanian SSR, and the Kaliningrad District in the USSR, the exploitable hydrocarbon deposits are found in the Vendian (Upper Proterozoic) and Middle (and Upper) Cambrian terrigenic complex having a mean total thickness of 200—400 m (Fig. 1.45). Besides the stratigraphic information provided by Soviet literature and cited by Volkolakov et al. (1977) very little drill-hole information is available from the marine area. However, the combination of land data with information from drillings, e.g., off Gotland and in the coastal region of southern Sweden, provides a limited but acceptable basis for the interpretation of the rather abundant both Swedish and Soviet geophysical data mainly in the form of continuous seismic profiles.

According to Volkolakov et al. (1977), the Vendian—Cambrian complex along the Baltian coast line and within the Gulf of Gdansk consists of sandstones with occasional siltstone layers. The western part of the complex, extending to Öland, consists mainly of clayey strata with only minor sandy and silty layers. Along the Scanian coast and in the southern part of Öland alum shales and bituminous limestones seem to abound. In the central part of the marine area Vendian—Cambrian strata consist of sandstones with many intermediary layers of siltstones and shales. Although this suggested lateral variation in the lithology of especially Upper and Middle Cambrian beds does somewhat limit the areal extent of petroperspective beds the lateral transition from sandstone to shale might form suitable traps for the accumulation of hydrocarbons. The overlying Ordovician shales and limestones are rather uniform across the entire sedimentary formation in the central and southern Baltic Sea and may be considered to constitute an adequate cap rock.

The argillaceous—calcareous, Ordovician—Silurian complex, well developed across the larger part of the central and especially the southern Baltic Proper contains several marginally petroperspective horizons of which the

* By Boris Winterhalter.

Upper Ordovician reef limestones (bioherms) seem to be the most attractive. These reef structures are known to be oil-bearing in the coastal regions of the Latvian SSR and the Lithuanian SSR. These structures have also yielded minor oil in drillings off the eastern shore of Gotland. Considering the nature of such rocks the oil deposits would generally be small and thus also of limited economic interest.

The potential interests in the Devonian sediments of the southeastern part of the Baltic Proper are limited to the coastal region. Permian and Mesozoic deposits, as a whole, which lie in the aquatorial part of the Baltic Depression, (see p. 37—39) are hardly attractive due to their rather limited thickness and areal extent.

The total sedimentary thickness within the Danish—Polish Depression in the southwestern Baltic Proper amounts to 3000 m off the Scanian coast, increasing to approximately 7000 m towards the SE. Sedimentologically and structurally, the Mesozoic strata seem to be rather attractive, but, so far, the

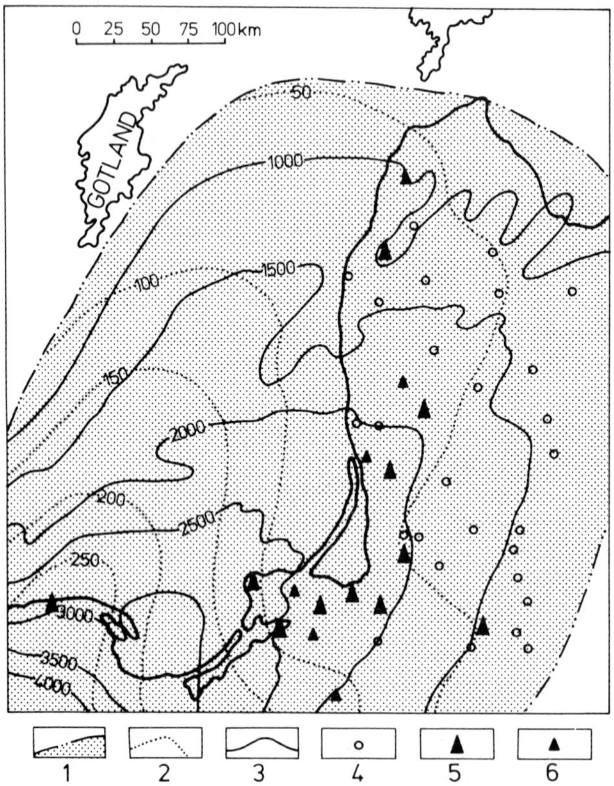

Fig. 1.45. Extent of oil- and gas-bearing Middle and Upper Cambrian sedimentary rocks. *1* = extent of petroperspective strata; *2* = Middle and Upper Cambrian isopachs; *3* = isopachs of overlying sedimentary rocks; *4* = drill holes; *5* = oil deposits; *6* = traces of oil. According to Volkolakov et al. (1977).

rather limited exploratory work within the area has not yielded oil gas deposits of commercially exploitable size. Although Cambrian bituminous shale in the form of erratics has been found sparingly along the southern coast of the Bothnian Sea and traces of hydrocarbons were detected in drill cores from the southwestern part of the Lower Paleozoic sequence (Fig. 1.3a) (P. Thorslund, pers. commun., 1978), it is evident that economically exploitable hydrocarbon reserves will hardly be found anywhere in the northern Baltic Sea.

Ferromanganese concretions

The worldwide interest in manganese geochemistry together with the commercial interests in oceanic deep-sea nodules has accelerated research also in questions dealing with precipitates of iron and manganese in the Baltic Sea. Of special interest in this respect are the papers by Manheim (1961, 1965), Winterhalter (1966, 1980), Gorshkova (1967), Winterhalter and Siivola (1967), Varencov (1973) and Varencov and Blažčišin (1976).

Occurrence and type of concretions. As stated by Varencov (1973) the formation of ferromanganese concretions is especially noteworthy in five regions, i.e. the southern and eastern parts of the Central Baltic Sea, the Gulf of Riga (Fig. 1.46), the eastern part of the Gulf of Finland and the Bothnian Bay. The nature of the deposits is very similar in the different parts of the Baltic Sea. However, locally the size and form of the concretions may vary considerably from spheroidal, with diameters up to 3—4 cm, and discoidal concretions to crusts of various thickness and form.

Relation between form and environment. The sedimentation during the last 7000 years up to the present has been characterized by a rather high amount of organic matter, both terrigenic and authigenic, being trapped in the bottom deposits of sedimentary basins. The decay of this material generally causes the formation of a reducing environment mobilizing eventual oxyhydrates of iron and especially of manganese that may happen to be buried in the sediment. Consequently, the occurrence of concretions is restricted to sea-floor areas of non-deposition or very weak intermittent sedimentation and erosion. In the latter case the thickness of the resulting sediment layer must not surpass 5—10 cm of readily reworkable deposit to ensure that non-reducing conditions prevail most of the time at the level of ferromanganese accretion. Bottom-dwelling fauna may play some role in the mixing of the sediment cover by aiding aeration.

The variations in the forms of the concretions are closely related to the type of bottom. Thus, spheroidal concretions (Fig. 1.47) occur on what might be called "soft bottom", i.e., in areas where a change in sedimentation conditions has led to the cessation of formerly active deposition of clay, silt

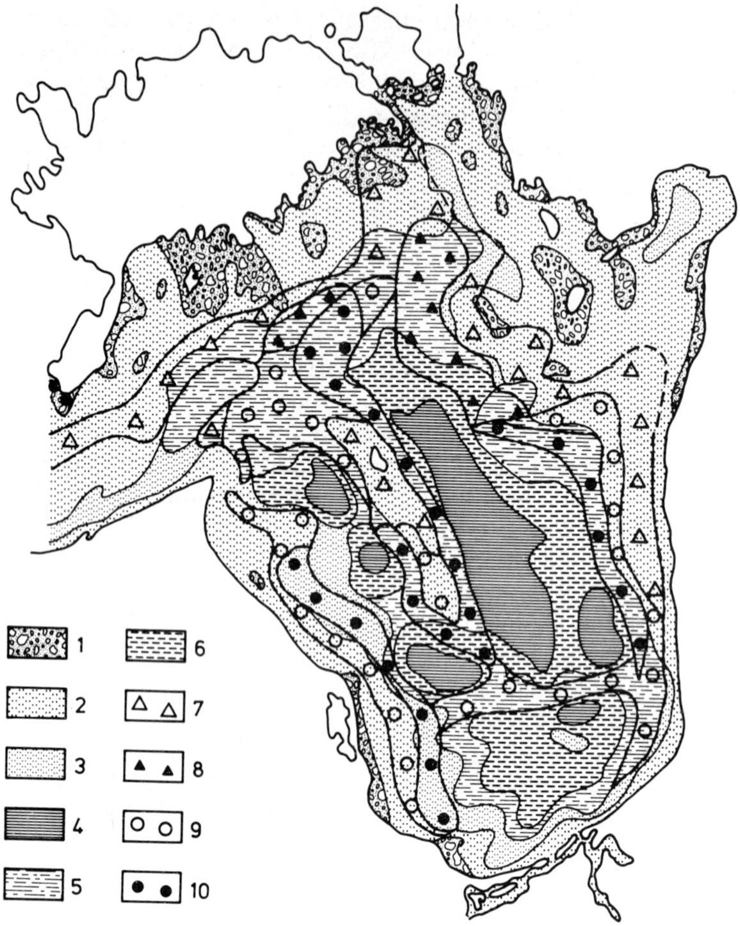

Fig. 1.46. Map showing the distribution of ferro-manganese concretions and their relation to the type of bottom sediments in the Gulf of Riga according to Varencov and Blažčišin (1976). *1* = till; *2* = sand; *3* = coarse silt; *4* = silty mud; *5* = clayey mud; *6* = silty clayey mud; *7* = sparse discoidal concretions and crusts; *8* = numerous discoidal concretions and crusts; *9* = sparse spheroidal concretions; *10* = numerous spheroidal concretions. Note the occurrence of concretionary material around the flanks of basins with recent sedimentation.

and mud, leaving only a thin more or less mobile layer on and in which concretions can form by concentric growth as spheroids.

The discoidal concretions (Fig. 1.48) generally form around a separate nucleus, e.g., a pebble or a fragment of concretionary material and occur on rather hard, often sandy or silty, bottoms. Also the rate of sedimentation must be nil or amount at most only to local deposition of some millimeters of mobile sediment during temporary slack-current conditions, to be once more removed with an increase in current velocity.

Fig. 1.47. Spheroidal concretions on a silty bottom partly covered by a thin mobile sediment layer. Diameter of the trigger weight compass of the automatic camera is 5 cm. The location is 64° 53′N and 22° 45′E and the water depth is 75 m.

The third group, consisting of slabs and crusts of concretionary material, covers a wide variety of subtypes. They do, however, constitute only a minor part of the total bulk of iron and manganese precipitated in concretions in the Baltic Sea. The greater part of this group consists of slabs of, e.g., glacial clay encrusted by a layer of oxyhydrates of iron and manganese only some millimetres thick (Fig. 1.49). The prerequisite is the existence of stiff glacial clays, generally varved, exposed to submarine erosion. Uniform crusts of concretionary material up to several millimetres thick have been observed at places on silty bottoms with non-sedimentation conditions prevailing (Win-

Fig. 1.48. Discoidal concretions from the Gulf of Finland formed around a pebble nucleus (*A*) and probably around a fragment of concretionary material (*B*). The diameters of both concretions are approximately 5 cm.

terhalter, 1966). Slabs and crusts are also known to coalesce into larger aggregates as a result of the change in current patterns affecting the environmental conditions.

Composition and internal structure of Baltic concretions. The three different types of concretions generally indicate specific environmental conditions. Thus, the threefold classification according to the form is also re-

Fig. 1.49. Crusts of concretionary material formed around lumps of late-glacial varved sediment indicating conditions of bottom erosion at a depth of 95 m in the northern Bothnian Sea. The ferro-manganese encrustation is generally only a few millimetres thick.

flected in the chemical composition especially in the interelement ratios, e.g., in the Mn/Fe ratio. The relation between the manganese and iron content and the type of concretion is shown in Fig. 1.50.

Considering that the Mn/Fe ratio in oxidized recent (post-glacial) sediments is generally about 0.1, portraying the availability of these metals dispersed originally in the water phase. In temporarily anoxic basins, e.g., in the Gotland Deep (cf. p. 101) this ratio is generally above unity (Niemistö and Voipio, 1974). Thus it appears that because the formation of concretions depends more or less on the supply of manganese and iron directly from the water phase, the crusts and slabs exhibit low ratios. The occurrence of spheroidal concretions, however, found embedded in soft sediment along the border of active sedimentary basins can be explained by the fact that manganese is easily remobilized due to the reducing conditions generally prevailing in recent sediments. The discoidal concretions, not as directly related to any specific environment as are the two other types, obviously must exhibit a wide variation in the Mn/Fe ratio (see Fig. 1.50).

The Mn/Fe ratio does not only vary in concretions from different localities, but even different parts of a single concretion may show considerable variation. This is due to the fact that the concretions as a rule consist of concentrically arranged alternating iron-rich and manganese-rich layers (Winterhalter and Siivola, 1967). The thickness of the layers may vary either in

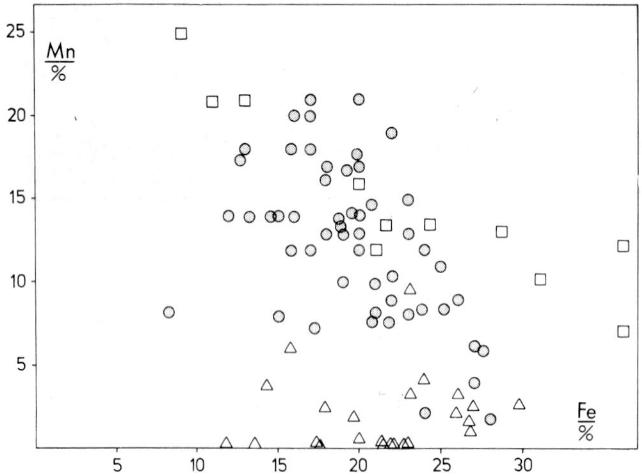

Fig. 1.50. Relationship between the iron and manganese contents in various types of concretions in the Baltic Sea. circles = spheroidal concretions; triangles = crusts and slabs; squares = discoidal concretions (see text).

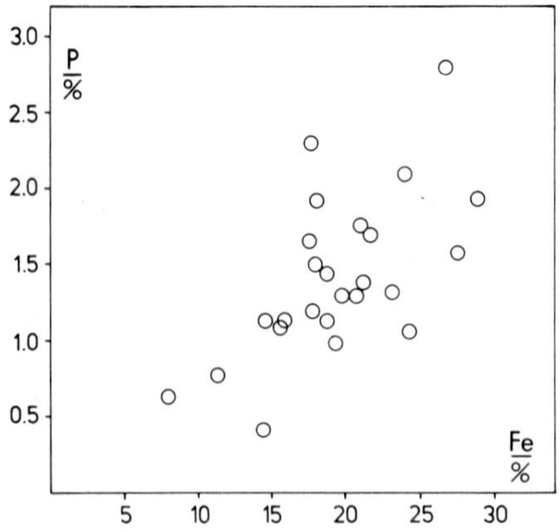

Fig. 1.51. Relationship between the iron and phosphorus contents in Baltic Sea concretions.

TABLE 1.III

Average concentrations of major and minor elements in concretions from various parts of the Baltic and from some lakes. The values are in ppm unless otherwise stated

Locality	Mn (%)	Fe (%)	P (%)	Cu	Ni	Co	Mo	Cr	Zn	Pb	V	B	ti	No of samples
Gulf of Bothnia (Winterhalter, 1980)	14.6	16.6	1.35[a]	80	260	140	330	110	200	40	100	80	1700	15
Gulf of Finland (Varencov, 1973)	13.3	19.7	1.24	10	40	100	—	20	110	10	70	—	2800	9
Gulf of Finland (Winterhalter, 1980)	12.8	17.7	1.65[b]	60	140	120	290	100	230	50	100	80	1400	25
Gulf of Riga (Varencov, 1973)	9.7	22.8	0.69	20	50	60	—	20	140	20	100	—	2900	19
Central Baltic (Manheim, 1965)	14.0	22.5	0.70	50	750	160	130	10	80	40	150	60	2200	24
SE Baltic (Varencov, 1973)	8.6	18.4	1.10	40	150	90	—	30	140	20	130	—	3100	7
Lake Eningi-Lampi, Carelian SSR (Varencov, 1972)	21.5	23.5	0.002	5	10	110	—	4	180	4	6	—	7400	8
5 Finnish lakes (Halbach, 1976)	9.7	30.9	0.4	—	70	100	—	—	—	—	10	—	—	5

a Mean of 19 samples for which Fe = 19.8% and Mn = 12.2%.
b Mean of 6 samples for which Fe = 15.3% and Mn = 12.5%.

different parts of a single layer or among different layers. In addition, the iron-rich layers are associated with a rather high amount of detrital matter. The iron-rich layers exhibit also phosphorus contents that are considerably higher than those observed in deep-sea nodules. The phosphorus content may amount to several per cent of the bulk dry weight of the concretion. A definite correlation with iron is present (Fig. 1.51).

Table 1.III gives a general picture of the trace element distribution in Baltic Sea concretions. As a rule the trace metals, e.g., Cu, Ni and Co, are far less enriched in Baltic Sea concretions than in deep-sea nodules. This is evidently due to the considerably faster growth rate in the Baltic Sea and the lower content of the trace metals in the water. The growth rate of concretions in the Baltic Sea can be estimated as 0.05—0.2 mm a^{-1}.

Amber, phosphorite and glauconite

Mineral exploration along the submarine slopes of the Sambian Peninsula (USSR) in the southeastern part of the Baltic Sea has led to the unearthing of commercially attractive deposits of amber, phosphorite and glauconite (Fig. 1.52).

The formerly rich amber deposits on the shores of the southeastern Baltic Sea are nearly exhausted. According to Blažčišin et al. (1976), considerable effort was made by Soviet research programs at the beginning of the 1970s to determine the submarine extent of the Paleogene glauconite silts ("light blue earth") known to be the source of amber. Four separate localities have so far been detected along the submarine slopes of the Sambian Peninsula. The outcropping glauconite beds (Fig. 1.52) are covered by only a thin layer (0.5—2.5 m) of Holocene sands. The thickness of "light blue earth" beds varies considerably being 2—10 m at Primore, and 5—11 m in the Bay of Pokrov according to Blažčišin et al. (1976). The highest concentrations of amber occur in the Bay of Pokrov. Here the average thickness of the beds has been estimated as 7 m, with an exploitable areal extent of 9 km^2 and of this an area of some 5 km^2 is virtually exposed at a depth of 10—17 m. The average concentration of amber has been estimated to 600—1500 gm^{-3}.

The amber-bearing "light blue earth" is underlain by a sandy glauconitic clay containing a large number of phosphorite lenses. The thickness of this clay bed varies from 1—6 m. Off the Cape of Taran, the deposit is covered by only a very thin layer of Quaternary sediments (0.5—1.5 m) at a depth of 7—9 m. Further south along the western submarine slope of the peninsula, the depth increases to 20—25 m. The richest deposits exist in the northern part of the area and have a phosphorite content of 500—700 kg m^{-3}, assaying 10—20% of P_2O_5 (Blažčišin et al., 1976).

Glauconite occurs in the "light blue earth" and in the subordinate "wild earth" beds closely related to both amber and phosphorite. The glauconite

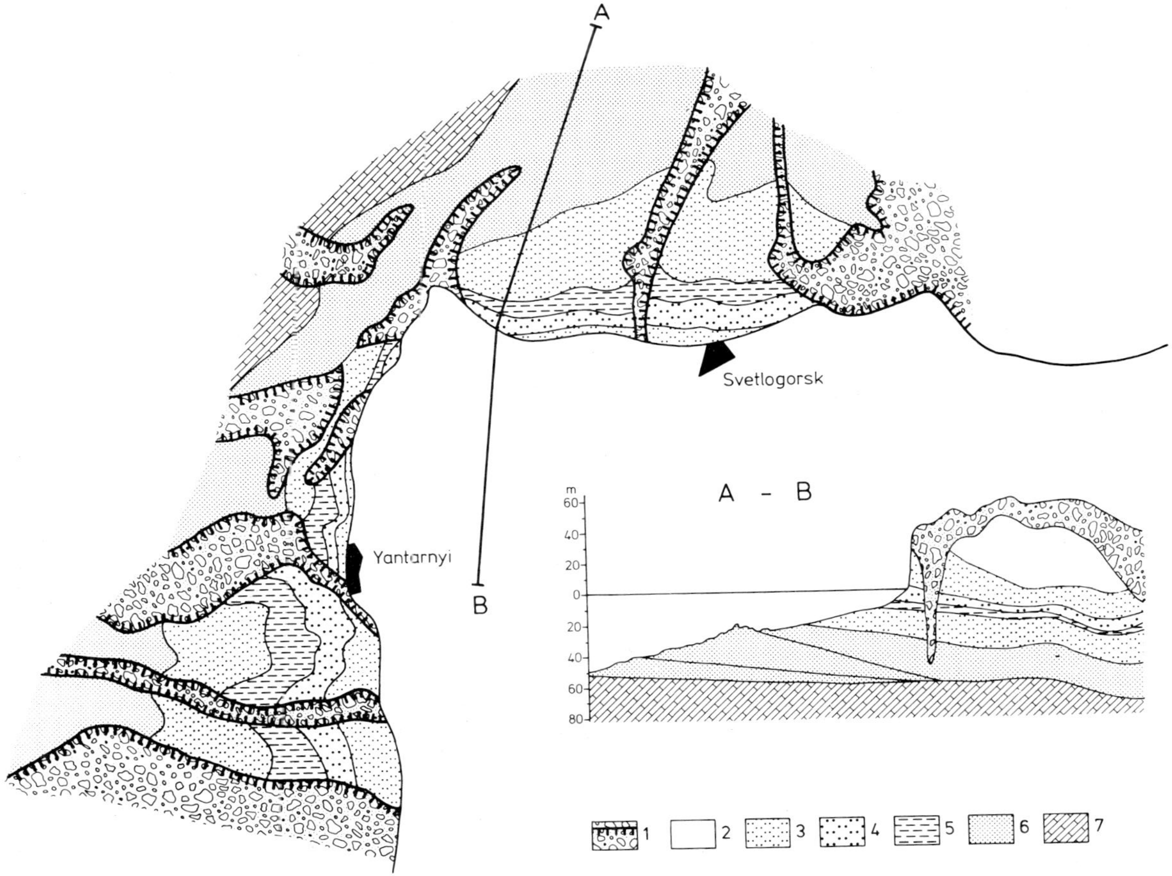

Fig. 1.52. Offshore geology of the Sambian Peninsula (USSR) in the southeastern Baltic Proper. Modified from fig. 18, in Blažčišin et al., 1976. *1* = Quaternary (till); *2* = Neogene; *3* = Eocone; *4* = "light blue earth"; *5* = "wild earth"; *6* = Lower Paleogene and Upper Cretaceous; *7* = Upper Cretaceous. For further explanation see text.

beds have an average thickness of 10—12 m and according to Blažčišin et al. (1976) they contain 35—40% of glauconite.

The fact that amber, phosphorite and glauconite occur close to one another makes their possible exploitation even more attractive. Whether a commercial utilization of these submarine deposits is already under way, has not so far been verified in available Soviet literature.

Sand and gravel

The Pleistocene glaciation of northern Europe left behind a multitude of evidence pertaining to the exarative power of the ice sheet. Some of the most spectacular monuments are the various forms of stratified drift being today exploited at an increasing pace to quench the hunger of the modern society for sand and gravel. Such deposits occur in the form of eskers, kames and various ice-marginal formations. Together with a geographically uneven distribution, the overexploitation of the land resources and the restrictive conservational measures taken by naturalists are producing a shortage of these essential commodities. This is becoming especially evident in many highly populated coastal areas of the Baltic Sea. Thus it will be necessary to turn to the many known submarine deposits of sand and gravel in the coastal areas of the Baltic Sea.

Today, probably the greatest activity in submarine exploitation of sand and gravel is taking place in the southwestern part of the Baltic Sea, especially in the Danish Straits by Denmark and in Öresund by Sweden. According to E. Heller (pers. commun., 1978) of the Geological Survey of Denmark 6.1 $\times 10^6$ m^3 of sand and gravel were recovered in 1975 from Danish waters constituting 22% of the total annual sand and gravel production.

Approximately 4.4 $\times 10^6$ m^3 of submarine sand and gravel have been recovered from the Swedish side of the Öresund during the ten-year period, 1966—1976.

In Finland, only limited use has been made of submarine deposits since large deposits on dry land have been available. However, the accelerating demand for construction purposes together with more stringent conservational requirements are working up a demand for submarine sand and gravel deposits also in Finland.

No exact figures of marine sand and gravel exploitation by the Socialist countries are available, but it is understood that it does play a role in the economy of the coastal states.

Placer deposits

Considering the regressive nature of the greater part of the Baltic Proper (see part B of this chapter) and the rather short time span available since deglaciation, economically exploitable placer deposits might probably be found only in the southernmost part of the Baltic Sea along and south of the zero-isobase (see Fig. 1.5). Thus, the coastal regions of the German Democratic Republic, Poland and the USSR may justify the exploration of such minerals as ilmenite, magnetite, rutile, zirkon, leucoxene and garnet.

Blažčišin et al. (1976) give the following total concentrations of placer minerals for the various coastal regions: the Soviet Union 60—70 kg m^{-3} with a maximum of 200 kg m^{-3}; and an average of 30—40 kg m^{-3} and a maximum of 70 kg m^{-3} for the coasts of the German Democratic Republic and Poland. The deposits so far explored are rather thin, at most a few meters thick, and of a limited areal extent. It seems to be rather unlikely that deposits warranting commercial exploitation will be found in other parts of the Baltic Proper, although it should be pointed out that very little work on placer minerals has so far been carried out especially within the westernmost part of the southern Baltic Proper.

REFERENCES

Aartolahti, T., 1972. On deglaciation in southern and western Finland. Fennia, 114: 1—84.

Alhonen, P., 1971. The stages of the Baltic Sea as indicated by the diatom stratigraphy. Acta Bot. Fenn., 92: 1—18.

Bergqvist., E., 1977. Postglacial land uplift in northern Sweden. Some remarks on its relation to the present rate of uplift and the uncompensated depression. Geol. Fören. Stockholm Förh., 99: 347—357.

Blažčišin, A.I., 1976a. Zur Stratigraphie spätquartärer Bodenablagerungen der mittleren Ostsee. Beitr. Meeresk., 38: 49—59.

Blažčišin, A.I., 1976b. Osnovnye himičeskie komponenty v donnyh osadkah (Main chemical components in bottom sediments). In: V.K. Gudelis and E.M. Emeljanov (Editors), Geologija Baltijskogo Morja (Geology of the Baltic Sea). Mokslas, Vilnius, pp. 255—287.

Blažčišin, A.I., Boldyrev, V.L. and Šuiskij, Ju. D., 1976. Drugie poleznye iskopameye (Other useful minerals). In: V.K. Gudelis and E.M. Emeljanov (Editors), Geologija Baltijskogo Morja (Geology of the Baltic Sea). Mokslas, Vilnius, pp. 349—357.

Boulanger, Ju., Deumlich, F., Entin, I., Joo, I., Kashin, L., Hristov, V., Lillienberg, D., Setunskaya, L., Vyskochil, P., Wyrzykowsky, T. and Zotin, M., 1975. Summary map of the recent vertical crustal movements for eastern Europe. In: Problems of Recent Crustal Movements. Proc. 4th Int Symp, Moscow, USSR, 1971, pp. 31—43.

Brangulis, A., Kala, E., Mardla, A., Mens, K., Pirrus, E., Sabaljauskas, V., Fridrihsone, A. and Jankauskas, T., 1974. Shema strukturnaja fazialnogo rajonyrovanija territorii Pribaltiki Vend i Kembrii. Izv. Akad. Nauk, Khim. Geol., 23(3): 218—225.

Dadlez, R., 1976. Zarys geologii podłoza kenozoiku w basenie nołudniowego Bałtyku (Outline of sub-Cainozoic geology in the South Baltic Basin). Biul. Inst. Geol. Warsaw, 285: 21—50 (in Polish, with English and Russian summaries).

De Geer, G., 1940. Geochronologia Suecica principles. K. Sven. Vetenskapsakad. Handl., III Ser., 18(6): 1—367.

Donner, J.J., 1978. The dating of the levels of the Baltic Ice Lake and the Salpausselkä moraines in South Finland. Commentat. Phys.-Math., Soc. Sci. Fenn., 48(1): 11—38.

Erlenkeuser, H., Suess, E. and Willkomm, H., 1974. Industrialization affects heavy metal and carbon isotope concentrations in recent Baltic Sea sediments. Geochim. Cosmochim. Acta, 38: 823—842.

Eronen, M., 1974. The history of the Litorina Sea and associated Holocene events. Commentat. Phys.-Math., Soc. Sci. Fenn., 44(4): 1—195.

Flodén, T., 1973. De jotniska sedimentbergarternas utbredning i Östersjön. Ymer Årsbok Stockholm, pp. 47—57.

Flodén, T. and Brännström, B., 1965. En Thumperprofil genom Landsortsdjupet. Geol. Fören. Stockholm Förh., 87(3): 337—346.

Fredén, C., 1967. A historical review of the Ancylus Lake and the Svea River. Geol. Fören. Stockholm Förh., 89: 239—267.

Fromm, E., 1943. Havsbottnens morfologi utanför Stockholms södra skärgård (Morphology of the sea bottom outside the southern part of the Stockholm archipelago). Geogr. Ann., 3—4: 137—169 (in Swedish, with an English summary).

Fromm, E., 1965. Beskrivning till jordartskarta över Norrbottens län nedanför lappmarksgränsen (Quaternary deposits of the southern part of the Norrbotten County). Sver. Geol. Unders. Ser. Ca, 39: 1—236 (in Swedish, with an English summary).

Gorbatschev, R., 1967. Petrology of Jotnian rocks in the Gävle Area. Sver. Geol. Unders. Ser. C, 621: 1—50.

Gorshkova, T.I., 1967. Marganec v donnyh otloženijah severnyh morej SSSR. In: Margancevye mestoroždenuja SSSR. pp. 117—134. Moscow.

Gripenberg, S., 1934. A study of the sediments of the North Baltic and adjoining seas. Merentutkimuslaitoksen Julk. Havsforskningsinst. Skr., 96: 1—231.

Gudelis, V.K., 1967. Morfogenetičeskie tipy beregov Baltijskogo morja (The morphogenetic types of the Baltic Sea coasts). Baltica, 3: 123—145 (in Russian, with English and German summaries).

Gudelis, V.K., 1973. Relef i četvertičnye otloženija Pribaltiki. Vilnius, 264 pp.

Gudelis, V.K., 1976. Lithochemical characteristics of the recent bottom sediments in the southeastern Baltic. Ambio, Spec. Rep., 4: 149—154.

Gudelis, V.K. and Litvin, V.M., 1976. Geomorfologija dna (Bottom geomorphology). In: V.K. Gudelis and E.M. Emeljanov (Editors), Geologija Baltijskogo Morja (Geology of the Baltic Sea). Mokslas, Vilnius, pp. 25—34.

Halbach, P., 1976. Mineralogical and geochemical investigation on Finnish lake ores. Bull. Geol. Soc. Finl., 48: 33—42.

Härme, M., 1961. On the fault lines in Finland. Bull. Comm. Géol. Finl., 196: 437—444.

Häsänen, E., 1977. Dating of sediments based on ^{210}Po measurements. Radiochem. Radioanal. Lett., 31(4—5): 207—214.

Hoppe, G., 1961. The continuation of the Uppsala Esker in the Bothnian Sea. Geogr. Annlr., 43(3—4): 329—335.

Hult de Geer, E., 1954. Skandinaviens geokronologi. Geol. Fören. Stockholm Förh., 76(2): 299—329.

Hult de Geer, E., 1957. Old and new datings of Swedish Ice Lakes and the thermals of Bölling and Alleröd. Geol. Fören. Stockholm Förh., 79(1): 93—100.

Hyyppä, E., 1966. The late-Quaternary land uplift in the Baltic sphere and the relation diagram of the raised and tilted shore levels. Ann. Acad. Sci. Fenn., Ser A 3, 90: 153—168.

Hyvärinen, H., 1975. Myöhäisjääkauden Fennoskandia — käsityksiä ennen ja nyt (Lateglacial paleogeography of Fennoscandia). Terra 87(3): 155—166 (in Finnish, with an English summary).

Ignatius, H., 1958. The rate of sedimentation in the Baltic Sea. C. R. Soc. Géol. Finl., 30: 135—144.

Ignatius, H. and Niemistö, L., 1971. Itämeren sedimentit ja sedimentaatio (Sediments and sedimentation in the Baltic). Luonnon Tutkija, 75(3—4): 72—80 (in Finnish, with an English summary).

Ignatius, H. and Tynni, R., 1978. Itämeren vaiheet ja piilevätutkimus (Baltic Sea stages

and diatom analysis). Turun Yliopiston Maaperägeologian Osaston Julk., 36: 1—26 (in Finnish, with an English summary).
Ignatius, H., Kukkonen, E. and Winterhalter, B., 1968. Notes on a pyrite zone in upper Ancylus sediments from the Bothnian Sea. Bull. Geol. Soc. Finl., 40: 131—134.
Ignatius, H., Kukkonen, H. and Winterhalter, B., 1970a. Marine geological map, Gulf of Bothnia, Bothnian Bay, Quaternary deposits. Geological Survey of Finland.
Ignatius, H., Kukkonen, H. and Winterhalter, B., 1970b. Marine geological map, Gulf of Bothnia, Bothnian Sea, Quaternary deposits. Geological Survey of Finland.
Ignatius, H., Niemistö, L. and Voipio, A., 1971. Variations of redox conditions in the recent sediments of the Gotland Deep. Geologi, 3: 43—46.
Jerbo, A., 1961. Den gyttjebandade leran i bottniska sediment. Geol. Fören. Stockholm Förh., 83(3): 303—312.
Jerbo, A., 1965. Bothnian clay sediments — a geological-geotechnical survey. Swed. State Railways, Bull., 11: 1—159.
Kögler, F.-C. and Larsen, B., 1979. The West Bornholm basin in the Baltic Sea: geological structure and Quaternary sediments. Boreas, 8: 1—22.
Kolp, O., 1965. Paläogeographische Ergebnisse der Kartierung des Meeresgrundes der westlichen Ostsee zwischen Fehmarn und Arkona. Beitr. Meeresk. 12—14: 1—59.
Kukkamäki, T.J., 1975. Report on the work of the Fennoscandian subcommission. In: Problems of Recent Crustal Movements. Proc. 4th Int. Symp. Moscow, USSR, 1971, pp. 25—30.
Kumpas, M., 1978. Distribution of sedimentary rocks in the Hanö Bay and S. of Öland, S. Baltic. Stockholm Contrib. Geol., 31(3): 95—103.
Laurén, L., Lehtovaara, J., Boström, R. and Tynni, R., 1978. On the geology and the Cambrian sediments of the circular depression at Söderfjärden, western Finland. Geol. Surv. Finl. Bull., 297: 1—81.
Lillienberg, D., Setounskaya, L., Blagoboline, N., Bylinskaya, L., Gorelov, S., Nikonov, A., Rozanov, L., Serebryannyi, L. and Filkine, V., 1975. L'analyse morphostructurale des movements verticaux actuels de la partie européenne de l'URSS. In: Problems of Recent Crustal Movements. Proc. 4th Int. Symp., Moscow, 1971, pp. 57—67.
Liszkowski, J., 1975. Recent movements of the earth's crust in Poland. Tectonophysics, 29: 1—4.
Logvinenko, N.V., Barkov, L.K. and Gontarev, E.A., 1978. Sostav i dinamika sovremennyh donnyh osadkov vostočnoj časti Finskogo zaliva (The composition and dynamics of the bottom sediments in the eastern part of the Gulf of Finland). Vestn. Leningr. Univ. 12, Geol. Geogr. Vyp., 2: 14—25 (in Russian, with an English summary).
Lundqvist, G., 1961. Beskrivning till karta över landisens avsmältning och högsta kustlinjen i Sverige (Outline of the deglaciation in Sweden). Sver. Geol. Unders. Ser. Ba, 18: 1—148 (in Swedish, with an English summary).
Magnusson, N., Lundqvist, G. and Granlund, E., 1957. Sveriges geologi. Svenska bokförlaget, Stockholm, 557 pp.
Manheim, F., 1961. A geochemical profile in the Baltic Sea. Geochim. Cosmochim. Acta, 25: 52—70.
Manheim, F., 1965. Manganese-iron accumulations in the shallow marine environment. Symp. Mar. Geochem. Narragansett Mar. Lab. Univ. Rhode Island, Occ. Publ., 3: 217—276.
Martinsson, A., 1956. Neue Funde kambrischer Gänge und Ordovizischer Geschiebe im südwestlichen Finnland. Bull. Geol. Inst. Uppsala, 36(5): 79—105.
Martinsson, A., 1958. The submarine morphology of the Baltic Cambro-Silurian area (Deep boring on Gotska Sandö. I). Bull. Geol. Inst. Univ. Uppsala, 38: 11—35.
Martinsson, A., 1974. The Cambrian of Norden. In: C.H. Holland (Editor), Cambrian of the British Isles, Norden and Spitzbergen. Lower Paleozoic Rocks of the World, Vol. 2. Wiley-Interscience, London, pp. 185—283.

Mörner, N.-A., 1977. Post and present uplift in Sweden: glacial isostasy, tectonism and bedrock influence. Geol. Fören. Stockholm Förh., 29: 48—54.

Mörner, N.-A., Flodén, T., Beskow, B., Elhammar, A. and Haxner, H., 1977. Late Weichselian deglaciation of the Baltic. Baltica, 6: 33—51.

Munthe, H., 1927. Studier över Ancylus sjöns avlopp (Studies in the outlets of Ancylus Lake). Sver. Geol. Unders. Ser. Ca, 346: 1—107 (in Swedish, with an English summary).

Niemistö, L. and Tervo, V., 1978. Preliminary results of heavy metal contents in some sediment cores in the northern Baltic Sea. Proc. XI Conf. Baltic Oceanogr., Rostock, 24—27 April, 1978. Vol. 2, pp. 653—672 (mimeogr.).

Niemistö, L., Tervo, V. and Voipio, A., 1978. Storage of iron and phosphorus in the sediments of the Bothnian Sea. Finn. Mar. Res., 244: 36—41.

Niemistö, L. and Voipio, A., 1974. Studies on the recent sediments in the Gotland Deep. Merentutkimuslaitoksen Julk. Havsforskningsinst. Skr., 238: 17—32.

Nilsson, E., 1968. Södra Sveriges senkvartära historia (The late-Quaternary history of southern Sweden). K. Sven. Vetenskaps Akad. Handl. Fjärde Ser., 12(1): 1—117 (in Swedish, with an English summary).

Olausson, E., Gustafsson, O., Melin, T. and Svensson, R., 1977. Current level of heavy metal pollution and eutrophication in the Baltic Proper. Geol. Lab. Univ. Göteborg. Medd. Maringeol. Lab. 9: 1—95 (mimeogr.).

Öpik, A.A., 1956. Cambrian (Lower Cambrian) of Esthonia. Proc. 20th Int. Geol. Congr., Mexico, The Cambrian System, 1: 97—126.

Panasenko, G.D., 1977. Zemletrjasenii Fennoskandii 1951—1970. Katalog. (Earthquakes in Fennoscandia 1951—1970. Catalogue). In: A.P. Lazareva (Editor), Materialy Mirovogo Centra B. Moskva, 111 pp.

Papunen, H., 1968. On the sulfides in sediments of the Bothnian Sea. Bull. Geol. Soc. Finl., 40: 51—57.

Penttilä, E., 1978. Earthquakes in Finland 1610—1976. Inst. Seismology, Univ. Helsinki. Rep. SI.: pp. 1—13.

Plissov, A.A., Gorijanskii, V. Ju., Vanderflit, E.K. and Sapoznikova, P.S., 1975. Novye dannye o rasščelenii Vend na severo-zapade russkoj platformy i ego granica v Kembri. In: A.Ja. Lunc (Editor), Geologija Kristalličeskogo Fundamenta i Usatočnogo čehla Pribaltiki. Riga, pp. 64—81.

Pratje, O., 1948. Die Bodendeckung der südlichen und mittleren Ostsee und ihre Bedeutung für die Ausdeutung fossiler Sedimente. Dtsch. Hydrogr. Z., 1(2—3): 45—61.

Pustelnikov, O.S., 1977. Balans osadočnogo materiala i skorosti sovremennogo osadkoobrazovanija v Baltijskom more (po dannym izučenija vzvesi). The balance of sediments and recent sedimentation rates in the Baltic Sea (according to the data of suspension studying). Baltica (Vilnius), 6: 155—160 (in Russian, with an English summary).

Rankama, K. (Editor), 1963. The Precambrian. The Geologic Systems 1. New York, N.Y. 280 pp.

Sauramo, M., 1923. Studies on the Quaternary varve sediments in southern Finland. Bull. Comm. Géol. Finl., 60: 1—164.

Sauramo, M., 1929. The Quaternary geology of Finland. Bull. Comm. Géol. Finl., 86: 1—110.

Sauramo, M., 1954. Das Rätsel des Ancylussees. Geol. Rundsch., 42(2): 197—233.

Sauramo, M., 1958. Die Geschichte der Ostsee. Ann. Acad. Sci. Fenn. Ser. A3, 51: 1—522.

Simonen, A., 1971. Das finnisches Grundgebirge. Geol. Rundsch., 60(4): 1406—1421.

Simonen, A., and Kouvo, O., 1955. Sandstones in Finland. Bull. Comm. Géol. Finl., 168: 57—87.

Suess, E. and Erlenkeuser, H., 1975. History of metal pollution and carbon input in Baltic Sea sediments. Meyniana, 27: 63—75.

Tarkiainen, E., Rinne, I. and Niemistö, L., 1974. On the chemical factors regulating the primary production of phytoplankton in the Baltic Proper. Merentutkimuslaitoksen Julk. Havsforskningsinst. Skr., 238: 39—52.

Thorslund, P., 1938. Deep-boring through the Cambro-Silurian at File Haidar, Gotland. Sver. Geol. Unders. Ser. C, 415: 1—57.

Thorslund, P., 1960. The Cambro-Silurian of Sweden. Sver. Geol. Unders. Ser. Ba, 16: 69—110.

Thorslund, P., 1970. Sommarens borrningsrapport: Ingen olja vid Finngrundet. Lufttrycket (Atlas-Copco), Stockholm, 8: 3.

Thorslund, P. and Axberg, S., 1979. Geology of the southern Bothnian Sea. Part 1. Acta Univ. Ups. Nova Acta Regiae Soc. Sci. Ups. Ser. VC, 8: 1—62.

Tornquist, A., 1913. Grundzüge der geologischen Formations- und Gebirgeskunde. Bornträger, Berlin, 296 pp.

Tulkki, P., 1977. The bottom of the Bothnian Bay, geomorphology and sediments. Merentutkimuslaitoksen Julk. Havsforskningsinst. Skr., 241: 1—89.

Tuominen, H.V., Aarnisalo, J. and Söderholm, B., 1973. Tectonic patterns in the Central Baltic Shield. Bull. Geol. Soc. Finl., 45(2): 205—217.

Tynni, R., 1978. Muhoksen muodostuman mikrofossiilitutkimuksen tuloksia. Geol. Surv. Finl. Tutkimusrap. Rep. Investig., 30: 1—18.

Varencov, I.M., 1972. Geochemical studies on the formation of iron-manganese nodules and crusts in recent basins. I. Eningi—Lampi Lake, Central Karelia. Acta Miner.-Petrogr., Szeged., 10: 363—381.

Varencov, I.M., 1973. Geochemical aspects of formation of ferromanganese ores in shelf regions of recent seas. Acta Miner.-Petrogr. Szeged 21(1): 141—153.

Varencov, I.M., and Blažčišin, A.I., 1976. Železo-margancevye konkrecii (Iron and manganese concretions). In: V.K. Gudelis and E.M. Emeljanov (Editors), Geologija Baltijskogo Morja (Geology of the Baltic Sea). Mokslas, Vilnius, pp. 307—348.

Veltheim, V., 1962. On the pre-Quaternary geology of the bottom of the Bothnian Sea. Bull. Comm. Geol. Finl., 200: 1—166.

Veltheim, V., 1969. On the pre-Quaternary geology of the Bothnian Bay area in the Baltic Sea. Bull. Comm. Geol. Finl., 239: 1—56.

Voipio, A. and Niemistö, L., 1979. Sedimentological studies and their use in pollution research. ICES C.M. 1979/C:46. 10 pp. (mimeogr.)

Volkolakov, F.K., Polivko, I.A., Agalcova, E.N. and Jakovleva, V.I., 1977. Geologičeskoe stroenie i neftegazonosnost akvatorialnoj časti Baltijskoj sineklizy. Izdatelstvo Zinatne, Riga, 136 pp.

Vorma, A., 1976. On the petrochemisty of Rapakivi-granites with special reference to the Laitila Massif, southwestern Finland. Bull. Geol. Surv. Finl., 285: 1—98.

Winterhalter, B., 1966. Pohjanlahden ja Suomenlahden rauta-mangaanisaostumia (Iron-manganese concretions from the Gulf of Bothnia and the Gulf of Finland). Geotek. Julk., 69: 1—77 (in Finnish, with an English summary).

Winterhalter, B., 1967. The Sylen and Solovjeva Shoals as observed by a diving geologist. Geol. Fören. Stockholm Förh., 89: 205—217.

Winterhalter, B., 1972. On the geology of the Bothnian Sea, an epeiric sea that has undergone Pleistocene glaciation. Bull. Geol. Surv. Finl., 258: 1—66.

Winterhalter, B., 1980. Ferromanganese concretions in the Baltic Sea. In: I.M. Varencov (Editor), International Monograph on the Geology and Geochemistry of Manganese. Hungarian Academy of Sciences, Budapest, 3: 227—254.

Winterhalter, B., and Siivola, J., 1967. An electron microprobe study of iron, manganese and phosphorus in concretions from the Gulf of Bothnia, northern Baltic Sea. C. R. Soc. Géol. Finl., 39: 161—172.

Chapter 2

HYDROLOGY OF THE BALTIC SEA

ULF EHLIN

HYDROMORPHOLOGY

The Baltic Sea has a meridional extension of more than 1500 km and a latitudinal extension of about 650 km. It consists of a series of basins of which the Bothnian Bay, the Bothnian Sea, the Gulfs of Finland and Riga and the Baltic Proper are the main bodies of water. The Belt Sea, including the Danish straits and the Sound, form together with the Kattegatt the transition zone of the Baltic Sea with the Atlantic Ocean (cf., also Chapter 1, part B, and Chapter 3).

The drainage basin covers an area which is 4.3 times as large as the area of the Sea itself (Fig. 2.1). The land in the drainage basin varies greatly geographically with the high mountain and forest areas in the northwestern part of the Scandinavian peninsula, the vast Finnish lake and forest areas in the northeast and the areas of agriculture in the southeast. Due to its long north—south extension, the drainage basin also varies greatly with respect to climatic conditions with long winters with snow and ice in the north, and short winters with less frequent snow in the south. Even the amount of precipitation varies, and the highest amounts fall in the western areas, especially in the mountains. The mean run-off from the drainage area to the different sea basins is shown in Table 2.I. The Gulf of Bothnia has the highest run-off with 12 $dm^3\ s^{-1}\ km^{-2}$. The Baltic Proper has only half of that run-off.

TABLE 2.I

Inflow of river water to the Baltic Sea and its different subbasins in $km^3\ a^{-1}$ (according to Mikulski, 1970, 1972). The numbers refer to the subbasins given in Fig. 2.1

Period	1 Bothnian Bay	2 Bothnian Sea	3 Gulf of Finland	4 Gulf of Riga	5 Baltic Proper	6 Belt Sea	7 Katte-gat	8 Baltic Sea area
1951-1960	99.9	82.8	115	32.6	110			441
1961-1970	101	83.8	110	26.7	112			433

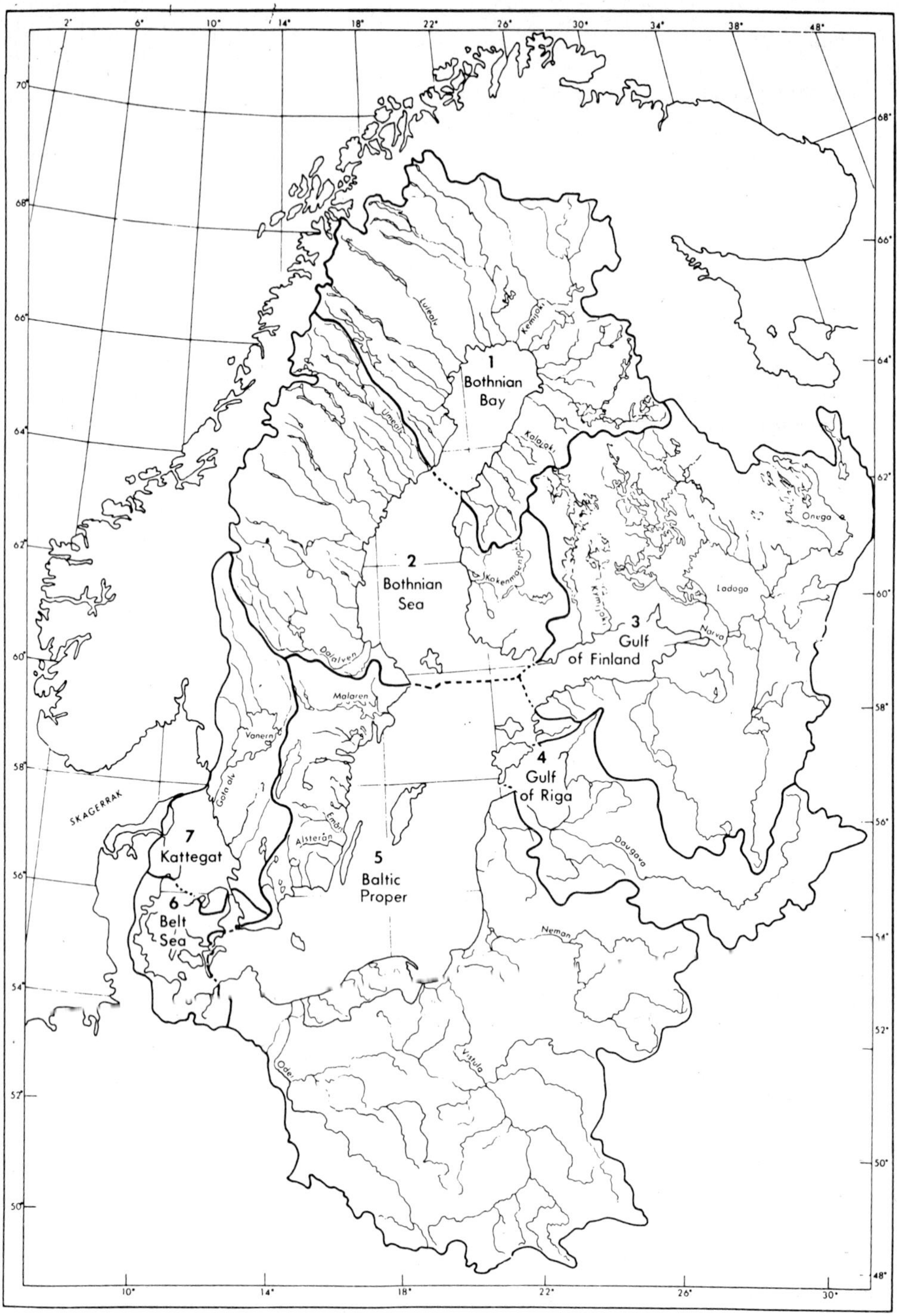

Fig. 2.1. Drainage basin and subregions of the Baltic Sea and its transition area (Falkenmark and Mikulski, 1974). Boundaries between Baltic Sea subregions (dashed lines). Boundaries between the corresponding drainage basins (thick lines).

RIVER INFLOW

Due to the geographical and climatological differences in the drainage basin the system of water courses transporting fresh water to the sea is different in the northern and southern regions. Many rivers of variable sizes discharge water into the Gulf of Bothnia from the Scandinavian mountains and from the forest areas on both sides of the Gulf. In the southeastern and southern areas, however, there are only a few main rivers flowing into the Baltic Sea (Fig. 2.1).

The largest rivers discharging into the Baltic Sea are the Neva flowing into the Gulf of Finland and the Vistula flowing into the Baltic Proper (Table 2.II).

There is a great seasonal variation in the river discharge into the Baltic Sea (Fig. 2.2). The maximum run-off occurs in spring during the thawing period

TABLE 2.II

Drainage area (A), discharge (Q) and run-off (q) of the largest rivers discharging into the Baltic Sea

River	A(km^2)	Q($m^3\ s^{-1}$)	q($dm^3\ s^{-1}\ km^{-2}$)
Neva	281,100	2600	9.25
Vistula	193,910	954	4.92
Daugava	87,900	688	7.83
Neman	98,200	674	6.86
Kemijoki	51,400	581	11.3
Luleälv	25,250	477	18.9

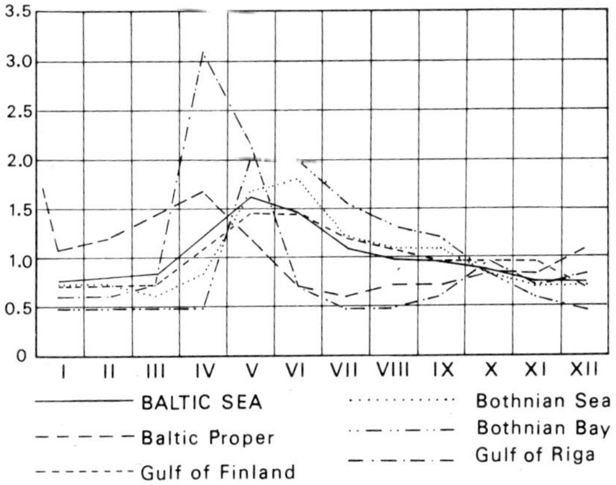

Fig. 2.2. Seasonal variation of river inflow (from Mikulski, 1970). Vertical axis shows mean monthly run-off as a quotient of mean annual run-off.

and the minimum during later summer or in midwinter. The delay in the thawing maxima while moving east- and northwards is also very pronounced. Thus, in the southwestern part of the Baltic Sea, the inflow of thaw water yielding the highest monthly totals occurs already in March, but in the middle part of the Baltic Proper in April. In the Gulf of Riga, a partial delay of the high-water inflow until May is observed, while the Gulf of Finland is characterized by the highest inflow in May and June. The situation is similar in the Gulf of Bothnia, where a considerable part of the high-water inflow takes place first in July. The annual minima of inflow appear in summer in the central part of the Baltic Sea and in autumn and winter in the northern parts.

Such regional differences in the times of the maxima discharge contribute to a rather steady supply of river water to the sea. As a result the outflow peak of brackish surface water from the Baltic Sea to the North Sea is diminished. The annual maximum inflow of river water to the Baltic Sea as a whole is observed from April to June. The corresponding minimum value, which is half of the maximum, occurs in winter from November to January.

The total annual inflow of river water to the Baltic Sea and its subbasins has been calculated by Mikulski (1970, 1972) for the periods 1951—1960 and 1961—1970 (Table 2.I). The total inflow differs for the two decades by 10 km^3 a^{-1} and is about 10% less than figures calculated by Brogmus (1952) for 1921—1930. The difference depends not only on better observations during more recent time but also on climatic variations resulting in a decrease in precipitation over the drainage basin.

Comprehensive run-off regulations for hydroelectric power purposes, especially in the large rivers of northern Sweden, have markedly changed the run-off by reducing its yearly extremes. The spring flood is thus kept in dammed storage in the rivers in order to make possible a substantially increased discharge during the following winter. The increase in March in the discharge from the River Luleälven can, e.g., in some years be as large as 60% of the unregulated flow (Ehlin and Zachrisson, 1974).

PRECIPITATION AND EVAPORATION

The determination of the amount of precipitation over the Baltic Sea area is difficult due to the fact that it has to be based mainly on data from shore stations since observations on islands, vessels or platforms are rare.

A special difficulty arises from the different national practices used for correcting the precipitation values for different errors, mainly the aerodynamic deficit, which will probably be rather large at wind-exposed locations frequent in the coastal zone.

The estimates of the annual mean precipitation over the Baltic Sea vary from 400 mm to 600 mm as shown in Table 2.III. Dahlström (1977 and

TABLE 2.III

Annual precipitation on the Baltic Sea estimated by various scientists

Author	Period	Precipitation (mm a^{-1})
Schulz (1938)	1921-1930	400-550
Simojoki (1949)	1886-1935	525
Brogmus (1952)	(40-50 years?)	471
Dahlström (1977) (preliminary results)	1931-1960	619

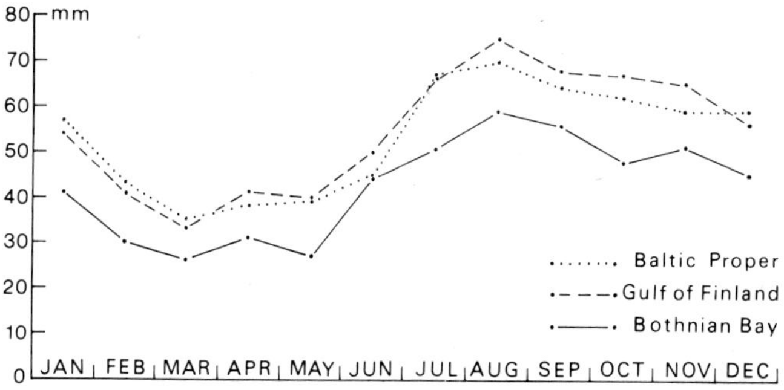

Fig. 2.3. Seasonal variation in precipitation (Dahlström, 1977).

1979) is the only author using corrected observation values. His preliminary result, 619 mm a^{-1}, is also considerably larger than previously published results. For the whole Baltic Sea area including the Danish Sounds and the Kattegatt the values of Dahlström give a total precipitation volume of 257 km^3 a^{-1}. If the transition areas are excluded, the precipitation volume is 229 km^3 a^{-1}.

The precipitation is, in general, lower during winter and spring and higher during summer and autumn (Fig. 2.3). It is also higher in the southern part of the system and in the coastal areas and lower in the northern and the central parts of the sea basins.

The evaporation from the sea surface is a very vaguely known element of the hydrological cycle. The lack of methods for direct measurement of evaporation has made it necessary to look for empirical formulas permitting an approximate estimation from climatological or aerological data. Only a few of these methods can, however, be applied to large water areas. The methods available are primarily the following three:

(1) The bulk aerodynamic method, which is based upon data of the vertical gradients of temperature and humidity and on wind data. The equation also contains an exchange coefficient, the determination of which de-

pends on calibration measurements. This method has been applied by Brogmus (1952).

(2) The aerological method (Palmén and Söderman, 1966), which requires accurate data of precipitation and a very good knowledge of the vertical moisture and wind distribution. Furthermore, it depends on reliable assumptions regarding the wind field between the aerological stations.

(3) The energy balance method, which gives the evaporation as a function of the difference between two large terms. This, of course, leads to difficulties relative to accuracy.

According to the calculations by Simojoki (1949), the areal mean evaporation for the Baltic Sea is 460 mm a^{-1}. Brogmus (1952) estimated the value as 471 mm a^{-1}, in fact identical with the figure he calculated for precipitation. The aerological method used by Palmén and Söderman (1966) looks promising. However, one difficulty, when applied to the Baltic Sea, is that very few aerological posts are distributed around the sea. In Fig. 2.4 values for the monthly evaporation from the Baltic Proper calculated by Palmén and Söderman, Simojoki and Brogmus are compared.

The study on the water balance of the Baltic Sea within the framework of the International Hydrological Program, IHP, incorporates new estimates on evaporation.

Preliminary results from calculations for 1975—1976 using the bulk aerodynamic method (Henning, 1979) show that evaporation during autumn months may even be larger than both precipitation and river run-off (Fig. 2.5). The annual evaporation is larger in the southern areas than in the northern ones due primarily to warmer water and often ice-free winters.

The difference between precipitation and evaporation can be expressed by

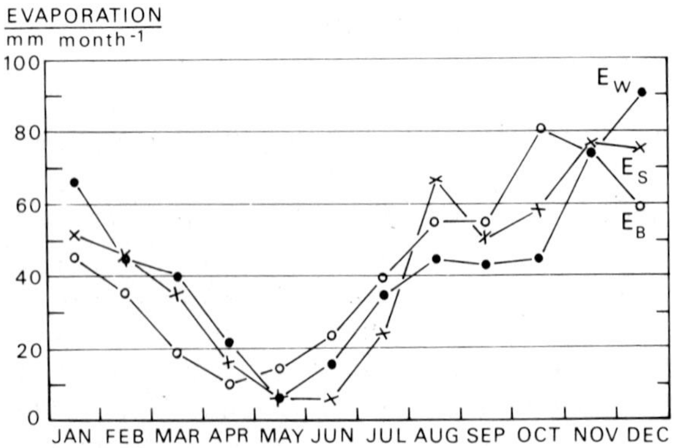

Fig. 2.4. Monthly evaporation from Baltic Proper (according to Palmén and Söderman, 1966). Evaporation computed for the period from October 1961 to September 30, 1962 (E_W) compared with evaporation according to Simojoki (1949) (E_S) and Brogmus (1952) (E_B).

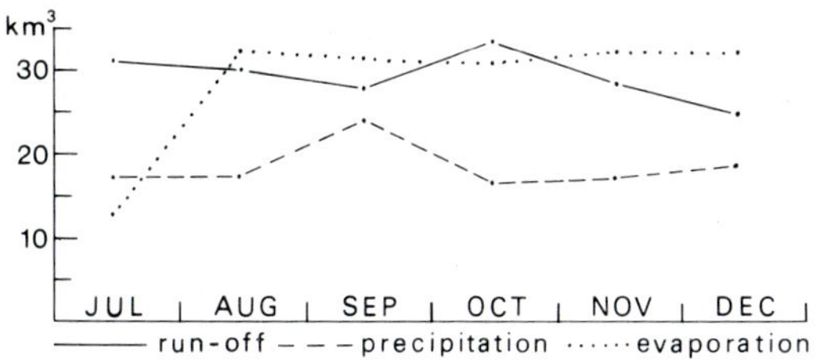

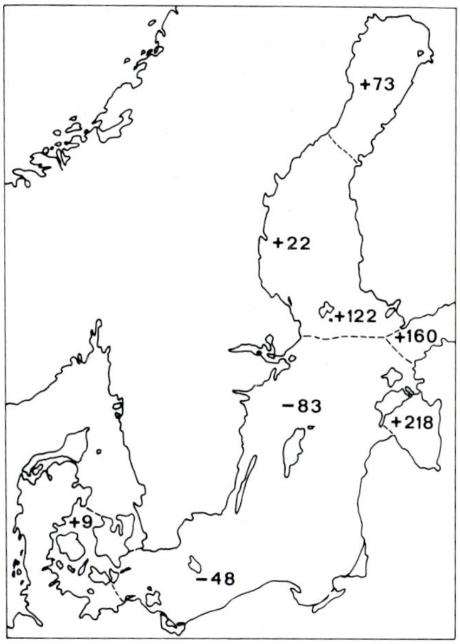

Fig. 2.5. Monthly values of run-off, precipitation and evaporation in the Baltic Sea area in 1975 (Mikulski, 1977; Dahlström, 1979; Henning, 1979).

Fig. 2.6. Vertical water balance of the different subregions (mm a^{-1}) according to Brogmus (1952).

the so-called vertical water balance, the distribution of which is shown in Fig. 2.6.

According to available data, the annual vertical balance seems to be about zero when averaged over the whole Baltic Sea area. It is, however, positive over the northern and eastern sub-regions (up to 200 mm a^{-1}) and negative over the Baltic Proper (down to -100 mm a^{-1}). This is the combined result of relatively low precipitation over open-sea areas and of higher evaporation in the south due to warmer water and often ice-free winters. During different seasons, however, evaporation departs markedly from precipitation, and significant irregularities of the vertical balance should also be expected during different years.

WATER TRANSPORT THROUGH THE DANISH SOUNDS

In addition to the precipitation and the inflow of fresh water from the drainage area, the Baltic Sea is fed by inflowing salt water from the North Sea (see Chapter 3, p. 139). Voluminous water inflows from the North Sea are mainly sudden and intensive but of short duration. The salinity of the inflowing water is distinctly lower than that of ocean water, due to the fact that the inflowing water partly consists of water that has flowed out of the Baltic Sea and become mixed with inflowing oceanic water in the Sounds and the Kattegatt. The salinity usually varies within a range of 15—25‰. The water inflow is an element extremely variable in time, both in its annual course and in the multi-year one. The volume and intensity of the inflows from the North Sea depend on the actual anemobaric situation and the differences in water level and salinity. The salt-water inflow is made up of two principal components: a frequently occurring deep-water stream, generated by the horizontal salinity gradient between the Baltic Sea and the North Sea, and an episodic, much more intensive inflow connected with persistent westerly winds. The latter form usually occurs during autumn and winter. However, the interval between large successive inflows of this kind can be several years (cf., Chapter 3, p. 143 and Chapter 4, p. 190).

Soskin (1963) assumed the transport through the Danish Sounds to be either completely outwards or completely inwards. He studied the general variation of mean monthly inflow for 1898—1944. His calculation is based on stream-velocity data from the light ships Lappegrund in the northern part of the Sound and Halsskov Rev in the central Great Belt, using correlations with the inflow as separately determined for 30 years, for which the inflow could be calculated from water balance data. The long-term mean inflow was 1187 m^3 a^{-1} during the whole period. The highest monthly inflow (120 km^3) occurs generally in January and November and the lowest monthly inflow in May (73 km^3). There are also very clear multi-year variations between a maximum of 1508 km^3 a^{-1} (1921) and a minimum of 983 km^3 a^{-1} (1937).

The inflow series reveals large differences between different 20-year periods: the 20-year annual mean for 1900—1919 was thus 1119 km^3 a^{-1}, or about 10% lower than for the following two decades when it amounted to 1238 km^3 a^{-1}. These inflow values deviate rather much from the long-term inflow, as estimated from conditions of water and salt balance (Knudsen relations, see p. 131).

The outflow through the Danish Sounds is principally determined by the sea-level differences between the Baltic Sea and the North Sea. The larger this difference is, the larger is the outflow. The winds also influence the outflow, repressing it by westerly winds (Wyrtki, 1954) and enforcing it by easterly winds. According to the studies of Soskin (1963), comprising the years 1898—1944, the strongest monthly outflow occurred in December and March (150 km^3) and the minimum monthly outflow in June (115 km^3). In June—October the outflow increased. The long-term variations revealed the highest outflow of 2082 km^3 a^{-1} in 1927 and the lowest of 1287 km^3 a^{-1} in 1942. The long-term annual average was 1660 km^3 a^{-1}.

For a few years in the middle of the 1970s Danish oceanographers studied water transport by using recording current meters placed in the Sounds.

The results from this study are so far not published in the final form but computations from the first part of the study show reasonable agreement with Soskin's (1963) result (Jacobsen, 1976).

The processes of salt-water inflow and fresh-water outflow in the Danish Sounds combine to a resultant water exchange between the Baltic Sea and the North Sea.

Attempts made hitherto to determine the water exchange and its elements through field measurements have generally been sporadic and used mainly for determining the volume of individual inflows from the North Sea to the Baltic Sea. In other instances the water exchange has been determined through indirect calculations using current data from light ships, water-balance relations, etc. The strong consequences in the dynamics of the Baltic Sea due to the water exchange through the Danish Sounds have caused considerable research in order to approach the water-exchange problem, often from a theoretical point of view. Svansson (1972) summarized the different approaches.

The long-term water exchange has been studied by means of the classical approach of Martin Knudsen (see Fig. 2.7).

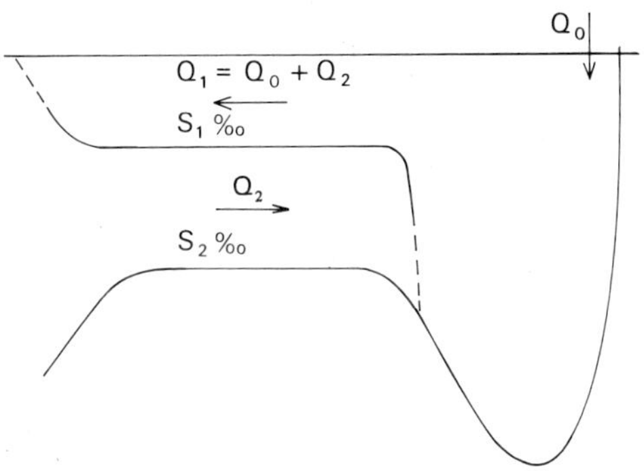

Fig. 2.7. Water exchange according to Knudsen (from Svansson, 1972). Schematic figure of an enclosed sea with fresh water supply Q_0 and salinity S_1 connected through a strait with an ocean with salinity S_2. Q_2 is the compensation transport.

This approach is based on the model of an enclosed sea with fresh-water discharge Q_0 and salinity S_1‰, connected through a strait with an ocean from which water with salinity S_2 enters the enclosed sea as a so-called compensation transport Q_2. It is thereby assumed that the connecting strait contains two different water layers, an upper layer of outgoing brackish water and a lower layer of ingoing ocean water. Assuming the net salt transport to be zero, i.e., stationary salinity conditions in the Baltic Sea, Knudsen's relations give the possibility to determine Q_1 and Q_2 when Q_0, S_1 and S_2 are known, as follows:

$$Q_1 = \frac{S_2}{S_2 - S_1} Q_0 ; \quad Q_2 = \frac{S_1}{S_2 - S_1} Q_0$$

Knudsen's relations thus do not give the possibility to calculate the net transport but rather the figures of the different components in the exchange equation.

Due to vertical mixing, the volume and salinity of the outgoing and ingoing water increase with increasing distance from the Baltic Sea. When comparing the results from calculations of the size of the transports made by different scientists, it is therefore necessary to notice which area in the Danish Sounds the calculations represent.

Brogmus (1952), using Knudsen's relations, obtains a net long-term water exchange of 472 $km^3\ a^{-1}$ through the Danish Sounds, this being the difference between an inflow of 472 $km^3\ a^{-1}$ and an outflow of 944 $km^3\ a^{-1}$ (characteristic salinities from Darsser Schwelle: $S_1 = 8.7‰$; $S_2 = 17.4‰$).

These flow values are considerably lower than the flows determined by Soskin (1963). This is mainly due to the fact that the Knudsen flows refer to an upstream section of the Danish Sounds, whereas Soskin's study was carried out further downstream and thus includes salt water that joined the outgoing current before reaching the Baltic Sea.

On a shorter time scale, the resultant water exchange is composed of perpetual to-and-fro movements, governed mainly by the weather conditions. Soskin (1963) gives the following description. The two-layered current system in the Danish Sounds in fact represents the general situation during calm weather, i.e., when an anticyclone is located over northern Europe, characterized by small pressure gradients between the North Sea and the Baltic Sea. Normally, however, the front separating Baltic Sea water from North Sea water, which during calm weather is located near the northern entrance to the Belts, oscillates between a position in the central part of the Skagerrak during persistent easterly winds, and a position near the sills in Öresund and in the Darsser area during westerly winds. This oscillation is

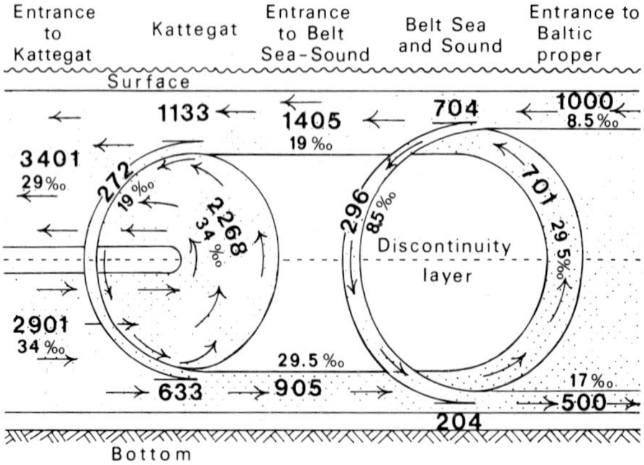

Fig. 2.8. Water exchange through Danish Sounds (Steemann-Nielsen, 1940).

caused by the movement of the water masses, implying enforcement of the outgoing currents during periods of strong westerly winds. The main transport through the Danish Sounds takes place in the upper 10—15 m layer. During the outward phase, relatively fresh water masses move into the Belt Sea, where they become mixed vertically with saltier masses. During the next period of inflow, these water masses return to the Baltic Sea but are forced downwards at the Drogden and Darsser sills. Fig. 2.8 summarizes these mixing and transport processes.

WATER STORAGE AND WATER EXCHANGE

The water reaching the Baltic Sea is stored there while waiting for its transport through the Sounds. The active storage capacity of the Baltic Sea between the highest and lowest monthly mean water levels is about 500 km^3. In other words, the basin has the capacity to store the average annual freshwater input. At any time, the balance of water inflow and outflow in the basin is recorded as volume changes. The salt water inflow and the outflow through the Danish Sounds depend strongly on the atmospheric circulation. Consequently, large irregularities of the volume changes are also characteristic. The storage change between two consecutive months can thus rise to 150 km^3 or even 200 km^3 (Wyrtki, 1954).

The total water turnover (the sum of inflow and outflow) is about 650 km^3 a^{-1} when only fresh water is taken into account, and about 1100 km^3 a^{-1} when also the long-term salt-water exchange is included. Averaged over the surface of the Baltic Sea, this corresponds to 1700 mm a^{-1} and 2800 mm a^{-1}, respectively.

When considered from a strictly hydrological viewpoint, the Baltic basin has a more stagnant than a through-flow character, due to the relative importance of the vertical terms in the water balance (Falkenmark and Mikulski, 1974).

REFERENCES

Brogmus, W., 1952. Eine Revision des Wasserhaushaltes der Ostsee. Kiel. Meeresforsch., 9(1): 15—42.

Dahlström, B., 1977. Estimation of precipitation for the Baltic Sea — preliminary result. Ad Hoc Meeting of the Pilot Study Group of Experts. Norrköping (mimeogr.).

Dahlström, B., 1979. Determination of areal precipitation for the Baltic. Sixth Meeting of Experts on the Water Balance of the Baltic Sea, Hanasaari Cultural Centre Near Helsinki, 30.1—2.2, 1979, Paper 11, 1 p., tables (mimeogr.).

Ehlin, U. and Zachrisson, G., 1974. Redistribution of runoff to the Baltic through river regulations in Sweden. Proc. 9th Conf. Baltic Oceanogr., Kiel, 17—20 April, 1974, pp. 265—274 (mimeogr.).

Falkenmark, M. and Mikulski, Z., 1974. Hydrology of the Baltic Sea — General background of the international project. Water Balance of the Baltic Sea — a Regional Cooperation Project of the Baltic Countries. International Hydrological Decade. Project Documents, Stockholm, Warszawa, 1: 1—51.

Henning, D., 1979. Baltic Sea evaporation. Pilot study year. Summary of interim results. Sixth Meeting of Experts on the Water Balance of the Baltic Sea, Hanasaari Cultural Centre Near Helsinki, 30.1—2.2, 1979, Paper 10 (mimeogr.).

Jacobsen, T.S., 1976. Preliminary transport calculations for Store Belt. Proc. 10th Conf. Baltic Oceanogr., Gothenburg, 2—4 June, 1976, Paper 28, 29 pp. (mimeogr.).

Mikulski, Z., 1970. Inflow of river water to the Baltic Sea in the period 1951—1960. Nord. Hydrol., 4: 216—227.

Mikulski, Z., 1972. The inflow of the river waters to the Baltic Sea in 1961—1970. Proc. 8th Conf. Baltic Oceanogr., Copenhagen, October 1972. 3 pp. (mimeogr.).

Mikulski, Z., 1977. River inflow to the Baltic Sea July 1975—December 1975. Fifth Meeting of Experts on the Water Balance of the Baltic Sea, Rostock, 23—27 May, 1977, 1 p. (mimeogr.).

Palmén, E. and Söderman, D., 1966. Computation of the evaporation from the Baltic Sea from the flux of water vapour in the atmosphere. Geophysica, 8(4): 261—280.

Schultz, S., 1938. Die Bilanz der Ostsee. VI Baltische Hydrologische Konferenz, Berlin, 21: 1—6.

Simojoki, H., 1949. Niederschlag und Verdunstung auf dem Baltischen Meer. Fennia, 71(1): 1—25.

Soskin, I.M., 1963. Mnogoletnie izmenenija gidrologičeskih harakteristik Baltijskogo morja. Leningrad, 159 pp.

Steemann-Nielsen, E., 1940. Die Produktionsbedingungen des Phytoplanktons im Übergangsgebiet zwischen der Nord- und Ostsee. Medd. Dan. Fisk.- o Havunders. Ser. Plankton, 3(4): 1—55.

Svansson, A., 1972. The water exchange of the Baltic. Ambio, Spec. Rep., 1: 15—19.

Wyrtki, K., 1954. Schwankungen im Wasserhaushalt der Ostsee. Dtsch. Hydrogr. Z., 7(3—4): 91—129.

Chapter 3

PHYSICAL OCEANOGRAPHY

GUNNAR KULLENBERG

INTRODUCTION

In this chapter the physical oceanographic conditions in the Baltic Sea will be discussed, including the salinity and temperature distribution and their seasonal and long-term variations; the density stratification and its variations; the mean and time-dependent circulation and its relation to wind and thermohaline effects; the vertical and horizontal transports generated by different mixing processes in the open sea and in the coastal boundary zone; the optical conditions and aspects of the heat budget and ice conditions.

The presentation will be mainly descriptive, and reference will be made to original articles with respect to theoretical developments. The physical oceanography of the Baltic Sea to a large extent is governed by the water balance, i.e. the fresh-water supply and the water exchange through the Danish Straits. But also the topographic conditions are of great importance, in particular the division of the Baltic Sea into several basins separated by more or less well-defined sills, and the shallow and narrow connections with the North Sea. These subjects are dealt with in separate chapters and will not be discussed here.

SALINITY AND TEMPERATURE DISTRIBUTIONS

Salinity

A dominating feature of the Baltic oceanography is the marked permanent salinity stratification characterized by limited variations in comparison with the considerable variations occurring in the Transition Area (subregions 6 and 7 in Fig. 2.1). This feature is of great importance for the chemical and biological conditions in the deep waters. The fresh-water supply to the Baltic Sea generates a brackish surface layer of outflowing water, and incoming subsurface flow forms layers of more saline deep waters and bottom waters. Although the circulation is weak, a cyclonic salinity distribution is evident (Fig. 3.1a, b) with the low-salinity surface water being concentrated along the Swedish coast and the high-salinity deep water flowing inwards primarily along the eastern coasts. Such a distribution is also present in the Gulf of Finland.

The inflow of deep water is continuous over the Darss Sill and successively into the Baltic Sea basins. This gives a continuous source of salt and oxygen to the deep water. Sometimes strong pulses of inflowing water are generated by special meteorological conditions. These major inflows usually cause a successive renewal of the bottom water in the Baltic Sea basins.

The fresh-water supply is mixed downwards by a combination of the wind-generated mixing and thermohaline convection during the fall and early winter. This generates an almost homohaline surface layer with insignificant

NOTATION

a_y	light absorption coefficient for yellow substance
c_d	drag coefficient
c_p	specific heat capacity
E	evaporation in mm per 24 hours
e_s	saturated water vapour pressure at temperature of sea surface
e_a	water vapour pressure at level *a*
$f = 2\omega \sin\varphi$	Coriolis parameter
F	heat flux
g	acceleration of gravity
H	water depth
K_H	vertical turbulent transfer coefficient for heat
K_z	vertical turbulent mixing coefficient
K_h	horizontal turbulent mixing coefficient
K_m	vertical transfer coefficient for momentum
p	pressure
q	horizontal current vector
Q_A	long-wave radiation from sea surface
Q_B	back radiation from sky
Q_K, Q_V	convective and latent heat exchange, respectively
Q_n	net radiation budget for sea surface
Q_R	radiation reflected from sea surface
Q_S	incoming radiation
Ri, Rf	Richardson and flux Richardson number, respectively
Ri_*	overall Richardson number
T	temperature
t	time
$u,v,w,$	velocity components along x,y,z axes, respectively
$\tilde{u}$, $\tilde{v}$	vertically averaged velocities
u_*	friction velocity
W_a	wind velocity at level *a*
x,y,z	coordinate axes, *z* positive downwards
$N = \left(\frac{g}{\rho}\frac{d\rho}{dz}\right)^{\frac{1}{2}}$	Brunt-Väisälä frequency
ω	angular velocity of rotation of the earth
φ	latitude
ρ, ρ_a	density of water and air, respectively
τ	wind stress at sea surface
ζ	sea level

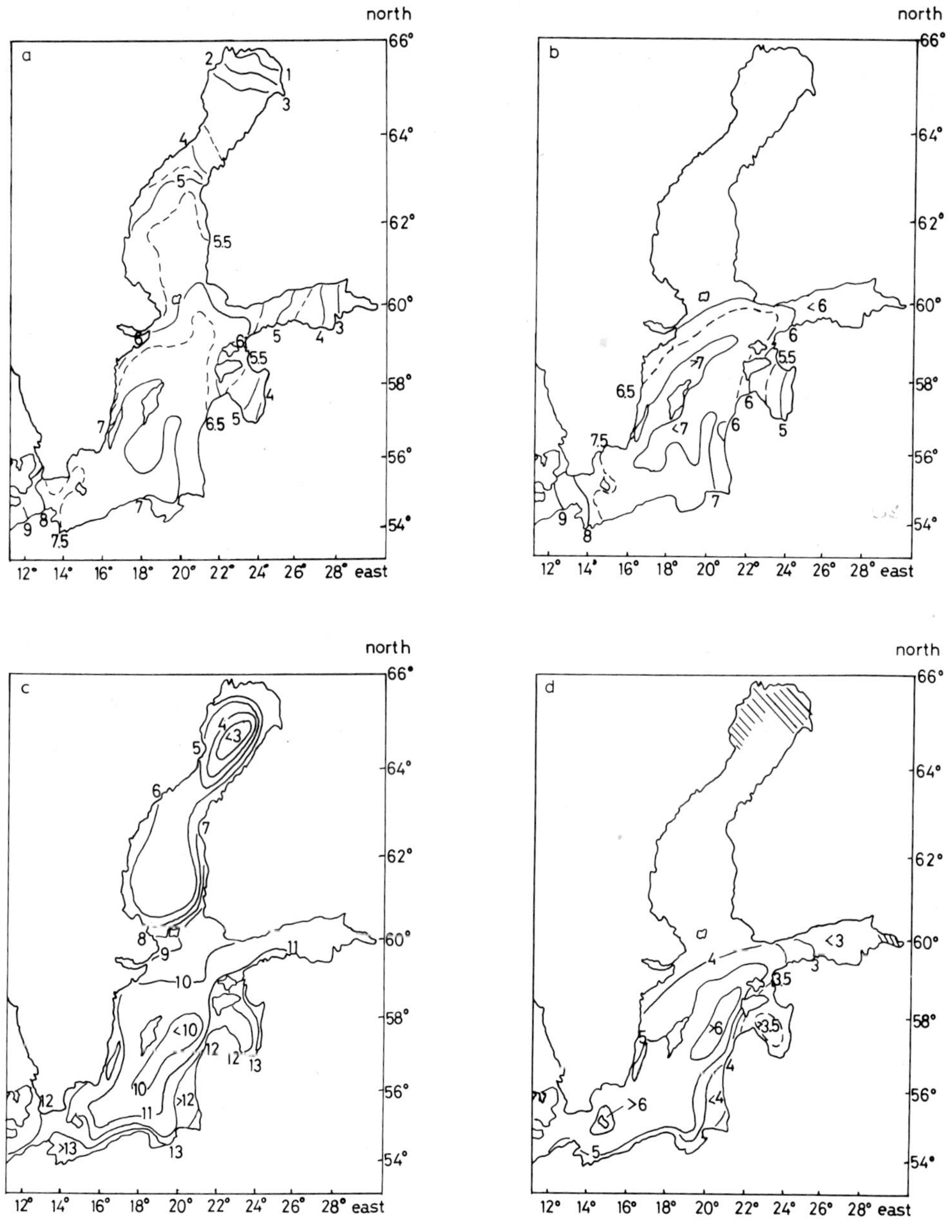

Fig. 3.1. Surface salinity ($S/‰$) distribution for June (a) and December (b). From Bock, 1971.

Surface temperature ($T/\ ^{\circ}C$) distributions for June (c) and December (d). From Lenz, 1971.

vertical salinity variations during the winter and weak variations during the summer. The surface layer is separated from the deep water by the primary halocline. The transition layer is about 10—20 m thick. The thickness of the surface layer varies from basin to basin. It is related to the sill depths between the basins, since cross-sectional areas large enough in the layer beneath the halocline must be available for the deep-water flow, and to the efficiency of the mechanical mixing in the surface layer. In the central Arkona Basin the depth of the halocline layer varies between 30 m and 40 m, and in the Bornholm Basin the depth is generally 40—50 m. In the Stolpe Channel the depth increases slightly, and in the large deep water reservoir of the Baltic Sea in the central Baltic Proper it varies between 60 m and 70 m, sometimes reaching a depth of 80 m in the northern part and in the Landsort Deep (e.g., Fonselius, 1969). Since the beginning of this century there has been a tendency towards a decrease of the halocline depth in the central Baltic Proper. However, prolonged series of observations may well show this to be a periodic feature.

The water between the Bothnian Sea and the Baltic Proper is exchanged through the Åland Sea because this is deeper and more open than the shallow Archipelago Sea between Åland and the Finnish mainland. The exchange is more or less continuous and is forced by both fresh-water supply and meteorological factors, the latter generating the most important transport (Hela, 1977). It is often a two-layer flow with southgoing transport in the surface layer. The inflowing water originates partly from the surface layer of the Baltic Proper which has small salinity and density variations. The density variations are further reduced by the mixing during the inflow which implies that the stratification in the Gulf of Bothnia is considerably weaker than in the Baltic Proper. The stratification is stable throughout the year, but the depth of the weakly developed halocline varies strongly, and in parts of the basin the halocline may be absent during certain periods of the year, especially during winter. During summer and fall a fairly shallow halocline may be present. Due to the efficiency of the water exchange between the Gulf of Bothnia and the Baltic Proper, the deep and bottom waters are renewed annually in most of the Gulf of Bothnia although the salinity stratification prevents thermohaline convection to the greatest depths. In the deepest parts of the Gulf of Bothnia the bottom water salinity is about 1.3‰ higher than the surface-water salinity.

The deep water entering the Bothnian Sea during summer originates from the 50—70 m-deep layer in the northern Central Basin and gradually fills up the Åland sea. During winter oxygen-rich surface-layer water from the Baltic Proper is forced by strong winds into the Åland Sea. This water penetrates northwards along depth contours into the Bothnian Sea (Palosuo, 1973), with velocities of the order of a few cm s^{-1}. The relatively warm and salt summer water also penetrates northwards along the Finnish coast and gives rise to a zone of warm water, particularly evident during the fall.

The annual inflow of water to the Bothnian Sea is of the order of 900 km^3, and the outflow is of the order of 1100 km^3 (Fonselius, 1971). In the Åland Sea the mean surface currents and deep currents are generally south- and northgoing, respectively, although recent current measurements have shown that reversals occur over time scales of the order of several months. These observations also indicate a considerably larger transport than that calculated by Knudsen's relations (Fonselius, 1971; Ehlin and Ambjörn, 1978; see also Chapter 2, p. 129).

The water entering the Bothnian Bay is derived primarily from the surface layer of the Bothnian Sea (Palosuo, 1973). During fall and winter the incoming water generally has a density high enough to allow penetration by convection to the deep waters and bottom waters of the Bothnian Bay (Palosuo, 1964). The annual water transport has been estimated at 300—400 km^3 of salt water entering the Bay and 400—500 km^3 leaving the Bay, which gives a residence time for the water in the Bay of about 3 years (Fonselius, 1971; Dahlin, 1976).

The deep-water salinity is in the Gulf of Bothnia 3—7‰, in the Gulf of Finland 5—9‰ and in the Baltic Sea proper 10—13‰. In Fig. 3.2 examples of vertical salinity profiles are given for the different Basins. In connection with particularly strong inflow from the Kattegat, for instance such as occurred in November—December 1951 (Wyrtki, 1954), the bottom water salinity in the Bornholm Basin may reach 20‰, and in the eastern and western Gotland Basins 14‰ and 11‰, respectively. During such inflows a secondary halocline may develop at a depth of 110—130 m (Voipio and Mälkki, 1972). Depending upon the intensity of the inflow the passage time of the deep water from the Arkona Basin to the Gotland Basin varies from 4 to 9 months (e.g., Francke and Nehring, 1973).

The residence time of the deep waters and bottom waters in the Baltic Proper depends primarily upon their densities. The water beneath the sills is mainly replaced through inflowing water which must have enough excess density and kinetic energy to force the old water away. The inflowing water will flow into the basin at its appropriate density level. After a major inflow the density of the new bottom water normally must decrease by vertical mixing before a new inflow can replace it. During these periods the motion in the bottom water is very weak, and they are therefore often called stagnation periods. The residence time of the bottom water varies considerably, from less than 2 years to about 5 years (e.g., Fonselius, 1969, 1976). In the Arkona Basin the water is renewed annually due to the combined action of inflowing Belt Sea water and thermohaline convection (Krause, 1969).

Temperature

The deep-water temperature fluctuations are largest in the southwestern parts of the Baltic Sea where the temperature of the inflowing Kattegat

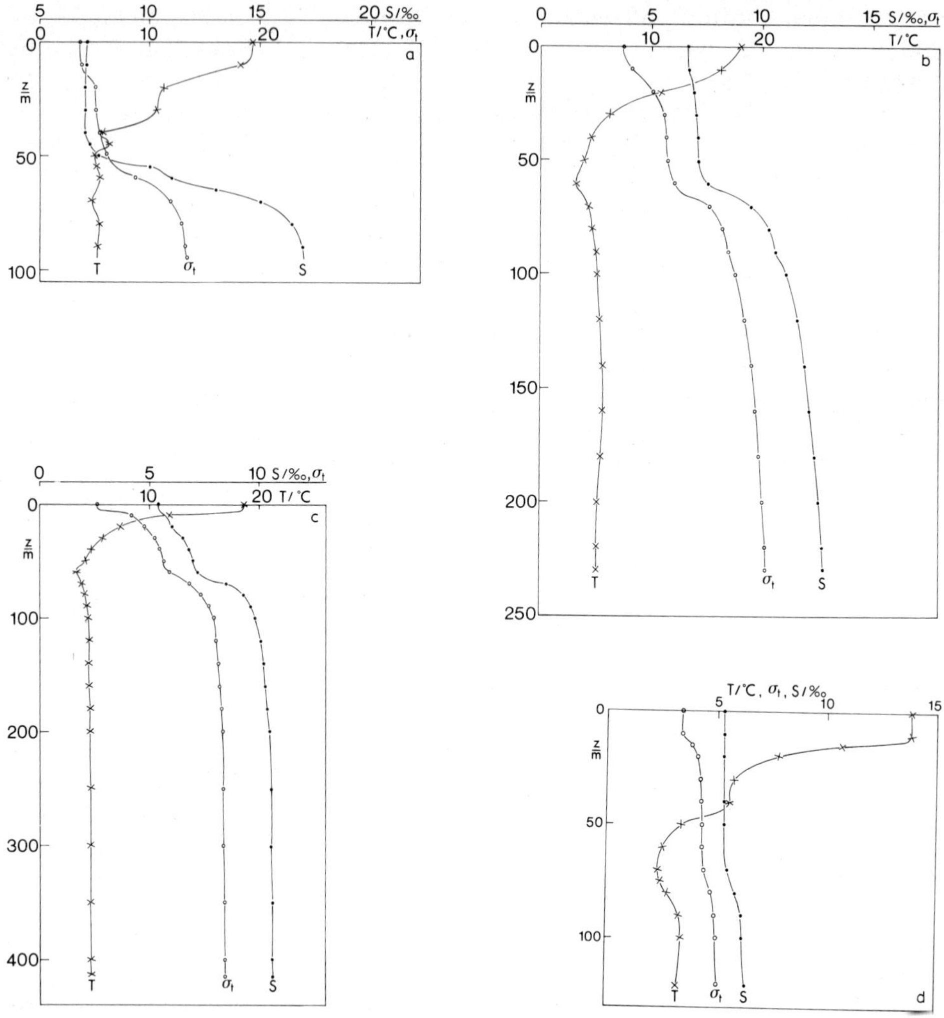

Fig. 3.2. Salinity (S/‰), temperature (T/°C) and density (σ_t) profiles from: (a) Bornholm Deep 55°19.5′N, 15°38.5′E, 4.8.1938; (b) Gotland Deep 57°21.5′N, 20°02.5′E, 27.7.1938; (c) Landsort Deep 58°38.5′N, 18°16.5′E, 28.7.1938; (d) Bothnian Sea 61°04′N, 19°35′E, 12.7.1938.

water, having a range from 2° to 14°C, determines the deep-water temperature. High temperature is normally combined with high salinity. In the central part of the Baltic Proper the temperature of the deep waters and the bottom waters does not fluctuate so much. This is due to the influence of the large water volume, the inflowing water being about 20% of the deep-water volume annually, and the mixing with water in and just above the halocline layer where the temperature is always low. The deep-water temperature increases from the halocline towards the bottom (Fig. 3.2), and a normal range is from 4° to 6° C. Sometimes strong or unusual inflows of Kattegat water may raise the deep-water temperature in the Gotland Deep above 6° C, for instance in 1952—1953, in 1971 and in 1976—1977 (Fonselius, 1977). At the end of 1976 the temperature of the bottom water in the Gotland Deep was 5.75° C, whereas in January 1977 it was 7.43° C. This increase (Fonselius, 1977) was related to an inflow of unusually warm water from the Kattegat in the autumn of 1976. The inflow also increased the deep-water salinity in the Gotland Deep from 12.46‰ to 13.28‰, the highest value observed there since 1962 (Fonselius, 1977). After such inflows both the temperature and the salinity of the bottom water will decrease through vertical mixing for a period of time which may last up to 5 years until renewal of the bottom water will take place.

The temperature distribution in the Baltic Sea (Fig. 3.1c, d) and its seasonal fluctuations display several characteristic features. During fall and early winter the homohaline surface layer becomes homothermal through a combination of thermohaline and mechanical wind-induced mixing. Although the surface water reaches its maximum density around 2.5° C the homohaline layer always obtains a lower temperature (see Fig. 3.2), in the southern parts sometimes reaching 0° C. This may be explained by an effective mechanical convection induced by the strong winds during fall and winter, often reaching storm forces and often with a duration up to a week.

The convective mixing reaches the top of the primary halocline where further mixing is suppressed by the strong stratification across the halocline layer. Due to further cooling of the surface water, a thin winter thermocline may develop. The surface layer above this thermocline is generally cooled to the freezing point or to a temperature slightly less than 0° C. The typical annual temperature variation is shown in Fig. 3.3 for the central Baltic Proper. The layer between the winter thermocline and the halocline layer remains homothermal during the winter. During spring warming the heat is initially transferred from the surface by thermocline convection. The warming is mostly so rapid that a thermocline is developed before the whole homohaline layer has received any heat input. A spring and summer thermocline therefore usually forms over most of the Baltic Sea area at depths between 15 m and 20 m. This implies that the so-called winter water formed during the preceding winter remains as a layer on top of the primary halocline where a very characteristic temperature minimum is observed throughout the year in most Baltic Sea temperature profiles (Fig. 3.2). It should, however, be noted that heat can penetrate to the deep water. For instance, in the Arkona and Bornholm Basins the conditions are very much influenced by advection, also in the deep and bottom waters. However, in the Gotland Basin advection plays a minor role. In the open central Baltic Proper the annual variation of the temperature in the different layers can to the first

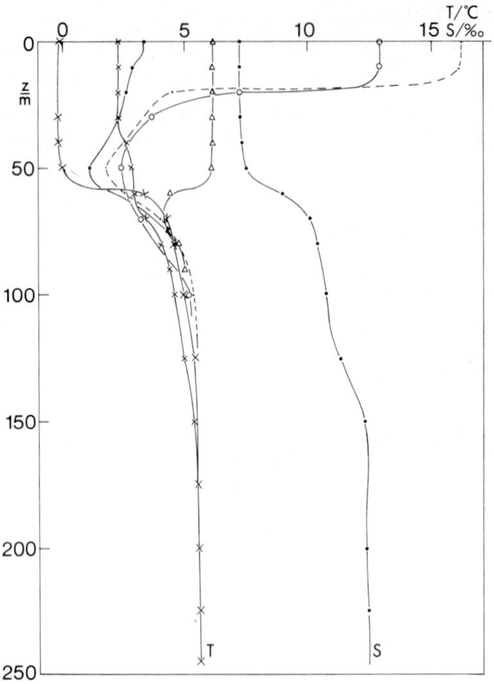

Fig. 3.3. Temperature ($T/\ ^\circ C$) profiles from different months showing development of seasonal thermoclines in the Gotland Deep (F81, 57°20′N, 19°59′E). For reference one salinity profile is also shown.

approximation be explained by heat exchange with the atmosphere and by the subsequent vertical mixing of warm or cold water into deeper water layers (e.g., Matthäus, 1977a). An annual variation of the temperature can be traced to depths of about 70 m. The variation is, of course, suppressed with increasing depth, and the temperature minimum occurs increasingly towards the end of the year. The halocline layer has a very marked effect, the annual variation below this layer being very small.

The summer thermocline may be very sharp, and the stability across the thermocline layer may thus effectively suppress vertical exchange between the surface layer and deeper layers. This implies that during this period the supply of nutrients to the euphotic zone is very limited. The euphotic zone is about 20 m deep in most parts of the Baltic Sea. Another secondary result of the suppressed vertical mixing is that during the summer the fresh-water supply becomes trapped in the warm surface layer above the thermocline. This further increases the stratification. The warm surface layer reaches its salinity minimum towards the end of July or about the same time as it reaches its temperature maximum. This shows that the salinity minimum is more due to the thermocline suppressing the mixing than to an increase in the fresh-water supply, since this reaches its maximum around May.

In the Gulf of Bothnia a summer thermocline normally also forms at a depth of 15 m to 20 m. However, since the salinity stratification in the Gulf is much weaker than in the Baltic Proper, the heat input gradually mixes towards deeper layers by wind-induced mixing during the fall. The heat penetration in the open Gulf reaches a depth of at least 100 m (e.g., Hela, 1966a). However, in the Bothnian Sea advective transport of heat by the inflowing water from the Baltic Proper may be relatively more important than advective heat transport in the central part of the Baltic Proper.

The winter water of the Gulf of Bothnia is not as clearly defined as in the Baltic Proper. In late summer the thermocline is generally situated at a depth around 20 m, and the thermocline layer may be rather thick due to successive heating. The layer becomes thinner and sharper when coinciding with a halocline, which is often the case in the Åland Sea and the northern parts of the Gulf.

The heating and cooling occur faster along the shallow coast line than in the open sea. Thus bands of warm and cold water occur generally along the coasts during summer and winter, respectively. These are clearly separated from the open-sea water and may suppress the exchange between the different zones. Summaries of the oceanographical conditions in the Gulf of Bothnia have been given by Ehlin and Ambjörn (1978) and by Dahlin (1978).

Long-term variations

The temperature and salinity of the Baltic Sea waters show important long-term variations, studied in particular by Ahlnäs (1962), Soskin (1963), Hela (1966b), Fonselius (1969, 1977) and Matthäus (1977b).

The salinity increase in both the surface water and the deep water during this century has been documented by several authors (Granqvist, 1952; Lindquist, 1959; Soskin, 1963; Hela, 1966a; Fonselius, 1969; Matthäus, 1977b). In the central Baltic Proper the salinity showed a marked decrease during the early 1930s, followed by an increase starting about 1938 which probably continued during the 1940s. The great salt-water inflow during 1951—1952 further increased the salinity (Fonselius, 1962), and during the subsequent stagnation period the salinity decreased almost continuously until 1959—1960 when a new major inflow occurred (Fonselius, 1969). Both the variability and the general trend of increasing salinity are demonstrated in Fig. 3.4, showing the 3-year gliding means of the salinities at two stations from the Baltic Proper. However, the fluctuations are as large as, or larger than, the net increase since 1908. The marked minima at 100 m around 1910, 1938 and 1960 are noticeable. There is reason to believe that such fluctuations are normal and will occur also in the future. Also in the northern part of the Baltic Proper and in the Gulf of Bothnia a general increase of

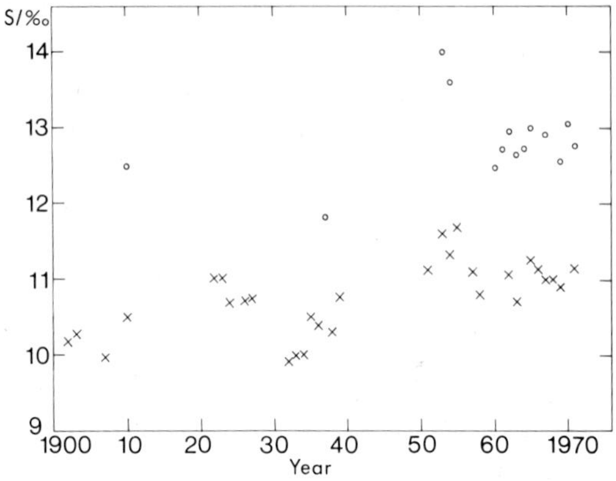

Fig. 3.4. Salinity ($S/‰$) variations in the Landsort Deep; circles: 200—240 m, crosses: 300 m.

the deep-water salinities has occurred as demonstrated by Fonselius (1969).

The most recent study of the secular variations of the surface layer salinity for the Baltic Proper has been conducted by Matthäus (1977b). He demonstrated an increase in the salinity of the surface water over the whole study area, from the Arkona Basin to the northern Baltic Proper in the range 0.5—0.9‰ from 1900 to 1975. The increase was found to be almost uniform, with a mean value in the range 0.007—0.012‰ per year.

The salinity variations will also influence the depth of the halocline layer, as shown by Fonselius (1969). In the central part of the eastern Gotland Basin the top of the halocline layer was located around a depth of 80 m in the beginning of this century, and about 1960 it was at a depth of around 60 m. It should, however, be noticed that considerable fluctuations of the depth of the halocline have taken place. Fonselius (1969) has shown that the depth was about 65 m in 1914—1920, about 80 m in 1942—1946 and about 60 m around 1968. These fluctuations are related to the salinity variations and to variations of the meteorological conditions driving the vertical mixing.

The depth of the halocline is of importance for the conditions in the deep waters, and it is of interest to define the upper boundary of the halocline layer in an unambiguous way.

Hela (1966a) used the method proposed by Tully (1958) of plotting the salinity vs. logarithmic depth to define this upper boundary. When plotted vs. logarithmic depth, the salinity of the permanent halocline in an estuary forms a straight sloping line whereas the salinity of the waters above and below forms vertical lines. Thus such a plot gives an unambiguous method for the determination of the depth of the upper boundary of the halocline layer and the salinity at that depth. However, as Hela (1966a) pointed out, the hydrographic conditions in the Baltic Sea differ from those of a normal estuary in several ways:

(1) the Baltic Sea boundary to the open ocean is a transition area with a shallow sill depth; in this respect the Baltic Sea is more like a fjord than an estuary.

(2) The width and the depth of the Baltic Sea vary considerably, and the basin is divided into a series of subbasins separated by more or less pronounced sills.

(3) Several rivers discharge into the Baltic Sea.

(4) The Gulf of Finland and the Gulf of Bothnia form a part of the Baltic Sea, with the former being a true estuarine embayment and the latter being divided into two separate basins.

In three areas of the Baltic Sea, Hela (1966a) found that the salinity on a log depth plot forms an almost straight line, namely in: (1) the westernmost part of the Baltic Sea including at times the western corner of the Bornholm Basin; (2) the easternmost part of the Gulf of Finland except when the summer thermocline reduces the vertical mixing of salt; and (3) the northernmost part of the Bothnian Bay.

However, in the Gotland Deep the salinity of the halocline layer appears not as a straight line but falls on the arc of the circle. Despite this the required salinity and depth of the upper boundary of the halocline layer can be determined. Hela (1966a) used a longitudinal hydrographic section between the Bornholm Sill and the mouth of the Gulf of Finland obtained in August 1956 and found the halocline depth to be around 60 m in the Gotland Basin and barely 50 m in the Bornholm Basin. The salinity of the upper boundary of the halocline layer was virtually the same, about 7.7‰, throughout the section.

In the Gotland Deep a secondary halocline was found at a depth of around 100 m. This was observed in the log plot through the appearance of two circular areas, the water below the permanent halocline thus consisting of the deep water down to a secondary halocline and the bottom water from there to the bottom.

Using the same technique for a section across the Bothnian Sea in June 1958, Hela (1966a) found the upper boundary of the permanent halocline there to be in the depth range from 50 m to 75 m. The boundary was deepest on the western side where the salinities in the whole water column also were lower than in the central parts. Between a depth of 10 m and 20 m there was a secondary halocline caused by the existence of a seasonal summer thermocline in combination with the river runoff. This halocline was best developed on the western side of the section. In the central parts of the Bothnian Sea the salinity variation from surface to bottom was 1.5‰ whereas it was 5‰ in the Gotland Deep and 8‰ in the Bornholm Basin.

Hela (1966a) went on to investigate the secular changes of the salinity in the upper water layer using the salinity of the upper boundary of the permanent halocline as a measure of the mean salinity of the upper water layer of the preceding winter. For this study Hela used five stations covering the open parts of the northern Baltic Proper, one station in the Bothnian Sea and one in the Bothnian Bay. Observations at these stations exist since about 1900 and the series are quite complete. Hela found a rather high correlation among individual stations as well as among the different basins. This shows that the salinity parameter used is reliable and that the secular changes of salinity in the basins are more or less analogous. The analyses, however, also confirmed the regional delay of salinity variations already found by Granqvist (1949, 1952). Salinity increases in the Gulf of Finland and the Gulf of Bothnia were consequently observed first in the eastern and southern parts, respectively. The consecutive decade mean values of the salinity of the upper layers of the northern Baltic Proper varied in the range from 7.00‰ to 7.65‰, with the lowest values occurring in the 1930s and the highest values around 1950. From there on till 1965 Hela found a slight decrease of the salinity. Matthäus (1977b) did not find this trend for the surface salinity when considering specifically the period 1952—1974.

Hela (1966a) used the results of Jensen (1937) to extend his curve backwards to about 1880. Jensen found that on the whole the 5-year gliding means of the surface salinity at Christiansø, in the Bornholm Basin and in the Kattegat correlated well with high values in the period 1880—1904, low values in the period 1895—1914, high values between 1915 and 1924 and low values, finally, from 1925 to 1935. Similar results were found by Neumann (1940) for the lightships Skagen Rev, Schultz' Grund and Gedser Rev. Hela's conclusion is that the salinity in the Baltic Proper was higher during the period 1880—1895 than around 1905. This seems to agree with observations in the Baltic Proper for 1877 (Ekman, 1893). Hela also compared his results with those of several other authors studying the salinity variations of the Baltic Sea. He found that the order of magnitude of the salinity increase calculated by different methods is the same. Some particular features of the salinity variations are worth noticing. Hela (1966a) referred to Lisitzin (1948) who, studying the mean salinities for different water layers at Utö, observed that the maximum values in a secular sense show a slow transfer from the bottom layers to the surface with a transfer time of about two years. Hela further noticed the correlation between the fluctuations found by him in the surface layers and those found by Fonselius (1962) for the depth of 200 m in the Gotland Basin. Both Hela's and Fonselius' presentations show that the minimum salinities occurred in the middle 1930s, the high salinities in the early 1950s and that subsequently the salinity decreased. The fluctuations in the bottom water are, however, three or four times as large as those in the surface layer.

Fonselius (1969) investigated the salinity variations in the deep waters of the northern Baltic Proper, the Åland Sea, the Bothnian Sea and the Bothnian Bay. The decreasing trend in the Gotland Deep after the major inflow in

1951—1952 lasted until about 1959 when a new inflow increased the salinity. In all areas Fonselius found a general trend of increasing salinity, with minima around 1910 and 1935.

Also the temperature of the Baltic Sea waters shows marked long-term variations which have been studied among others by Soskin (1963) and Fonselius (1969). During the twentieth century the temperature of the ocean water has increased, and this has also included an increase of the water temperature in the Baltic Sea (Soskin, 1963). The deep water temperature in the Landsort Deep shows both the fluctuation generated by the major inflow and the long-term trend of increasing temperature (Fonselius, 1969). Since around 1890 the temperature there has increased from about 3.8° C to 5.0° C around 1965. The warmest water, with a temperature of 5.5° C, was found there around 1955 due to the major inflow into the Baltic Sea in 1951—1952. In Fig. 3.5 temperature variations at an open-sea station in the Baltic Proper are shown.

Nilsson and Svansson (1974) investigated the long-term variations of the surface salinity and temperature in the Gulf of Bothnia using observations from several lightships. They found a trend of increasing temperature since the early part of the twentieth century and an increase of the salinity of about 0.5‰ since around 1930 to 1960. Before that period the salinity showed a decreasing trend.

Causes of the long-term salinity variations

The source of the salinity is the inflowing Kattegat water and it seems natural to seek the explanation for the long-term salinity variations in variations of the exchange of water between the Baltic Sea and the North Sea. However, the salinity variations may also be related to variations in the fresh-water supply, mainly river runoff. The connection between the fluctuations of river runoff and salinity has been investigated by Soskin (1963), Fonselius (1969) and Kaleis (1976), among others. The results show a fairly evident correlation between the fluctuations in the sense that periods of low salinity tend to coincide with periods of high-river runoff and vice versa, although with an evident time lag. Kaleis (1976) investigated three USSR rivers representing 28% of the total runoff to the Baltic Sea and found a clear connection between periods of high and low runoff and low and high annual mean salinities in the water column (0—95 m) in the Bornholm Deep. In 1882—1898 the runoff was low which agrees with the expected high salinity for that period.

The water exchange between the Baltic Sea and the North Sea is strongly influenced by the meteorological conditions in the North Sea—Baltic Sea area, and therefore it is reasonable to expect a correlation between salinity fluctuations and fluctuations in the meteorological conditions. Jensen (1937) and Hupfer (1962) tried to interpret long-term salinity variations in

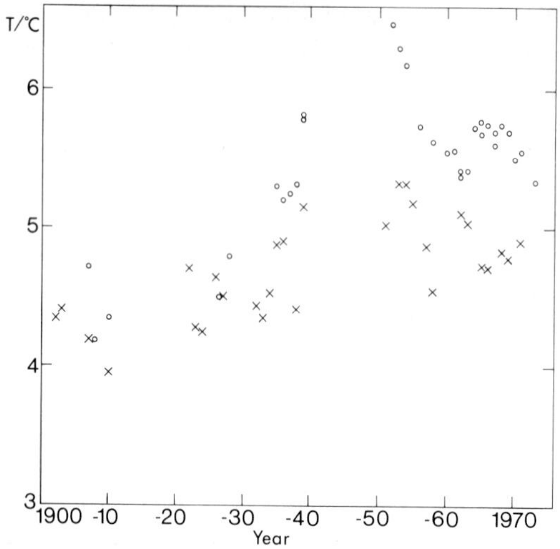

Fig. 3.5. Temperature ($T/\ ^{\circ}C$) variations in the Landsort Deep (crosses) at 300 m and the Gotland Deep (circles) at 200 m.

the Transition Area on the basis of meteorological changes. Hupfer's results clearly showed the importance of the meteorological conditions and furthermore coupled the variations to simultaneous changes in the general atmospheric circulation over Europe. Dickson (1971, 1972) concluded that the main cause of long-term salinity variations in the European shelf seas is a persistent pressure anomaly pattern over the mid-latitudes of the North Atlantic Ocean. He related high surface salinities occurring with 3—4-year intervals in the European shelf seas to this anomalous atmospheric circulation system, and he further noted that major inflows into the Baltic Sea tend to occur during periods of high salinity in the shelf seas. On the basis of the hypothesis that high-salinity conditions in the shelf seas are necessary for a major inflow into the Baltic Sea, Dickson (1973) made a successful prediction of such an inflow. Hupfer (1975) found a close connection between the salinity at a depth of 15 m at the Lappegrund lightship and the easterly wind component over the North Sea. Kullenberg (1977) found a fair degree of correlation between the deep water salinity variations in the Gotland Deep and the wind fluctuations at Gedser Rev lightship.

All these studies clearly show the significance of meteorological conditions and it seems safe to conclude that the salinity fluctuations in the Baltic Sea are connected with variations of meteorological conditions over northern Europe. This conclusion is not at odds with the result that the salinity fluctuations appear to be connected with variations of river runoff, since the variations in turn most likely are related to changes in atmospheric circulation, as discussed e.g., by Soskin (1963) and Fonselius (1969). However, there exist no studies of the long-term changes of atmospheric circulation which can be directly applied to this problem, and additional studies are therefore needed.

DENSITY STRATIFICATION AND ITS VARIABILITY

From the above discussion it appears that, except in very limited areas, the water column in the Baltic Sea is stably stratified throughout the year. The stratification is related both to the seasonal changes of temperature and river runoff and the permanent salinity layering. The seasonal variation of the stratification in the almost homohaline layer above the permanent halocline is governed by the annual temperature changes.

For the halocline layer and the deep water the stratification is determined mainly by the salinity and its essentially long-term fluctuations. The major inflows of deep water are clearly reflected in the density variations (e.g., Fonselius, 1969). The most important effect of the stable density stratification is its effect on the vertical exchange between the various layers. Since the stratification will suppress the mixing depending on the degree of stability it is of great interest to investigate the stability in the water column and its long-term variations. It is primarily the stability across the halocline layer which is of interest. Fonselius (1969) calculated the stability ($E = 1/\rho \cdot \Delta\rho/\Delta z$) across the halocline layer using observations from the 100 m and 150 m depth levels in the Landsort Deep and the Gotland Deep. He found an increase in the stability in the Landsort Deep during this century, in particular when only values from the summer months June, July and August were considered. In the Gotland Deep there were large fluctuations of stability, evidently coupled to major inflows, but there was no significant trend towards an increase of the stability.

Instead of calculating the stability between 100 m and 150 m, one may investigate the maximum stability across the halocline layer. Some caution is needed since it is not always possible to define the layer of maximum stability in an unambiguous way. However, as an indication of the stability variations the maximum stability across the halocline layer is given in Fig. 3.6a, b for the Gotland Deep and the Landsort Deep, respectively. No marked trends occur. It appears that there have been no drastic long-term changes of the stability across the halocline layer. The stability is essentially determined by the salinity differences between the surface water and the deep water. Since the long-term salinity variations in both layers tend to correspond, since the salinity source is the inflowing Kattegat water and since the long-term salinity increase occurs in both layers there is no reason to expect any marked long-term change in the stability. The salinity variations appear to be coupled to the meteorological conditions prevailing over northern Europe. It may be expected that the stability fluctuations are related to these condi-

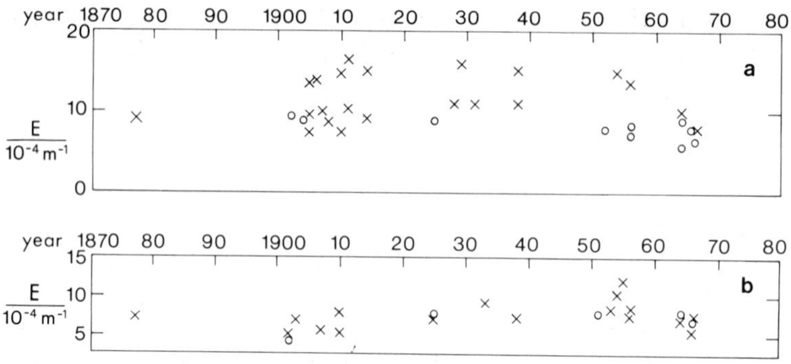

Fig. 3.6. Variations of stability (E) over the halocline layer in: (a) Gotland Deep and (b) Landsort Deep. crosses: summer season; circles: winter season.

tions as well, as indicated by the correlation between the stability in the 50—100 m-deep layer in the Gotland Deep and the wind at Gedser Rev (Kullenberg, 1977).

CIRCULATION

Mean circulation

The permanent circulation in the Baltic Proper is very weak and is clearly related to the excess fresh-water supply. The current velocities are of the order of a few cm s^{-1} in the surface layer and slightly less than 1 cm s^{-1} in the deep water. Primarily the horizontal salinity distribution with a marked NE—SW inclination of the isohalines, but also to a certain extent the temperature distribution, show the long-term average circulation to be cyclonical. In the Gulf of Finland and the Gulf of Bothnia the mean circulation is also cyclonical with a velocity of the order of 1 cm s^{-1}. There is one large gyre covering the Bothnian Sea and another covering the Bothnian Bay (Witting, 1912; Palmén, 1930). These gyres are clearly depicted in the mean monthly surface temperature distribution (Böhnecke and Dietrich, 1951). The influence of the Coriolis effect on the mean circulation is significant.

The mean motion in the surface layer is slightly more persistent along the west coast than along the east coast due to the combined effect of outgoing river runoff and the Coriolis effect. The mean circulation contains a weak vertical shear. Although storms over the Baltic Sea are frequent and often persistent, the mean winds are generally weak and the mean circulation in the Baltic Sea appears to be mainly estuarine and thermohaline.

Time-dependent motion

The fluctuating part of the motion predominates, and it is primarily induced by varying meteorological conditions. The tidal motion in the Baltic Sea is very slight, of the order of 1 cm s^{-1}, whereas the wind-induced motion can be up to two orders of magnitude as large.

There exist some early studies of the fluctuating motions in the Baltic Sea. Witting (1912) and Palmén (1930) made basic investigations of the currents in the Gulf of Finland and the Gulf of Bothnia, also including parts of the northern Baltic Proper. They used observations obtained at Finnish lightships over several years. Palmén managed to show the direct influence of the wind on the surface-layer velocities, in fairly good agreement with the theory of Ekman. Palmén found that the surface current was almost directly proportional to the local wind and it could reach a velocity of about 30 cm s^{-1}.

Already during the 1920s and 1930s, current measurements were carried out from anchored research vessels and in some instances by moored instruments in the southern and central parts of the Baltic Proper. On the basis of such measurements, Gustafsson and Kullenberg (1936) presented current measurements for the summers of 1931, 1932 and 1933 covering the southern and central parts of the eastern and western Gotland Basins. In 1932 one station with recording meters at depths of 10 m, 17 m and 28 m was placed at 56°44′N, 19°37′E for 10 days. During the same time a number of anchor stations were occupied in order to investigate the phase difference between the currents. During the first 3 days of the observations the sea was calm, then the wind velocity was about 15 m s^{-1} from SW—W for about 4 days followed by varying winds. The recordings from the moored instruments at the depths of 10 m and 17 m showed very clearly rotating currents. Unfortunately, the instrument at the depth of 28 m did not work satisfactorily. The effect of the strong wind in the middle of the recording was seen as a phase-shift of 140° in the rotating current vector, but after the storm the angular velocity became the same as before. The currents at the depths of 10 m and 17 m were very similar. However, the results from the anchor stations showed that below the thermocline the currents were different. During the calm period they showed no rotational component, but after the storm a weak rotation of the current was evident, with a clear phase difference relative to the currents above the thermocline.

The aim of the 1933 programme was to carry out simultaneous observations at several stations. At each station one recording instrument was moored in the top layer which was 15—17 m thick. Four different combinations were used covering scales of about 35 km and 70 km in the NNE and ESE directions, including comparisons between simultaneous observations east and west of Gotland. During the observations the wind varied from weak to W and NNW with a velocity of 10—14 m s^{-1}. Rotational currents were traced in the records and their period was determined using intervals when the wind velocity was less than 7 m s^{-1}. The period of rotation was found to agree well with the inertial period, the difference being at most 10 minutes. The analyses of the horizontal extent of the inertial motions showed that both the amplitude and the phase varied considerably. The phase variation was equivalent to about 1/3 period over a distance of about 35 km. The phase difference also varied with time. An analysis of nearby coastal water level records covering the period of current measurements showed that the water level oscillations contained no inertial period. Gustafsson and Kullenberg concluded that the inertial motions are local phenomena and are damped before they reach the coast.

The investigation of the inertial motions was continued during a cooperative international expedition with four research vessels in the summer of 1939 (Kullenberg and Hela, 1942). The observations were carried out along an almost east-west line at 56°20′N with five stations, at first about 11 nautical miles apart and later about 29 nautical miles apart. Besides the research vessels measuring at the depths of 15 m and 30 m, one mooring was laid with two instruments at the depth of 12 m and 15 m. During the period of observation, 23 July—12 August, the wind velocity was less than 10 m s^{-1} except on 25—27 July with northerly winds with a velocity exeeding 15 m s^{-1}, and on 31 July—2 August with southwesterly winds with a velocity greater than 15 m s^{-1}. Observations from anchored vessels were only carried out during periods of weak winds. The surface mixed layer was 16—17 m thick separated from deeper layers by a well-defined thermocline at a depth of 17—20 m.

The analysis of the current measurements showed that inertial oscillations generally occurred. The mean period determined from the moored instruments was only a few minutes shorter than the theoretical one. At the same station the phase difference was 2° between the depths of 12 m and 15 m, whereas at the second moored station it was 15°, and the measurements did not permit an accurate determination of the period. The observations from the anchored vessels showed that inertial motions were present at the depths of both 15 m and 30 m, i.e., below the thermocline layer. The phase difference varied between 190° and 40°, mostly being greater than 90°. In some instances the amplitude at the depth of 30 m was larger than the amplitude at the depth of 15 m. The amplitudes showed a considerable range, from about 3 cm s^{-1} to about 20 cm s^{-1}.

The observations were also used to investigate the horizontal extent of the inertial motions. The two moored stations were separated 16 pendulum days in time and about 25 nautical miles in a north—south direction. Despite this the records showed a phase difference of only 9° at the depth of 12 m and 23° at the depth of 15 m. This shows a considerable inertia in the system and indicates that the rotating system was fairly large. The simultaneous observations at different locations also show good coherence. Between the eastern and western parts there was a phase difference at the depth of

15 m of 33°, or 89 minutes. Two stations about 20 nautical miles apart showed almost identical phase over a measuring period of a week. Kullenberg and Hela concluded that the surface layer water masses covered by the east—west line about 180 km long were in a state of inertial motion with practically the same phase but considerable variation in amplitude. The station closest to the eastern coast, about 19 km from the coast line, did not show any inertial oscillations, whereas they were present at the station 40 km from the coast line.

As regards the conditions at the depth of 30 m, Kullenberg and Hela (1942) did not consider the occurrence of inertial motions definitely proved, although the indications based on the circular form of the flow line were quite strong. Finally, it should be mentioned that an analysis for other periods was also carried out using records from the moored instruments. The semi-diurnal tide M_2 was found with an amplitude of 1—2 cm s^{-1}, but no diurnal tide could be traced.

Thus a fruitful series of Baltic Sea studies prior to World War II was terminated and it was not revived again until well after the war.

Subsequent observations, mainly by means of moored current meters, have revealed that inertial oscillations occur in most parts of the Baltic Sea and also in the deeper layers at some distance from the coast (e.g., Kowalik and Taranowska, 1967; Kielmann et al., 1969, 1973; Mälkki, 1975). These investigations show that the response of the open Baltic Sea is mainly barotropic for wind periods lasting longer than about 50 hours and that the main baroclinic response occur for wind periods in the range 10—40 hours. The largest variability is also found in that range, and periods shorter than about 10 hours are rather quickly supressed. In the current spectra the inertial peak usually dominates. The energy of the baroclinic mode has been observed to be about one order of magnitude larger than the energy of the barotropic mode (Kielman et al., 1973). Even for periods around 100 hours the barotropic mode only contains about 50% of the total energy. This means that one should be careful in treating the Baltic Sea as homogeneous; such models are at best crude approximations.

A common picture brought out by all the current studies is that the meteorological conditions have a considerable influence on the fluctuating motion. In several instances simultaneous observations of meteorological conditions and currents have been carried out. It is generally found that the air pressure and the wind velocity spectra have approximately the same shape, although the pressure spectrum has a mean slope of —1.6 for frequences above about 10^{-2} h^{-1}, whereas the wind spectrum then has a mean slope of —0.8 (e.g., Krauss, 1974a).

A comparison between the wind and current spectra shows them to be in general rather similar. The current spectrum has often a mean slope of —0.8 (Kielmann et al., 1973; Krauss, 1974a). At the inertial period there is generally a very marked resonance peak. In the top layer the fluctuating motion is directly proportional to the wind stress, as already found by Palmén (1930). In the deep layers the motion seems to be more related to the divergence and the curl of the wind stress. The fluctuating velocities can be of the order of 50 cm s^{-1} and it seems to be clear that the main energy source for the motion in the Baltic Sea is the wind. The tidal forces are at least an order of magnitude smaller (Hollan, 1969).

Prominent features of the time-dependent motion in the Baltic Sea are different kinds of waves, seiches and internal waves. Kielmann et al. (1969) found that the kinetic energy in slightly more than half of the water column was proportional to $N^{-2/3}$ and in the rest (deeper parts) of the water column proportional to N, where N is the Brunt-Väisälä frequency. This result may suggest an essentially turbulent motion in the top layer and motion dominated by internal waves in the deep layer. Attempts have been made to fit current observations to linear internal wave models. Krauss (1974a) found a rather good agreement between filtered current observations and the sum of the four basic internal wave modes. The observations covered a period of about two months in the Arkona Basin, and the 10—18 h period range was used in the computations. Although the model fit was good for the whole water column, the differences between the model and the observations were greater for the top layer than for the near bottom layers as should be expected. Long internal wave motion has also been studied by Hollan (1969) who used observations from moored current meters at 5 different depths covering the whole water column in the Gotland Deep. Hollan showed that wind forcing could generate internal waves in the whole water column with periods close to the inertial period. The periods appeared to depend upon the horizontal extent of the wind disturbance, and it was concluded that mainly large-size strong disturbances could generate long internal waves in the whole area. Hollan (1969) found that long waves with a period of 13.5 hours could explain most of the fluctuations of the motion in the area. The amplitude of the waves varied from 20—25 cm s^{-1} near the surface to 5—7 cm s^{-1} near the bottom, and the waves showed phase differences up to 180° over vertical distances of 50 m. Since strong meso-scale wind disturbances over the central part of the Baltic Sea are fairly frequent, one may expect that such motion frequently occurs in the area. On a theoretical basis, Hollan (1969) argued that the motion will cover the whole Gotland Deep. Consequently, these wave motions may have a considerable influence on the stratification and the distribution of matter in the area.

Other important long-period motions in the Baltic Sea are the seiches. They have been studied by, among others, Neumann (1941), Krauss and Magaard (1962), Magaard and Krauss (1966) and Krauss (1974b). The theoretical calculations of Neumann, using one-dimensional theory, were corroborated by Krauss and Magaard, but when Magaard and Krauss analysed a data set for one year from 27 tide gauges they did not find any energy concentrations at the main theoretical periods, namely 39.1 h, 27.5 h and

19.3 h. These periods were found only at 10—20% of the stations analysed, whereas peaks in the range 23—24 h occurred at more than 50% of the stations. Despite this it appears that seiches are important phenomena in the Baltic Sea and that the discrepancy lies largely in the shortcomings of one-dimensional theory (Krauss, 1974b). Krauss developed a two-dimensional theory and computed the seiches as an initial value problem for the systems: (1) southwestern Baltic Proper — Gulf of Finland with the Gulf of Bothnia closed; (2) the whole Baltic Sea with the Gulf of Bothnia being a part of the system; and (3) southwestern Baltic Proper — Gulf of Bothnia with the Gulf of Finland closed, with and without the Coriolis effect included.

The theory gave results in basic agreement with the observations of Magaard and Krauss (1966), since periods around 32 h and 22 h are predicted. The 32.5 h generally occurs in sea level spectra of the Baltic Sea. Periods in the range 50—60 h have also been observed in the Baltic Sea system (Magaard and Krauss, 1966; Kielmann et al., 1973; Nielsen, 1973). This period has then been interpreted as a higher harmonic of the wind-force 120-hour oscillation, but in the calculations of Krauss it emerges as the basic Eigen-oscillation of the Baltic Sea system. It should be noted that periods of 27 h and 39 h also have been demonstrated (e.g., Neumann, 1941; Kielmann et al., 1973), but Krauss (1974b) does not regard them as representing Eigen-oscillations but rather as resulting from a superposition of such oscillations.

It should be noted that the results of Krauss to a certain extent differ from earlier results, and it may be concluded that further work on this problem is required in order to reach a satisfactory understanding of the processes involved.

In the Bothnian Bay the water level variations are more pronounced than in other parts of the Baltic Sea, reaching values of about 1 m during winter, with a period of around 2 days, i.e., rather close to the appropriate seiche period (e.g., Lisitzin, 1967). In the Bothnian Sea variations with the same period occur but with considerably smaller amplitude.

In the open main part of the Baltic Proper the amplitudes of the sea level oscillations are of the order of 10 cm. They become larger both towards the entrances and in the Gulf of Bothnia and the Gulf of Finland where they may reach a value of 100 cm.

Svansson (1972) made extensive calculations of the sea level variations in the Baltic Sea using a canal model and dividing the area into a large number of segments. He found a fair agreement between calculated and observed variations.

The coastal boundary layer

Considering the dominating influence of the wind on the motion in the Baltic Sea, it appears reasonable to pay particular attention to the coastal boundary layer since the wind forcing there will generate special effects due to the presence of the solid boundary. Also from a practical point of view, the coastal zone is of a special interest. Already in the 1920s and 1930s, the effect of the coast on the temperature distribution was discussed (e.g., Mäe, 1928; Palmén, 1930).

The response of a system like the Baltic Sea to transient forcing has been studied theoretically by Walin (1972a). His results suggest that the baroclinic response could be essentially limited to a narrow coastal boundary zone, about 5—10 km wide. In this zone the response can be quite vigorous while the response further out is of another type. Walin (1972b) used daily observations in a section perpendicular to the coast in southern Sweden during the summer months to demonstrate this effect experimentally. In a 5—10 km wide coastal zone the vertical excursion of the isotherms was large, sometimes comparable with the total depth, whereas at a distance of 10 km from the coast the isotherms did not deviate appreciably from the vertical structure typical of the season.

Shaffer (1975, 1977) pursued these studies in the coastal region south of Stockholm. He demonstrated the capacity of the wind-forcing to generate large vertical fluxes in the coastal boundary zone. Both upwelling and downwelling appear to be important processes. Although the very large fluxes sometimes found by Shaffer may be related to the special canyon type topography of his study area, his results clearly show the importance of the coastal boundary layer for vertical transfer. The coastal zone is important in most water bodies, from the oceans to confined systems such as the Great Lakes in North America where the coastal boundary layer has been extensively studied (e.g., Csanady, 1967, 1977).

Mälkki (1975) used observations from moored current meters at four stations over a period of 80 days in July—September to study the current variability in a region at Landsort. The area probably covered a transition zone from the coastal boundary layer to the open sea system, the coast line being 10—20 km away. He found clear rectification effects of the coast. The main part of the fluctuating energy was contained in the low frequency range of the longshore current component. During the calm summer months the most predominant periodic feature was the inertial oscillation, with average current velocities exceeding 10 cm s^{-1}. Towards autumn it became less distinct, and at the end of September no inertial motion was detected. The inertial waves above and below the thermocline were in opposite phases, in agreement with the results of Kullenberg and Hela (1942). However, Mälkki (1975) also found large horizontal phase differences which he considered an effect of the relative proximity of the coast. The coast will distort the wind field and will also increase the frictional damping. Mälkki formulated a model and found that the damping was rather large. Since the inertial motions vary with the wind stress in time and space (Krauss, 1972), an inhomogeneous wind field may distort that type of motion. Mälkki made a careful attempt at the evaluation of the various terms in both the momentum and the energy equations. In the former he found a balance between the pressure gradient force, the Coriolis effect and the frictional force. The energy considerations indicated that the fluctuating motion was feeding energy to the mean motion.

These studies clearly show the complexity of the coastal boundary layer. More studies are needed to clarify the processes and, in particular, to investigate the coupling between the coastal zone and the open sea.

Finally, a small scale detailed study of the near shore zone is mentioned. The study included observations extended over an approximately 6 km long section perpendicular to the coast of Poland at Lubiatowo, of winds, currents, waves, diffusion, suspended matter, temperature and heat budget. The results were presented by Druet et al. (1976).

Theoretical considerations

A brief account will be given of some basic theoretical approaches used in or developed for the study of the Baltic Sea circulation.

Palmén (1930), to a certain extent following and developing the work of Witting (1912), studied the relation between wind-induced current and the Ekman theory (Ekman, 1905). Palmén also investigated the coupling between the wind-generated motion and stratification with special reference to the Gulf of Finland. He elucidated the combined effects and in particular stressed the generation of a vertical circulation through the compensating horizontal flows in the semi-enclosed basin. Considering the quasi-stationary balance between the frictional force, the pressure gradient force and the Coriolis effect, Palmén derived the following equation for the vertical current gradient in the longitudinal direction of the Gulf (x positive in current direction, z positive downwards):

$$\frac{\partial \rho u}{\partial z} = -\frac{g}{2\omega \sin\phi} \cdot \frac{\partial \rho}{\partial z} \operatorname{tg}\psi \qquad (3.1)$$

where ψ is the slope of the isopycnals in the steady state. The expression was applied to the Gulf of Finland. He showed that strong westerly winds may bring the deep water towards the surface along the coast of Finland, thus suggesting the possible role of upwelling. Palmén also applied Bjerknes' (1901) circulation theorem to a number of different wind conditions and was able to explain the main features of the density distribution in a cross-section for easterly and westerly winds, respectively. The vertical current distribution was calculated using eq. 3.1. In order to obtain the absolute current distribution, Palmén used the stationary gradient equation for the surface current velocity and the water level difference. The comparison between observed and computed currents was fairly good for the surface current but not so good for the subsurface layers. However, the available current observations were not ideal for the comparison. Palmén also discussed the often occurring sudden temperature drops observed along Baltic Sea coast lines, referring to Mäe (1928) who had observed such drops also during coast-parallel winds forcing the water away from the coast. Palmén concluded that the dominating process generating temperature drops, e.g., in areas like the Gulf of Finland and the Gulf of Bothnia, was the transverse circulation generated by combined effects of the coast line, the wind and the Coriolis effect. He demonstrated the effect of coastal upwelling with an illustrative example of wind, air and water temperature observations at two stations (Hangö, Odinsholm) across the western part of the Gulf of Finland.

During the last 2 or 3 decades, the theory of time-dependent motion in oceans and lakes has been developed, and in the 1960s interest has been focused on the development of the time-dependent motion along the coastal boundary. Csanady (e.g., 1967, 1971) made several important studies of this zone in the Great Lakes in North America, particularly in Lake Ontario, which have many similarities with the Baltic Sea. Walin (1972a) investigated the general hydrographic response to transient meteorological disturbances with special application to the Baltic Sea. He developed a theory on the basis of scale analysis and boundary-layer technique. The basic equations were linearized around a state of rest with a specified stratification, and the analysis was limited to an intermediate range of time scales, large compared to the inertial period, but small compared to the time scale for diffusion. The ratio of the Brunt-Väisälä frequency N to the Coriolis parameter f turned out to be an important external parameter for the types of motion considered by Walin. He found that the ratio of the horizontal to the vertical scale of the motion, L/H, is the order of N/f. For the Baltic Sea this implies that the response to a large scale disturbance does not vary with depth. The response forced from the coast, however, has the length scale L_c with the prescribed depth H_0, giving $t_c \sim 5$ km for typical Baltic Sea conditions. Observations also indicate larger values of the width (e.g., Shaffer 1977).

The theoretical studies of Palmén (1930) and Walin (1972a) to a certain extent emphasize the vertical response and the effect of the coastal boundary. Hollan (1969) investigated the response of the interior to both small scale and meso- to large-scale wind disturbances. He found that both could generate internal wave motion. Small scale disturbances, from about 0.1 km to a few km, could generate waves of periods from some minutes up to an hour, with a wave-length of the same order as the disturbance. This could be due to local variations in the mean current field or to local changes in meteorological conditions. Large scale disturbances, extending over a distance of more than 50 km, may generate internal waves with a period just below the inertial period. In this case, the required external forcing could be specified as resulting from the action of strong winds.

Hollan's (1969) analysis of the small scale waves is based on the small scale approximation for internal waves (e.g., Krauss, 1966) solving the problem for the case of a constant vertical density gradient. The result is proportional to N^{-2}, and the smallest possible wave period is $2\pi/N$. Hollan first investigated the long-period oscillations by means of the theory of Fjeldstad (1958), neglecting external forcing. In order to study the forcing he then developed a theory including the horizontal external forcing, which he expanded in a special way. The results are not easily summarized but appear to conform largely with his observations.

Stigebrand (1976) has shown that mixing generated by internal waves breaking along a sloping bottom is of great importance for the conditions in the deep waters of fjords, with special application to the Oslo Fjord. It is possible that a similar mechanism is significant for deep water mixing in the Baltic Sea. Also in the open sea, breaking internal waves most likely play a role in generating vertical transfer.

The long period large scale seiches of the Baltic Sea were investigated by Krauss (1974b) who used the vertically integrated equations of motion:

$$\frac{\partial \tilde{u}}{\partial t} = -f\tilde{v} - g\frac{\partial \zeta}{\partial x} - r\tilde{u} + \frac{\tau^{x}}{\rho H} \qquad (3.2a)$$

$$\frac{\partial \tilde{v}}{\partial t} = +f\tilde{u} - g\frac{\partial \zeta}{\partial y} - r\tilde{v} + \frac{\tau^{y}}{\rho H} \qquad (3.2b)$$

$$\frac{\partial \zeta}{\partial t} = -\left[\frac{\partial H\tilde{u}}{\partial x} + \frac{\partial H\tilde{v}}{\partial y}\right] \qquad (3.2c)$$

where $\tilde{u}$, $\tilde{v}$ are vertically averaged current components. The bottom friction coefficient r equals 5×10^{-6} s^{-1} on the basis of Neumann's (1941) observations. Similar damping values have been found by Kullenberg and Hela (1942) and Mälkki (1975) for inertial oscillations.

Krauss integrated the equations numerically, treating the Baltic Sea as a basin closed at Femahrnbelt and the Sound. The same equations were used to compute the stationary drift currents, giving stationary currents after 120 hours.

One incentive for studying the circulation in a semi-enclosed area like the Baltic Sea is the necessity of obtaining reliable values for fluxes of various substances (e.g., salt, nutrients, oxygen), both horizontally and vertically between the basins and the different layers. For this purpose the present knowledge of the circulation of the Baltic Sea is very meager. Many attempts have been made to construct box models with two or three layers (e.g., Fonselius, 1969) and using continuity conditions starting from the Knudsen equations to calculate inflows and outflows. This approach can only give reliable results when long-time series of observations are used.

Recent attempts have been made to estimate the fluxes to and from the Baltic Sea as well as inside the area on the basis of some known external forcing functions. Welander (1974) used the outside salinity, the fresh-water supply and the meteorologically driven barotropic transport as forcing functions. He found that a single steady state can exist and that variations in the salinity or the fresh-water supply tend to increase, respectively decrease, the salinity within the system.

Sarkisyan et al. (1976) calculated the transport and current velocities at 15 levels using observed average density and wind fields. Their results showed the importance of the combined effect of baroclinicity and of bottom relief. The baroclinicity and the direct wind effect were found to be decisive factors for the stationary circulation. However, the authors cautioned that the data base for the calculations was not ideal. Walin (1977) attempted to give a general description of the estuarine system using salt as the only independent hydrographic variable. He gave expressions for fluxes both between isohalines and across a given isohaline. This method will make possible the use of the long-term hydrographic observations in a better way than is possible in the box model. Finally, Pedersen (1977) used a hydraulic approach to calculate the flux from the Bornholm Basin into the Gotland Deep through the Stolpe Channel. He started from the conditions prevailing in the eastern Bornholm Basin and worked upstream to the Darss Sill. Combining the hydraulic relations with the long-term observations of salinity profiles at Christiansø he was able to determine both the mean volume flux and its expected range. His results are in fair agreement with actual observations at the Bornholm Sill (Petrén and Walin, 1975). Thus it is possible to predict also the residence time for the deep water in the Bornholm Basin, which Pedersen found to be of the order of months, again in reasonable agreement with observations.

MIXING CONDITIONS

Small-scale motion

The development of the wind-generated surface waves in the Baltic Sea is limited by the fetch which does not allow a fully developed sea to occur for wind forces above 7 Bf. The largest possible fetch is between the Åland Sea and the southern coast, about 300 nautical miles. For a wind force of 7 Bf, the expected wave characteristics after 24 hours duration are: mean wave height 4.5 m, mean wave length 80 m, mean period 8.7 s (Magaard, 1974). Systematic investigations of the wave conditions have been carried out in the western parts of the Baltic Sea (Magaard, 1974).

Internal small scale motion is considerably affected by the stratification. This suppresses the convective vertical motion except in the homohaline layer during parts of fall and winter. Otherwise, most of the vertical motion occurs in the form of internal waves. Short internal waves in the thermocline with a frequency close to the local Brunt-Väisälä frequency were observed already by Neumann (1946, see also Dietrich et al., 1957, p. 319). These were standing cellular type waves with a period of 45 s. Short period internal waves have later been studied in the Arkona Basin in the pycnocline around a depth of 30—40 m by Krauss et al. (1973). They used 20 thermistors with a separation of 20 cm attached to a bottom mounted subsurface tower to record the oscillations. In addition, observations of the sound scattering layer related to the pycnocline were carried out from a vessel. The spectra of the temperature fluctuations generally showed a marked peak close to the local Brunt-Väisälä period which mostly varied between 30 s and 60 s. The energy level in the peak was typically increased by a factor of 10—50 relative to the background. Spectra not containing this peak had lower energy density at the higher frequency end of the spectrum than those spectra where the peak occurred. The slope of the spectra varied in the range −2.3 to −1.0.

Hollan (1969), besides finding near inertial motion, also observed short internal waves with periods in the range 0,1—1 h in the central Gotland Deep. These short waves occurred as packets with a duration of up to several hours and amplitudes of about 3 cm s^{-1}, with periods of weak intensity in between. The vertical extent of the waves was roughly 100 m, sometimes even 200 m, showing that the whole water column was affected. The waves were apparently generated by surface disturbances with an extent from 0.1 km to a few kilometres, the waves having the same extent initially. However, by radiation the locally concentrated kinetic energy became distributed over a larger area.

These investigations show that small internal waves occur frequently both in the open Baltic Sea and closer to the coast. Since the stratification is generally strong, the internal wave occurrence is important when considering internal mixing both in the deep water and across the pycnocline layers.

In the coastal boundary layer, special circulation patterns occur which are very important for the mixing. The wind-induced vertical circulation was mentioned previously and can clearly serve as a window for vertical transfer across the pycnocline layer. In the nearshore zone covering also the surf zone, other processes occur caused by both winds and waves. The currents in this zone were studied by Wojewódzki et al. (1976) as part of the "Lubiatowo 74" experiment. The current measurements were made from 4 m water depth to a total water depth of 25 m, about 6 km from the shore, covering a time period of about 1.5 months. The data showed that the currents at bottom depths of 17 m and 25 m were dominantly shore parallel and were strongly influenced by the shore parallel wind component. The current and wind spectra had a similar shape. At a bottom depth of 4 m, in the wave surf zone, the currents were generally stronger than further offshore and were clearly affected also by other factors than the wind. For strong winds the current velocities at the depths of 4 m, 17 m and 25 m were similar. Also in the surf zone the current direction was generally shore parallel. It seems likely that the wave transport affected the conditions in the surf zone, but the authors considered it necessary to carry out more detailed experiments before the processes in this complicated zone could be elucidated.

In conclusion, although the mixing in the surf zone is strong, the transport is often alongshore, the offshore transport being confined to limited regimes, such as rip currents. The processes governing the exchange with the offshore zone have not yet been clarified. It is a primary task for Baltic Sea research to study the exchange between the nearshore zone and the open sea.

Vertical and horizontal mixing

The vertical mixing and transfer between separate layers are of main interest. The mixing of the Baltic Sea is primarily affected by the stable stratification and the energy input from the wind and through solar radiation. These processes may operate together to erode the stratification by mechanical energy and free convection, or they may work in opposite directions. Salinity stratification is always present and tends to suppress, in particular, vertical mixing.

The rate of vertical mixing can be studied by means of various tracers, such as heat, salinity, oxygen or a dye tracer. Both the salinity and temperature distribution in the layer above the halocline show that this layer is vertically well mixed during some periods of the year. The rate of mixing, however, varies strongly throughout the year. Observations of temperature as a function of time and depth have been used to calculate a vertical transfer coefficient for heat K_H, (Simojoki, 1946; Hela, 1966a; Piechura, 1972; Kremser and Matthäus, 1973; Matthäus 1977a; Lundberg, 1964 (cited by Matthäus, 1977a)). Several methods which have been developed to calculate K_H from temperature measurements were discussed by Kremser and Matthäus (1973). They also developed a method based on the equation:

$$F = -c_p K_{II} \frac{\partial T}{\partial z} \tag{3.3}$$

The water column was divided into n layers, down to a depth where the yearly temperature variation is insignificant, assuming only vertical heat transfer.

This approximation may be accepted for parts of the open Gotland Deep as regards the mean transport (e.g., Hollan, 1969). The method allows calculation of K_H as a function of depth and time, and Matthäus (1977a) presented results from 10 stations in the open Baltic Sea covering the Arkona and Bornholm Basins and the Gotland Deep. The mean annual temperature profile was used, and the water column was divided into layers 10 m thick. The temperature profile was approximated by a Fourier series, and daily mean values for the respective depth layers were determined.

Since advection may be expected to be important throughout the year in the deeper layers of both the Arkona Basin and the Bornholm Basin the calculations were limited to depths of 30 m and 50 m, respectively.

In the open Gotland Deep, however, observations indicated no significant influence by either advection or upwelling phenomena. The mean annual temperature variation was uniform down to a depth of 80 m, the amplitude decreasing with increasing depth and the temperature maximum being reached successively later in the year. It may be concluded that the mean annual temperature variation in the open Baltic Proper to a first approximation results from vertical heat exchange in the surface layers, the heat being transferred downwards by turbulent mixing.

The results of Matthäus (1977a) will be summarized here. In the whole area the values of K_H were largest in the surface layer during October—November and in February. The maximum value was slightly greater than 100 cm^2 s^{-1} in the Arkona Basin, and the minimum value was greater than 10 cm^2 s^{-1}. Also in the depth layers 30—40 m and 40—50 m, values up to 40 cm^2 s^{-1} were observed. Matthäus concluded that the high mixing rates during these periods are due to the generally strong winds, with a wind force of 6 Bf during 20—30% of the time (Defant, 1972). In addition, the heat exchange with the atmosphere acts to generate free convection during these months. By March the winter stratification has been produced, suppressing the vertical exchange and giving low values of K_H. By April the heating sets in and generates an intermediate period of fairly effective exchange, with relatively high values of K_H in the range 10—100 cm^2 s^{-1} in the top 10 m and 3—30 cm^2 s^{-1} in the 20—30 m deep layer. In the period from April—May to August—September the values of K_H in the surface layer were generally in the range 2—10 cm^2 s^{-1}, decreasing with depth down to about 50 m and being generally below 1 cm^2 s^{-1} in the subsurface layer.

Below 50 m the values of K_H, with few exceptions, were less than 1 cm^2 s^{-1} at all the stations.

Matthäus (1977a) explained regional differences between the Arkona Basin and the Gotland Deep by the difference in wind fetch and, partly, wind strength. In the Arkona Basin the fetch is small, and the winds are slightly weaker than in the Gotland Deep, corresponding to maximum exchange values up to or slightly greater than 100 cm^2 s^{-1} in the Arkona Basin, but up to 150 cm^2 s^{-1} or more in the Gotland Deep. These differences are also present in the subsurface layers. In the western Gotland Deep the fetch is smaller and the exchange correspondingly smaller although K_H reaches values exceeding 100 cm^2 s^{-1}; however, K_H decreases rapidly with depth. The values for the Baltic Sea are generally comparable with values found in other shelf areas, e.g., in the North Sea (Weidemann, 1973; Talbot and Talbot, 1974).

For the northern part of the Baltic Sea, Simojoki (1946) and Hela (1966a) used temperature observations at Finnish lightships and coastal stations to calculate K_H. The values fall generally within the same range as those found by Matthäus (1977a). The minimum values determined by Hela (1966a) were of the order of 1 cm^2 s^{-1}, whereas Simojoki (1946) found values of the order of 0.1 cm^2 s^{-1}. At most stations Hela (1966a) found an increase of K_H with depth from the minimum layer of the pycnocline. This behaviour was not confirmed by Matthäus for the southern Baltic Sea areas. Hela used temperature observations for the period 1948—1957. Advection and upwelling were neglected in the calculations. Hela concluded that advective effects were probably important in the surface layer but less important at middle depths.

In conclusion, the use of heat penetration to calculate vertical exchange coefficients is attractive for several parts of the Baltic Sea, giving a fair coverage of the area in space and time, and yielding consistent values which conform with those found by other methods.

Matthäus and Kremser (1976) also modified their method of calculating K_H in order that it could be applied to the transfer of oxygen. Thereby the air—sea exchange of oxygen was taken into account as well as various sources and sinks of oxygen in the water column. Values for the vertical exchange coefficient for oxygen were calculated from observations in the Gotland Deep. The values were similar to those found for K_H, and the mean annual variation showed generally the same behaviour. The calculations were carried down to the 40—50 m layer, where values up to 10 cm^2 s^{-1} were found.

An alternative way is to use dye tracers to investigate vertical mixing for short periods of time, of the order of hours and days. In the Baltic Sea this has been attempted by Brosin (1972, 1974a, b) and Kullenberg (1971, 1972, 1974a, 1977). The latter injected the dye rhodamine B into the thermocline layer or into the halocline layer, tracing the dye in situ, and calculating diffusion coefficients from the observed concentration distributions. The vertical diffusion coefficients were in the range 0.01—0.2 cm^2 s^{-1} for depth layers between 25 m and 55 m in the Arkona Basin and in the Bornholm Basin. The experiments were carried out in calm weather conditions in May, August and December in different years, and the stability N^2 in the layers varied from 2.8×10^{-3} to 3.3×10^{-5} s^{-2}. The results for a depth of about 30 m in the Bornholm Basin are in very good agreement with the corresponding mean value calculated by Matthäus (1977a). It appears that the stability (stratification as given by N^2) affects the vertical transfer considerably, although no satisfactory empirical relationship can as yet be given, and it also appears that the wind has a strong influence on the transfer in the open sea.

Kullenberg (1977) made a preliminary investigation in order to decide whether the winds over the Baltic Sea can supply the energy needed for the annual vertical transfer between the deep water and the homohaline layer. Using mean vertical mixing values determined by means of dye experiments in the southwestern Baltic Sea, it was possible to find the range of energy consumed for vertical mixing down to a depth of 60 m. The energy available from the wind for vertical mixing was calculated on the basis of general results obtained by Denman and Miyake (1973), Denman (1973) and Kullenberg (1976), among others, and was found to be almost the same as the energy consumed by vertical mixing. This result suggests that the wind energy is important also in the Baltic Sea, and Kullenberg therefore correlated wind observations made at Gedser Rev lightship with time series of salinity and oxygen determinations at some stations in the deep water in the Gotland Deep. A fair degree of correlation was found in observations covering the period 1900—1975. The same result was reached regarding the stability in the 50—100 m deep layer in the Gotland Deep. These results, although preliminary, show the importance of wind for the conditions in the Baltic Sea. Besides influencing the exchange between the North Sea and the Baltic Sea, the wind also influences the mixing. The periodical renewal of the bottom water is partly governed by the vertical mixing conditions for which the mixing in the open Baltic Sea plays a significant role. The deep water mixing is partly governed by the entrainment caused by the inflowing deep water.

However, special processes occur in the coastal zone which can produce a very effective vertical exchange between surface water and deep water. Shaffer (1977) used essentially the approach described by Walin (1977) to calculate vertical diffusive and advective fluxes from observations of salinity, temperature and currents during about one year in a small canyon south of Landsort in the Stockholm archipelago. He found that across the halocline in the canyon, the advective flux per unit area could be up to two orders of magnitude as large as the mean flux per unit area in the whole Baltic Sea. The diffusive flux per unit area in the canyon could be one order of magnitude as large as in the open Baltic Sea. The mean exchange coefficients across the halocline layer were about 1 cm^2 s^{-1}, and Shaffer's corresponding diffusive flux was 12×10^{-6} kg salt per m^{-2} s^{-1}. Using the exchange coefficients from the dye experiments and the mean salt gradients across the halocline layer, Kullenberg (1977) found a flux of 0.6×10^{-6} kg salt per m^2 s^{-1} for the open Baltic Sea. From other data, Shaffer (1977) calculated a flux of 0.8×10^{-6} kg salt per m^2 s^{-1} for the open Baltic Sea. The area of the Baltic Sea at the depth of the halocline layer is about 10^5 km^2 (Ehlin et al., 1974), and the coastal length is about 10^3 km for the Baltic Sea proper. With an effective width of the coastal boundary layer of 5—10 km, this gives a ratio of the integrated diffusive fluxes of about unity, indicating that both the open sea and the coastal zone are of importance for vertical mixing. As regards the large advective flux found by Shaffer, it seems hard to believe this to be relevant for the coastal zone in general. The high flux value might be caused by special topographic conditions in the area studied by Shaffer, and more studies of this kind in different areas are necessary.

In the coastal zone the forcing is due to the wind, and also in the open sea the wind energy appears to be very important. This is true for most oceanic areas. However, in the Baltic Sea with its low salinity water, it is conceivable that the free convection generated by cooling and heating during parts of the fall, winter and early spring, respectively, is also important. The difficulty is to separate this effect from the wind effect, especially since strong winds are common during the same periods. As yet, the data required for a detailed study of this problem are not available. An example of the influence of the wind on the vertical structure is shown in Fig. 3.7, based on observations in the southern Gotland Basin during September before the cooling period had started. The profiles were obtained in situ with a conductivity-temperature-depth sensor immediately before and after a period of strong winds.

The horizontal mixing in the Baltic Sea has been investigated by means of a limited number of experiments with a dye tracer (Kullenberg, 1972, 1974a, 1977; Brosin, 1974a; Schott et al., 1978) and in the coastal zone using drifters (Brosin, 1974b). For the thermocline and halocline layers in the Arkona Basin and the Bornholm Basin, Kullenberg found horizontal diffusion velocities as defined by the theory of Joseph and Sendner (1958), in the range from 3.5 m h^{-1} to 7.2 m h^{-1}, whereas Schott et al. (1978) found

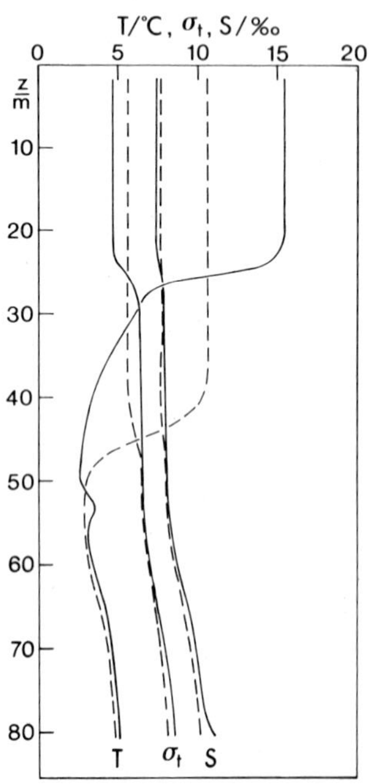

Fig. 3.7. Profiles of salinity ($S/‰$), temperature ($T/\,°C$) and density (σ_t) in the southern Gotland Basin obtained by temperature—conductivity—depth recorder (CTD), before (full drawn) and after (dashed) a storm with winds up to 30 m s^{-1} and duration of several days.

values from 9.7 m h^{-1} to 21 m h^{-1} from experiments made at the surface. Compared with the dispersion in the oceanic surface layer (e.g., Okubo, 1971), the values in the Baltic Sea are generally smaller by a factor of 5—10, which is quite reasonable.

The variation of diffusion with length scale is generally also found to be slightly different from that in the open ocean. Using results from subsurface dye experiments, Kullenberg (1977) found the horizontal diffusion coefficient K_h to be proportional to the horizontal length scale l_h, obtaining $K_h \propto l_h^{0.9}$. Similar results were obtained by Brosin (1974a, b) for the near surface layer. Brosin studied the dispersion of a cluster of current crosses in the top 1—2 m on two occasions in the fall about 10—20 km from the coast. He obtained the result $K_h \propto l_h^{0.8}$. Brosin (1974a, b) also studied the dispersion in the near-shore zone by tracking current crosses optically from the shore.

The present knowledge about the mixing processes is limited. Probably several energy sources are important in the Baltic Sea, not only wind and surface heat exchange, but also lateral inflow affecting the deep water circulation and mixing. The intermediate deep water layer just beneath the halocline layer appears to be considerably more active than the deeper lying layers (e.g., Kaleis, 1976). The small scale motion is probably important for the mixing. The layered temperature structure often observed in the Baltic Sea in or below the halocline layer is noteworthy (e.g., Wüst and Brogmus, 1955; Kullenberg, 1974b). The layers often have a thickness between 1 m and 10 m and may be caused by both inflowing water spreading at its appropriate density level and small-scale vertical mixing processes such as breaking internal waves with subsequent shearing. A considerable part of the horizontal mixing in the pycnocline layers can be explained by the vertical shear effect (Kullenberg, 1974a).

Knowledge of the mixing rates is of great practical interest with reference to the overall ecological conditions and to local pollution problems. Several local studies in the nearshore zone have been conducted for obtaining background information towards predicting the effects of local sources of e.g., heat and sewage.

OPTICAL PROPERTIES, HEAT BALANCE AND ICE CONDITIONS

Optical properties

There exist few investigations of the inherent and apparent optical properties of the Baltic Sea waters. It is pertinent to start with a brief account of the inherent properties, i.e., absorptance, scatterance and attenuance. These are given by the coefficients of absorption a, scattering b and attenuation c; for an account of definitions, relationships and measuring techniques, reference is given to Jerlov (1976).

An early systematic study of factors influencing the attenuation was carried out by Jerlov (1955) in a section along the western side of the Baltic Sea up to Sundsvall in the Bothnian Sea. The transparency was measured in situ at 380 nm and 655 nm. In the Baltic Sea the attenuation is very much affected by the selective absorption by yellow substance, i.e., humus-like substances in the water.

Jerlov (1955) also studied the particle distribution by means of Tyndall measurements, finding a very high load of particles in the deep waters of the Bothnian Sea. The particle content generally increased logarithmically towards the bottom below the pycnocline layer, which suggests deposition of slowly settling material under influence of horizontal flow. The high particle content was coupled to the large river runoff carrying a considerable load of suspended and dissolved matter. In salt water much of this material flocculates, starting with the smallest grains. The aggregates settle and much of the material is deposited rather close to the river mouths. Some flocculation and settling apparently go on also in the open sea (Jerlov, 1955).

Also in the central part of the Baltic Sea, Jerlov (1955) found similar particle distributions with increasing loads towards the bottom although this phenomenon was more pronounced in the Bothnian Sea. Another typical feature was a minimum of particles occurring in the upper part of the halocline layer at a depth of around 50 m. As regards the yellow substance, Jerlov (1955) found a distribution pattern conforming with the general circulation picture of outflowing surface water through the Åland Sea and further along the coast of Sweden. In the Bothnian Sea, values of $a_y = 2\ m^{-1}$ were found, about 5 times as high as the Kattegat values.

The inherent properties were further studied by Højerslev (1974) in a section from the Sound to Landsort, and by Lundgren (1976) at selected stations along the same section.

Højerslev measured the absorption at a number of wavelengths in the range 372 nm to 633 nm down to a depth of 40 m, the attenuation coefficient in the red (655 nm), green (525 nm) and ultraviolet (380 nm) parts of the spectrum, down to a depth of 100 m.

The absorption was dominated by the yellow substance, the particulate matter being of less importance. The scattering coefficient in the red varied between 0.10 m^{-1} and about 0.30 m^{-1}, with the largest values in the surface layer down to a depth of about 20 m and in the bottom layer roughly 5—10 m thick. The smallest values were observed at intermediate depths (30—60 m) in the western Gotland Deep.

The attenuation distribution generally followed the scattering distribution but displayed a more detailed structure in the UV part of the spectrum. The attenuation was smallest in the green and largest in the UV, with intermediate values in the red, generally.

Højerslev (1974) also measured the fluorescence in the blue-green part of the spectrum (490 nm), finding no systematic variation with salinity variations. The fluorescence decreased towards the south and generally increased with depth.

The volume scattering function has been studied by Kullenberg (1969) who made in situ measurements down to a depth of 70 m at three stations in the Bornholm Basin, in the red (655 nm, 633 nm), green (525 nm), blue-green (488 nm) and blue (440 nm) parts of the spectrum. The scattering functions are strongly peaked in the forward direction, with a minimum around a scattering angle of 100°—120° and a slight increase for larger angles. The results indicated a larger scattering in the green than in the other parts of the spectrum, but more data are needed to confirm this. Integrating the scattering functions, the scattering coefficients obtained were in the range 0.7—0.1 m^{-1}, with the largest values near the surface layer. The minimum values were found at intermediate depths, increasing again towards the bottom. The bottom boundary layer was about 5 m thick.

Jerlov (then Johnson) and Liljequist (1938) initiated studies of the underwater radiant energy conditions by measuring the radiance distribution at 6 stations in the central and western Baltic Sea in the blue, green and red parts of the spectrum, down to a maximum depth of 40 m.

The maximum transmission was observed in the green part of the spectrum around 525 nm, and for the deeper strata there was an indication of the direction of the maximum intensity to approaching the direction of the vertical. The approach towards the asymptotic radiance distribution was further studied for green light by Jerlov and Nygård (1968, see also Jerlov, 1976, p. 121). At a depth of 100 m in the central Baltic Sea, the distribution appears to be close to the asymptotic state, but the directional change of the light field is not completed even at that depth.

As regards the irradiance, spectral measurements in the central Baltic Sea show that there is a large shift of the transmittance peak with depth towards about 550 nm and that the ultraviolet is extinguished already at a depth of 5 m. This is mainly explained by the abundance of yellow substance in the water (Jerlov, 1976, p. 128). Højerslev (1974) measured the irradiance in the green (525 nm) simultaneously with the quanta irradiance integrated over the range 350—700 nm, in the Bornholm Basin, western Gotland Deep and northern Baltic Proper. The 1% levels of the green light were in the depth range 25—30 m, and the 1% quanta levels (i.e., integrated over the spectral range 350—700 nm) were at a depth of 20—22 m in the whole area. Højerslev (1974) also measured the spectral irradiance at 8 wavelengths distributed in the interval from 371 nm to 693 nm. The Baltic Sea water acts as a green filter, and the maximum transmittance was found around 555 nm, in agreement with previous results. The ratio of upwelling to downwelling irradiance never exceeded 1%, with the maximum value falling around 533 nm. At 371 nm the daylight was attenuated more than at 693 nm.

Jerlov (1975) compared irradiance measurements at virtually the same position in the Baltic Sea obtained in 1937—1938, 1963 and 1973 during the spring and early summer. He found that the irradiance transmission about 525 nm was slightly higher in 1973 than in 1937 and concluded that the amount of particulate matter in the surface layer has at least not increased since 1937.

The colour of the relatively turbid Baltic Sea water is a mixture of various colour components and the purity of the colour is therefore low (Jerlov, 1976, p. 166). The colour is affected by scattering due to water molecules and particles and by absorption due to water and dissolved substances; in areas like the Baltic Sea the yellow substance plays an important role. A useful colour index for the near surface water is defined by the ratio of upwelling irradiance in the blue (around 450 nm) and the green (around 520 nm) (Jerlov, 1974).

Høјerslev (1974) measured this ratio at many locations in the Baltic Sea, finding values around 0.5. The ratio in the western Mediterranean Sea south-west of Sardinia is around 3.0 (Jerlov, 1974). Further studies of the irradiance and, in particular, of the irradiance fluctuations due to waves and meteorological conditions have been carried out by Dera and Olszewski (1967) in the southern part of the Baltic Sea.

The optical properties are often sensitive indicators of water conditions with respect to the amount of various dissolved substances and the load of suspended matter (cf., Chapter 4, p. 211). The problem is, as usual, mainly to relate an observed change correctly to its cause. However, as more information on the optical conditions is obtained along with other properties of water, both biological and chemical as well as physical, this tool may become more used. The optical properties are also of great interest in relation to remote sensing techniques.

Heat balance

The heat balance of the Baltic Sea is of interest with reference to the climatic conditions of the countries bordering the sea. Early studies of the heat balance were made by Witting (1918), Wallerius (1932) and Jerlov (1940) on the basis of measurements from lightships and research ships.

Jerlov used material collected on cruises during different seasons in 1938 and 1939 and covering large parts of the Baltic Proper. Simultaneous observations from 3 ships were obtained during the international expedition in July 1939. Observations of the temperature distribution from the surface to the bottom were made at intervals of about 30 nautical miles covering the whole Baltic Proper.

For each station, Jerlov calculated the heat content down to the depths of 50 m, 70 m, 100 m and 150 m (or bottom).

The stations were grouped together for each latitude and mean values were calculated for each latitude, from 55° N to 60° N, for May (1938), August (1938), March (1939) and July (1939). The mean values showed a decrease of the heat content down to a depth of 50 m with increasing latitude, but there was no tendency towards an east—west variation. The heat contents in the depth intervals 50—70 m and 70—100 m were then calculated.

The results for the 70—100 m depth interval showed that the heat content in the southern Baltic Sea, essentially in the Bornholm Basin, was slightly larger and varied more with time than in other parts of the area investigated. In the central and northern parts, the variations in space were small. In the 50—70 m depth interval, the heat content increased by about 25% from March to August showing that the heat variation penetrates into or just below the halocline layer, whereas in the depth interval 70—100 m the heat content was virtually constant, showing that this layer does not contribute significantly to the annual heat exchange. The results are in good agreement with later studies on the annual temperature variation in the Baltic Proper (e.g., Matthäus, 1977a). Jerlov (1940) also calculated the heating during the periods 15 May—15 August, 1938, and 17 March—26 July, 1939, down to the depths of 50 m, 70 m and 100 m. These calculations showed that the heating was uniform for the whole Baltic Sea and that only a few percent of the accumulated heat penetrated below the 50 m level. Jerlov gave the total heat budget for the Baltic Proper as 1275—1670 MJ m^{-2} (305—400 Mcal m^{-2}) and for the Gulf of Bothnia and the Gulf of Finland 1190—1570 MJ m^{-2} (285—375 Mcal m^{-2}) and 1340—1690 MJ m^{-2} (320—405 Mcal m^{-2}), respectively.

His final results for the Baltic Proper gave the magnitude of the heat budget as 1990 MJ m^{-2} (475 Mcal m^{-2}) and 2650 MJ m^{-2} (640 Mcal m^{-2}) for 1938 and 1939, respectively. The predominant factors were incoming radiation and heat accumulation in the water mass. Finally, considering the Baltic Sea as a factor affecting the climate, Jerlov found that about 40% of the heat accumulated during the warm seasons was given up to the atmosphere during the cold season. This has a considerable influence on the climate, implying relatively mild winters along the coasts of the Baltic Proper.

Heat budget studies were taken up again after World War II by Simojoki (1946, 1949), Hela (1951), Brogmus (1952), Palmén (1963), Hankimo (1964), Pomeranec (1964), Hupfer (1967) and Sturm (1968) who studied various aspects of the problem.

Sturm (1968, 1970a) investigated the heat balance in the Transition Area and in the southern Baltic Sea using, in particular, long-term observations made at the lightship Fehmarn Belt. He calculated, for an interval of a decade, mean values of the terms in the budget given by the equation:

$$Q_n = Q_S - Q_R - Q_A + Q_B + Q_K + Q_V \qquad (3.4)$$

At Fehmarn Belt the surface water receives the largest heat input in June and releases the largest amount in November. The shift from the negative to the positive balance occurs at the end of March and from the positive to the negative balance in early September. The results show that the advection of heat by the subsurface current in the Baltic Sea is of great importance for the overall budget. Consequently, the inflowing North Sea water, which is anomalously warm for its latitude (Dietrich, 1950), has a significant influence on the heat budget in the southern Baltic Sea. Sturm (1970a) calculated the advective heat transport as the rest-term in the heat budget. Hupfer (1967) also found a large advective transport of heat in the coastal zone of the southern Baltic Sea.

Sturm (1970b) further considered the advective heat exchange between the southern and northern parts of the Baltic Sea. The deep water current yields an important contribution also to the heat budget of the central and northern Baltic Proper during the late spring and the early summer. Major inflows also affect the heat balance in a more aperiodic way, e.g., in 1934, 1938, 1948, 1952 and 1959. The importance of advection for the surface layer is also demonstrated by the mean annual heat balance given by Pomeranec (1964; cited by Sturm, 1970b) with negative values of Q_n (positive advection balance) on the eastern side up to the Gulf of Finland and positive values, i.e., negative advection balance, on the western side. This picture agrees well with the general mean circulation picture.

The investigations of Simojoki (1946) of the temperature and salinity distribution at Bogskär, at the entrance to the Bothnian Sea, showed an inflow of warm water from the central Baltic Sea in the depth interval 20—65 m in late April—May.

Simojoki estimated that about 9% of the annual heat turnover in the northern Baltic Sea (i.e. the Gulf of Bothnia) was related to advective inflow from the Baltic Proper. This is related to the inflow of relatively warm and salt water which enters the Gulf of Bothnia along the bottom from the Baltic Proper during summer and fall. Sturm (1970b) found that the corresponding figure for the Fehmarn Belt area was about 10%.

In order to study the annual heat budget in the Bothnian Sea, Hankimo (1964) investigated the vertical exchange of heat using observations made at the lightship Finngrundet. He also compared evaporation calculated by Palmén (1963) with evaporation calculated by means of Jacobs' (1942) equation:

$$E = k(e_s - e_a)W_a \tag{3.5}$$

where k is a dimensional coefficient, e_a and W_a are water vapour pressure and wind velocity at level a, originally 6 m. The evaporation and flux of sensible heat over the Baltic Proper were calculated for the period December 1961—May 1962.

For the Baltic Proper the evaporation was determined using the equation:

$$E = 0.114(e_s - e_{10})W_{10} \qquad \text{mm} \cdot \text{day}^{-1} \qquad (3.6)$$

and for Finngrundet:

$$E = 0.127(e_s - e_4)W_4 \qquad \text{mm} \cdot \text{day}^{-1} \qquad (3.7)$$

whereby the value of k has been adjusted so that the unit of E is mm · day^{-1} and the effect of salinity on e_s has been neglected, since the surface salinity in the area is only about 6 ‰. In order to calculate the total short-wave radiation, the reflected radiation and the back radiation, the equations given by Laevestu (1960) were used, with observations from lightships and a merchant ship passing weekly the route Helsinki-Copenhagen.

For the Baltic Proper, Hankimo's results showed the evaporation and the flux of sensible heat to be the largest in December, viz., 103.5 mm and 19 540 J cm^{-2}, respectively. The values of Palmén (1963) were generally somewhat smaller and the total for the 6 months was about 7% smaller than the total found by Hankimo. He concluded that Palmén's values were a few percent too small and that the coefficient 0.114 in eq. 3.7 gave generally good results. The evaporation was lowest in April and May, and the total energy flux was directed from the atmosphere to the sea.

At Finngrundet the smallest amount of evaporation (1 mm) occurred in May and the largest (87 mm) in December, the annual amount being 495 mm. The total energy exchange by convection processes was directed from the sea to the atmosphere in all months except May and June. The total budget showed that the sea had received a net input of heat during the year. The temperature observations, however, showed that the sea had lost energy during the year. Hankimo therefore concluded that Laevestu's (1960) equation for calculating the total short-wave radiation and the reflected radiation gave too large values. The measured values at nearby coastal stations were about 25% smaller than the calculated values.

Using corrected values Hankimo obtained a net energy flux of 420 J cm^{-2} (100 cal cm^{-2}) per year from the sea. According to temperature measurements, the heat storage of the water column had decreased by about 20 kJ cm^{-2} (5000 cal cm^{-2}) during the year. Hankimo concluded that advective effects and computational errors might account for the imbalance. The annual fluctuations of the heat storage in the Bothnian Sea, as represented by Finngrundet, from the maximum in September to the minimum in March was approximately 170 kJ cm^{-2} (41 kcal cm^{-2}), which is comparable with the corresponding value of 190 kJ cm^{-2} (44.5 kcal cm^{-2}) found by Jerlov (1940) for the Baltic Proper.

Further studies of problems concerning the heat budget and thermal regime in different parts of the Baltic Sea have subsequently been carried out by Palosuo (1971), Sturm et al. (1976) and Matthäus (1977a, c).

Ice conditions

The ice conditions in the Baltic Sea, particularly in the northern parts, are of great importance for shipping. Therefore elaborate warning, forecasting and ice-breaking services have been established.

Studies aiming at improving the prediction methods and the understanding of the conditions are being conducted, particularly in Finland and Sweden.

Ice atlases have also been prepared, e.g., by the German Hydrographic Institute (Deutsches Hydrographisches Institut, 1956) using material collected during the period 1900—1950.

The freezing and ice cover depend mainly on the meteorological conditions, and large fluctuations occur from year to year, depending upon the high-pressure situation over the North Sea, Scandinavia and the USSR.

The number of consecutive days with negative air temperature which is required for freezing varies considerably and has been used to predict the formation of ice (e.g., Kühnel, 1967). Palosuo (1966) prepared a chart of ice coverage probability based on data from 1931—1960.

The Bothnian Bay is normally completely covered by ice by January. Normally, complete ice coverage also occurs in the coastal zone down to the Åland Sea, along the Gulf of Finland and its inner parts, and along the Gulf of Riga.

In the central parts of the Baltic Proper the probability of occurrence of ice is less than 25%. In the open Bothnian Bay the freezing commonly starts early in January, in the coastal zones of the Bothnian Sea in middle January and further south by middle February (Palosuo, 1966). The Bothnian Bay is covered by ice for 100—150 days. The Bothnian Sea becomes covered by ice five winters out of ten.

The ice in the coastal zones is normally fast whereas the ice in the open sea moves influenced by winds and currents. The thickness of the ice (mean 80—50 cm in the Bothnian Bay) therefore varies considerably in the open area, and open zones and ice walls occur frequently. Relatively broad open lanes often occur along the coasts.

Recently attempts have been made to investigate the properties of ice and the dynamics of ice drift in some detail, using modern observation techniques including satellite remote sensing (e.g., Udin and Omstedt, 1976; Blomquist et al., 1976). One reason for these studies is the requirement for better data for developing and testing forecasts of the ice situation based on numerical modelling. The observations showed good correlation between wind conditions and ice drift. The main forces determining the motion of the ice were found to be the wind stress and the water stress, but the balance of forces was also influenced by the Coriolis effect and the internal ice stress. Using these results, Udin and Ullerstig (1976) developed a two-dimensional numerical model for predicting ice conditions on the basis of meteorological conditions. They used mass and momentum equations and assumed the ice to be viscous when the relative ice coverage was low but to behave as a plastic substance when the ice coverage became large. The model gave reasonable predictions for a number of different occasions.

Calculations of ice drift have also been carried out by Valli and Leppäranta (1975) who used a development of Doronin's (1970) model. They found reasonable agreement between calculated and observed data. The model is being used for predicting ice conditions.

REFERENCES

Ahlnäs, K., 1962. Variations in salinity at Utö 1911—1961. Geophysica, 8(2): 135—149.

Bjerknes, V., 1901. Cirkulation relativ zur Erde. K. Sven. Vetenskapakad. Handl., 10: 1—20.

Blomquist, Å., Pilo, C. and Thompson, T., 1976. Sea Ice-75, Summary rep. Styrelsen för Vintersjöfartsforskning, Forskningsrapp. Winter Navigation Research Board, Res. Rep., 16(9) 1—26.

Bock, K.-H., 1971. Monatskarten des Salzgehaltes der Ostsee dargestellt für verschiedene Tiefenhorizonte. Ergänzungsh. Dtsch. Hydrogr. Z., Reihe B(4°), 12: 1—147.

Böhnecke, G. and Dietrich, G., 1951. Monatskarten der Oberflächentemperatur für die Nord- und Ostsee und die angrenzenden Gewässer. Dtsch. Hydrogr. Inst., No. 2336: 1—18, Hamburg.

Brogmus, W., 1952. Eine Revision des Wasserhaushaltes der Ostsee. Kieler Meeresforsch., 9(1): 15—42.

Brosin, H.J., 1972. Untersuchungen zur horizontalen turbulenten Diffusion in den Gewässern um Rügen. Beitr. Meeresk., 30/31: 35—40.

Brosin, H.J., 1974a. Untersuchungen zur mittelmasstälichen horizontalen Diffusion mit Driftbojen in den Gewässern um Rügen. Beitr. Meeresk., 34: 5—8.

Brosin, H.J., 1974b. Photogrammetric investigation on turbulent diffusion with discrete particles. Rapp. P.-V. Réun. Cons. Int. Explor. Mer, 167: 222—224.

Csanady, G.T., 1967. Large scale motion in the Great Lakes. J. Geophys. Res., 72: 4151—4162.

Csanady, G.T., 1971. Baroclinic boundary currents and long edgewaves in basins with sloping shores. J. Phys. Oceanogr., 1: 92—104.

Csanady, G.T., 1977. The coastal jet conceptual model in the dynamics of shallow seas. In: E.D. Goldberg, I.N. McCave, J.J. O'Brien and J.H. Steele (Editors), The Sea. Wiley-Interscience, New York, N.Y. 6: 117—144.

Dahlin, H., 1976. Hydrokemisk balans för Bottenhavet och Bottenviken. Vannet in Norden, 1: 62—63.

Dahlin, H., 1978. Översikt av Bottniska vikens vattenkemi och materialbalans. Finnish-Swedish Seminar of the Gulf of Bothnia, Vaasa, Finland, March 8th—9th, 1978. 18 pp., appendices (mimeogr.).

Defant, F., 1972. Klima und Wetter der Ostsee. Kieler Meeresforsch., 28: 1—30.

Denman, K.L., 1973. A time-dependent model of the upper ocean. J. Phys. Oceanogr., 3(2): 173—184.

Denman, K.L. and Miyake, M., 1973. Upper layer modification at ocean station Papa: observation and simulation. J. Phys. Oceanogr., 3(2): 185—196.

Dera, J. and Olszewski, J., 1967. On the natural irradiance fluctuations affecting photosynthesis in the sea. Acta Geophys. Pol., 15(4): 351—364.

Deutsches Hydrographisches Institut, 1956. Atlas der Eisverhältnisse der Deutschen Bucht und der westlichen Ostsee. Hamburg.

Dickson, R.R., 1971. A recurrent and persistant pressure-anomaly pattern as the principal cause of intermediate-scale hydrographic variation in the European shelf seas. Dtsch. Hydrogr. Z., 24(3): 97—119.
Dickson, R.R., 1972. The beginnings of a new Baltic inflow? -ICES C.M. 1972/C:10. 9 pp. (mimeogr.).
Dickson, R.R., 1973. The prediction of major Baltic inflows. Dtsch. Hydrogr. Z., 26(3): 97—105.
Dietrich, G., 1950. Kontinentale Einflüsse auf Temperatur und Salzgehalt des Ozeanwassers. Dtsch. Hydrogr. Z., 3(1—2): 33—39.
Dietrich, G., Kalle, K., Krauss, W. and Siedler, G. (Editors), 1957. Allgemeine Meereskunde. Eine Einführung in die Oceanographie. Bornträger, Berlin, 492 pp.
Doronin, Ju. P., 1970. Metodike rasčeta spločennesti i drejfa l"dov. Trudy, Arkt. Antarkt. Naučno-Issled. Inst., 291: 5—17 (in Russian). On a method of calculating the compactness and drift of ice flows. Aidjex Bull., 3: 22—39 (in English).
Druet, Cz., Hupfer, P. and Shadrin, I. (Editors), 1976. Properties and transformation of hydrodynamical processes in coastal zone. Pr. Morsk. Inst. Ryb. Gdyni, Rep., 2a: 1—253.
Ehlin, U. and Ambjörn, C., 1978. Bottniska vikens hydrografi och dynamik. Finnish-Swedish Seminar of the Gulf of Bothnia. Vaasa, Finland, March 8th—9th, 1978. 54 pp. (mimeogr.).
Ehlin, U., Mattisson, I. and Zachrisson, G., 1974. Computer based calculations of volumes of the Baltic area. Proc. 9th Conf. Baltic Oceanogr., Kiel, 17—20 April, 1974, pp. 115—128.
Ekman, F.L., 1893. Den svenska hydrografiska expeditionen år 1877. I. K. Sven. Vetenskapsakad. Handl., 25(1): 1—72.
Ekman, V.W., 1905. On the influence of the earth's rotation on ocean currents. Ark. Mat. Astron. Fys., 2:1—11.
Fjeldstad, J.E., 1958. Ocean currents as an initial problem. Geofys. Publ., 20(7): 1—24.
Fonselius, S.H., 1962. Hydrography of the Baltic deep basins. Fish. Board Sweden, Ser. Hydrogr. Rep. 13: 1—41.
Fonselius, S.H., 1969. Hydrography of the Baltic deep basins III. Fish. Board Sweden, Ser. Hydrogr. Rep., 23: 1—97.
Fonselius, S.H., 1971. Om Östersjöns och speciellt Bottniska vikens hydrografi. Vatten, 27(3): 309—324.
Fonselius, S.H., 1976. On the nutrient variability in the Baltic. Ambio, Spec. Rep., 4: 17—26.
Fonselius, S.H., 1977. An inflow of unusually warm water into the Baltic deep basins. ICES C.M. 1977/C:15, 3 pp., figs. (mimeogr.).
Francke, E. and Nehring, D., 1973. Physical and chemical variations in the eastern part of the Gotland Basin in 1969/70. Oikos Suppl., 15: 14—20.
Granqvist, G., 1949. The increase of the salinity along the coast of Finland since 1940. Fennia, 71(2): 1—14.
Granqvist, G., 1952. Harmonic analysis of temperature and salinity in the sea off Finland and changes in salinity. Merentutkimuslaitoksen Julk. Havsforskningsinst. Skr., 152: 1—29.
Gustafsson, T. and Kullenberg, B., 1936. Untersuchungen von Trägheitsströmungen in der Ostsee. Svenska Hydrogr.-Biol. Komm. Skr. Ny Ser. Hydrogr., 13: 1—28.
Hankimo, J., 1964. Some computations of the energy exchange between the sea and the atmosphere in the Baltic area. Finn. Meteorol. Office Contrib., 57: 1—26.
Hela, I., 1951. On the energy exchange between the sea and the atmosphere in the Baltic area. Ann. Acad. Sci. Fenn. Ser. A 1(97): 1—48.
Hela, I., 1966a. Secular changes in the salinity of the upper waters of the northern Baltic Sea. Commentat. Phys.-Math., Soc. Sci. Fenn., 31(14): 1—21.

Hela, I., 1966b. Vertical eddy diffusivity of waters in the Baltic Sea. Geophysica, 9(3): 219—234.

Hela, I., 1977. Synoptic survey of the hydrography of the Åland Sea. Merentutkimuslaitoksen Julk. Havforskningsinst. Skr., 241: 101—110.

Højerslev, H.K., 1974. Inherent and apparent optical properties of the Baltic. Rep. Inst. Phys. Oceanogr. Univ. Copenhagen, 23: 1—40.

Hollan, E., 1969. Die Veränderlichkeit der Strömungsverteilung im Gottlandbecken am Beispiel von Strömungsmessungen im Gottland-Tief. Kieler Meeresforsch., 25: 19—70.

Hupfer, P., 1962. Meeresklimatische Veränderungen im Gebiet der Beltsee seit 1900. Veröff. Geophys. Inst. Univ. Leipzig, 17(4): 355—512.

Hupfer, P., 1967. Die thermischen Verhältnisse in der ufernahen Zone des Meeres — dargestellt am Beispiel der Ostsee bei Zingst. Habilitationsschrift, Leipzig (unpubl. manuscript).

Hupfer, P., 1975. Marine climatic fluctuations in the Baltic Sea since 1900. Z. Meteorol., 25: 85—93.

Jacobs, W.C., 1942. On the energy exchange between the sea and the atmosphere. J. Mar. Res., 5: 45—47.

Jensen, A.J.C., 1937. Fluctuations in the hydrography of the transition area during 50 years. Rapp. P.-V. Réun. Cons. Int. Explor. Mer, 102(1): 3—49.

Jerlov (Johnson), N.G., 1940. Östersjöns värmeekonomi. Svenska Hydrogr.-Biol. Komm. Skr. Ny Ser., Hydrogr., 15: 1—18.

Jerlov, N.G., 1955. Factors influencing the transparency of the Baltic waters. Göteborgs K. Vetensk. o VitterhSamh. Handl. Ser. B 6(14): 1—19.

Jerlov, N.G., 1974. Significant relationships between optical properties of the sea. In: N.G. Jerlov and E. Steemann Nielsen (Editors). Optical Aspects of Oceanography. Academic Press, Aberdeen, 77—94.

Jerlov, N.G., 1975. Long period changes in the optical properties of the Baltic. J. Cons. Int. Explor. Mer, 36(2): 188—190.

Jerlov, N.G., 1976. Marine Optics. Developments in Oceanography, 14, Elsevier Amsterdam, 231 pp.

Jerlov, N.G. and Liljequist, G., 1938. On the regular distribution of submarine daylight and on the total submarine illumination.-Sven. Hydrogr.-Biol. Komm. Skr. Ny Ser. Hydrogr., 14: 1—15.

Jerlov, N.G. and Nygård, K., 1968. Inherent optical properties from radiance measurements in the Baltic. Københavns Universitet, Inst. Fys. Oceanogr. Rep., 1: 1—7.

Joseph, J. and Sendner, H., 1958. Über die horizontale Diffusion im Meere. Dtsch. Hydrogr. Z., 11(2): 49—77.

Kaleis, M.V., 1976. Present hydrographic condition in the Baltic. Ambio, Spec. Rep. 4: 37—44.

Kielmann, J., Krauss, W. and Magaard, L., 1969. Über die Verteilung der kinetischen Energie im Bereich der Trägheits- und Seichesfrequenzen der Ostsee im August 1964. Kieler Meeresforsch., 25(2): 245—254.

Kielmann, J., Krauss, W. and Keunecke, K.H., 1973. Currents and stratification in the Baltic Sea and the Arkona Basin during 1962—1968. Kieler Meeresforsch., 29(2): 90—111.

Kowalik, Z. and Taranowska, S., 1967. Horizontal large scale turbulence in the Baltic Sea. Cah. Océanogr., 19(4): 295—310.

Krause, G., 1969. Ein Beitrag zum Problem der Erneuerung des Tiefwassers im Arkona-Becken. Kieler Meeresforsch., 25(2): 268—271.

Krauss, W., 1966. Das Spektrum der internen Bewegungsvorgänge der Ostsee im Periodenbereich von 0.5 bis 7 Stunden. Kieler Meeresforsch., 22(2): 28—34.

Krauss, W., 1972. Wind-generated internal waves and inertial-period motions. Dtsch. Hydrogr. Z., 25(6): 241—250.

Krauss, W., 1974a. Interne Wellen. In: L. Magaard and G. Rheinheimer (Editors), Meereskunder der Ostsee. Springer, Berlin, pp. 77—84.
Krauss, W., 1974b. Two-dimensional seiches and stationary drift currents in the Baltic Sea. ICES Special Meeting on Models of Water Circulation in the Baltic. Paper 10, 32 pp. (mimeogr.).
Krauss, W. and Magaard, L., 1962. Zum System der Eigenschwingungen der Ostsee. Kieler Meeresforsch., 18(2): 184—186.
Krauss, W., Koske, P. and Kielmann, J., 1973. Observations on scattering layers and thermoclines in the Baltic Sea. Kieler Meeresforsch., 29(2): 85—89.
Kremser, U. and Matthäus, W., 1973. Grundlagen und Methoden zur Berechnung mittlerer vertikaler Wärmeaustauschkoeffizienten in der Ostsee. Gerlands Beitr. Geophys., 82(2): 128—134.
Kühnel, I., 1967. Die Eisvorbereitungszeiten für die Ostsee östlich der Linie Trelleborg-Arkona und für den Finnischen und Rigaischen Meerbusen sowie für die südlichen Randbezirke der Bottensee. Dtsch. Hydrogr. Z., 20(1): 1—6.
Kullenberg, B. & Hela, I., 1942. Om tröghetssvängningar i Östersjön. Sven. Hydrogr.-Biol. Komm. Skr. Ny Ser. Hydrogr., 16: 1—14.
Kullenberg, G., 1969. Light scattering in the central Baltic. København Universitet, Inst. Fysisk Oceanogr., Rep., 5: 1—12.
Kullenberg, G., 1971. Vertical diffusion in shallow waters. Tellus, 23(2): 129—135.
Kullenberg, G., 1972. Apparent horizontal diffusion in stratified shear flow. Tellus, 24(1): 17—28.
Kullenberg, G., 1974a. Some observations of the vertical mixing in the Baltic. Proc. 9th Conf. Baltic Oceanogr., Kiel, 17—20 April, 1974, pp. 129—137 (mimeogr.).
Kullenberg, G., 1974b. An experimental and theoretical investigation of the turbulent diffusion in the upper layer of the sea. København Universitet. Inst. Fysisk. Oceanogr., Rep., 25: 1—272.
Kullenberg, G., 1976. On vertical mixing and the energy transfer from the wind to the water. Tellus, 28(2): 159—165.
Kullenberg, G., 1977. Observation of the mixing in the Baltic thermo- and halocline layers. Tellus, 29(6): 572—587.
Laevastu, T. 1960. Factors affecting the temperature of the surface layer of the sea. Commentat. Phys. Math., Soc. Sci, Fenn., 25(1): 1—136.
Lenz, W., 1971. Monatskarten der Temperatur der Ostsee dargestellt für verschiedene Tiefenhorizonte. Ergänzungsh. Dtsch. Hydrogr. Z., Reihe B(4°), 11: 1—148.
Lindquist, A., 1959. Studien über das Zooplankton der Bottensee II. Inst. Mar. Res. Lysekil, Ser. Biol. Rep., 11: 1—136.
Lisitzin, E., 1948. On the salinity in the northern part of the Baltic. Fennia, 70(5): 1—24.
Lisitzin, E., 1967. Day-to-day variation in sea level along the Finnish coast. Geophysica, 9(4): 259—275.
Lundberg, O.R., 1964. Die Bestimmung des Koeffizienten der vertikalen Temperaturleitfähigkeit durch Änderungen der Wassertemperatur in der Ostsee. Tr. Gos. Okeanogr. Inst., 81: 94—105 (in Russian, with a German summary).
Lundgren, B., 1976. Spectral transmittance measurements in the Baltic. København Universitet, Inst. Fys. Oceanogr., Rep., 30: 1—38.
Mäe, H., 1928. Über die Temperatursprünge in der Ostsee. Sber. Akad. Wiss. Wien, IIa, 137(1—2): 1—44.
Magaard, L., 1974. Wasserstandsschwankungen und Seegang. In: L. Magaard and G. Rheinheimer (Editors), Meereskunde der Ostsee. Springer, Berlin, pp. 67—76.
Magaard, L. and Krauss, W., 1966. Spektren der Wasserstandsschwankungen der Ostsee im Jahre 1958. Kieler Meeresforsch., 22(2): 155—162.
Mälkki, P., 1975. On the variability of currents in a coastal region of the Baltic Sea. Merentutkimuslaitoksen Julk. Havsforskningsinst. Skr., 240: 3—56.

Matthäus, W., 1977a. Mittlere vertikale Wärmeaustauschkoeffizienten in der Ostsee. Acta Hydrophys., 22(2): 73—92.
Matthäus, W., 1977b. Zur säkularen Veränderlichkeit des Oberflächensalzgehaltes in der offenen Ostsee. Beitr. Meeresk., 39: 37—49.
Matthäus, W., 1977c. Zur mittleren jahreszeitlichen Veränderlichkeit der Temperatur in der offenen Ostsee. Beitr. Meeresk., 40: 117—155.
Matthäus, W. and Kremser, U., 1976. Die Berechnung mittlerer vertikaler Austauschkoeffizienten in der Ostsee auf der Grundlage von Sauerstoffkonzentrationswerten. Beitr. Meeresk., 37: 111—136.
Neumann, G., 1940. Mittelwerte längerer und kürzerer Beobachtungsreihen des Salzgehaltes bei den Feuerschiffen im Kattegat und in der Beltsee. Ann. Hydrogr., 69: 373—386.
Neumann, G., 1941. Eigenschwingungen der Ostsee. Arch. Dtsch. Seewarte Marineobserv., 61(4): 1—59.
Neumann, G., 1946. Stehende zellulare Wellen im Meere. Naturwissenschaften, 33(9): 1—282.
Nielsen, A., 1973. Water level and current spectra from the Great Belt 1970. Københavns Universitet, Inst. Fys. Oceanogr., Rep. 22: 1—45.
Nilsson, H. and Swanson, A., 1974. Long-term variations of oceanographic parameters in the Baltic and adjacent waters. Medd. Havsfiskelab. Lysekil 174: 1—11, appendices.
Okubo, A., 1971. Oceanic diffusion diagrams. Deep-Sea Res., 18: 789—802.
Palmén, E., 1930. Untersuchungen über die Strömungen in den Finnland umgebenden Meeren. Commentat. Phys.-Math., Soc. Sci. Fenn., 5(12): 1—94.
Palmén, E., 1963. Computation of the evaporation over the Baltic Sea from the flux of water vapor in the atmosphere. I.A.S.H. Comm. Evap., 62: 244—252.
Palosuo, E., 1964. A description of the seasonal variations of water exchange between the Baltic Proper and the Gulf of Bothnia. Merentutkimuslaitoksen Julk. Havsforskningsinst. Skr., 215: 1—32.
Palosuo, E., 1966. Ice in the Baltic. In: H. Barnes (Editor), Oceanogr. Mar. Biol. Annu. Rev., 4: 79—90.
Palosuo, E., 1971. Water exchange and the rate of cooling in the Gulf of Bothnia. ICES, C.M. 1971/C, 28, 14 pp. (mimeogr.).
Palosuo, E., 1973. Bottenhavet och Bottenviken — två olika bäcken av Bottniska viken. Terra, 85(3): 139—148.
Pedersen, F.B., 1977. On dense bottom currents in the Baltic deep water. Nord. Hydrol. 8(5): 297—316.
Petrén, O. and Walin, G., 1975. Some observations of the deep flow in the Bornholm Strait during the period June '73—December '74. Göteborgs Universitet, Oceanogr. Inst., Rep., 12: 1—30.
Piechura, J., 1972. The density structure and mixing of southern Baltic waters. Proc. 8th Conf. Baltic Oceanogr., Copenhagen, October 1972, Paper 4, 5 pp (mimeogr.).
Pomeranec, K.S., 1964. Die Wärmebilanz der Ostsee. Trudy Gos. Okeanogr. Inst., 82: 87—109 (in Russian, with a German summary).
Sarkisyan, A.S., Stashkevich, A. and Kowalik, Z., 1976. Diagnostic computation of the summer circulation in the Baltic Sea. Oceanology, 15(6): 653—656.
Schott, F., Ehlers, M., Hubrich, L.M. and Quadfased, D., 1978. Small-scale diffusion experiments in the Baltic surface-mixed layer under different weather conditions. Dtsch. Hydrogr. Z., 31: 195—215.
Shaffer, G., 1975. Baltic coastal dynamics project — the fall downwelling regime off Askö. Contrib. Askö Lab., Univ. of Stockholm, 7: 1—61 (mimeogr.).
Shaffer, G., 1977. Calculations of cross-isohaline salt exchange in a coastal region of the Baltic. Göteborgs Universitet, Inst. Oceanogr., Rep. 24: 1—26, appendices.

Simojoki, H., 1946. On the temperature and salinity of the sea in the vicinity of the Bogskär Lighthouse in the northern Baltic. Commentat. Phys.-Math. Soc. Sci. Fenn., 13(7): 1—24.
Simojoki, H., 1949. Niederschlag und Verdunstung auf dem Baltischen Meer. Fennia, 71(1): 1—25.
Soskin, I.M., 1963. Mnogoletnie izmenenija gidrologičeskih harakteristik Baltijskogo morja. 159 pp. Leningrad.
Stigebrandt, A., 1976. Vertical diffusion driven by internal waves in a sill fjord. J. Phys. Oceanogr., 6(4): 486—495.
Sturm, M., 1968. Untersuchungen der Wärmebilanz der südlichen Ostsee im Bereich des Feuerschiffes "Fehmarnbelt". Tellus, 20(3): 485—494.
Sturm, M., 1970a. Zum Wärmehaushalt der Ostsee im Bereich der südlichen Beltsee (Fehmarnbelt). Beitr. Meeresk., 27: 47—61.
Sturm, M., 1970b. Zu Fragen der horizontalen Wärmeaustausches zwischen der Nord- und Ostsee. Monatsber. Dtsch. Akad. Wiss. Berlin, 2(4): 267—286.
Sturm, M., Francke, E. and Matthäus, W., 1976. A few aspects of the thermal regime in the Baltic during the summer of 1975. Proc. 10th Conf. Baltic Oceanogr., Göteborg, 2—4 June, 1976, Paper 4, 11 pp. (mimeogr.).
Svansson, A., 1972. Canal models of sea level and salinity variations in the Baltic and adjacent waters. Fish. Board Swed. Ser. Hydrogr. Rep., 26: 1—72.
Talbot, J.W. and Talbot, G.A., 1974. Diffusion in shallow seas and English coastal and estuarine waters. Rapp. P.-V. Réun. Cons. Int. Explor. Mer, 167: 93—110.
Tully, J.P., 1958. On structure, entrainment and transport in estuarine embayments. J. Mar. Res., 17: 523—535.
Udin, I. and Omstedt, A., 1976. Sea Ice-75, Dynamical report. Styrelsen för Vintersjöfartsforskning. Forskningsrapp. Winter Navigation Research Board. Res. Rep., 16(8): 1—64.
Udin, I. and Ullerstig, A., 1976. A numerical model for forecasting the ice motion in the Bay and Sea of Bothnia. Styrelsen för Vintersjöfartsforskning. Forskningsrapp. Winter Navigation Research Board. Res. Rep., 18: 1—40.
Valli, A. and Leppäranta, M., 1975. Calculation of ice drift in the Bothnian Bay and the Quark. Styrelsen för Vintersjöfartsforskning Winter Navigation Research Board, Forskningsrapp., 13: 1—14, figs.
Voipio, A. and Mälkki, P., 1972. Variations of the vertical stability in the northern Baltic. Merentutkimuslaitoksen Julk. Havsforskningsinst. Skr., 23: 3—12.
Walin, G., 1972a. On the hydrographic response to transient meteorological disturbances. Tellus, 24(3): 169—186.
Walin, G., 1972b. Some observations of temperature fluctuations in the coastal region of the Baltic. Tellus, 24(3): 187—198.
Walin, G., 1977. A theoretical framework for the description of estuaries. Tellus, 29(2): 128—136.
Wallerius, D., 1932. Östersjövattnets värmeinnehåll. Sven. Hydrogr.-Biol. Komm. Fyrskeppsunders., 1932: 47—62 (English summary).
Weidemann, H., 1973. The ICES diffusion experiment RHENO 1965. Rapp. P.-V. Réun. Cons. Int. Explor. Mer, 163: 1—111.
Welander, P., 1974. Two-layer exchange in an estuary basin with special reference to the Baltic Sea. J. Phys. Oceanogr., 4: 542—556.
Witting, R., 1912. Zusammenfassende Übersicht der Hydrographie des Bottnischen und Finnischen Meerbusens und der nördlichen Ostsee nach den Untersuchungen bis Ende 1910. Finnl. Hydrograph.-Biol. Unters., 7: 1—82.
Witting, R., 1918. Hafsytan, geoidytan och landhöjningen utmed Baltiska hafvet och vid Nordsjön. Fennia, 39(5): 1—346.

Wojewódzki, T., Hupfer, H.A. and Shadrin, J., 1976. Currents in the surf zone. Based on data of experiment "Lubiatowo '74". Pr. Morski Inst. Ryb. Rep. Gdyni, 2a: 75—88.

Wüst, G. and Brogmus, W., 1955. Ozeanographische Ergebnisse einer Untersuchungsfahrt mit Forschungskutter "Südfall" durch die Ostsee Juni—Juli 1954 (anlässlich der totalen Sonnenfinsternis auf Öland). Kieler Meeresforsch., 11(1): 3—21.

Wyrtki, K., 1954. Der grosse Salzeinbruch in die Ostsee im November und Dezember 1951. Kieler Meeresforsch., 10(1): 19—25.

Chapter 4

CHEMICAL OCEANOGRAPHY

KLAUS GRASSHOFF and AARNO VOIPIO

ANOMALIES IN THE COMPOSITION OF BALTIC SEA WATER

The first evidence of interest in the chemical oceanography of the Baltic Sea appears to be the early studies of the chemical composition of sea water. For instance, in 1837, P.A. v. Bonsdorff of the University of Helsinki carried out several determinations of calcium in sea water. However, the time was not ripe for a systematic study of the relationships between the different components of the Baltic Sea water, especially the so-called major constituents, Cl, SO_4, Br, B, F, Na, Mg, Ca, K and Sr. Information was needed on the constant relation between the chloride concentration, or more exactly, the total amount of halides excluding fluoride expressed as chlorides, and the total salt content of the different sea waters.

Knudsen (1903) established a relationship between salinity and chlorinity in sea water, viz.:

$$S‰ = 1.805\ Cl‰ + 0.0300\ ,$$

this equation being based on 13 different samples, only 8 of which were taken from the Baltic Sea and the transition area. The constant term takes into account the mean anomalies of the Baltic Sea water which derive from the dilution of ocean water with river water. More recently, salinity has been redefined on the basis of the relative conductivity of sea water (Cox et al., 1962), and the relation between salinity and chlorinity has been internationally defined by the expression (Wooster et al., 1969):

$$S‰ = 1.80655\ Cl‰^{*}$$

Among the more than 100 sea water samples from all oceans and seas used as the basis for this equation, only a few were from the Baltic Sea, and it has been found that in waters with a salinity below 20‰ the deviations from the relationship may exceed 0.02‰*.

The proportions of the conservative elements in sea water are usually expressed relative to chlorinity, and anomalies are defined relative to the proportions of these elements in ocean water. Because of the unusual hydrographical features of the Baltic Sea, it might be expected that these elements would show significant anomalies. More than 200 rivers annually discharge about 450 km^3 of fresh water from drainage areas in different geological en-

* Because of the principle difficulties with a double definition of salinity in terms of relative conductivity and of chlorinity, salinity and chlorinity have been decoupled and can only be converted at 35.000‰ S and 19.374‰ Cl, respectively (UNESCO, 1979).

vironments. In addition, the water added to the Baltic Sea by precipitation (about 200 km^3 a^{-1} ; cf., Chapter 2) has a chemical composition which differs significantly from that of distilled water. The situation is further complicated by the presence of dissolved organic substances having considerable complexing properties, e.g., humic substances, and by variations in the redox conditions. When these factors are considered, it is surprising that the relative proportions of the major elements in Baltic Sea water are so similar to those in ocean water.

An extensive discussion on the anomalies of the constituents of the Baltic Sea water has recently been published by Grasshoff (1975). This discussion is summarized here with some additional comments.

Sodium. Kremling (1969) has found an average Na/Cl ratio of 0.5547 ± 0.0021, which value is not significantly different from that of 0.5555 (Culkin, 1965) for open ocean water. Indeed, no marked differences would be expected as the Na/Cl ratio for river water is of the same order of magnitude.

Potassium. The K/Cl ratio is not quite constant in the samples collected from different parts of the Baltic Sea (Kremling, 1969). The lowest values have been found in the southern Baltic Sea, while the highest have been recorded in the Åland Sea and in the Gulf of Bothnia. However, the mean value of 0.0206 is the same as that reported for oceans (Cox and Culkin, 1966).

Calcium. It has been proved by numerous workers (Von Brandt, 1936; Gripenberg, 1937; Wittig, 1940; Rohde, 1966; Kremling, 1970) that of all the major elements calcium is the one whose ratio to chlorinity in the Baltic Sea water most markedly exceeds the ratio in ocean water. This is the result of the high Ca/Cl ratio of the river water discharged into the Baltic Sea. The Ca/Cl ratio recorded near the mouths of several rivers shows marked elevation compared with that for oceans, i.e., 0.02125 (Cox and Culkin, 1966). Also in the open sea areas both the slope and constant term in the regression equation for Ca and Cl vary notably in surface water with a chlorinity below 4.5‰. This results partly from regional differences, partly from scattering of the analytical results. However, the calcium content in the intermediate and deep water of the Baltic Proper is relatively constant in relation to chlorinity. Some selected values for the relationship between Ca and Cl have been compiled in Table 4.I.

Magnesium. In contrast to calcium, magnesium does not show marked regional variations in its relation to chlorinity. This was first shown in 1957 by Voipio, who obtained a ratio of 0.06695, which is rather similar to that found in oceans, i.e., 0.06680 (Cox and Culkin, 1966). However, more recent analyses indicate that the Baltic Sea value evidently slightly exceeds the oceanic Mg/Cl ratio (Nehring and Rohde, 1966; Rohde, 1967; Kremling, 1969, 1970) i.e., 0.0672.

TABLE 4.I

Linear relation of the calcium content (Ca), given in grams of calcium per kilogram sea water, to chlorinity (Cl ‰), expressed by the equation Ca = A Cl‰ + B

Sampling year	Area	Number of samples	Cl range	A	B	Source
1935	northern Baltic Proper	6	> 4.5‰	0.0204	0.0226	Gripenberg (1937)
1935	northern Baltic Proper and Gulf of Finland	10	< 4.5‰	0.0238	0.0083	Gripenberg (1937)
1935	Gulf of Bothnia	19	< 4.5‰	0.0258	0.0019	Gripenberg (1937)
1967	southwestern and central Baltic Proper	12	< 4.5‰	0.0216	0.0174	Kremling (1969)
1967	southwestern and central Baltic Proper	20	> 4.5‰	0.0200	0.0247	Kremling (1970)
1968	southwestern and central Baltic Proper	15	< 4.5‰	0.0201	0.0234	Kremling (1970)
1968	southwestern and central Baltic Proper	13	> 4.5‰	0.0204	0.0218	Kremling (1970)
	Baltic Sea (general)		4.5‰	0.0228	0.0127	Trzosińska (1968)

Sulphate. The available data suggest that the Baltic Sea waters with a salinity higher than 8‰ do not show sulphate anomalies (Kremling, 1970). However, the relatively large standard deviations indicate local deviations from the oceanic SO_4/Cl ratio, i.e., 0.1400 (Morris and Riley, 1966). The results obtained by Kremling for chlorinities 4.5‰ were 0.1410 and 0.1405 for 12 and 15 samples collected in 1967 and 1968, respectively.

The formation of hydrogen sulphide in the deep basins of the Baltic Sea should theoretically cause a decrease in the SO_4/Cl ratio as most of the sulphide sulphur originates from the dissolved sulphate (cf., Kwiecinski, 1965). It is, however, very unlikely to be observed in the SO_4/Cl ratio. The maximum concentration of hydrogen sulphide found has been about 40 m mol m^{-3} and moreover, most of the hydrogen sulphide is evidently formed in the interstitial water of the sediments (cf. Section on p. 205, so that the deficiency can hardly be visible in sulphate concentrations of about 10,000 m mol m^{-3}.

Earlier analyses (Thompson et al., 1931) for waters of the Bothnian Bay and the Gotland Deep gave SO_4/Cl ratios of 0.1412—0.1419 for the surface water and 0.1404 for the deep water. Zarinš and Ozolinš (1935) have reported values of 0.1410 for the Gotland Deep. Kwiecinski (1965) obtained a mean ratio of 0.1413 for 26 samples collected from different depths in various areas of the central parts of the Baltic Sea. Trzosińska (1967) found a mean ratio of 0.1436 for 26 samples for the Bornholm Basin, the Arkona Basin and the Bay of Gdansk, but some of her individual values were considerably higher. The main cause of the positive sulphate anomaly of the surface water is certainly dilution with river water, which has a SO_4/Cl ratio of 0.3—0.6 (Rubey, 1951).

Bromide. Kremling (1969, 1970) has found a slight negative anomaly for the Br/Cl ratio, which is evidently caused by the low ratio in the river water component of the Baltic Sea water (see Table 4.II).

TABLE 4.II

Range of mean ratios between ionic concentrations (g kg^{-1}) and chlorinity (‰) in the Baltic Sea 1966—1968 according to Kremling (1970)

Ion	Baltic Sea	Ocean $\frac{g\ kg^{-1}}{‰}$	Excess (+) Deficiency (—)
Na^+	0.5547	0.5555	0
K^+	0.0203—0.0209	0.0206	0 (but local anomalies exist)
Ca^{2+} ($+Sr^{2+}$)	< 4.5‰ Cl: = 0.0201 Cl‰ + 0.0234 = 0.0216 Cl‰ + 0.0174	0.02166	+
	> 4.5‰ Cl: = 0.0200 Cl‰ + 0.0247 = 0.0204 Cl‰ + 0.0218		+
Mg^{2+}	0.0671—0.0672	0.0669	(+)
Alkalinity	0.303—0.342 $\left[\frac{eq\ m^{-3}}{‰}\right]$	0.123 $\left[\frac{eq\ m^{-3}}{‰}\right]$	+
$SO_4{}^{2-}$	< 4.5‰ Cl: 0.1405—0.1410	0.1400	
	> 4.5‰ Cl: 0.1400—0.1403		0 (but local anomalies exist)
Br^-	= 0.00328 Cl‰ + 0.0005 = 0.00348 Cl‰ + 0.0004	0.00347	—
H_3BO_3	B (mg kg^{-1}) = 0.220 Cl‰ + 0.15	0.230 $\left[\frac{mg\ kg^{-1}}{‰}\right]$	+
F^-	spring 1966: 7.24×10^{-5} spring 1967: 7.33×10^{-5} autumn 1967: 8.45×10^{-5} spring 1968: 7.12×10^{-5}	6.70×10^{-5}	+

Boron. Both Gripenberg (1961) and Kremling (1970) have reported a positive anomaly for the B/Cl ratio. Gripenberg gave a constant ratio of 0.264, while Kremling found that the boron content, given in g kg^{-1}, could be expressed as 0.220 + 0.15 Cl‰. Dyrssen and Uppström (1974) gave a relationship of 0.227 + 0.050 Cl‰.

Fluoride. The F/Cl ratios of Baltic Sea waters vary significantly from one area to another and with depth. In addition, seasonal variations have been observed. Thus, values in the autumn have been found to be significantly greater than those in the spring. The most striking differences have been observed in surface water at stations in the central parts of the Baltic Sea (Kremling, 1969, 1970). The F/Cl ratios during autumn 1967 were larger than the ratio in ocean water (cf., Greenhalgh and Riley, 1963). In Baltic Sea water of chlorinity of 5‰, the average decrease of the fluoride content between autumn and spring amounted to about 0.07 mg kg^{-1}. There are three possible explanations for these changes:

(1) They may be produced by seasonal variations in the F/Cl ratio of the water entering the Baltic Sea. Values of 0.605—0.090 g F m^{-3}, with an average of 0.166 g F m^{-3}, have been reported for Swedish rivers (Kullenberg and Sen Gupta, 1973). However, it is not possible to explain positive deep water anomalies in this way.

(2) As suggested by Sillén (1961), fluoride may be taken up by phytoplankton along with silicate and may replace O^{2-} or OH^- groups and be subsequently released during decomposition.

(3) They may be caused by biological uptake of fluoride. Some mysids, crabs and snails are able to bind considerable amounts of calcium fluoride (cf., Lowenstam and McConnell, 1968).

A significant reduction in the F/Cl ratio has been observed in stagnant water (Kullenberg and Sen Gupta, 1973), suggesting that fluoride may be lost from water to the sediment in association with magnesium ions or with the formation of apatite (cf., Arrhenius, 1963), at least during periods of stagnation in the deeper parts of the Baltic Sea.

The average cation and anion concentrations for all parts and depths of the Baltic Sea in 1966—1968 are summarized in Table 4.II.

The observations cited above show that the composition of Baltic Sea water is clearly affected by river water discharged to the sea. The positive anomalies seem to be more common than the negative ones. This indicates that the constant term in the Knudsen formula is real. Kremling (1970) has proved this by summing up the masses of the cations and anions in his samples and calculating the relation of the sum to the chlorinity. His coefficient agrees excellently with that found by Knudsen, while his constant term is slightly higher (Table 4.III), which is not surprising in view of his more refined and precise analytical techniques and far larger number of samples.

TABLE 4.III

Relationships between S‰ (cations + anions) and Cl‰ in the Baltic Sea from 1966—1968 according to Kremling (1970)

Sampling year	Number of samples	Range of Cl‰	Correlation	Standard deviation (equation)
1966	38	3.196—10.981	S‰ = 1.802_8 Cl‰ + 0.041	$S_a = \pm 0.006_5$ $S_b = \pm 0.000_6$
1967	36	3.669— 9.734	S‰ = 1.802_5 Cl‰ + 0.040	$S_a = \pm 0.005_5$ $S_b = \pm 0.000_6$
1968	34	1.959— 8.193	S‰ = 1.802_5 Cl‰ + 0.039	$S_a = \pm 0.006_0$ $S_b = \pm 0.000_8$

THE DISTRIBUTION OF DISSOLVED OXYGEN

In the Baltic Sea, a permanent oxygen deficiency exists below the permanent halocline. The main reasons for this are the hydrographical features of the Baltic Sea. The estuarine circulation with limited inflow of more saline water, and a positive water balance, caused by the excess of inflowing fresh water over the annual evaporation, result in a pronounced stratification, i.e., a large gradient in the permanent halocline and in the other pycnoclines as well (cf., Section on p. 135).

In the aquatic ecosystem, a part of the biogenic material sinks and is gradually decomposed. If the density stratification is not regularly broken up by sufficient convection or advective down-welling, a clear decrease in the oxygen saturation percentage will be caused by the decomposition of biogenic autochthonous matter or other biogenic material of allochthonous origin. Therefore, one may assume that an oxygen deficiency has prevailed at least from time to time in extensive deep areas since the Baltic Sea attained its present hydrographic features (cf., Section on p. 205).

In the deep areas of the Baltic Proper, e.g., in the Gotland Deep, there are often three main layers during the wintertime, and four in summer when the summer thermocline develops. Between the surface and the thermocline, the oxygen content is always in near equilibrium with the atmosphere. The oxygen content corresponds to the saturation content or temporarily even exceeds it during periods of intense primary production. At such times, the concentration of carbon dioxide decreases, and the values of pH increase. In the layer between the thermocline and the permanent halocline there is no continuous supply of oxygen, and consequently the oxygen concentration in this layer decreases during summertime down to 70—90% of saturation (see Fig. 4.1).

In winter, thermal convection annually "refills" the oxygen reservoir of the water column practically down to the depth of the permanent halocline. However, the time of mixing is so rapid that the whole layer may often remain slightly undersaturated.

The permanent halocline forms an effective barrier to convection (cf., Section on p. 135). However, certain processes, e.g., shear stress between the differently moving layers, cause some vertical circulation. This is especially pronounced in areas where the halocline meets the bottom. Similarly, although the oxygen transport occurring during thermal convection does not penetrate the halocline, some feeble exchange of oxygen obviously also takes place through this barrier. This transport downwards is probably small in comparison with that connected with the inflows of water through the Danish Sounds, even though some recent calculations (Gargas et al., 1977) seem to suggest that the amount of oxygen penetrating the permanent halocline might almost equal the amount carried into the Baltic Sea deep water by inflows through the Danish Sounds.

The water between the permanent halocline and the bottom is usually divided into two layers by a secondary halocline (cf., Fig. 4.1). The layer between the primary and secondary halocline receives new water irregularly but

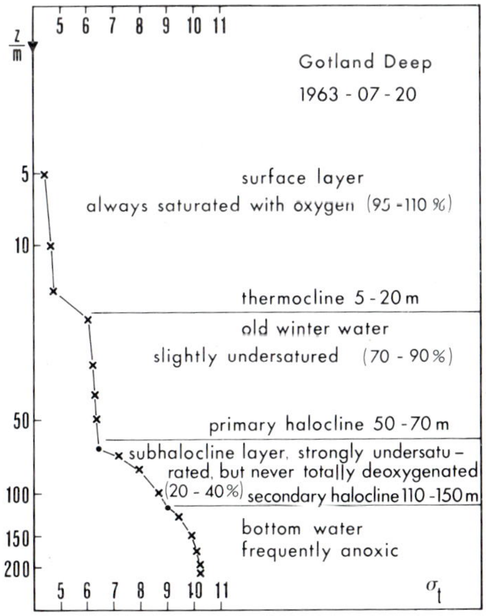

Fig. 4.1. Stratification of water in the Baltic Proper. z denotes depth in metres, in logarithmic scale, and σ_t is the relative density —1.

fairly often from the inflows through the Danish Sounds. However, the transport of these water masses from the Sounds to the central parts of the Baltic Proper takes several months, during which the oxygen is partly consumed. In addition, mixing with old water and local oxygen consumption result in a considerable oxygen saturation deficit, although in this layer in the central parts of the Baltic Proper saturation is seldom less than about 20 or 30% of the surface saturation.

The quality and quantity of the inflowing water depend largely on the meteorological conditions, and the frequency with which the aerated water masses enter from the open ocean is linked with local climatic fluctuations (cf., Chapter 3). Especially irregular and unpredictable is the inflow of water saline enough to renew the layer below the secondary halocline. Under unfavourable conditions, the oxygen may be totally stripped from the deep water, and hydrogen sulphide may form.

The vertical distribution of oxygen outlined above is typical of the deep basins in the Baltic Proper and the Gulf of Finland. In the shallow areas, the layer most liable to stagnation is absent and the most severe result of oxygen deficiency, i.e., the formation of hydrogen sulphide, seldom occurs, except in isolated basins near pollution centres.

The hydrography of the Gulf of Bothnia differs from that of the other parts of the Baltic Sea. The amount of the more saline water component annually flowing into this area is larger in proportion to the volume of the basin than the amount entering the Baltic Proper. Furthermore, since it is mainly surface water of the Baltic Proper, its salinity is only slightly higher than that of water in the Bothnian Sea and thus no sharp halocline will form. Under these circumstances the thermal convection occurring in winter takes place nearly throughout the water column. Consequently, the oxygen level seldom decreases below 60—80% of saturation.

It was pointed out above that an oxygen deficiency has presumably existed in the deep layers of the Baltic Sea as long as the present form of water circulation has prevailed. However, there have been very pronounced variations in the extent of this phenomenon for the last hundred years.

Fonselius (1969a, b, 1977) has devoted special attention to the trends in the dissolved-oxygen conditions in the Baltic Sea and has followed the fluctuations occurring since the first reliable data became available, i.e., since the beginning of the twentieth century.

In the Gotland Deep, stagnation leading to the formation of hydrogen sulphide was first observed in 1931 (Granqvist, 1932; Kalle, 1943). However, a rapid decrease in the oxygen content had already been recorded by Pettersson (1894) in the Landsort Deep in the 1890s. In 1900—1914 the oxygen content in bottom water in the Gotland Deep was always above 1.43 g m^{-3} (1 ml l^{-1}). During the 1920s, the oxygen content at that depth was generally below 1 ml l^{-1}, but around 1928—1929 an increase in the oxygen content of the water below the halocline was observed. This was caused by the intrusion of high-salinity water, which seems to have remained in the basin for some years before stagnant conditions became established in the bottom water. The stagnant conditions came to an end in 1933 (Fonselius, 1969a, b).

From 1952 to 1961, a prolonged period of stagnation led to a pronounced decrease in the oxygen content of the water below the halocline in the central parts of the Baltic Proper. The bottom water in the Gotland Deep was finally replaced by new saline water during the summer of 1961. This stagnation was evidently caused by an unusually large inflow of water of comparatively high density in November and December 1951 (Wyrtki, 1954). The inflow led to a sharp increase in the salinity of the deep water throughout the Baltic Sea. In the Gotland Deep, the increase at 200 m was 1—2 ‰, and the temperature rose by about 1° C.

After the intrusion of new water in 1961, another stagnation period began and eventually led to the development of anoxic conditions during 1963 (Fig. 4.2). The deep water was renewed again in January 1964, and a third period of reduced deep-water exchange ensued. This alternation between intrusion of new bottom water and stagnation has been studied intensively by Fonselius (1962, 1967, 1969a, b), who was able to show that on an average the dissolved-oxygen concentration has tended to decrease in the central, northern and eastern parts of the Baltic Proper. Fig. 4.3 shows that the oxygen content of deep water in the Gotland Deep was about 2.86 g m^{-3} (2 ml l^{-1}) at the beginning of the twentieth century and has decreased to values close to zero at the present time (Matthäus, 1977). This general fall has been interrupted from time to time by the inflow of saline water with a higher oxygen content. Certain signs of a decreasing trend of the oxygen content have also been found in the Gulf of Bothnia (see Fig. 4.4).

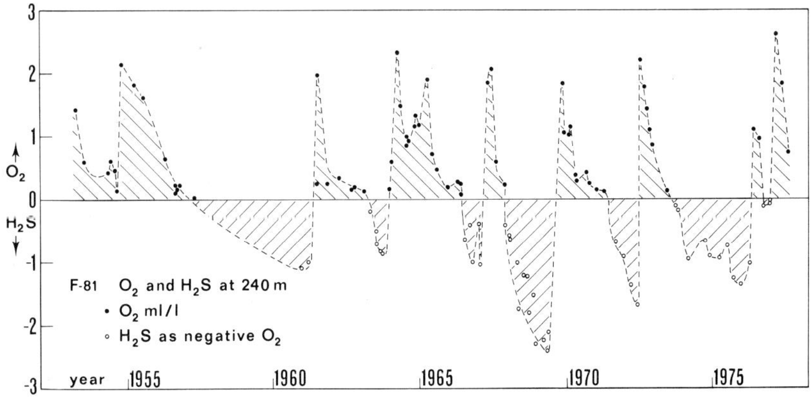

Fig. 4.2. Periods of oxic and anoxic conditions in the Gotland Deep (Station F81) from 1950 to 1977 (according to Fonselius, 1977). Oxygen content is given in ml O_2 l^{-1}. 1 ml O_2 l^{-1} is equivalent to 1.43 g O_2 m^{-3}. Hydrogen sulphide is given as negative oxygen corresponding to the amount of oxygen needed for oxydizing the hydrogen sulphide.

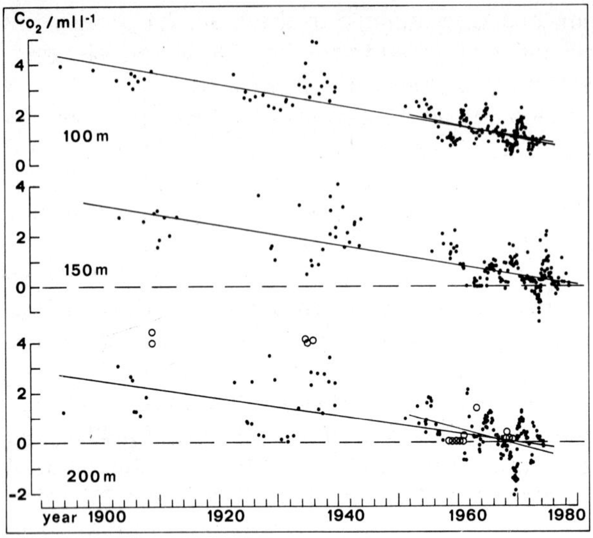

Fig. 4.3. Long-term trends in the oxygen content of the deep water in the Gotland Deep (Matthäus, 1977).

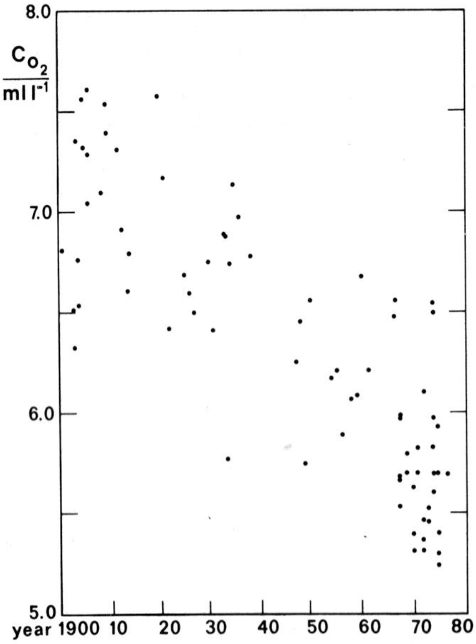

Fig. 4.4. Oxygen content (see legend for Fig. 4.2) in water samples collected from the depths of 175 m and 200 m in the northwestern part of the Bothnian Sea (F24 and Us2) in the twentieth century.

Both medium- and short-term fluctuations are superimposed on the secular changes in the oxygen content of the deeper waters in the Baltic Sea basins. The fluctuations are greatest in the southwestern parts of the Baltic Proper and the oxygen content of the deep water also shows considerable variations in the Bornholm Basin (e.g., Ruppin, 1912; Soskin, 1963; Fonselius, 1967, 1968, 1969a, b; Matthäus, 1973). The magnitude of the medium-term fluctuations is also shown in Fig. 4.2. It also appears that there have been numerous minor intrusions of oxygenated water.

The oxygen and hydrogen sulphide isopleths for a typical intrusion of new water into the Gotland Deep during the International Baltic Year (1969—1970) are shown in Fig. 4.5. The inflow of saline water displaces the old bottom water upwards, and the hydrogen sulphide which it contained is oxidized. The initial oxygen content of the intruding water was reduced from more than 1.43 g m^{-3} (1 ml l^{-1}) to 0.71 g m^{-3} (0.5 ml l^{-1}), and after 7 months the oxygen content of the water had been reduced to less than 0.71 g m^{-3} (0.5 ml l^{-1}).

The short-term variations of the hydrographical parameters are superimposed on the medium-term fluctuations. They are probably caused by inhomogeneity in the distribution of non-conservative parameters, oscillations of the water masses and residual net transport of deep water. The interconnection between water transport and short-term fluctuations of chemical parameters has been studied by Hollan (1969a, b). Figure 4.6 shows the short-term variations of the oxygen content observed in water below the halocline in the Gotland Deep in September 1967 and May 1968. The magnitude of the short-term changes suggests that if medium- and long-term variations are to be identified with certainty, it will be necessary to make a large number of observations extending over long periods. In some layers the magnitude of the hourly and daily oxygen variations may even be similar to the magnitude of the long-

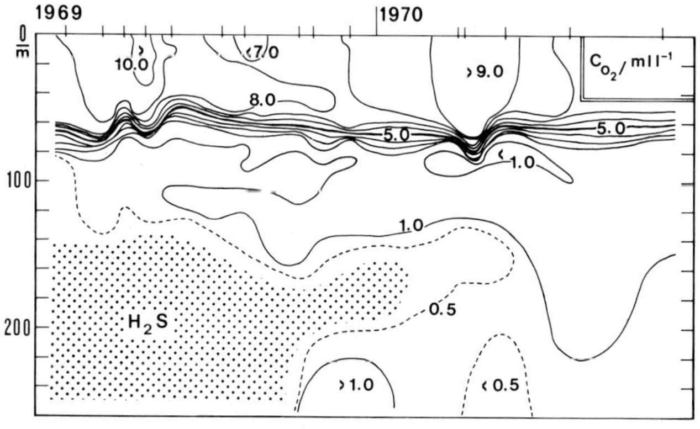

Fig. 4.5. Medium-term variations of the oxygen content (see legend for Fig. 4.2) in the Gotland Deep from January 1969 to November 1970 (according to Nehring and Francke, 1971). The presence of H_2S indicated by the stippled area.

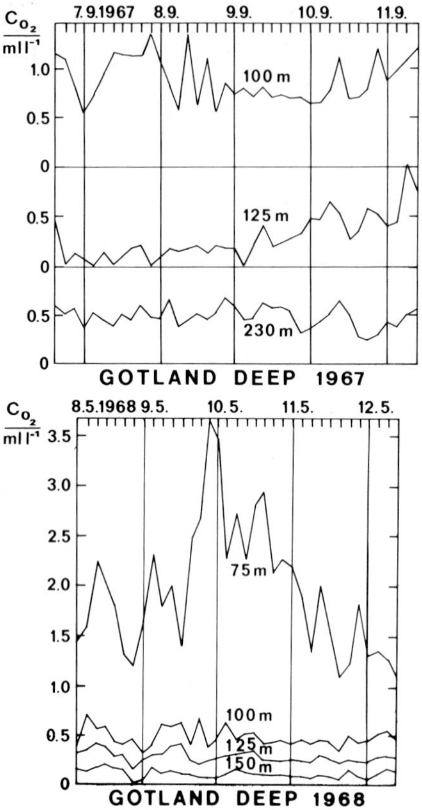

Fig. 4.6. Short-term variations of the oxygen content (see legend for Fig. 4.2) at various depths in the Gotland Deep in September 1967 and in May 1968 (according to Gieskes and Grasshoff, 1969).

term variations. Such variations in the oxygen content can often be correlated with variations in other parameters, both conservative and nonconservative (Gieskes and Grasshoff, 1969).

NUTRIENTS

During the 1920s the oceanographers round the Baltic Sea prepared themselves for studies on the distribution and circulation of nutrients, and on the relations between different nutrients and biological production (e.g., Buch, 1932; Kalle, 1932). The first investigations dealt with ammonium, nitrate and phosphate ions. Fairly soon nitrite and silicate were included, but only since the 1960s have extensive studies been carried out on the total content of phosphorus and nitrogen. Recently, sporadic studies have also been made on urea and hydroxylamine.

The vertical distribution of phosphate and other nutrients above the halocline follows closely the thermal layering and convection. During the spring bloom, phosphate is rapidly consumed and the content decreases close to zero. Phosphate is often the limiting nutrient, especially in the Bothnian Bay. Below the summer thermocline the phosphate-phosphorus content is usually near 0.3 m mol m^{-3} the whole year round, a uniform level being maintained by the autumnal decomposition of biogenic matter and by vertical mixing due to thermal convection. However, near coastal areas, e.g., in the northern Baltic Proper and at the head of the Gulf of Finland, vertical circulation also introduces phosphate to the surface layers. This mechanism is especially pronounced at a time when the stagnation and nutrients accumulation in the deep water are interrupted, and old water is displaced upwards. In addition to the internal circulation of the Baltic Sea water masses, the input of phosphorus from land-based sources also affects the regional distribution of total phosphorus. Surface water samples collected in July—August during the years 1966—1975 from the Bothnian Bay, the Bothnian Sea and the Gulf of Finland show that the levels of total phosphorus are about 0.15, 0.25 and 0.35 m mol m^{-3}, respectively. The last-mentioned value corresponds rather closely to the level in the Baltic Proper in late summer.

The distribution of dissolved phosphate in a longitudinal section is shown in Fig. 4.7. A rapid concentration increase can be seen in the permanent halocline. In the Gulf of Bothnia, the permanent halocline is absent, and no strong vertical phosphate gradient can be found (Fig. 4.8).

The permanent halocline is a diffusion barrier for both oxygen and other dissolved substances. Therefore the content of phosphate-phosphorus above the halocline is usually equal to, or less than 0.3 m mol m^{-3}. Enormous amounts of phosphate accumulate below the halocline, especially during the stagnation periods. PO_4—P concentrations up to 9 m mol m^{-3} have been observed in the Gotland Deep. The accumulation of phosphates arises partly from redissolution of phosphates deposited on the bottom. This mobilization

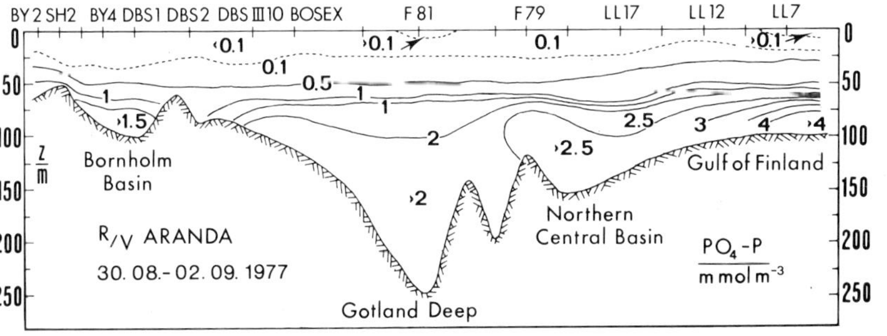

Fig. 4.7. Distribution of dissolved phosphate in a longitudinal section from the Gulf of Finland to the Bornholm Basin. The sampling stations visited indicated by their codes above the figure.

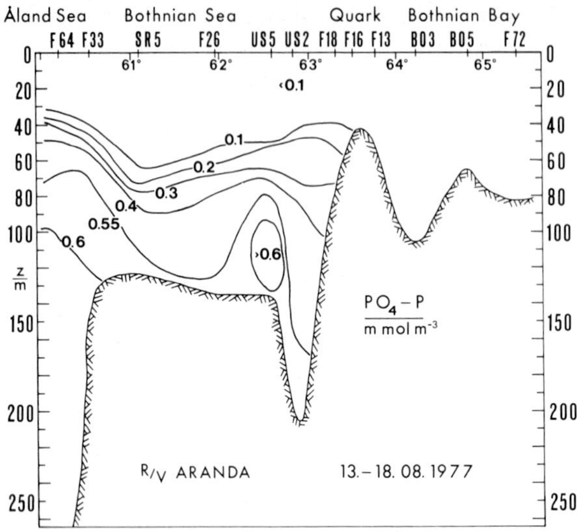

Fig. 4.8. Distribution of dissolved phosphate in a longitudinal section in the Gulf of Bothnia. The sampling stations visited indicated by their codes above the figure.

probably takes place mainly in a relatively loose layer between the water and the bottom sediments (see Section on p. 205).

In addition to reduced water exchange in the deep basins and the possible redissolution of phosphates from the surface layer of sediments, there is a third potential mechanism causing the accumulation of phosphates in the deep water layer. At least a part of the phosphorus liberated in the upper layers may be adsorbed on the sedimenting material, e.g., clay particles or diatom frustules, or sedimented as iron phosphate, finally becoming desorbed or dissolved in the deep layers, where the pH and redox potential are lower (cf., Voipio, 1969). This mechanism may contribute to the increase of the AOU:P ratio* in the surface layer observed by Sen Gupta (1973). It is also in good agreement with the well-known decrease of this ratio in anoxic deep layers (Fonselius, 1967; Gieskes and Grasshoff, 1969).

Besides the seasonal and sporadic variations, the phosphate and total phosphorus contents of the surface water seem to show a generally increasing trend (cf., Fonselius, 1972, 1976b; and Fig. 4.9). Indications of an increasing total phosphorus content have also been observed in the surface water of the Gulf of Finland (cf., Voipio, 1977). However, it should be kept in mind that the time during which changes in the total phosphorus content have been followed is still rather short and may not cover the total range of normal seasonal

---

* The value of the apparent oxygen utilization (AOU) is found by subtracting the actual oxygen concentration observed from that of the saturation concentration. P is the content of dissolved (released) phosphate.

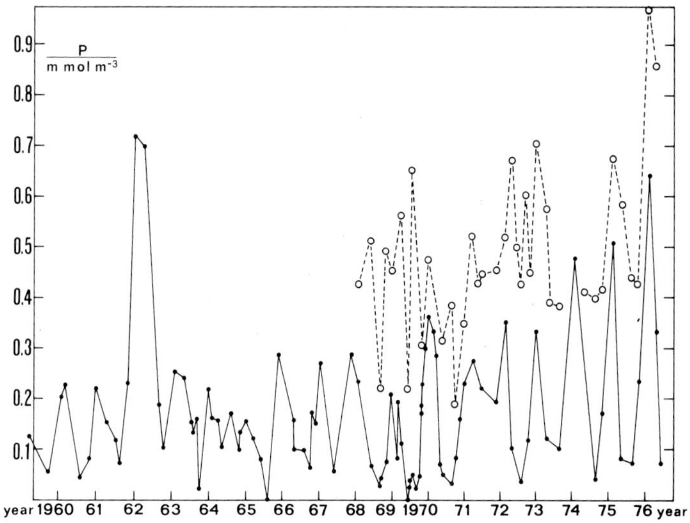

Fig. 4.9. Variations of the content of total phosphorus (dashed line) and phosphate (solid line) in surface water of the Gotland Deep in 1959—1976 (according to Fonselius, 1976b).

variations. The recent increase in the frequency of sampling may have yielded extreme values that previously passed unrecorded, and which give somewhat misleading trends.

The circulation in the marine environment of nitrogen, in its different chemical forms, and its relation to biological production, has long been recognized as one of the key questions, not only in marine chemistry, but also in biological and microbiological oceanography. However, until the last ten or fifteen years, the systematic study of these problems has been hindered by the lack of adequate analytical methods (e.g., Sen Gupta, 1973).

Observations of the surface layer indicate that nitrate disappears almost completely (NO_3—N < 0.1 m mol m^{-3}) from the water column above the thermocline during the spring bloom of plankton algae, mainly diatoms. As detectable amounts of phosphate are often still present, except in the Gulf of Bothnia, it has also been suggested that nitrogen is the principal factor limiting the primary production of phytoplankton (Sen Gupta, 1973). All the nitrate generated from organically bound nitrogen and from nitrogen compounds with a lower oxidation level, and brought into the photic layer by thermal convection, is rapidly consumed in April—June. Subsequently, the primary production seems to be regulated by nitrogen compounds set free in the surface layers or deposited from the atmosphere. At this stage ammonia becomes the main nitrogen source (e.g., Niemi, 1975). There is evidence that the phytoplankton population even prefers ammonia nitrogen to nitrate nitrogen (Tarkiainen et al., 1974) in midsummer.

In late summer large populations of blue-green algae frequently appear, some of which are known to be able to fix considerable amounts of molecular nitrogen (Rinne et al., 1976; Rinne, 1977). These blooms are enhanced by excess phosphorus (Horstmann, 1975). In autumn, when the water becomes cooler, production slows down and decomposition exceeds consumption, so that regeneration of nitrates commences. No seasonal variations occur in the deep layers. There the nitrate concentration depends on the accumulation of nutrients and on the redox conditions. Nitrates are always absent in water containing hydrogen sulphide.

The ammonia content of the surface water is probably controlled by the uptake of phytoplankton or, less probably, by its oxidation. In the open sea the content of ammonia nitrogen in the oxygen-containing waters usually lies well below 1.0 m mol m^{-3}. During the production period it decreases to some 0.2 or 0.3 m mol m^{-3}. In the oxygen-containing deeper waters the content of ammonia nitrogen remains low, but in the anoxic water its content rises to about 5 m mol m^{-3} and values up to 9 m mol m^{-3} have been observed.

The nitrite content is usually very low throughout the water column as long as oxygen is present. However, nitrite maxima may be noted even in the presence of oxygen. One of these sometimes occurs when the first spring bloom develops in water with a relatively high nitrate content, usually at depths of about 15—20 m. The second nitrite maximum may be the result of nitrification. It usually occurs in a thin layer near the secondary halocline (Fig. 4.10), above the layer with increasing nitrate content. Nitrite contents

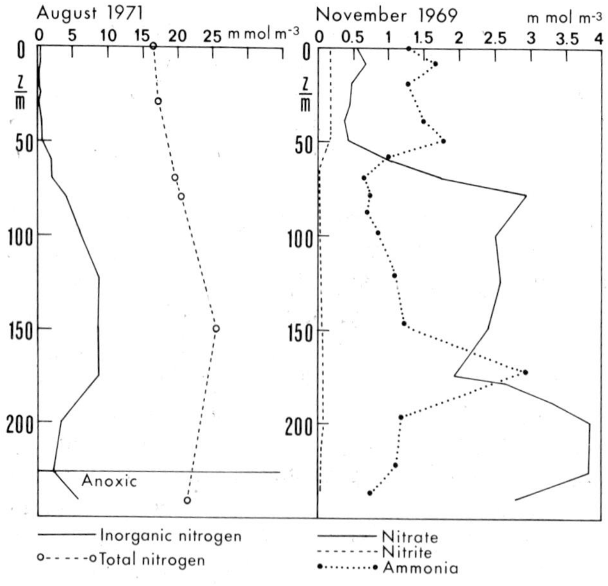

Fig. 4.10. Vertical distribution of nitrogen compounds in the Gotland Deep (according to Sen Gupta, 1973).

of 1—2 m mol m^{-3} NO_2—N are not unusual. Regular measurements of total nitrogen have not been made until recent years. The level usually varies from 10 to 25 m mol m^{-3} (Fig 4.10). From 70% to 90% of the total nitrogen seems to be bound organically (Sen Gupta, 1973).

In the open Baltic Proper there is always an accumulation of organic matter in the lower part of the permanent halocline, and total nitrogen has often a maximum in this layer. In deeper waters, the total nitrogen increases with depth during prolonged anoxic conditions, partly because biogenic material accumulates when the water exchange is reduced and partly because anoxic conditions retard decomposition. Unlike phosphorus, total nitrogen is not drastically decreased by reoxygenation. At the initial stage of an anoxic regime, the total nitrogen content may decrease in the upper part of the anoxic layers (cf., Sen Gupta, 1973). This is perhaps the result of denitrification to the molecular nitrogen.

It has been suggested that nitrate, nitrite and ammonia may not be the only nitrogen compounds of importance in the marine ecosystem. Because of analytical difficulties only occasional attempts have been made to detect other nitrogen compounds. Hydroxylamine and urea have been found in significant amounts in the surface waters. Sen Gupta (1973) found 0.2—0.3 m mol hydroxylamine-N m^{-3} in surface waters of the central parts of the Baltic Proper in November 1969; the content decreased with increasing depth. There is evidence that urea may play an important role as a potential source of nitrogen for primary producers (e.g., McCarthy, 1972; Savidge and Hutley, 1977). Koroleff (1974) found contents as high as 2 m mol urea-N m^{-3}; but the number of samples examined was too small for any clear correlations with other parameters to be discerned. However, these data suggest that urea is likely to be an essential source of nitrogen for phytoplankton, particularly when other sources have been exhausted.

The ratios among carbon, nitrogen and phosphorus in the different parts of the marine environment are often compared with the classical atomic ratios in plankton, i.e., C:N:P = 106:16:1 (cf., Fleming, 1940). It has been assumed that the ratios of the elements in mineralized form, i.e., after decomposition of biogenic matter, should be similar to their ratios in plankton. In the open ocean, the nitrate:phosphate ratio approaches the planktonic value (16:1) before the annual start of phytoplankton production (Redfield et al., 1963).

There are rather few direct determinations of the C:N:P ratio in Baltic Sea plankton. Voipio (1973) studied samples collected in 1967—1970 with a 150 μm mesh net and found the ratio 101:19:1 (mean of 12 samples). The samples were dried with the "supernatant" sea water, i.e., without washing, and, as Voipio pointed out, this may have given too high a value for nitrogen. Koroleff (see Sen Gupta and Koroleff, 1973) analyzed samples collected in 1973 and found a N:P ratio of 13.3. He washed his samples with distilled water, which may have released certain substances from the dying plankton organisms. Ehrhardt (1969) determined the C:N ratio of the particulate organic matter retained by a 1 μm pore filter and found a value of 8.8. His value is thus slightly higher than the ratio of 6.6 for the oceanic plankton, while that of Voipio (5.3) is smaller.

Sen Gupta (1973) calculated that the ratio of the apparent oxygen utilization, (AOU), (cf., p. 196) to the phosphate phosphorus content, (P), (AOU:P) is 360 (in atoms). Subtracting from this value the amount of oxygen needed to oxidize 13 nitrogen atoms from the amino level to the nitrate level, i.e., 26 oxygen molecules, he obtained a value of 154 oxygen molecules, which yields a C:P ratio of 154. Correspondingly, for the C:N:P ratio he gave the value 154:13:1 (see also Sen Gupta and Koroleff, 1973). However, there are some factors which may cause minor errors in the estimation of both AOU and P. Sen Gupta and Koroleff mentioned that the AOU values might be slightly too high due to undersaturation of oxygen during the period of thermal convection. The values used refer mainly to the layer of "old winter water" formed by thermal convection. This undersaturation may be about 5—10% of the saturation value. Moreover, the potential scavenging of phosphates by material being deposited, referred to above (p. 196), may decrease the phosphate concentration. Both these factors tend to decrease the AOU:P ratio.

However, the ratio between the mineralized forms of nitrogen and phosphorus in the water phase is generally rather low. Using a large material collected during different seasons in 1968—1970, Sen Gupta and Koroleff (1973) obtained a value of 4 (in atoms) for the ratio of NO_3—N to PO_4—P. This value rises to about 5 when NO_2—N and NH_3—N are included. As possible reasons for this low ratio the authors gave the use of urea as nitrogen source and denitrification processes leading to the formation of molecular nitrogen.

Furthermore, when considering the nutrients in water, it must be kept in mind that the typical total content of nitrogen varies from 15 to 20 m mol m^{-3} and that of phosphorus from 0.3 to 1 m mol m^{-3}. This corresponds to a N:P ratio of 15:50. It should also be noted that information on the real residence or turn-over times of the nutrient elements in biogenic material is minimal.

Although some evidence has been given that the C:P ratio may be higher in Baltic Sea plankton than in oceanic plankton, the authors believe that the most uncertain feature in the element composition of plankton is the level of nitrogen. However, for many purposes, e.g., ecological modelling, the rounded values of 100:15:1 may be used as a plausible approximation of the ratio C:N:P. Taking into account various analytical difficulties and other potential sources of error, there does not seem to be sufficient reason to assume that this deviates greatly from the generally accepted oceanic values.

The relationship between the carbon and nitrogen contents in the Baltic Sea sediments has been studied, among others, by Gripenberg (1934), Niemistö and Voipio (1974), and Pecherzewski (1974). The contents seem to be closely correlated. For instance, the values found by Niemistö and Voipio

(1974), for two cores from the Gotland Deep, from which 49 and 69 samples were taken, yielded correlation coefficients of 0.91 and 0.95, respectively. The ratio of organic carbon to total nitrogen varied between 6.6 and 11.0, and the mean value of 9.0 was practically the same as that (8.9) reported by Gripenberg (1934). Another set of 141 samples from the various localities in the Gulf of Bothnia, the Gulf of Finland and the northern Baltic Proper showed a mean ratio of 9.6 and a correlation coefficient of 0.96 (Niemistö and Tervo, 1978). The overall mean value observed by Gripenberg for the same areas was 10.0.

Data are also available regarding the ratio of organic carbon to phosphorus. Recalculation of the results of Varencov and Blažčišin (1976) gives a mean C:P ratio of 98 for the soft sediments. The results of Niemistö and Voipio (1974) yield ratios varying between 89 and 123 for samples from the uppermost 10 cm layer. The latter samples were taken from an environment with clearly reducing conditions ($E_h < -100$ mV) and can hardly contain significant quantities of inorganic phosphorus.

The element composition of the organic substances in sediments resembles rather closely that of the other biogenic matter in the marine environment. The only definite difference seems to be the rather high C:N ratio.

Silicate is brought into the Baltic Sea mainly by the rivers discharging from the Precambrian bedrock areas, i.e., chiefly from Finland and Sweden. Average concentrations of about 60 m mol m^{-3} have been reported for rivers flowing through catchment areas rich in granite (Hofman-Bang, 1904). Quaternary deposits usually have significantly lower silicate contents. Consequently, water in rivers from the European continent has a considerably lower silicate content than water in Fennoscandian rivers.

The water flowing in through the Danish Sounds usually has a silicate silicon content of only a few m mol m^{-3}. But the high-salinity water from the deep layers of the Kattegat entering during exceptionally heavy influxes may contain silicate silicon up to 20 m mol m^{-3}.

In general, the silicate content of the Baltic Sea surface water decreases from the inner parts of the Gulf of Finland and the Gulf of Bothnia towards the Baltic Proper. As a consequence of the uptake of silicate by diatoms, the content may vary considerably. The diatom bloom occurs after the melting of the ice in the headward parts of the bays and gulfs and a second, but much smaller maximum may occur in the autumn (cf., Niemi, 1975). Thus silicon shows strong seasonal variations in the surface water affected by the availability of nitrogen compounds and phosphates. Silicate itself has never been observed to be a factor limiting primary production in the open Baltic Sea.

Figure 4.11 shows the distribution of silicate along a longitudinal section through the Baltic Sea in August 1977. The highest silicate content in surface water, sometimes reaching 50 m mol m^{-3}, is found in the Gulf of Bothnia (Voipio, 1961). Because of the low silicate content of the Neva and Narva, rivers in the USSR, the content in surface water may be expected to decrease

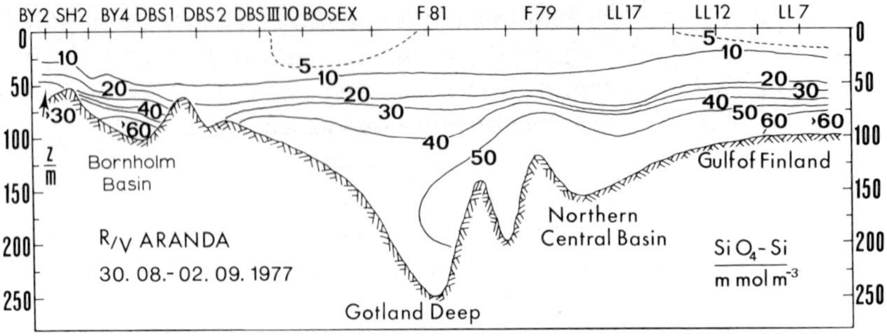

Fig. 4.11. Distribution of dissolved silicate in a longitudinal section from the Gulf of Finland to the Bornholm Basin. The sampling stations visited indicated by their codes above the figure.

from the west to the east in the Gulf of Finland. The upwelling of silicate-rich deep water at the entrance of the Gulf of Finland may also contribute to the silicate content in this Gulf (Niemi, 1975). The distribution of silicate at the mouth of a river emptying in an archipelago is shown in Fig. 4.12 which also shows that silicate is a useful parameter for studying mixing processes in an estuary because of the large differences between the silicate contents of the open-sea and river water.

Although the silicate content of surface water is comparatively uniform down to the permanent halocline, particularly after the winter mixing, it increases sharply beneath the halocline. Depending on the area and the depth of the basin, contents of 60—80 m mol m^{-3} are found in the deep water of the Baltic Proper. Silicate accumulates during periods of stagnation. It has been observed that the fine structure of the skeleton of diatoms dissolves more rapidly under anoxic conditions and that the frustules are completely destroyed by the time they reach the sea floor when reducing conditions prevail, e.g., in the Landsort Deep. This may be due to the reduction of iron (III), which under oxidizing conditions preserves the skeletons of diatoms from attack. In addition, large amounts of silicate are dissolved from the bottom sediment when the overlying water is deoxygenated (W. Balzer, 1978). The accumulation of dissolved silicate in the stagnant deep water is possible because sea water undersaturated with respect to silicate anion (Rankama and Sahama, 1955; Krauskopf, 1956). Biological processes are probably responsible for the removal of silicate from the water, and because they affect the transport of silicate from the surface water to the deep water and to the sediments, variations of the silicate content of the deep water can be used to evaluate horizontal and vertical mixing in the deep water in the Baltic Sea. The silicate distribution can also be used to examine the local occurrence of stagnant environment and reduced water exchange (cf., Niemi, 1975).

Fig. 4.12. Distribution of silicate in a delta area of Kemijoki River, discharging into the northernmost part of the Bothnian Sea, 1959 (according to Voipio, 1961).

TRACE METALS

In addition to the improvement of general knowledge of the chemistry of the sea, the following reason exists for the study of trace elements in marine environments. Firstly, many of the trace metals are biologically active, either enhancing or depressing biological production. Secondly, in certain instances it may be possible to exploit the mineral resources of the sea. However, in the Baltic Sea such exploitation is rather small and is limited to the sediments (cf., Section on p. 105).

Consideration of the enhancement and inhibition of biological production is hampered by several fundamental difficulties. Very little is known about the fraction which is biologically active. Most probably it consists of not only the ionic forms of metals but also of at least some of the chelated ions. It is also evident that the total amount of the various metals cannot be uniformly active. Great interest is thus focused on the chemical form of the heavy metals, especially in the Baltic Sea, which receives significant contributions from land-based sources in river water and direct discharges of waste waters.

Moreover, the analytical procedures from sampling to the final measurement involve many difficult problems. How can samples be collected representing only dissolved, either ionic or complex, forms of the heavy metals? How can the dissolved substances be separated from those adsorbed or chemically bound to biogenic and mineral particles of different sizes? How does the chemical form change during the pretreatment of the samples and, finally, how do the variations in the matrices, e.g., in salinity or lithological properties of sediments, interfere with the measurement stage of the analysis?

These difficulties are commonly encountered in chemical oceanography but are especially severe in the study of trace metals. Furthermore, the work necessitates several rather arbitrary choices of procedures, and these may also differ among the analysts. For instance, Koroleff (1968) filtered his iron samples through a 0.5 μm membrane filter and Kremling (1973) through a 0.4 μm filter, while Schmidt (1976) used unfiltered samples.

As early as 1947, Koroleff published some data on the total manganese content in northern Baltic Sea waters. These contents vary mainly between 2 and 10 mg m^{-3}, Koroleff (1968) gave also data for iron. The iron content in the soluble, i.e., filtered, fraction usually varies from some 2 mg m^{-3} to 10 mg m^{-5} in the Baltic Proper and in the Gulf of Bothnia, respectively (cf., Morozov et al., 1974). In anoxic waters the iron content in the soluble fraction evidently reaches the solubility of FeS, or about 70 mg m^{-3}. Under these conditions practically all iron is present in the filtered fraction, except in water samples from the nearbottom, often turbid waters. The manganese content can also be rather high in anoxic water; F. Koroleff (pers. commun., 1978) has recorded total manganese contents of several hundred mg m^{-3}. The total iron content can be fairly high even in waters containing oxygen, but it is usually below 10 mg m^{-3} in the surface water of the Baltic Proper (Koroleff, 1968; Schmidt, 1976). In the bottom waters the iron content seems to be related to the turbidity, but in the surface waters the highest total iron contents are recorded in the Bothnian Bay, where humic substances are most abundant (Koroleff, 1968; V. Tervo, pers. commun., 1978).

Sen Gupta (1972), Kremling (1973) and Brügmann (1974) have obtained fairly consistent results for the zinc content of water. The background value of soluble zinc in the uncontaminated and unfiltered water samples is 5—10 mg m^{-3}. The same authors have also studied the cadmium, copper and lead contents. The cadmium content was of the order of 0.2 mg m^{-3} (cf., Schmidt, 1976). The copper content is usually 2—5 mg m^{-3}. These contents are very similar to those in the open ocean (Kremling, 1973).

The present level of lead, viz., 0.5—1.0 mg m^{-3}, has been assumed to represent slight contamination, e.g., by air-borne pollution. However, the analytical difficulties are very pronounced (cf., ICES, 1977).

Sen Gupta (1972) and Morozov et al. (1974) have also studied the concentrations of nickel and cobalt. The levels for both elements are usually below 2 mg m^{-3}, that for cobalt being often lower than nickel.

Some determinations of the arsenic content have been made. For instance, in 1975 F. Koroleff (pers. commun., 1978) recorded up to 0.02 m mol^{-3} in the surface water of the Baltic Proper and about 0.03 m mol m^{-3} in deep waters. In the Bothnian Bay the level was clearly higher: 0.04 m mol m^{-3} in surface water and 0.06 m mol m^{-3} in deeper waters.

Most of the investigations cited above provide background information on the "normal" levels of the elements studied. Only Sen Gupta (1972) has considered their relation to production but his data are insufficient for any general conclusion.

Weigel and Kremling (1975) have studied the Pb, Cu, Cd, Fe and Zn in seston, i.e., in living and non-living particulate matter in water. Their results indicate that the heavy-metal concentrations in seston, excluding iron, are only small fractions of the total contents of those metals in a certain water column. On the basis of later studies of Kremling and Petersen (1978), it has been estimated that 2.3% of zinc, 4.3% of cadmium and 13.7% of copper are extracted from sea water by phytoplankton (Bornholm Basin, April 1975). They estimated also the concentration factors for these elements, which are of the order of 10^4.

Much intensive study has been recently devoted to trace metal concentrations in the sediments (see also Section below). In the sediment phase, iron, manganese and titanium no longer rank as trace metals, their concentrations being too high, a few per cents of the dry matter as regards iron and sometimes equally high for manganese (Niemistö and Voipio, 1974; Varencov and Blažčišin, 1976). The content of titanium is usually one order of magnitude lower. Emeljanov (1976) has given ample data on the distribution of trace metals of different kinds in the uppermost sediments. The data for soft bottom sediments are presented together with some other data in Table 4.IV.

SEDIMENT—WATER INTERACTIONS

One of the basic factors controlling the chemistry of sea water is the interaction between sediments and water. In this interaction three main aspects can be distinguished: bacterial activity, effects of the redox potential, and chemical equilibria. In addition to these, biogeochemical processes, physical processes, such as wave action, erosion by currents and mechanical mixing by moving bottom animals must be taken into account. Another important para-

TABLE 4.IV

Selected data of the content of some heavy metals in soft sediments. The values of Olausson et al. are labelled "mean values for off-shore areas". Those of Hallberg are for a core taken from the Gotland Deep and cut in 223 transverse sections of 2 mm each. The data from Suess and Erlenkeuser refer to a core from the Bornholm Basin.

Element	Number of samples	content mg^{-1} of dry weight			Reference
		min.	mean	max.	
Ba	70	20	972	2910	Emeljanov, 1976
Cd	125	0.05	0.73	5.64	Niemistö and Tervo, 1978
					Olausson et al., 1977
	21	0.5	1.1	2.2	Suess and Erlenkeuser, 1975
Co	121	2	18	36	Niemistö and Tervo, 1978
	21	4.4	6.2	9.5	Suess and Erlenkeuser, 1975
Cr	72	19	94	252	Emeljanov, 1976
	141	11	39	50	Niemistö and Tervo, 1978
Cu	18	12	44	65	Emeljanov, 1976
		28	72	214	Hallberg, 1974
	141	7	37	64	Niemistö and Tervo, 1978
			12		Olausson et al., 1977
	21	25	45	65	Suess and Erlenkeuser, 1975
Hg	141	0.01	0.13	0.68	Niemistö and Tervo, 1978
			0.12		Olausson et al., 1977
Mo	82	1	13	108	Emeljanov, 1976
	186	13	103	615	Hallberg, 1974
Ni	82	14	37	60	Emeljanov, 1976
	121	10	33	47	Niemistö and Tervo, 1978
			14		Olausson et al., 1977
	21	34	52	88	Suess and Erlenkeuser, 1975
Pb	133	10	31	100	Niemistö and Tervo, 1978
			17		Olausson et al., 1977
	21	13	64	105	Suess and Erlenkeuser, 1975
Sn	63	2.4	5.5	6.6	Emeljanov, 1976
V	83	10	102	300	Emeljanov, 1976
	12	35	85	128	J. Launiainen and R. Danielson unpubl. results

TABLE 4.IV (Continued)

Element	Number of samples	concentent mg^{-1} of dry weight			Reference
		min.	mean	max.	
Zn	21	49	145	268	Emeljanov, 1976
		79	161	619	Hallberg, 1974
	141	46	163	477	Niemistö and Tervo, 1978
			80		Olausson et al., 1977
	21	96	167	270	Suess and Erlenkeuser, 1975
Zr	67	80	267	600	Emeljanov, 1976

meter, affecting both the biogeochemical and physical processes, is the bottom configuration.

The definition of the sediment—water interface deserves consideration. It is much less similar to a geometric plane than is the sea—atmosphere interface. For the present purpose it could in many cases be best defined as an intermediate layer between turbid water and the fairly well consolidated bulk sediment. At the same time it is necessary to remember that several of the processes dealt with in this section take place at any (physical) interface of water and particulate matter, while some of them evidently need rather bulky surroundings existing only on the sea floor before the reaction products can be found.

The diffusivity of ions and molecules in the intermediate layer is only slightly lower than in free water, owing to its high content of interstitial water (e.g., Manheim, 1970). This property enhances the interaction between the intermediate layer and the water phase, and several substances which would be permanently trapped in the sediment bulk can become dissolved from the intermediate layer.

It is difficult to determine the real thickness of the intermediate layer. It depends greatly both on the hydrodynamic conditions under which the deposition of particulate matter takes place and on the quantity of the sinking material. The extent of this layer is perhaps best characterized by the dry matter content of the sediment. Figure 4.13 indicates that below the few uppermost centimetres of sediment there exists a rather clear gradient in the dry matter content and in porosity, i.e., the ratio of the volume of water to the total volume of sediment.

In his classical work, Mortimer (1942) has shown that phosphorus becomes mobilized from the deposited particulate matter in lake sediments when the redox potential decreases below + 230 mV in relation to the normal hydrogen electrode. According to him this is mainly caused by liberation of phosphate ions from iron (III) phosphate or desorption of those ions from some other

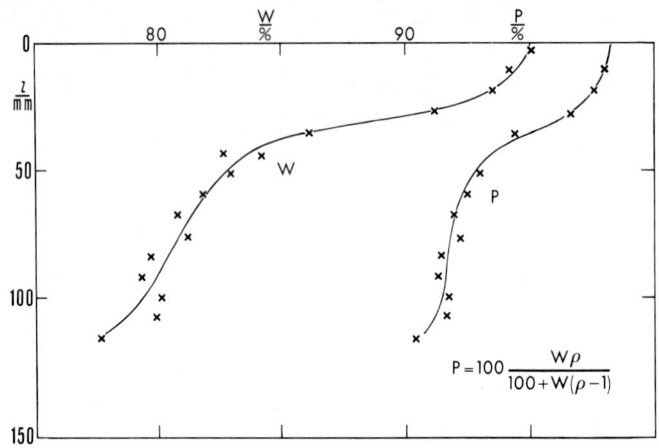

Fig. 4.13. Distribution of porosity and water content in a core from the Gotland Deep, collected in 1971. Porosity (*P*) is defined by equation (1) where *W* is the water content of the sediment and ρ the density of the sediment (2.7 g cm^{-3}). Data from Niemistö and Voipio (1974).

iron (III) compounds when iron (III) is reduced to iron (II). In lakes the iron (II) content increases in water when the phosphate content increases. In the Baltic Sea the increase of iron is not as clear as in lakes because of the parallel reduction of sulphate to sulphide, which combines with iron (II) to form iron (II) sulphide. The distribution of iron in relation to phosphates and redox conditions has received relatively little attention in the Baltic Sea, but the results of Koroleff (1968) show that under reducing conditions in the Gotland Deep the iron content in water increases with decreasing oxygen content.

It has been assumed that phosphatic materials dissolve from the sediment in anoxic water whose pH decreases to near 7.0 (Fonselius, 1967; Koroleff, 1968; Gieskes and Grasshoff, 1969). A log—log plot of pH vs. the concentration of phosphate shows that a correlation between pH and the phosphate content exists down to a pH value of 7.1. This probably indicates the effect of carbon dioxide liberated by the decomposition of organic material during which phosphates are formed. However, an increase in the hydrogen ion concentration also favours the desorption of phosphates from particulate matter (cf., e.g., Voipio, 1969). Below the pH value of 7.1, the phosphate content rises sharply. Hallberg and his co-workers (1972) have given experimental evidence indicating that phosphate dissolves from sediments under anaerobic conditions especially if they contain large amounts of organic material. The process seems to be reversible, because the phosphate content of the supernatant water rapidly decreases when the water is reaerated.

The processes mentioned above evidently take place in an interface layer between water and sediment. It is difficult to say whether this usually thin layer consists of water with a very high content of particulate matter or of extremely soft sediment in which the water content perhaps exceeds 90—95%. In the more consolidated sediment the material is much less mobile. Even the most anoxic sediments in the Baltic Sea have a phosphorus content of about 0.09% of dry sediment, while the C:P ratio is usually very similar to that in plankton material (cf., Section on p. 194). Below the sediment—water interface no simple correlation seems to exist between the redox potential and the phosphorus content of the sediment (Niemistö and Voipio, 1974).

As mentioned above, the deposition of iron and some other heavy metals is regulated by changes in the oxygen content in the water phase, heavy metals being less soluble under oxic than anoxic conditions. Consequently, it can be expected that such changes in the oxygen content will be reflected in the vertical distribution of heavy metals in sediments, provided that sedimentation has not been disturbed, for instance, by significant water turbulence or movements of bottom animals. The iron content in the sediment cores taken from the Gotland Deep does in fact show some vertical variation, but greater variation can be seen in the manganese content (cf., Niemistö and Voipio, 1974).

The redox conditions in the sediment phase also play a very pronounced role. The postglacial sediments in the Baltic Sea contain large amounts of organic matter (see e.g., Niemistö and Voipio, 1974; Pustelnikov, 1976, 1977; see also Section on p. 101). The decomposition of the organic matter decreases the redox potential and promotes the remobilization of certain heavy metals from the sediment. Changes in the carbon dioxide system may also cause mobilization, followed by diffusion towards the water phase. However, the formation of sulphide ions increases the entrapment of some metals as sulphides, e.g., FeS, CuS (Papunen, 1968; Hallberg, 1974; Niemistö and Voipio, 1974). Moreover, iron sulphide can be transformed to pyrite or marcasite (Ignatius et al., 1968).

The pH of sediments being around 7, manganese (II) sulphide is more soluble than iron (II) sulphide. Manganese is therefore expected to be more mobile in an anoxic than in an oxic environment. Consequently, the stratification of iron and manganese in sediments, and especially their ratio, should reflect the variations of the redox conditions in the water phase. Evidence that this is the case has been given by Niemistö and Voipio (1974). Hallberg (1974) has studied the sediments in the same area (Gotland Deep), examining the ratio of copper to zinc and the ratio of the sum of copper and molybdenum to zinc in the light of the solubilities of their sulphides.

While some correspondence exists between the variation of these ratios studied by Niemistö and Voipio, and by Hallberg, some discrepancies are also observed. Some of them may be caused by the inhomogeneity of the different cores. Some evidently reflect additional interfering chemical processes. For instance, manganese is often trapped in the Baltic Sea sediments as manganous calcium carbonate, whose chemical composition corresponds fairly closely to that of rhodochrosite (Manheim, 1961; Niemistö and Voipio, 1974). The stratification of manganese in sediment is thus regulated not only by the redox conditions but also by the presence of organic matter in such quantity that its decomposition will yield enough carbon dioxide for the formation of carbonate. Furthermore, several organic compounds most probably form complexes with metal ions, but these processes are inadequately known.

The decomposition of biogenic organic matter in an aquatic environment is an oxidation process yielding carbon dioxide. When the dissolved oxygen has been consumed from the interstitial water of the sediments, heterotrophic microbes can utilize the oxygen of the sulphate ions when they use the biogenic matter as a carbon source. Since sulphate is one of the major constituents of sea water, the sulphate in water with a salinity of 10‰ gives an oxygen supply about 50 times as great as the supply of dissolved molecular oxygen in water in equilibrium with the atmosphere. It is therefore easy to understand that the methane formation frequently occurring in lake sediments with a high content of biogenic organic matter has not been observed in the Baltic Sea sediments.

The production rate of hydrogen sulphide is kinetically of the zero order down to sulphate concentrations of some 2 mM and is clearly correlated to the magnitude of the carbon supply (Bågander, 1977b). The interstitial water in sediments rich in biogenic matter becomes soon saturated with hydrogen sulphide. Nor can the metal ions combine with excess amounts of sulphide ions. Therefore most of the hydrogen sulphide formed must escape to the water phase. Bågander, for instance, has observed cases in which about 80% of the hydrogen sulphide produced was transported from the sediment into the water phase.

The observations cited above agree well with the reports of other authors indicating that the number of sulphate-reducing bacteria is very high (10^5—10^6 cm^{-3}) in the sediment phase (Seppänen and Voipio, 1971; Bansemir and Rheinheimer, 1974. Further on, the number of those microbes in water containing hydrogen sulphide is usually low, often smaller than in water rich in oxygen, where hydrogen sulphide is evidently formed in the anoxic microenvironment inside the particles of biogenic matter and becomes rapidly oxidized outside.

The importance of sulphate-reducing bacteria in the decomposition of biogenic organic matter in sediments is enhanced by the fact that hydrogen sulphide, being toxic to most organisms, reduces the number of other organisms able to compete for biogenic matter as their carbon source. However, the activity of the sulphate-reducing bacteria seems to depend fairly closely on temperature (Bågander, 1977b).

The hydrogen ion concentration of the sediments increases with the decreasing redox potential and the value of pH is about 7 when hydrogen sulphide is present. The hydrogen ion concentration is mainly regulated by the carbon dioxide system. Consequently, when carbon dioxide is used during the bacterial oxidation of hydrogen sulphide, the value of pH rises above 8

(Bågander, 1977a). This change corresponds well to the behaviour of pH in the water phase.

In shallow water, the oxidation of sulphides is greatly enhanced by the presence of photosynthetic sulphur bacteria. The oxidation evidently takes place in two steps, first to elemental sulphur and then to sulphate (see Bågander, 1977b).

DISSOLVED ORGANIC MATTER

The study of the organic substances in the Baltic Sea was probably originally undertaken for two main reasons. Firstly, the wish to discover why the sea water has a colour and even different colours, depending on the place and time of observation (cf., Kalle 1938). The earliest published material relating to the colour of the Baltic Sea is evidently that presented by Witting in 1912. Secondly, it was found that the colour of the water itself interfered with the coloured compounds formed during the photometric determinations, e.g., of silicate (cf., Kalle, 1937).

The Baltic Sea water is clearly more yellow than ocean water. In the northernmost part of the Gulf of Bothnia it is even yellow-brownish. Since it has been virtually impossible to determine the actual composition of the mixture of the various organic compounds present in Baltic Sea water, Kalle introduced (1937) the concept of "Gelbstoff" or "yellow substance" for this mixture. While in ocean water this mixture contains mainly the decomposition products of autochthonous biological matter, in the Baltic Sea it has two additional sources. Even under natural conditions the rivers discharging into the Baltic Sea contain much humic substances (Aschan, 1900, 1906; Odén, 1922). At present the municipal and industrial waste waters, especially those of the pulp and paper industry, discharge huge amounts of organic substances into the sea.

Until recent decades, the content of the yellow substance was mainly reported as absorptivity figures. These figures often give a fairly good indication of the relative content of the coloured organic substances in various water masses, but they do not give any indication of the actual content of the organic carbon in water. As an example, the relative concentration values along a Baltic Sea transect are given in Fig. 4.14 (Jerlov, 1955, See also Section on p. 167).

The lack of quantitative data of the content of dissolved organic carbon compounds (DOC) is evidently due to analytical difficulties. The principal methods used have been wet combustion, e.g., by potassium dichromate, or dry or wet combustion of organic substances followed by IR measurement of the amount of carbon dioxide formed. It seems that no comprehensive intercalibration study of the different methods or different instruments has yet been carried out. Moreover, some of the determinations are in fact of total organic carbon (TOC). However, the contribution of particulate organic carbon is very small, usually much less than 10% of DOC.

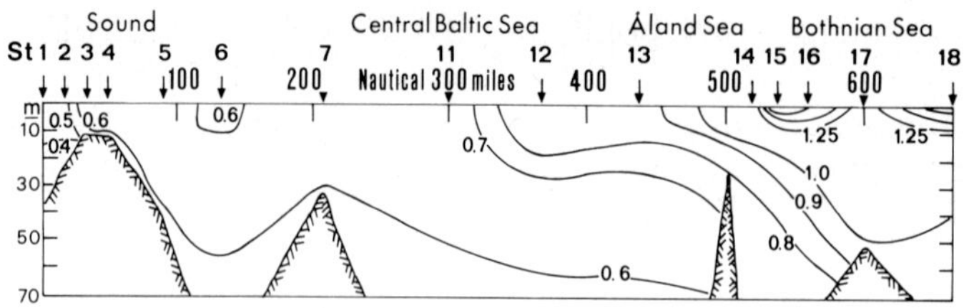

Fig. 4.14. Vertical distribution of the yellow substance along a longitudinal section from the Sound to the Bothnian Sea. The contents are given only in relative units (according to Jerlov, 1955).

The first extensive studies of the content of DOC are evidently those of Kay (1954) and of Skopintsev and his coworkers (1959). Kay reported carbon contents between 2.0 and 4.6 g m^{-3}. Skopintsev's data are from various seas, including the Baltic Sea and vary between 3 and 4 g m^{-3}. Szekielda (1968, 1971) found DOC values of 1.17—6.26 g m^{-3} in samples collected in 1962. Ehrhardt's (1969) values vary between 3.0 and 4.7 g m^{-3}, while Fonselius (1972) gave a rough mean value of 5 g m^{-3}. Later Swedish studies by S. Carlberg (pers. commun., 1978) indicate that the values should be somewhat lower. Jurkovskis and Luke (1974) reported values from 3.0 g m^{-3} to 11 g m^{-3} and Pecherzewski and Lawacz (1975) from 7 g m^{-3} to 13 g m^{-3}. These authors and also Szekielda found great seasonal and local differences in DOC. The Finnish levels vary from 4 g m^{-3} to 7 g m^{-3} in samples collected in 1971—1975 (V. Tervo, pers. commun., 1978). In waters polluted by wastes from the paper and pulp industry, the values can be considerably higher (see e.g., Lundstedt 1970).

Bladh (1972) and Carlberg (e.g., 1977) have studied the correlation among the yellow substance, total organic carbon (TOC) and oxidability or chemical oxygen demand (COD). According to these authors, the correlation between COD and the yellow substance is often better than that between TOC and the yellow substance. The results indicate that the partial breakdown of coloured organic substances is a much faster process than the final decomposition of dissolved organic substances.

In addition to the quantitative studies on the distribution of DOC, there are some papers dealing with miscellaneous topics related to the dissolved organic substances in the Baltic Sea water.

For instance, Zsolnay (1975) has used a modification of the biological oxygen demand determination method to measure easily cycling "labile carbon". According to him, about 1 g m^{-3} is decomposed during 3 months in dark bottles at 25° C. Brown (1975) has estimated that about 1% of the DOC is of high molecular weight (> 10,000). Salo and Saxén (1974) have made preliminary studies on the complexing effect of humic substances on some radionuclides.

Only recently have some studies been published on the chemical composition of DOC (see e.g., Josefsson, 1973; Josefsson et al., 1977). Nyquist (1976) has studied the distribution of humus and lignin sulphonates in the Baltic Sea using fluorescence measurements.

REFERENCES

Arrhenius, G., 1963. Pelagic sediments. In: M.N. Hill (Editor), The Sea, Vol. 3. Interscience Publishers, New York, London, pp. 655--727.

Aschan, O., 1900. Den organiska färgande substansen in Vanda ås vatten. Teknikern, 10, 11 pp.

Aschan, O., 1906. Humusämnena i de nordiska inlandsvattnen och deras betydelse, särskildt vid sjömalmernas daning. Bidr. Känn. Finl. Nat. Folk, 66: 5—176.

Balzer, W., 1978. Untersuchungen über Abbau organischer Materie und Nährstoff-Freisetzung am Boden der Kieler Bucht beim Übergang vom oxischen zum anoxischen Milieu. Rep. S.F.B. 95, Univ. Kiel, No. 36.

Bågander, L.E., 1977a. In situ studies of bacterial sulfate reduction at the sediment-water interface. Ambio, Spec. Rep., 5: 147—155 (Russian summary).

Bågander, L.E., 1977b. Sulfur Fluxes at the Sediment—Water Interface — an in situ Study of Closed Systems, E_h and pH. Diss. Dept. Geol., Univ. Stockholm, Microbial Geochem. Publ. 1, 11, 6, 14 and 18, 90 pp.

Bansemir, K. and Rheinheimer, G., 1974. Bakteriologische Untersuchungen über die Bildung von Schwefelwasserstoff in einer Vertiefung der inneren Kieler Förde. Kieler Meeresforsch., 30(2): 91—98.

Bladh, J.-O., 1972. Measurements of yellow substance in the Baltic and neighbouring seas during 1970—1972. Medd. Havsfiskelab. Lysekil, 138: 1—3, appendices.

Brown, M., 1975. High molecular-weight material in Baltic seawater. Mar. Chem., 3: 253—258.

Brügmann, L., 1974. Die Bestimmung von Spurelementen im Meerwasser unter Verwendung einer stationären Quecksilberelektrode. Acta Hydrochim. Hydrobiol., 2: 123—138.

Buch, K., 1932. Untersuchungen über gelöste Phosphate und Stickstoffverbindungen in den nordbaltischen Meeresgebieten. Merentutkimuslaitoksen Julk. Havsforskningsinst. Skr., 86: 1—28.

Carlberg, S.R., 1977. A study of the distribution of major organic constituents, measured as organic carbon, oxidability and yellow substance, in Baltic waters. Paper submitted to American Institute of Biological Sciences. Symposium "Concepts in Marine Organic Chemistry". Edinburgh, Scotland, 6—10 September, 1976. Mar. Chem.

Cox, R.A. and Culkin, F., 1966. Sodium, potassium, magnesium, calcium and strontium in sea water. Deep-Sea Res., 13: 789—804.

Cox, R.A., Culkin, F., Greenhalgh, R. and Riley, J.P., 1962. The chlorinity, conductivity and density of sea-water. Nature (London), 193(4815): 518—520.

Culkin, F., 1965. The major constituents of sea water. In: J.P. Riley and G. Skirrow (Editors), Chemical Oceanography. Academic Press, London, New York, 1: 121—161.

Dyrssen, D.W. and Uppström, L.R., 1974. The boron/chlorinity ratio in Baltic Sea water. Ambio, 3(1): 44—46.

Ehrhardt, M., 1969. The particulate organic carbon and nitrogen and the dissolved organic carbon in the Gotland Deep in May 1968. Kieler Meeresforsch., 25(1): 71—80.

Emeljanov, E.M., 1976. Malye i rassejannye elementy v osadkah (Small and trace elements in sediments). In: V.K. Gudelis and E.M. Emeljanov (Editors), Geologija Baltijskogo Morja (Geology of the Baltic Sea), Mokslas, Vilnius, pp. 288—306.

Fleming, R.H., 1940. The composition of plankton and units for reporting population and production. Proc. Sixth Pacific Sci. Congr. Calif., 1939, pp. 535—540.
Fonselius, S.H., 1962. Hydrography of the Baltic deep basins. Fish. Board Swed. Ser. Hydrogr. Rep., 13: 1—41.
Fonselius, S.H., 1967. Hydrography of the Baltic deep basins. II. Fish. Board Swed. Ser. Hydrogr. Rep., 20: 1—31.
Fonselius, S.H., 1968. On the oxygen deficit in the Baltic deep water. Lecture given at the VI:th Conference of the Baltic Oceanographers in Sopot June 1968. 10 pp., figs.
Fonselius, S.H., 1969a. On the stagnant conditions in the Baltic. Abstracts of Gothenburg Dissert. Science, 14: 1—17.
Fonselius, S.H., 1969b. Hydrography of the Baltic deep basins. III. Fish. Board Swed. Ser. Hydrogr. Rep., 23: 1—97.
Fonselius, S.H., 1972. On biogenic elements and organic matter in the Baltic. Ambio, Spec. Rep., 1: 29—36.
Fonselius, S.H., 1976a. On the nutrient variability in the Baltic. Ambio, Spec. Rep., 4: 17—25.
Fonselius, S.H., 1976b. On phosphorus in Baltic surface water. Medd. Havsfiskelab. Lysekil, 206: 1—3, appendices.
Fonselius, S.H., 1977. An inflow of unusually warm water into the Baltic deep basins. ICES. C.M. 1977/C:15. 3 pp., figs. Medd. Havsfiskelab. Lysekil, 229. 3 pp., figs.
Gargas, E., Dahl-Madsen, K.I., Schroeder, H. and Rasmussen, J., 1977. Dynamics of Baltic ecosystems and causes of their variability. Tech. Rep. Water Quality Institute, Denmark. 35 pp., appendices.
Gieskes, J.M. and Grasshoff, K., 1969. A study of the variability in the hydrochemical factors in the Baltic Sea on the basis of two anchor stations September 1967 and May 1968. Kieler Meeresforsch., 25(1): 105—132.
Granqvist, G., 1932. Croisière thalassologique et observations en bateaux routiers en 1931. Merentutkimuslaitoksen Julk. Havsforskningsinst. Skr., 81: 1—38.
Grasshoff, K., 1975. The hydrochemistry of landlocked basins and fjords. In: J.P. Riley and G. Skirrow (Editors), Chemical Oceanography. Academic Press, London, New York, San Francisco, (2nd ed.) 2: 455—631.
Greenhalgh, R. and Riley, J.P., 1963. Occurrence of abnormally high fluoride concentrations at depth in the oceans. Nature (London), 197: 371—372.
Gripenberg, S., 1934. A study of the sediments of the North Baltic and adjoining seas. Fennia, 60(3): 1—231.
Gripenberg, S., 1937. The calcium content of Baltic water. J. Cons. Int. Explor. Mer, 12: 293—304.
Gripenberg, S., 1961. Alkalinity and boric acid content of Barents Sea water. Rapp. P.-V. Réun. Cons. Int. Explor. Mer, 149: 31—37.
Hallberg, R.O., 1974. Paleoredox conditions in the eastern Gotland Basin during the recent centuries. Merentutkimuslaitoksen Julk. Havsforskningsinst. Skr., 238: 3—16.
Hallberg, R.O., Bågander, L.E., Engvall, A.-G. and Schippel, F.A., 1972. Method for studying geochemistry of the sediment water interface. Ambio, 2: 71—72.
Hofman-Bang, O., 1904. Studien über schwedische Fluss- und Quellwässer. Diss., Univ. Uppsala, 59 pp.
Hollan, E., 1969a. Die Veränderlichkeit der Strömungsverteilung im Gotland-Becken am Beispiel von Strömungsmessungen im Gotland-Tief. Kieler Meeresforsch., 25(1): 19—70.
Hollan, E., 1969b. Eine physikalische Analyse kleinräumiger Änderungen chemischer Parameter in den tiefen Wasserschichten der Gotlandsee. Kieler Meeresforsch., 25(2): 255—267.
Horstmann, U., 1975. Eutrophication and mass production of blue-green algae in the Baltic. Merentutkimuslaitoksen Julk. Havsforskningsinst. Skr., 239: 83—90.

ICES, 1977. Studies of the pollution of the Baltic Sea. II. Report on the baseline study of the level of contaminating substances in the living resources of the Baltic 1974/75, and the intercalibration exercise connected with it. ICES, Coop. Res. Rep., 63: 22—98.

Ignatius, H., Kukkonen, E. and Winterhalter, B., 1968. Notes on a pyritic zone in upper Anculys sediments from the Bothnian Sea. Bull. Geol. Soc. Finl., 40: 131—134.

Jerlov, N.G., 1955. Factors influencing the transparency of the Baltic waters. Medd. Oceanogr. Inst. Göteborg, 25: 1—19.

Josefsson, B., 1973. Determination of Dissolved Organic Compounds in Natural Waters. Diss., Gothenburg, 150 pp.

Josefsson, B., Lindroth, P. and Östling, G., 1977. An automated fluorescence method for the determination of total amino acids in natural waters. Anal. Chim. Acta, 89: 21—28.

Jurkovskis, A.K. and Luke, M.P., 1974. The results of bichromate and permanganate methods application in analytical investigations of the Baltic Sea water. ICES. C.M., 1974/C: 18. 10 pp. (mimeogr.).

Kalle, K., 1932. Phosphatgehaltuntersuchungen in der Nord- und Ostsee im Jahre 1931. Ann. Hydrol. Marit. Meteorol., 60(1): 6—17.

Kalle, K., 1937. Meereskundliche chemische Untersuchungen mit Hilfe des Zeisschen Pulfrich-Photometers. VI. Mitteilung. Die Bestimmung des Nitrits und des "Gelbstoffs". Ann. Hydrol. Marit. Meteorol., 65(1): 276—282.

Kalle, K., 1938. Zur Problem der Meereswasserfarbe. Ann. Hydrol. Marit. Meteorol., 66(1): 1—13.

Kalle, K., 1943. Die grosse Wasserumschichtung im Gotland-Tief vom 1933—34. Ann. Hydrol. Marit. Meteorol., 71(4/6): 142—146.

Kay, H., 1954. Untersuchungen zur Menge und Verteilung der organischen Substanz im Meerwasser. Kieler Meeresforsch., 10(2): 202—214.

Knudsen, M., 1903. Gefrierpunkttabelle für Meerwasser. Publ. Circonst. Cons. Int. Explor. Mer, 4—5: 11—13.

Koroleff, F., 1947. Determination of manganese in natural waters. Acta Chem. Scand., 1: 503—506.

Koroleff, F., 1968. A note on the iron content of Baltic water. ICES. C.M. 1968/C:34. 4 pp., appendices (mimeogr.).

Koroleff, F., 1974. On the determination of urea in seawater and some preliminary data from the Baltic. 9th Conf. of the Baltic Oceanographers, Kiel, 17—20 April, 1974. Paper 25a. 1 p. (mimeogr.).

Krauskopf, K.B., 1956. Factors controlling the concentrations of thirteen rare metals in sea water. Geochim. Cosmochim. Acta, 10(1): 1—32.

Kremling, K., 1969. Untersuchungen über die chemische Zusammensetzung des Meerwassers aus der Ostsee vom Frühjahr 1966. Kieler Meeresforsch., 25(1): 81—104.

Kremling, K., 1970. Untersuchungen über die chemische Zusammensetzung des Meerwassers der Ostsee II. Frühjahr 1967—Frühjahr 1968. Kieler Meeresforsch., 26(1): 1—20.

Kremling, K., 1973. Voltammetrische Messungen über die Verteilung von Zink, Cadmium, Blei und Kupfer in der Ostsee. Kieler Meeresforsch., 29(2): 77—84.

Kremling, K. and Petersen, H., 1978. The distribution of Mn, Fe, Zn, Cd and Cu in Baltic seawater; a study on the basis of one anchor station. Mar. Chem., 6:155—170.

Kullenberg, B. and Sen Gupta, R., 1973. Fluoride in the Baltic. Geochim. Cosmochim. Acta, 37: 1327—1337.

Kwiecinski, B., 1965. The sulphate content of Baltic water and its relation to the chlorinity. Deep-Sea Res., 12: 797—804.

Lowenstam, H.A. and McConnell, D., 1968. Biologic precipitation of fluorite. Science, 162: 1496—1498.

Lundstedt, K., 1970. Jämförelse mellan permanganattal och halt av organiskt kol i några olika ytvatten. Vatten, 26(2): 126—134.

McCarthy, J.J., 1972. The uptake of urea by natural populations in marine phytoplankton. Limnol. Oceanogr., 17(5): 738—748.

Manheim, F.T., 1961. A geochemical profile in the Baltic Sea. Geochim. Cosmochim. Acta, 25: 52—70.

Manheim, F.T., 1970. The diffusion of ions in unconsolidated sediments. Earth Planet. Sci. Lett., 9: 307—309.

Matthäus, W., 1973. Zur Hydrographie der Gotlandsee II. Der mittlere Jahresgang der Temperatur in Oberflächennähe. Beitr. Meeresk., 32: 105—114.

Matthäus, W., 1977. General trends in the development of the oxygen regime in the deep water of the Baltic. ICES. C.M./1977/C:16. 7 pp., figs. (mimeogr.).

Morozov, N.P., Demina, L.L., Sokolova, L.M. and Prohoryčeva, N.P., 1974. Transient and heavy metals occurring in the water and hydrobionts of the Baltic basin. Ecological aspects of chemical and radioactive pollution of aquatic medium 100: 32—36 (in Russian, with an English summary).

Morris, A.W. and Riley, J.P., 1966. The bromide/chlorinity and sulphate/chlorinity ratio in sea water. Deep-Sea Res., 13: 699—705.

Mortimer, C.H., 1942. The exchange of dissolved substances between mud and water in lakes. J. Ecol., 30(1): 147—201.

Nehring, D. and Francke E., 1971. Hydrographisch-chemische Veränderungen in der Ostsee seit Beginn dieses Jahrhunderts und während des Internationalen Ostseejahres 1969/70. Fisch.-Forsch., 9(1): 35—42.

Nehring, D. and Rohde, K.-H., 1966. Weitere Untersuchungen über anomale Ionenverhältnisse in der Ostsee. Beitr. Meeresk., 20: 10—33.

Niemi, Å., 1975. Ecology of phytoplankton in the Tvärminne area, SW coast of Finland. II. Primary production and environmental conditions in the archipelago and the sea zone. Acta Bot. Fenn., 105: 1—68.

Niemistö, L. and Tervo, V., 1978. Preliminary results of heavy metal contents in some sediment cores in the northern Baltic Sea. Proc. XI Conf. Baltic Oceanogr., Rostock, 24—27 April, 1978, Vol. 2: 653—672 (mimeogr.).

Niemistö, L. and Voipio, A., 1974. Studies on the recent sediments in the Gotland Deep. Merentutkimuslaitoksen Julk. Havsforskningsinst. Skr., 238: 17—32.

Nyquist, G., 1976. Bestämning av humus och ligninsulfonaten i Östersjön och Bottniska viken genom fluorescens-mätningar. Sven. Havsforskningsfören. Årsb. 1976, 7 pp., figures.

Odén, S., 1922. Die Huminsäuren II. Steinkopff, Dresden, Aufl. 33.

Olausson, E., Gustafsson, O., Mellin, T. and Svensson, R., 1977. The current level of heavy metal pollution and eutrophication in the Baltic Proper. Medd. Maringeol. Lab., 9: 1—28, appendices.

Papunen, H., 1968. On the sulfides in the sediments of the Bothnian Sea. Bull. Geol. Soc. Finl., 40: 51—57.

Pęcherzewski, K., 1974. Zawartość i rozmieszczenie C_{org} w powierzchniowej warstwie osadów dennych południowego Bałtyku (The content and distribution of organic C in the superficial layer of bottom sediments in southern Baltic). Zeszyty Naukowe Wydziału Biologii i nauk o ziemi. Oceanografia, 2: 5—21 (in Polish, with English summary).

Pęcherzewski, K. and Lawacz, W., 1975. Wstepne wyniki badan nad iloscia C_{org} rozpuszczone go i czasteczkowego w wodach południowego Bałtyku (Preliminary results of investigations on the quantity of dissolved and particulate C_{org} in the waters of the South Baltic). Zeszyty Naukowe Wydziału Biologii i Nauk o Ziemi. Oceanografia, 4: 25—43 (in Polish, with English summary).

Pettersson, O., 1894. Redogörelse för de svenska hydrografiska undersökningarne åren 1893—94 under ledning af G. Ekman, O. Pettersson och A. Wijkander. I. Östersjön. Bihang till K. Svenska Vet.-Akad. Handlingar 19 II(4): 1—14.

Pustelnikov, O.S., 1976. Organic matter in suspension and its supply to the bottom of the Baltic Sea. Oceanology 15(6): 673—675.
Pustelnikov, O.S., 1977. The balance of sediments and recent sedimentation rates in the Baltic Sea (according to the data of suspension studying). Baltica, 6: 160—172.
Rankama, K. and Sahama, T.G., 1955. Silicon. In: K. Rankama and T.G. Sahama (Editors), Geochemistry. Chicago, pp. 551—556.
Redfield, A.C., Ketchum, B.H. and Richards, F.A., 1963. The influence of organisms on the composition of sea water. In: M.N. Hill (Editor), The Sea, 2. Wiley, New York, London, pp. 26—77.
Rinne, I., 1977. Nitrogen fixation by blue-green algae in the Baltic Sea. Fifth Symposium of the Baltic Marine Biologists, Kiel, August 29—September 4, 1977, 14 pp. (mimeogr.).
Rinne, I., Melvasalo, T., Niemi, Å. and Niemistö, L., 1976. Information on Finnish research on nitrogen fixation by blue-green algae in the Baltic Sea. (Preliminary report.) International Symposium on the Environmental Role of Nitrogen Fixing Blue-Green Algae and Asymbiotic Bacteria. Uppsala, Sweden, September 20—24. 5 pp. (mimeogr.).
Rohde, K.H., 1966. Untersuchungen über die Calcium- und Magnesiumanomalie in der Ostsee. Beitr. Meeresk., 19: 18—31.
Rohde, K.H., 1967. Untersuchungen über die Calcium-Chlor- und Magnesiumanomalie in der Ostsee. Beitr. Meeresk., 20: 34—42.
Rubey, W.W., 1951. Geologic history of sea water. Bull. Geol. Soc. Am., 62: 1111—1147.
Ruppin, E., 1912. Beitrag zur Hydrographie der Belt- und Ostsee. Kiel. Wiss. Meeresuntersuch. N.F., 14: 205—272.
Salo, A. and Saxén, R., 1974. On the role of humic substances in the transport of radionuclides. Institute of Radiation Physics. Helsinki, Rep., SFL-A20: 1—31.
Savidge, G. and Hutley, H.T., 1977. Rates of remineralization and assimilation of urea by fractionated plankton population in coastal waters. J. Exp. Mar. Biol. Ecol., 28: 1—16.
Schmidt, D., 1976. Determination of cadmium, copper and iron in sea water of the western Baltic. ICES. C.M. 1976/C:9. 6 pp., appendices (mimeogr.).
Sen Gupta, R., 1972. On some trace metals in the Baltic. Ambio, 1(6): 226—230.
Sen Gupta, R., 1973. A study on nitrogen and phosphorus and their interrelationships in the Baltic. Univ. Gothenburg, Inst. Oceanogr., 82 pp., appendices (mimeogr.).
Sen Gupta, R. and Koroleff, F., 1973. A quantitative study of nutrient fractions and a stoichiometric model of the Baltic. Estuarine Coastal Mar. Sci., 1: 335—360.
Seppänen, H. and Voipio, A., 1971. Some bacteriological observations made in the northern Baltic. Merentutkimuslaitoksen Julk. Havsforskningsinst. Skr., 233: 43—48.
Sillén, L. G., 1961. The physical chemistry of sea water. In: M. Sears (Editor), Ocenography. AAAS, Washington, D.C., pp. 549—581.
Skopintsev, B.A., 1959. Organic matter of sea water. Reprints International Oceanogr. Congr. American Association for the Advancement of Science, pp. 953—954.
Soskin, I.M., 1963. Mnogoletnie izmenenija gidrologičeskih harakteristik Baltijskogo morja. Leningrad, 159 pp.
Suess, E. and Erlenkeuser, H., 1975. History of metal pollution and carbon input in Baltic Sea sediments. Meyniana, 27: 63—75.
Szekielda, K.-H., 1968. Vergleichende Untersuchungen über den Gehalt an organischem Kohlenstoff im Meerwasser und dem Kaliumpermanganatverbrauch. J. Cons. Int. Explor. Mer, 32(1): 17—24.
Szekielda, K.-H., 1971. Organisch gelöster und partikulärer Kohlenstoff in einem Nebenmeer mit starken Salzgehaltsschwankungen (Ostsee). Vie Milieu, Suppl., 22: 579—612.
Tarkiainen, E., Rinne, I. and Niemistö, L., 1974. On the chemical factors regulating the primary production of phytoplankton in the Baltic Proper. Merentutkimuslaitoksen Julk. Havsforskningsinst. Skr., 238: 39—52.
Thompson, T.G., Johnston, W.R. and Wirth, H.E., 1931. The sulfate-chlorinity ratio in ocean waters. J. Cons. Int. Explor. Mer, 6: 246—251.

Trzosińska, A., 1967. Metoda Knudsena-Sörensena w zastosowaniu do badania zasolenia wody południowego Bałtyku (Knudsen-Sörensen's method applied to research on water salinity of southern Baltic Sea). Przegl. Geofiz., 12(3—4): 367—381 (in Polish, with an English summary).

Trzosińska, A., 1968. Distribution of the calcium and magnesium ions content in the Baltic water. Proc. VI Conf. Baltic Oceanographers Sopot, 6—8 June, 1968.

UNESCO, 1979. Ninth report of the joint panel on oceanographic tables and standards. Unesco, Paris, September 11—13, 1978. UNESCO Tech. Pap. Mar. Sci., 30: 1—32.

Varencov, I.M. and Blažčišin, A.I., 1976. Železo-margancevye iskopaemye. Iron and manganese concretions. In: V.K. Gudelis and E.M. Emeljanov (Editors), Geologija Baltijskogo Morja (Geology of the Baltic Sea). Mokslas, Vilnius, pp. 307—348.

Voipio, A., 1957. On the magnesium content in the Baltic. Suom. Kemistil. B, 30: 84—88.

Voipio, A. 1961. The silicate in the Baltic Sea. Ann. Acad. Sci. Fenn. A II, 106: 1—15.

Voipio, A., 1969. On the cycle and balance of phosphorus in the Baltic Sea. Suom. Kemistil. A, 42: 48—54.

Voipio, A., 1973. Typen ja fosforin kokonaispitoisuudet meriympäristömme näytteissä (The level of total phosphorus and nitrogen in environmental samples from the northern Baltic Sea). Limnologisymposion 1969, Helsinki, pp. 59—68 (in Finnish, with an English summary).

Voipio, A., 1977. Itämeri eilen, tänään, huomenna (The Baltic Sea — yesterday, today and tomorrow). Kem.-Kemi, 4(1—2): 5—9 (in Finnish, with an English summary).

Von Bonsdorff, P.A., 1837. Vorläufige Bemerkungen über die Bestandtheile des Meerwassers, besonders in Rücksicht auf den Gypsgehalt desselben. Ann. Phys. Chemie, 40: 133—137. Herausgegeben von J.C. Poggendorff, Leipzig.

Von Brandt, A., 1936. Zum Kalkgehalt der Ostsee. Ergebnisse der ozeanographischen Untersuchungen des Fischeri-Instituts der Universität Königsberg Pr. im Jahre 1935—36 in der mittleren Ostsee. I. J. Cons. Int. Explor. Mer, 11(1): 314—342.

Weigel, H.-P. and Kremling, K., 1975. A new method for the determination of trace metals in seston by flameless atomic absorption spectrometry. ICES. C.M. 1975/E:39. 14 pp. (mimeogr.).

Winterhalter, B., 1980. Ferromanganese concretions in the Baltic Sea. In: I.M. Varencov (Editor), International Monograph on the Geology and Geochemistry of Manganese. Hungarian Academy of Sciences, Budapest, 3: 227—254.

Wittig, H., 1940. Über die Verteilung des Kalziums und der Alkalinität in der Ostsee. Meereskundl. Arbeiten Univ. Kiel, 70.

Witting, R., 1912. Zusammenfassende Übersicht der Hydrographie des Bottnischen und Finnischen Meerbusens und der nördlichen Ostsee nach den Untersuchungen bis Ende 1910. Finl. Hydrogr.-Biol. Unters., 7: 1—82, tables.

Wooster, W.S., Lee, A.J. and Dietrich, G., 1969. Redefinition of salinity. Deep-Sea Res., 16: 321—322.

Wyrtki, K., 1954. Der grosse Salzeinbruch in der Ostsee im November und Dezember 1951. Kieler Meeresforsch., 10(1): 19—25.

Zarinš, E. and Ozolinš, J., 1935. Untersuchungen über die Zusammensetzung des Meerwassers im Rigaschen Meerbusen und an der Lettländischen Küste des Baltischen Meeres. J. Cons. Int. Explor. Mer, 10(3): 275—301.

Zsolnay, A., 1975. Total labile carbon in the euphotic zone of the Baltic Sea as measured by BOD. Mar. Biol., 29: 125—128.

Chapter 5

BIOLOGICAL OCEANOGRAPHY

GUY HÄLLFORS, ÅKE NIEMI, HANS ACKEFORS, JULIUS LASSIG and ERKKI LEPPÄKOSKI

A. INTRODUCTION*

The location of the Baltic Sea in the northern high latitudes affects the structure and function of the Baltic ecosystem. There are also large regional differences in surface salinity (about 0.5—24 ‰) and chemical properties (see Chapter 4). However, the largest part of the Baltic Sea surface water belongs to the β-mesohaline salinity range (5—10 ‰), with the 10 ‰-surface isohaline in the Belt Sea area and the 5 ‰-surface isohaline in the Quark of the Gulf of Bothnia and in the inner part of the Gulf of Finland (Fig. 3.1). The Baltic deep water below the permanent halocline with a salinity exceeding 10 ‰ reaches the Gulf of Finland, but there is no permanent salinity stratification in the Gulf of Bothnia. The salinity conditions in the Baltic Sea are very stable. No marked tide-induced currents occur in the area.

From a geological point of view (see Chapter 1) the Baltic Sea is a young sea. During different marine and fresh-water stages there has been a succession of ecosystems since the last glaciation. Mainly limnic and marine organisms with a wide osmotic tolerance have been able to survive. The number of species is small. Few genuine brackish-water species have evolved in the Baltic Sea as compared to the geologically old, brackish Pontic—Caspian Sea area (Zenkevitch, 1963).

The productive system of the Baltic Sea consists of the trophogenic layers of the pelagial system and the littoral system, the latter of which is also of great importance in the northern part of the area because of the long and relatively shallow coast. The location of the Baltic Sea in high latitudes causes a pronounced maximum of light in summer. A difference in latitude of 10° gives rise to essential differences in the length of the growing period which is about 9—10 months in the south (Edler, 1977) but only 4—5 months in the north (Alasaarela, 1979). The duration of the ice cover is of decisive importance for the seasonal development of the vegetation and ecosystem functions and exhibits large year-to-year variations. In general, ice covers the northern part of the Baltic Sea for about 4—5 months, but sea ice occurs sometimes also in the southern and western parts of the Baltic Sea

* By Guy Hällfors and Åke Niemi.

(Palosuo, 1966). The nutrient level is another essential factor regulating production. The nutrient input to the productive systems comes from three sources: (1) natural outflow from land; (2) the waste-water load; and (3) the upwelling of phosphorus-rich Baltic deep water. The last source is absent in the Gulf of Bothnia, which, especially the Bothnian Bay, has markedly lower nutrient levels than the Baltic Proper (see Chapter 3).

The main consuming system, heterotrophic pelagic and benthic organisms, is dependent on organic matter sinking from the trophogenic layer. The transport of organic matter from the phytal zone out to the deep basins of the open sea must also be considered an essential input to the benthic system. The Baltic Sea is a typical detritus-based system (Jansson, 1972). The further biogeochemical circulation causes mobilized nutrients to return to the trophogenic pelagic and littoral system in connection with upwelling of deep water in coastal areas.

The Baltic Proper and the Gulf of Finland have a semistagnant character, with periodic depletion of oxygen below the permanent halocline, which kills the zoobenthos over large areas. This phenomenon does not occur in the Gulf of Bothnia (Andersin et al., 1978a). Because of apparent differences in hydrographic conditions (see Chapter 3) and in oxygen and nutrient levels (see Chapter 4), it is obvious that there should be marked regional differences in the seasonal succession and production of planktonic and benthic communities in the different parts of the Baltic Sea.

B. VEGETATION AND PRIMARY PRODUCTION*

Phytoplankton, general

The study of Baltic Sea phytoplankton was started already at the end of the 19th century by Hensen (1887). In the beginning attention was mainly devoted to the taxonomy, floristics and ecology of the common species. Already at the beginning of this century Lohmann (1908) tried to estimate the total phytoplankton biomass. Early records of cell and colony numbers were based on net or centrifuge samples. The results are thus subject to large errors and are of little value for quantitative production estimates.

The Utermöhl method (Utermöhl, 1958) made it possible to obtain cell counts with a much greater accuracy. By converting the cell counts into cell volume or biomass (e.g., Melvasalo et al., 1973; Smetacek, 1975) a more meaningful measure was obtained. By calculating the plasma volume, values are obtained which may be converted into carbon or energy equivalents (see Smetacek, 1975; Edler, 1977) for use in production studies. It is, however, difficult to measure the plasma volume of many species correctly due to the

* By Guy Hällfors and Åke Niemi.

presence of vacuoles or pusules or the complex and often variable shape of cells.

None of these methods gives a reliable picture of the actual primary production, as the relationship between biomass and production, and the assimilation number tend to vary with the external conditions (e.g., temperature, light, nutrients) and the physiological state of the organisms. The chlorophyll content is a much used parameter being relatively fast and simple to determine, but it will not always give a good measure of phytoplankton biomass.

Buch (1948, 1954) calculated the rate of carbon assimilation from changes in the orthophosphate, carbon dioxide and oxygen content of the productive layer during the vernal bloom in the coastal waters of the northeastern Baltic Proper, obtaining reasonable values (cf., Bagge and Niemi, 1971).

Only the introduction of the ^{14}C-method by Steemann Nielsen (1952) made it possible to estimate the primary production relatively accurately and fast at any time and during a variety of external conditions. The first ^{14}C-measurements in the Baltic Sea were made from Swedish light ships by Michanek (1970). Nowadays ^{14}C-fixation is measured all over the Baltic Sea, both in coastal areas and from research vessels (e.g., Lehmusluoto, 1968; Hübel, 1968; Bagge and Lehmusluoto, 1971; Bagge and Niemi, 1971; Sen Gupta, 1972; Hobro and Nyqvist, 1972; Renk, 1973; Renk et al., 1975; Ackefors and Lindahl, 1975a, b; Kaiser and Schulz, 1975; Niemi, 1975; Meskus, 1976; Lindahl, 1977a, b; Edler, 1977; Wulff, 1977; Lassig et al., 1978; Alasaarela, 1979; Ilus, 1979). The heterogeneous material is compiled in Fig. 5.1 and Table 5.I, to give some indication of the level of primary production in different parts of the Baltic Sea. Unfortunately, the values obtained by different laboratories are not exactly comparable as various modifications of the ^{14}C-method are being used. Recommendations for a standardization of the method for measuring primary production in the Baltic Sea have been worked out by the Baltic Marine Biologists organization (Dybern et al., 1976), but so far they have not been unanimously adopted.

Phytoplankton production, succession and regulating factors

Several papers have been published on phytoplankton of different parts of the Baltic Sea (cf., compilation of literature in Niemi, 1973 and 1975), though there are only a few general surveys (cf., Pankow, 1976). In recent years, a few new species have been described (cf., Hällfors and Niemi, 1974; Thomsen, 1977, 1979; Hasle, 1978). Much remains to be done on the taxonomy and ecology of Baltic phytoplankton (cf., Hällfors, 1979). Though apparently being of relatively great importance, the nanoplankton fraction is excluded from this discussion as its qualitative and quantitative composition is very incompletely known. Observations for instance in the Gulf of Finland

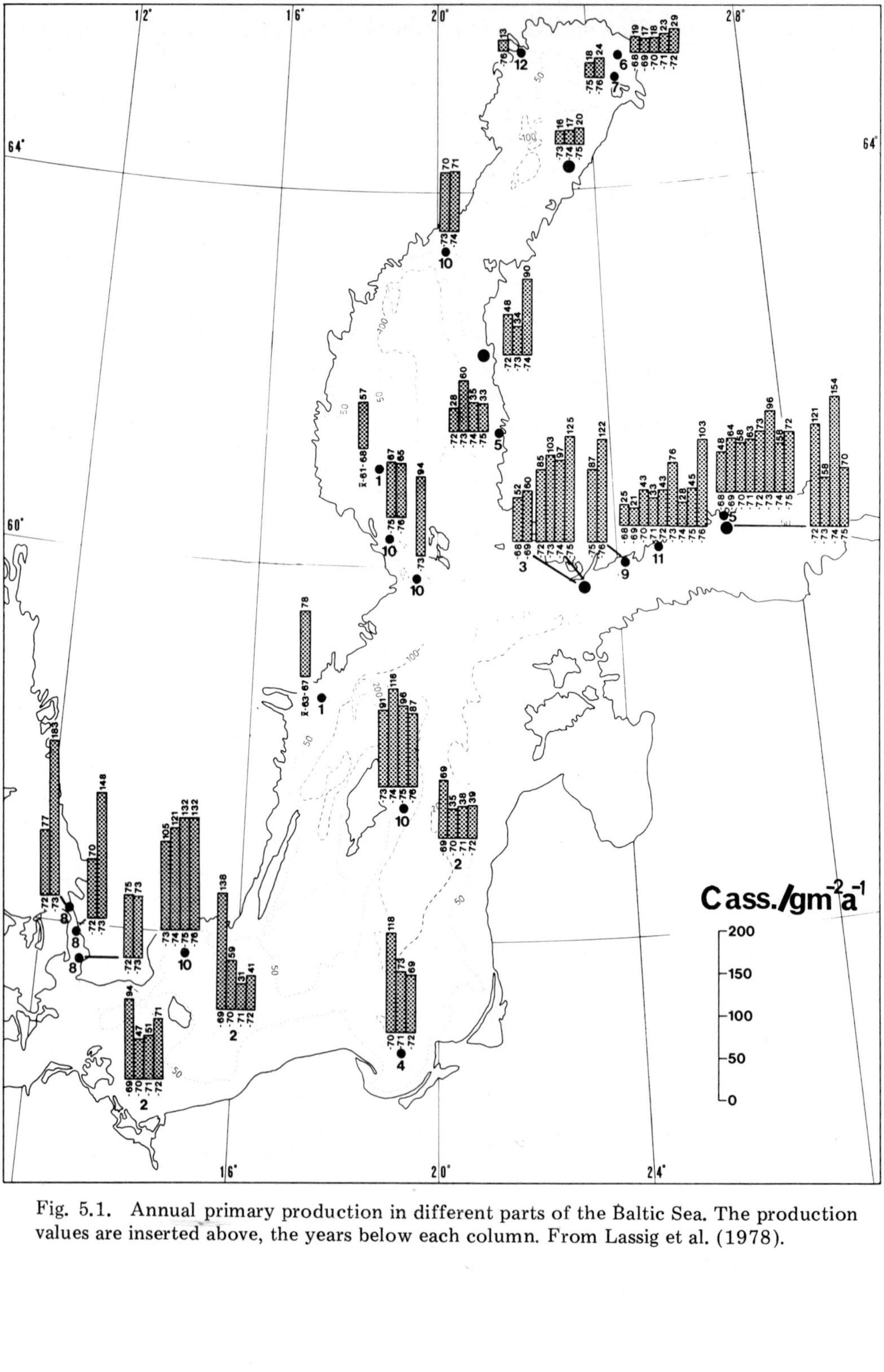

Fig. 5.1. Annual primary production in different parts of the Baltic Sea. The production values are inserted above, the years below each column. From Lassig et al. (1978).

TABLE 5.I

Mean daily production of assimilated carbon (mg m^{-2} d^{-1}) during the summer months in different parts of the Baltic Sea, from Lassig et al. (1978)

The Sound	Baltic Proper		Gulf of Finland		Bothnian Sea		Bothnian Bay	
420—520	Hanö	640	Tvärminne	390	Finngrundet	270	Ulkokalla	130
	Gdańsk Deep	400	Kopparnäs	680	Eurajoki	220	Hailuoto	76
	Gotland Sea	510	Lovisa archipelago	310	Kaskinen	280	Ulkokrunni	140
	Hävringe	340	Orrengrund	380	Quark	430		

by Niemi (1975) show the great importance of small flagellates in the primary production. Preliminary observations by G. Hällfors (unpubl.) in the same area indicate that species mainly of Cryptophyceae, Prymnesiophyceae, Chrysophyceae and Prasinophyceae dominate the centrifuge plankton in numbers of cells for much of the year.

In the central and northern parts of the Baltic Proper, in the Gulf of Finland and the Gulf of Bothnia, where the surface salinity ranges from 5‰ to 7‰, phytoplankton production is negligible in winter, because a well-developed ice-cover overlain by snow prevents enough light from reaching the water. When there is no or very little snow on the ice, diatoms have been observed growing on the underside of the ice, giving rise to a yellow-brown ice-layer (Niemi, 1973). The same phenomenon has been observed in the Belt Sea by Hickel (1969) and is also familiar in Arctic and Antarctic waters (Horner, 1976). Occasionally a bloom of the cold-stenothermal *Chrysochromulina birgeri* has been observed in, above and immediately below a transparent ice-cover in the southern archipelago of Finland (Hällfors and Niemi, 1974).

The central part of the northern Baltic Proper is often not frozen at all. During the winter phosphate, nitrate and silicate are present in far above limiting concentrations for production. But insufficient light intensity and an effective mixing below the critical depth prevents a phytoplankton net production (cf., Kaiser and Schulz, 1975; Niemi, 1975), although a small phytoplankton biomass consisting especially of *Chaetoceros danicus* and *Aphanizomenon flos-aquae* is present (Edler, 1979).

When enough light reaches the surface water, the vernal phytoplankton development will commence, being chiefly caused by marine 'cold-water' diatoms (*Achnanthes taeniata, Chaetoceros holsaticus, C. wighami, Melosira arctica, Nitzschia (Fragilariopsis) cylindrus, N. frigida, Skeletonema costatum, Thalassiosira baltica, T. lacustris*) and dinoflagellates, especially *Gonyaulax catenata* (see, e.g., Niemi, 1973, 1975; Lindahl, 1977a, b; Edler, 1979). In coastal areas the fresh-water diatom *Diatoma elongatum* is dominant during the late stage of the bloom.

The vernal phytoplankton bloom in the southern and central parts of the Baltic Proper generally occurs in the second half of April; in the northern part of the Baltic Proper and in the Gulf of Finland in early May, somewhat later in the Bothnian Sea, but not until June in the northern Bothnian Bay. Owing to meteorological fluctuations, however, year-to-year variations are great. The highest biomass values are commonly reached when a weak thermocline has developed, giving rise to a surface layer which in general coincides with the euphotic layer.

The course of the vernal bloom is regulated by the availability of light and the development of a thermal stratification which prevents the plankton from mixing (circulating) below the critical depth (Sverdrup, 1953; see also Kaiser and Schulz, 1975; Niemi, 1975). This bloom consumes all the measurable nitrate and phosphate from the mixed layer. Observations at the entrance to the Gulf of Finland (Niemi, 1975) and off Landsort in the northern part of the Baltic Proper (Hobro et al., 1975) show that the bloom starts to decline when the nitrate is used up, although there is still some phosphate left. The phytoplankton uses inorganic nitrogen and phosphorus in the weight ratio 7—8:1 (Cooper, 1937; Parsons et al., 1977). In the surface water of the northern Baltic Proper and the Gulf of Finland in late winter this ratio is about 4:1, which seems to indicate a deficit of inorganic nitrogen (cf., Section on p. 194).

With few grazers present during the vernal stage, a large portion of the produced organic matter sinks out of the productive layer to the consuming part of the pelagial and to the benthos. About 1/3 of the vernal production of organic matter reaches the bottom at a depth of 45 m in the northern part of the Baltic Proper (Jansson, 1978) constituting the main energy input to the benthic system.

When thermal stratification has developed and a low production stage has commenced after the vernal bloom, there are no or only traces of phosphate and nitrate (and nitrite) left in the euphotic layer (see Fig. 5.2). A few mg m^{-3} of ammonia are always measured. The euphotic layer mostly coincides with the mixed layer. The concentration of silicate-silicon is at its annual minimum (~ 250—150 mg m^{-3}; see Voipio, 1961; Niemi, 1975), but probably will never be a limiting factor for phytoplankton production. The concentration of total phosphorus has decreased below 10 mg m^{-3} (e.g., Niemi, 1975). At this stage phytoplankton production is mainly sustained by the recycling of ammonia and phosphate, while the effectivity of recycling is probably regulated by temperature, by the excretion by the increased stock of zooplankton, and by bacterial metabolism.

The low production stage which in the northern Baltic Proper, the Gulf of Finland and the Bothnian Sea falls on June and early July is characterized by a small biomass of phytoplankton which now is also regulated by the zooplankton. The chief primary producers consist of small flagellates with a high turn over rate. The euphotic layer is deep owing to good transparency.

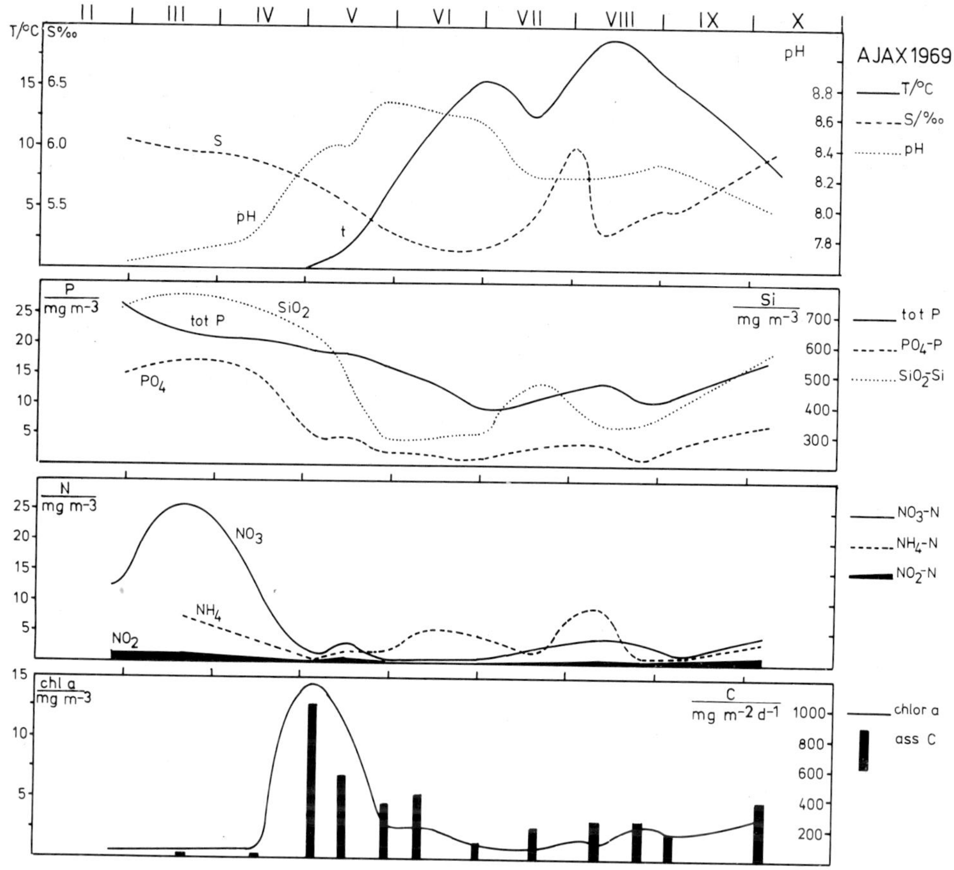

Fig. 5.2. Seasonal fluctuations of phytoplankton primary production, chlorophyll-a, nutrients and hydrographic properties in 1969 at Ajax at the entrance to the Gulf of Finland (mean values of the euphotic layer, from Niemi, 1975).

Probably very little organic matter produced during this stage sinks to the bottom. The phytoplankton community consists of different groups. In addition to diatoms (*Chaetoceros wighami, C. danicus*), dinoflagellates *Protoperidinium brevipes, Gonyaulax triacantha, Dinophysis acuminata, D. norvegica, Prorocentrum balticum*) and blue-green algae (*Gomphosphaeria lacustris* var. *lacustris* and var. *compacta*) there is a great abundance of small species of Cryptophyceae, Chrysophyceae (*Dinobryon balticum, D. petiolatum*), Euglenophyceae (*Eutreptia, Eutreptiella*), Prasinophyceae (*Pyramimonas*), Chlorophyceae (*Scenedesmus, Oocystis*) (Niemi, 1973, 1975), and Prymnesiophyceae (*Chrysochromulina* spp.) (S. Hällfors and G. Hällfors, unpubl.). In coastal areas of the Gulf of Finland and of the Gulf of Bothnia occasional blooms of the chrysophycean *Uroglena americana* and the dinoflagellate *Heterocapsa triquetra* have been observed (G. Hällfors and Å. Niemi, unpubl.).

Later in summer from July onward this low production stage is terminated by a nitrogen-fixing community of blue-green algae dominated by *Aphanizomenon flos-aquae*, *Nodularia spumigena* and *Anabaena lemmermannii*. From July on until early autumn, these species may give rise to blooms, sometimes even as late as in October (Niemi, 1975). In the central and southern Baltic Proper especially *Nodularia spumigena* develops sizeable blooms (Horstmann, 1975; Öström, 1976).

Blooms of blue-green algae are usually preceded by calm and stable weather conditions. The algae characteristically accumulate at the surface or in the uppermost metres of water. Mass occurrences of blue-green algae are probably facilitated by upwelling of phosphate-rich deep water. However, before and during a bloom there is practically no phosphate and nitrate in the euphotic layer (Buch, 1948). In spite of this nutrient-deficient condition, blooms are possible because of the following capacities of the participating species:

(a) They are able to store considerable amounts of phosphorus in their cells to be used for growth when the phosphorus in the environment is depleted (Stewart and Alexander, 1971; Whitton, 1973).

(b) Possessing gas vacuoles, they tend to rise to the surface during periods of intense growth.

(c) They are able to fix molecular nitrogen, and are thus not dependent on mineral nitrogen in the environment.

The nitrogen fixed is released to the surface water as different compounds, partly already during the bloom (cf., Jones and Stewart, 1969), partly when the algae finally decompose. Moreover, some phosphorus is probably brought to the surface layer with blue-green algae rising from below the thermocline. Thus the blooms would tend to increase the nutrient content of the surface water (Öström, 1976). The details of these events are still little known, as is the whole nitrogen cycle of the Baltic Sea. Preliminary nitrogen fixation studies in the open Baltic Sea have been carried out by Rinne et al. (1978), and in coastal waters by Hübel and Hübel (1974), Brattberg (1977) and Lindahl et al. (1978).

The growth of these bloom-forming blue-green algae is enhanced by the addition of phosphate (cf. Melin and Lindahl, 1973; Tarkiainen et al., 1974; Horstmann, 1975; Rinne and Tarkiainen, 1978). Unusually great production of blue-green algae in the Gulf of Finland in October 1969 was probably caused by frequent upwelling of deep water (Niemi, 1975, 1979). Apparently such a fertilization of the surface layer will increase the biological production also at higher trophic levels of the ecosystem.

In late summer and early autumn cryptomonads and large centric diatoms (*Chaetoceros danicus*, *Actinocyclus octonarius*, *Coscinodiscus granii*) become more abundant. However, marked autumnal diatom blooms are rare. In October the surface water already contains enough phosphate and nitrate for a vigorous growth of diatoms. The autumnal turn-over, however, mixes the water column below the critical depth for net production, namely, to the permanent halocline (in the Bothnian Sea to the bottom), thus preventing the mass development of diatoms except during exceptionally favourable meteorological conditions (Niemi and Ray, 1977).

From November on, phytoplankton production in the open sea is negligible due to poor (limiting) light conditions and effective mixing. The nutrient concentrations gradually increase until the start of the next vernal bloom.

In the Bothnian Bay the phytoplankton maximum does not develop until June and consists chiefly of brackish-water diatoms (*Chaetoceros wighami*, *Melosira arctica*, *Skeletonema costatum*, *Thalassiosira baltica*) and fresh-water diatoms (*Diatoma elongatum*, *D. vulgare*) and dinoflagellates (*Gonyaulax catenata*). The limnic element is important in the Bothnian Bay phytoplankton. Later in summer plankton production is dependent on hydrographic factors. The production is low already in October (Meskus, 1976; Alasaarela, 1979). Dense blooms of nitrogen-fixing blue-green algae have not been observed in the open sea. This is probably caused by the excess of nitrate-nitrogen relative to phosphate as reported by Pietikäinen et al. (1978) and Alasaarela (1979).

The yearly primary production of carbon in the Bothnian Bay is relatively small, about 20 g m^{-2} (see Fig. 5.1), due to low concentrations of nutrients, especially of phosphorus, as pointed out already by Buch (1932). In the Gulf of Bothnia, especially in the Bothnian Bay, there is a deficit of phosphate compared to nitrate (Voipio, 1976; Dahlin, 1978; Pietikäinen et al., 1978; Alasaarela, 1979). The N:P ratios indicate that shortage of phosphate in the Bothnian Bay area has a more restricting influence on the phytoplankton production than in the Baltic Proper (Voipio, 1976). The short growing period and the displacement of the vernal phytoplankton maximum toward midsummer is a feature typical of Arctic conditions (Lassig and Niemi, 1975).

The southern part of the Baltic Proper from Gotland to the sill of Darss with a surface water salinity of about 7—8‰, has a phytoplankton element dominated by the same species as in the sea areas north of Gotland. A vernal bloom is caused by *Achnanthes taeniata*, *Chaetoceros wighami*, *C. holsaticus*, *Melosira arctica*, *Skeletonema costatum*, *Thalassiosira baltica* (see, e.g., Rothe, 1941; Edler, 1975). The early summer is among other species characterized by *Dinobryon balticum* and later by *Chaetoceros danicus* and by the all-over occurring blue-green algae *Aphanizomenon flos-aquae* and *Nodularia spumigena*, while the diatoms *Actinocyclus octonarius* and *Coscinodiscus granii* become more important in late summer. The true marine element is more conspicuous in late summer and autumn with dinoflagellates of the genus *Ceratium* (*C. tripos* v. *subsalsa; C. fusus*) *Dinophysis acuta*, *Prorocentrum micans* and diatoms of the genus *Coscinodiscus* (Ringer, 1973).

In the southern part of the Baltic Sea the phytoplankton has its vernal maximum already in April (Nikolaev, 1957; Lindahl, 1977a, b). In these areas the seasonal succession typical of the northern Baltic Proper, viz., vernal maximum, summer minimum and a late summer—autumn blue-green algal maximum, is not so clear. Several peaks of diatoms, dinoflagellates, blue-green algae or nanoplanktonic species may occur.

The Darss sill constitutes an inward border for the marine phytoplankton element (Kell, 1973). In years with strong inflow of saline water, the marine phytoplankton element is more abundant in the whole southern part of the Baltic Sea (Ringer, 1973).

In the Baltic Sea west of the sill of Darss the number of α-mesohaline and polyhaline species increases drastically. Important genera include *Chaetoceros*, *Coscinodiscus*, *Rhizosolenia*, *Protoperidinium* and *Ceratium*. Several species are brought with inflowing Kattegat water into the Belt Sea, but are not able any more to build up new populations in this area due to too low salinity.

In the Arkona Basin and in the Belt Sea the phytoplankton development shows a more complicated pattern than in the central and northern parts of the Baltic Sea. Hydrographical and meteorological factors probably affect the seasonal development of phytoplankton more strongly. In the Belt Sea up to nine different maxima have been registered (Smetacek, 1975). Also in the Sound the phytoplankton succession is very variable with several production peaks (Edler, 1977).

In March a vernal phytoplankton bloom commences, consisting mainly of *Chaetoceros* spp., *Skeletonema costatum*, and *Detonula confervacea. Rhizososolenia* spp., *Protoperidinium* spp., *Ceratium* spp. indicate the onset of the summer stage. The summer stage is characterized mainly by the dominance of blue-green algae, *Aphanizomenon flos-aquae* and *Nodularia spumigena* sometimes forming heavy blooms if the water temperature is at least 17—18° C. Besides, there is a succession of many additional species, e.g., *Dinobryon balticum* and *Protoperidinium pellucidum* at an early summer stage, followed by among others, species of the genera *Ceratium*, *Protoperidinium*, *Dinophysis*, *Rhizosolenia*, and *Chaetoceros* etc. The autumnal stage is characterized by *Ceratium* spp. and large centric diatoms, e.g., *Actinocyclus octonarius* and *Coscinodiscus granii*, i.e., species with a large volume/surface ratio.

In the Belt Sea the outflowing Baltic surface current of low salinity is characterized by species typical of the central Baltic Sea: *Aphanizomenon flos-aquae*, *Nodularia spumigena*, *Coelosphaerium kuetzingianum*, *Chaetoceros danicus*, *C. wighami* etc. When Kattegat water of high salinity has flowed into the Danish Sounds the phytoplankton shows more really marine features.

Phytoplankton in near-shore areas

In unpolluted coastal waters, for instance in the archipelagos (Niemi, 1972), the phytoplankton differs but little from that of the open sea. In estuaries and eutrophicated coastal waters, however, its succession and production are markedly different, being strongly influenced by the fresh-water discharge regime, especially in areas where river water is rich in nutrients. Where river outflow gives rise to a permanent salinity stratification markedly above the critical depth, phytoplankton production continues until late autumn (Niemi, 1975, 1978).

In eutrophicated areas around pollution centres of the northern parts of the Baltic Sea, the phytoplankton is characterized by the dominance of a dense *Oscillatoria agardhii* assemblage in summer and early autumn. The vernal diatom bloom is here not followed by a low production stage, but the *Oscillatoria agardhii* assemblage begins to develop as the vernal diatom bloom declines. The production of this plankton community is far higher than the vernal production (Melvasalo, 1971; Niemi, 1972). Thus the production maximum is displaced toward late summer. Even in the northern part of the Bothnian Bay the *Oscillatoria agardhii* assemblage has been found well developed in eutrophicated estuarine waters of the Oulujoki River (Alasaarela and Siira, 1976).

Along the southern coast of the Baltic Sea the semi-enclosed, isolated water bodies, the "Boddengewässer" or "Haffs" are characterized by more or less lowered salinity and of eutrophication. A phytoplankton assemblage (Greifswalder Bodden) suggesting such conditions was described as early as 1908 by Abshagen, and later many studies on similar waters have been carried out (Trahms, 1939; Hübel, 1968; Schnese, 1973).

The phytoplankton of the near-coast waters is, in general, a mixture of fresh-water, brackish water and marine elements. Water bodies of different degree of isolation have their characteristic phytoplankton assemblages according to, among others, salinity and eutrophication (Schnese, 1973).

Benthic vegetation, general aspects

The southern and southeastern coasts of the Baltic Proper are in general unsheltered. The bottom consists mainly of mobile sediments (sand and gravel, silt in sheltered areas) which provide little substrate for haptophytic plants. Hard substrates are limited to boulders and mussel shells and artificial substrates such as moles and pile-work. Northwards the complexity of the coastal morphology and available substrates gradually increase, reaching their maximum in the extensive archipelagos of the northern part of the Baltic Proper. Here bedrock and boulders predominate at least in exposed localities, while sediment bottoms mostly are confined to relatively sheltered areas and therefore usually are rather stable. In the Gulf of Bothnia shore complexity is again reduced, the bedrock is covered by glaciogenic sediments. Boulders and stones predominate on exposed shores and sand and silt on sheltered shores.

The number of macroscopic plant species declines with decreasing salinity rapidly eastwards and northwards in the southern part of the Baltic Proper due to the disappearance of marine species (see Fig. 5.3; see also Wærn, 1952, p. 8) which is but little compensated for by brackish-water and fresh-water species. Thus the number of physiognomically dominating algal species is reduced from 154 at the western coast of Sweden (Schwenke, 1974) to 31 in the southwestern archipelago of Finland (24 according to Schwenke, 1974). Only in the innermost parts of the gulfs and in estuaries does the species number actually grow again due to the increase of fresh-water elements (cf., Levring, 1940; Luther, 1951a; Wærn, 1952). In the Bothnian Bay marine species are of little importance.

Several marine algae exhibit a gradual reduction in size with decreasing salinity (pauperization; cf., Fig. 5.3). Along with size reduction there is in some cases a reduction in the reproductive cycle. Most red algae lose their ability to form sexual reproductive organs, and reproduce by asexual spores or vegetatively. For most brown and green algae conclusive data on the reproductive cycle of their Baltic Sea populations are lacking. The sexual process is rare or lost also in many fresh-water algae growing in brackish water (e.g., *Cladophora aegagropila*, *C. glomerata*, Oedogoniaceae, Zygnemaceae).

A number of algal species growing in the littoral zone on tidal shores are found only in the sublittoral in the Baltic Proper (e.g., *Fucus vesiculosus*, *F. serratus*, *Ahnfeltia plicata*). This submergence (see Fig. 5.3) is probably not a pure brackish-water phenomenon, but seems to be caused by several more or less interconnected factors. Thus Wærn (1949) prefers to use the more general term 'downward process'. The most conspicuous example is *Fucus vesiculosus*, which in the Baltic Proper is excluded from the littoral by desiccation during periods of long-lasting low water levels occurring mainly in spring, and by ice-scouring in winter. On shores exposed to pack-ice, the upper limit of *Fucus vesiculosus* may be at a depth of several metres. In estuarine areas freshwater outflow may also cause a downward shift of the upper limit of this species. The occurrence of *Fucus vesiculosus* in the sublittoral in the Baltic Sea has been interpreted as an evolution of deep-water ecotypes in the absence of competing species. On oceanic coasts such ecotypes seem to be absent (Wærn, 1952). However, according to Menge (1975, see Rasmus et al., 1977) the lower limit is not physiological, but is determined by biological interactions.

The seasonal variations especially in temperature and light intensity induce a seasonality in growth and reproduction of the benthic vegetation which is most explicitly expressed in the succession of annuals. The algal periodicity was outlined already by Svedelius (1901), and Levring (1940) provided additional information.

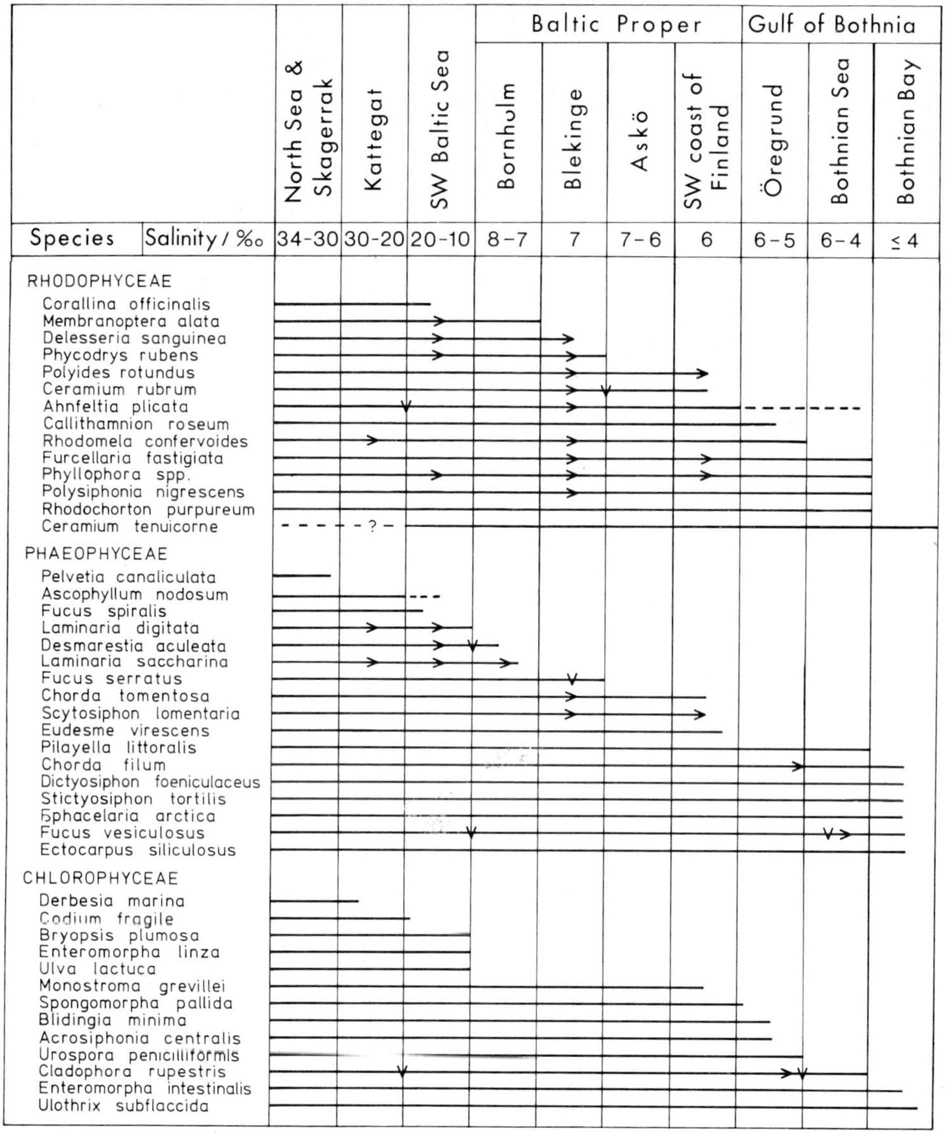

Fig. 5.3. Distribution of some representative marine algae in the Baltic Sea. > = pauperization, V = submergence. Compiled from various sources.

In the hydrolittoral and upper sublittoral of the northern part of the Baltic Proper the following seasonal groups may be discerned: (1) vernal species (*Acrosiphonia centralis*, *Monostroma grevillei*, *Ulothrix subflaccida*); (2) early summer species (*Chorda tomentosa*, *Dictyosiphon chordaria*, *Eudesme virescens*, *Scytosiphon lomentaria*); (3) high summer species (*Chor-*

da filum, *Cladophora glomerata*, *Dictyosiphon foeniculaceus*, *Ectocarpus siliculosus*, *Entermorpha* spp., *Spongomorpha pallida*, etc.); and (4) species occurring throughout the year but having their maximum development in spring and autumn (*Ceramium tenuicorne*, *Pilayella littoralis*, *Urospora penicilliformis*). Most perennials and higher aquatic plants have a period of intense vegetative growth in late spring and early summer (May—July) and a reproductive phase in late summer and early autumn (August—October). In the lower sublittoral this seasonality is more or less abated by the relatively constant environment.

Especially in the hydrolittoral, algal succession and dynamics are affected by several other, more or less irregularly varying, abiotic factors, so that hardly two years are quite alike. One may roughly speak of "green algal years" with *Cladophora glomerata* dominating and "brown algal years" with abundant filamentous brown algae (cf., Wallentinus, 1974).

Long-term fluctuations in the sublittoral involve changes in the abundances of, i.e., *Ectocarpus siliculosus*, *Fucus vesiculosus*, *Polysiphonia nigrescens*, *Rhodomela confervoides* and *Sphacelaria arctica* (G. Hällfors, unpubl.). The causes are, however, still little understood. Antagonistic abundance variations of the mussel *Mytilus edulis* and the red alga *Furcellaria fastigiata* have been interpreted in terms of salinity fluctuations and competition for space (H. Luther, pers. commun., 1972).

Zonation of the benthic vegetation

The benthic vegetation reaches down to a maximum depth of about 30 m in the Belt Sea area (Kiel Bight; Schwenke, 1964), to about 18—25 m in the northern Baltic Proper (Ravanko, 1968; Trei, 1975; Wallentinus, 1976a) and to about 10 m in the Bothnian Bay (P. Kangas, pers. commun., 1978).

The absence of a suitable substrate for attachment locally restricts the depth range especially in the southern and southwestern parts of the Baltic Sea and in the Gulf of Bothnia where mobile soft bottoms prevail. For instance, in the Kiel Bight below about 20 m the bottom consists of mud which is practically devoid of vegetation (Schwenke, 1966). Even in the archipelagos of the northern part of the Baltic Proper the depth range decreases inwards in the archipelagos due to the combined effects of decreasing transparency of the water, increased sedimentation and the ascent of soft bottoms without macroscopic plants into the euphotic zone.

Strong wave action excludes the vegetation from sediment bottoms due to the instability of the substrate. On less exposed shores the various plant species show different reactions to water movements (Luther, 1951a). The algal vegetation of rocky shores is affected mainly with respect to relative abundance and morphology of the species. The species composition is less variable. The biomasses tend to be inversely related to the intensity of wave action (A.-M. Jansson, 1974; Rönnberg, 1975; Hällfors et al., 1975; G. Hällfors, unpubl.). Near their inner border some marine species are favoured by increased water movement, e.g., *Fucus serratus* in Blekinge (Levring, 1940) and *Zostera marina* in Tvärminne. The reasons may be physiological, water movement compensating for low salinity, or they may have a complex ecological background (Luther, 1951b).

In the following some essential average traits of the phytal zonation in: (1) the southern part of the Baltic Sea and the Belt Sea; (2) the northern part of the Baltic Proper; and (3) the Bothnian Bay will be briefly described. The terminology relating to the littoral zones follows Du Rietz (1950; see also Wærn, 1952), distinguishing: (a) the geolittoral between the lowermost truly terrestrial plants and the mean (summer) sea level; (b) the hydrolittoral between the mean (summer) sea level and the lowest sea level; and (c) the sublittoral, the permanently submerged belt, from the lowest sea level (usually coinciding with the uppermost *Fucus vesiculosus* on sheltered rocky shores) down to the end of the vegetation.

The borders of the littoral zones are not fixed to absolute levels, but are rather a function of wave movement. On strongly exposed shores especially the hydrolittoral is widened upwards and downwards. A similar effect is caused by the swell along passages with frequent ferry traffic (Rönnberg, 1975).

Southern part of the Baltic Proper and the Belt Sea. Natural hard bottoms in the southwestern part of the Baltic Proper (see Fig. 5.4) are at best represented by accumulations of stones and boulders, or, more often, by stones and boulders interspersed on bottoms of sand and gravel (Schwenke, 1966, 1969). This leads to a complex situation with a mosaic of hard-bottom and soft-bottom vegetation. The Belt Sea is a transition area between the North Sea and the Baltic Proper, with many algal species at or near their inner border of occurrence. *Fucus vesiculosus* reaches high up into the hydrolittoral (Schwenke, 1969).

In the inshore areas of the southern coast of the Baltic Sea, e.g., the German "Boddengewässer" (Overbeck, 1965; Lindner, 1975) and the Bay of Gdańsk (Kornaś and Medwecka-Kornaś, 1949; Kornaś, 1959) the marine flora is already much impoverished and the vegetation and zonation are considerably simplified compared to the southwestern Baltic Sea and the Belt Sea.

Northern part of the Baltic Proper. The detailed account of Wallentinus (1976a, 1979) of the vegetation and flora in the Askö area is broadly of general validity for the archipelagos of the northern part of the Baltic Proper. The bedrock, mainly granite and gneiss, lies in most places bare and provides an excellent substrate for the attachment of haptophytic algae. The upper part of the shore is often rather smooth ice-polished rock, blocks increasing in frequency with the depth and in crevices. Figure 5.5 represents the average (idealized) conditions on a semi-exposed rocky shore in the outer archipelago zone.

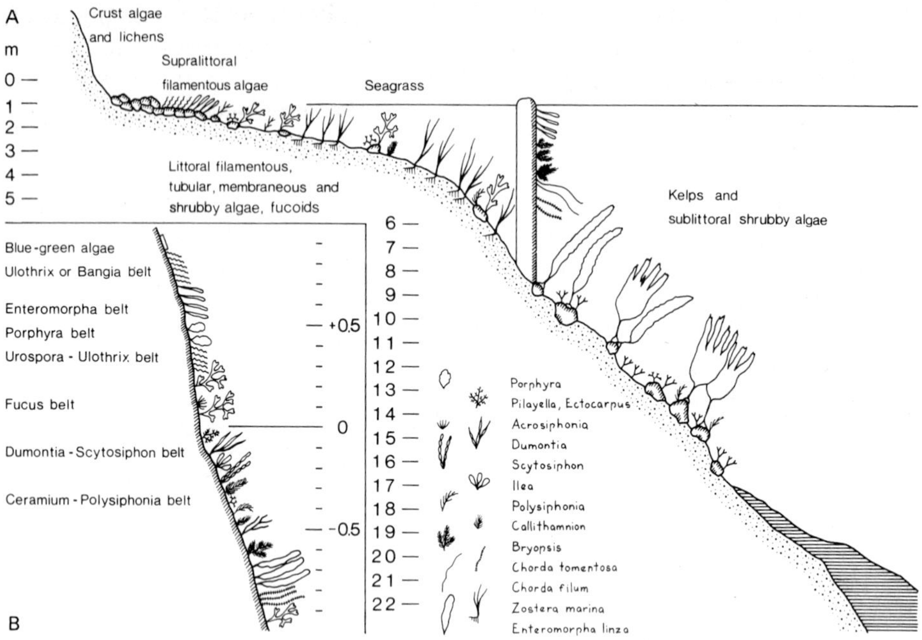

Fig. 5.4. Summarized vegetation profiles from the western Baltic Sea. A. total benthic vegetation. B. littoral vegetation on hard substrates. Modified after Schwenke (1966).

On semi-exposed soft bottoms the bottom material is mainly sand with scattered stones. *Zostera marina* forms the last patches of real seagrass meadows. With decreasing water circulation the density of the eelgrass vegetation is much reduced (Lappalainen et al., 1977).

On sheltered shores the bottom consists of silt and mud. In the most sheltered localities a reed-belt borders the shore. Otherwise the hydrolittoral has a patchy growth of small rhizophytes, e.g., *Eleocharis acicularis, Potamogeton filiformis, Ruppia maritima, Zannichellia palustris* and charophytes. In the sublittoral *Zostera* is replaced by a dense vegetation of *Potamogeton* spp. and *Myriophyllum spicatum*, i.e., limnic species favoured by a high electrolyte content of the water. Locally, up to 0.3 m thick accumulations of loose-lying *Fucus vesiculosus* cover the bottom. At a depth of 6—9 m the vegetation often ends with a peculiar community of loose-lying red algae, mainly *Phyllophora* spp. and some *Ahnfeltia plicata*, all much pauperized, and with only the uppermost tips sticking out of the mud.

Gulf of Bothnia. In the Bothnian Sea there is a sudden increase in the abundance of freshwater species (cf., Wærn, 1952). *Cladophora aegagropila* and the moss genus *Fontinalis* are the most prominent newcomers on hard substrates, while *Potamogeton vaginatus* attains some importance on sublittoral soft bottoms. *Fucus vesiculosus* is excluded from the upper sublittoral

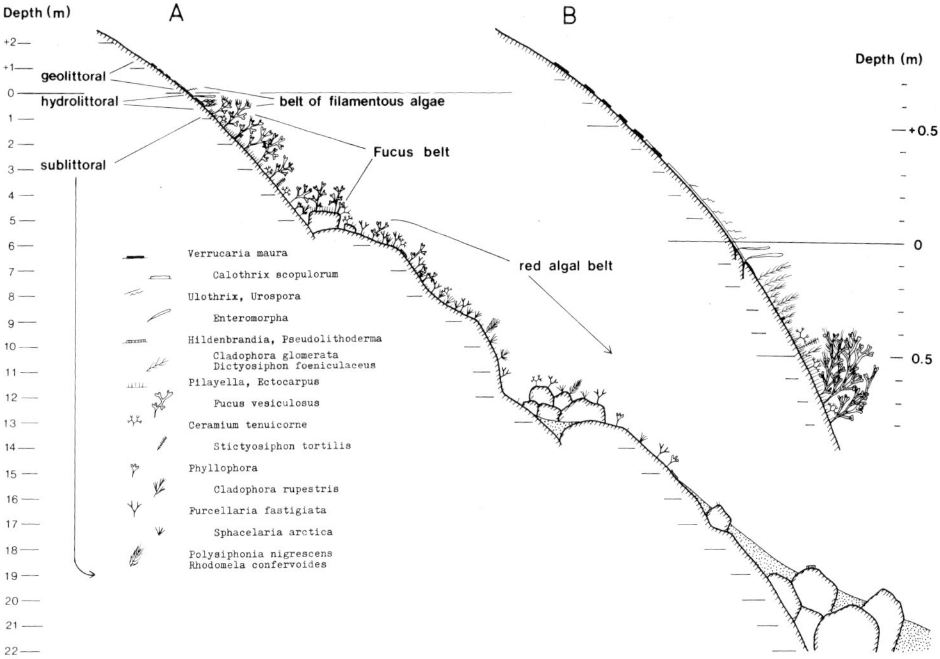

Fig. 5.5. Idealized profile from a semi-exposed rocky shore in the northern Baltic Proper. A. total benthic vegetation. B. littoral and uppermost sublittoral enlarged. Details not to scale. Original.

(above about 4 m) except in eutrophied water. Most marine species become sparse or disappear altogether.

In the Bothnian Bay conditions are relatively well known in the Krunnit area (Hällfors, 1976). On exposed shores (Fig. 5.6) the bottom consists mainly of stones and boulders. The hydrolittoral and upper sublittoral are dominated by *Cladophora glomerata* in summer. A perennial vegetation of *Cladophora aegagropila* and *Fontinalis* spp. starts at a depth of 2 m which is the normal lower limit of influence of the sea ice. In areas with pack-ice scouring *Fontinalis* is usually absent.

The hydrolittoral soft bottom is a mosaic of dense pads of mainly *Eleocharis acicularis* and/or *Potamogeton pusillus* coll. which alternate with eroded sandy patches bearing a thin growth of *Chara aspera* and *Zannichellia palustris*. Accessory species, *Potamogeton filiformis*, *Limosella aquatica*, *Subularia aquatica*, *Callitriche autumnalis*, *Elatine hydropiper* and *Alisma gramineum* spp. *wahlenbergii* are sparsely mixed with the dominants. Filamentous algae, chiefly *Cladophora glomerata*, are entangled in the rhizophytes. *Rivularia atra* is abundant on stones. In the sublittoral *Potamogeton perfoliatus* and *P. vaginatus* are the main species, and *Chara aspera* forms a thin bottom layer.

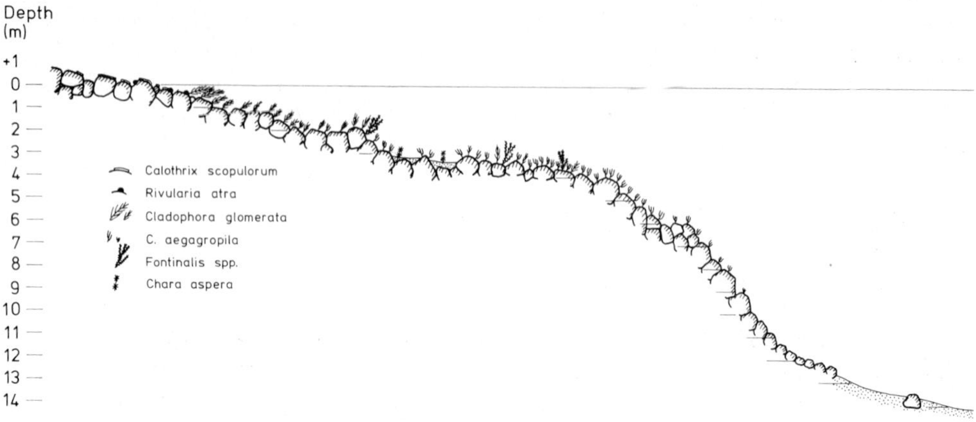

Fig. 5.6. Idealized profile from an exposed hard bottom in the Bothnian Bay (Krunnit area). Horizontal scale contracted to approx. 1/5—1/10. Original.

Available information from the Luleå archipelago (Wulff et al., 1977) indicates similar although more sheltered conditions, and a lower salinity than at Krunnit. This gives the flora an even more limnic character with *Isoëtes lacustris* and *Potamogeton gramineus* as new prominent species.

Littoral primary production

The production of the benthic vegetation is of considerable importance in the Baltic Sea where the littoral zone constitutes a relatively large part of the bottom. According to data given by Olsson (1971) bottoms at depths between 0 and 5.5 m constitute 3.3×10^{10} m^2, i.e., 8.9% of the total area (0—11 m, 6.0×10^{10} m^2, 16.1%; 0—16.5 m, 8.5×10^{10} m^2, 22.9%; 0—22 m, 11.0×10^{10} m^2, 29.6%).

With the exception of a few early studies (Bursa et al., 1939, 1948; Segerstråle, 1944; Kornaś et al., 1960), all relevant quantitative investigations have been made since the latter half of the 1960s. Several investigations have been made on the structure and standing stock of various communities in different areas (e.g., Ravanko, 1972; Luther et al., 1975; Hällfors et al., 1975; Rönnberg, 1975; Trei, 1975; Lappalainen et al., 1977; A.-M. Jansson and Kautsky, 1977; Wulff et al., 1977). Such biomass studies are of limited value, however, for the estimation of benthic primary production, unless simultaneous measurements of the metabolism of the community and, preferably also of the constituent species are made. Presently a wealth of such data is being collected (see, e.g., Schramm, 1973; Elmgren and Ganning, 1974; A.-M. Jansson, 1974, 1975; Wallentinus, 1975, 1976b, 1978; King and Schramm, 1976; B.-O. Jansson and Wulff, 1977; Guterstam, 1977; Schramm and Guterstam, in press) in an integrated effort to construct an ecosystem model of the Baltic Sea and its various subsystems (B.-O. Jansson, 1972, 1978; Schwenke et al., 1975).

Available results are still too few for any far-reaching conclusions to be made. Thus far published data for undisturbed areas in the northern part of the Baltic Proper indicate a gross primary production of carbon of 1.5—2.5 g m^{-2} d^{-1} in hydrolittoral *Cladophora glomerata* communities and 5.2—7.4 g m^{-2} d^{-1} in the uppermost *Fucus vesiculosus* vegetation in June—August (Elmgren and Ganning, 1974; A.-M. Jansson, 1974; B.-O. Jansson and Wulff, 1977), the conversion to carbon being based on a PQ value of 1.3 (see Guterstam, 1977).

On the basis of very limited background data B.-O. Jansson (1972) estimated the annual (net) production in the whole Baltic Sea littoral between the depths of 1 m and 6 m at 1.7×10^{12} g carbon. The figure is probably low. The annual (gross) production of the 0—1 m depth zone was estimated at 1.44×10^{12} g carbon by Elmgren and Ganning (1974).

Microscopic epiphytes (periphyton), mainly diatoms, make a significant contribution to the primary production (A.-M. Jansson, 1974). Except the qualitative results of Rautiainen and Ravanko (1972), little is known about the community structure and dynamics of periphyton in the Baltic Sea.

Trophic status of the Baltic Sea

The production of organic matter of the Baltic Sea shows marked regional variation which is chiefly caused by differences in nutrient levels and the length of the growing period. The level of the annual phytoplanktonic primary production of carbon is about 15—30 g m^{-2} in the Bothnian Bay, about 60 g m^{-2} in the Bothnian Sea, about 100 g m^{-2} in the Baltic Proper and the Gulf of Finland, and even higher in the Bornholm and Arkona Basins and in the Sound (Lassig et al., 1978).

If the area of the Baltic Sea including the Belt Sea is taken as 379,000 km^2, the annual net production of carbon in the pelagial amounts roughly to about 30×10^{12} g. The annual net primary production of the phytal is estimated at about 4×10^{12} gC. The total net primary production in the Baltic Sea thus amounts to about 34×10^{12} gC a^{-1}. If the annual zooplankton (excluding microzooplankton) production is taken as 10—20 gC m^{-2} (see Section on p. 238) in the Baltic Proper, and considerably less in the Gulf of Bothnia, the annual zooplankton production in the Baltic Sea will total about 4.4×10^{12} gC, which is about 15% of the phytoplankton production. For the zoobenthos we have not obtained enough information to calculate the production.

These values may be compared with the total catch of fish, which is about 900,000 tons fresh weight per year (cf., Sjöblom and Parmanne, 1978). This corresponds to about 7.6×10^{10} g carbon (conversion factor from fresh weight to dry weight is 0.2, from dry weight to ash-free dry weight 0.84, and from ash-free dry weight to carbon 0.5). Thus about 0.2% of the primary production is utilized by man.

In general the Baltic Sea has been considered an oligotrophic area, but lately the opinion has changed (e.g., Fonselius, 1972). The native archipelago people in Finland and Sweden assert that hydrolittoral filamentous algae in unpolluted archipelago areas have increased during the last 30—40 years. It is, however, not possible with present methods to obtain conclusive information as to a possible increase of biological production in the Baltic Sea because barely comparable production measurements have been made for only about 17 years. According to B.-O. Jansson (1978), the Baltic Sea is currently being driven out of a previously steady state by long-term processes which can be summarized as eutrophication and oceanization.

C. ZOOPLANKTON*

Introduction

The Baltic Sea, the largest brackish water area in the world, with low and stable salinity conditions, provides, in many respects, a rather unique habitat for its fauna and flora. It is also a very young sea from the geological point of view (see Section on p. 54). Since the last Glacial Period, fresh-water and brackish-water epochs have alternated, and the most recent brackish-water regime has only existed since 4000—5000 B.C. At the beginning of the present period the salinity was higher than today. As the evolutionary processes generally need a longer time for the formation of species, very few endemic species have evolved. This fact and the physiological difficulties of most aquatic organisms to adapt their populations to a brackish water of a salinity of 6—8‰, have promoted a fauna of low diversity.

Some 200,000 different species of plants and animals live in the seas around the world, of which about 85% are animals. In the southern Baltic Proper, only 145 macroscopic animals occur and further to the north in the southern Bothnian Sea no more than 52 (Zenkevitch, 1963). A lot of marine groups, such as Porifera and Echinodermata are not able to live in the Baltic Sea, and many others like Bryozoa and Tunicata have very few representatives.

Only 2% of marine animal species are pelagic and about half of those are permanent members of the zooplankton community. Consequently, the number of planktonic species is very small or just 1% of the known marine species. However, the number of temporary plankton species is very high. A rough estimate indicates that about 80% of marine evertebrates of 100,000 species have larval stages appearing as planktonic organisms during each re-

* By Hans Ackefors.

production cycle (Thorson, 1957). In tropical areas, 85—90% of the bottom community species have a long pelagic life, while very few species in coastal areas of Arctic and Antarctic Oceans or deep sea areas have pelagic larvae.

In boreal marine sea areas, corresponding to the Baltic Sea, about 2/3 of the species have pelagic larval stages. However, in the Baltic Sea only about 10% of the bottom invertebrates have planktonic larvae (see Ackefors and Hernroth, 1972; Hernroth and Ackefors, 1979). In total, a new estimate has shown that 130 species of bottom invertebrates are present in the southern Baltic Proper, not including occasionally found rare species (Forsman, 1972). This means that the meroplankton species are less than 15 in the area concerned. Together with the holoplankton species, the number of plankton species are not more than 40—50. About 40 of these species are listed in Table 5.II. If the fish species with pelagic larval stages are added, the number of animal plankton species does not exceed 60.

The low diversity of plankton species in the Baltic Proper is demonstrated by the low number of copepods, which normally make up the majority of holoplankton species as well as the bulk of the biomass in most sea areas. The copepods are estimated to be represented by 800 species in all oceans and seas together. In the Baltic Proper only 8 species of copepods occur regularly, but the production and the standing stock of the few copepod species is rather high (Hernroth and Ackefors, 1979).

Sampling technique

Different types of plankton nets — Nansen net, Hensen net, Judy net, l-m ringnet, etc. — have been used since the beginning of this century in the Baltic Sea. In 1971 Marine Biologists (BMB), founded in 1968, established six Working Groups to consider the sampling technique in various biological investigations. The aim was to recommend proper sampling methods and, if possible, to get an international agreement for using the same or at least similar gears. The reports from the Working Groups were published in 1976 (Dybern et al., 1976). For zooplankton sampling the fauna was split into three different size fractions; the microzooplankton (< 200 μm), the mesozooplankton (200 μm—1000 μm) and the macrozooplankton (> 1000 μm). The first group was sampled with a 5 l water sampler and the other groups were sampled mainly with nets. A WP 2 net with 90 μm mesh-size was proposed for ecological studies and the same net with 200 μm mesh-size for biomass studies (cf., Unesco, 1968). For more detailed ecological studies of the mesozooplankton a 23 l sampler was recommended (Ackefors, 1971a). A Bongo net was proposed for the sampling of macrozooplankton and fish larvae.

Simultaneous sampling with both new and old types of nets was carried out from Swedish research vessels in order to compare the efficiency of the old and new methods. The Nansen net had an efficiency of only about 50%

TABLE 5.II

Distribution of zooplankton in the northern (N), middle (M) and southern (S) parts of the Baltic Proper according to investigations in 1968—1972

Specimen	Abundant			Common			Sparse			Occasional		
	N	M	S	N	M	S	N	M	S	N	M	S
Sarsia tubulosa												X
Aurelia aurita		(X)	(X)	X	X	X						
Cyanea capillata								(X)	(X)		X	X
Pleurobrachia pileus (larvae)								X	X	X		
Keratella quadrata quadrata							(X)	(X)	(X)	X	X	X
K. qu. platei										X		
K. cruciformis eichwaldi										X		
K. cochlearis recurvispina												X
Synchaeta spp. (6 species)	X	X	X									
Pygospio elegans									X			
Harmothoë sarsi				X	X	X						
Bosmina coregoni maritima	X	X	X									
Podon intermedius				X	X	X						
P. leuckarti						X	X	X				
Pleopsis polyphemoides	(X)	(X)										X
Evadne normanni			X	X	X							
Calanus finmarchicus												X
Limnocalanus macrurus							X				X	X
Acartia bifilosa	X	(X)			X	X						
Acartia longiremis		(X)	X	X	X							
Euretemora sp.	X				X				X			
Centropages hamatus	X	X	X									
Temora longicornis	X	X	X									
Pseudocalanus m. elongatus	X	X	X									
Oithona similis									X	X	X	
Gastropoda							X	X	X			
Mytilus edulis						X	X	X				
Macoma baltica							X	X	X			
Cardium glaucum							X	X	X			
C. hauniense							X	X	X			
Mya arenaria							X	X	X			
Oikopleura dioica									(X)			
Fritillaria borealis	X	X	X									
Sagitta elegans baltica						(X)						
Sagitta setosa												X

compared with the WP 2 net, used since 1975/1976 by most countries in the Baltic area (Hernroth, in press, b). Due to this fact, old values, estimated from samplings with the Nansen net, can now be recalculated for comparison with results from recent samplings with the WP 2 net. This was done with all results used in this paper. Biomass values were estimated according to Ackefors (1972).

International oceanographic stations for plankton and hydrographical investigations (e.g., S and F stations, Fig. 5.7) have been visited regularly by research vessels for many years. In addition to these a station net suitable for monitoring studies has also been used (Fig. 5.7). Some of the stations have been visited every fortnight for primary and secondary production studies

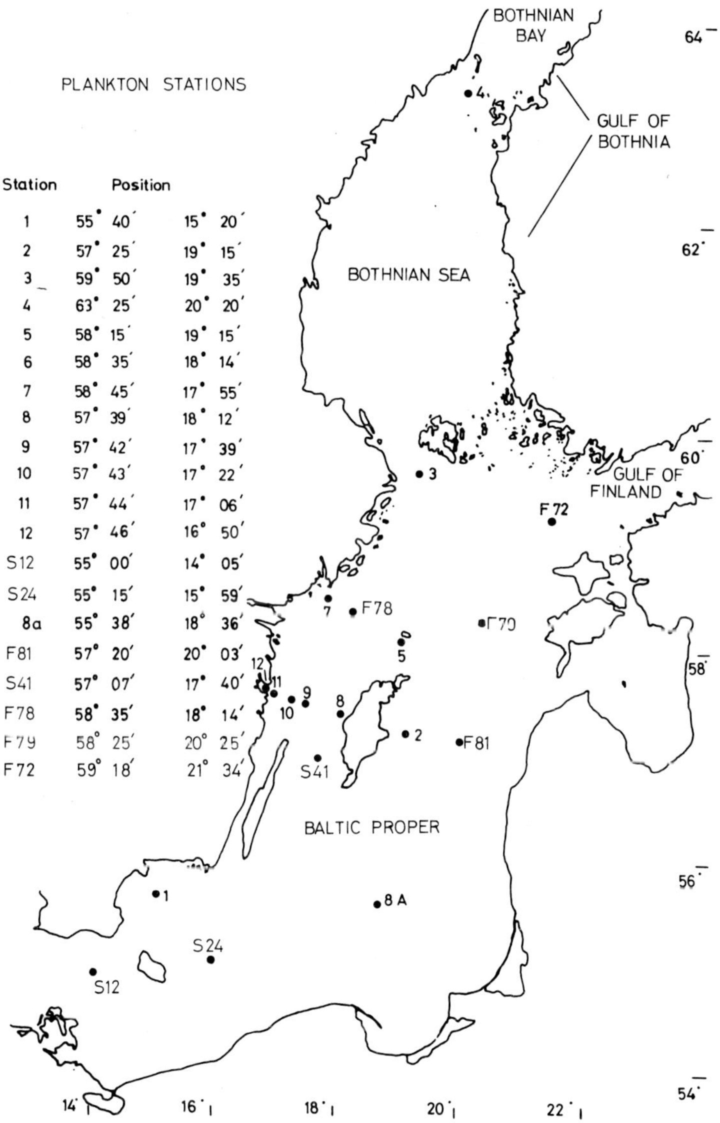

Station	Position	
1	55° 40′	15° 20′
2	57° 25′	19° 15′
3	59° 50′	19° 35′
4	63° 25′	20° 20′
5	58° 15′	19° 15′
6	58° 35′	18° 14′
7	58° 45′	17° 55′
8	57° 39′	18° 12′
9	57° 42′	17° 39′
10	57° 43′	17° 22′
11	57° 44′	17° 06′
12	57° 46′	16° 50′
S12	55° 00′	14° 05′
S24	55° 15′	15° 59′
8a	55° 38′	18° 36′
F81	57° 20′	20° 03′
S41	57° 07′	17° 40′
F78	58° 35′	18° 14′
F79	58° 25′	20° 25′
F72	59° 18′	21° 34′

Fig. 5.7. Plankton stations in the Baltic Sea visited by Swedish investigators (1—12) and the international oceanographic stations for plankton and hydrographical investigations (S 12—F 72).

(Lindahl, 1977a, b; Ackefors and Lindahl, 1979), while the majority of stations have been visited only 2—7 times a year (Hernroth and Ackefors, 1979).

Composition of the fauna

Because of the unique salinity conditions in the Baltic Sea a fauna comprising fresh-water, brackish-water and marine organisms has evolved. Brackish water of a salinity of 6—8‰ creates, however, critical conditions for most aquatic organisms. The number of species occurring in this salinity interval is very low. In water with a salinity less than 6‰ the number of fresh-water species increases and in water with a salinity above 8‰ the number of marine species increases (Remane, 1940). This implies that most marine organisms are excluded from the Baltic Sea due to their requirement for higher salinity. Only some euryhaline marine species tolerate brackish water with a salinity of 6—8‰. Those organisms are, however, the most important species in the zooplankton community of the Baltic Sea. Some of them, e.g., the copepods *Temora longicornis* and *Pseudocalanus m. elongatus*, are able to tolerate salinities down to 6‰, while others can even tolerate a salinity as low as 2‰, e.g., the cladocerans *Evadne nordmanni* and *Pleopsis polyphemoides* (Ackefors, 1969a, 1971b).

Only three endemic brackish-water species appear in the Baltic Sea, viz., *Bosmina coregoni maritima*, *Keratella quadrata platei* and *K. cochlearis recurvispina*, which is explained by the fact that the Baltic Sea is a geologically young sea (Segerstråle, 1957, 1962). Another important brackish-water species is *Limnocalanus macrurus*, which occurs mainly in the Gulf of Bothnia (Bothnian Sea plus Bothnian Bay) and in fresh-water lakes.

There are a small number of species of bottom evertebrates as well in the Baltic Sea, and only some of those species have pelagic larvae. The most common larvae are those of the bristle worm *Harmothoë sarsi* and the bivalve *Mytilus edulis*. The low diversity of plankton species is thus a characteristic feature of the plankton fauna of the Baltic Sea. Another characteristic feature is that the size of nearly all species is just about 1 mm or less.

The most abundant species in the Baltic Proper are six copepods, viz., *Temora longicornis*, *Pseudocalanus minutus elongatus*, *Centropages hamatus*, *Acartia longiremis*, *A. bifilosa* and *Eurytemora* sp.; rotifers of the genus *Synchaeta*; two cladocerans, *Bosmina coregoni maritima* and *Evadne nordmanni*; one larvacean *Fritillaria borealis* (Fig. 5.8). These species together make up 90—95% of the biomass not including the medusa *Aurelia aurita* (see Figs. 5.9—5.11). Due to a sampling technique, unsuitable for the capture of *Aurelia aurita*, it is not included in the biomass calculations. However, during summer and autumn *A. aurita* may be very abundant and owing to its size will easily exceed the biomass of all other species together on weight basis.

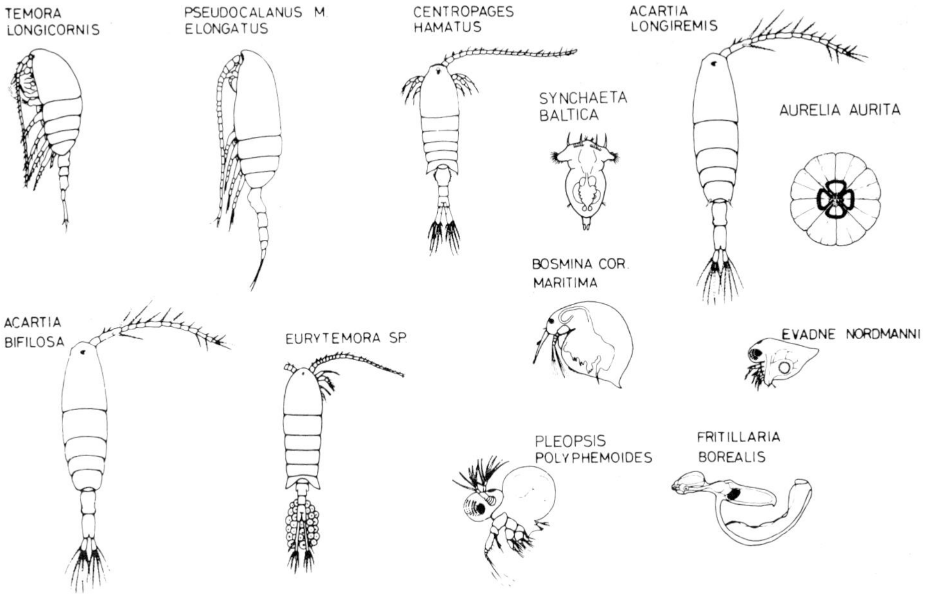

Fig. 5.8. Most important meso- and macrozooplankton species in the Baltic Proper. (According to Ackefors and Hernroth, 1972.)

In coastal areas of the Baltic Proper the rotifers of the genus *Synchaeta* and the cladoceran *Pleopsis polyphemoides* along with the copepod *Acartia bifilosa* are very important during certain periods in spring and summer. In offshore conditions these species are replaced in importance by the copepods *Temora longicornis*, *Pseudocalanus m. elongatus* and the larvacean *Fritillaria borealis*.

In the Bothnian Sea three copepods are very important, viz., *Limnocalanus macrurus*, *Acartia bifilosa* and *Eurytemora* sp. (Lindquist, 1959). But the rotifers of the genera *Sychaeta* and *Keratella* as well as the cladocerans *Bosmina cor. maritima*, *Evadne nordmanni* and *Podon* spp. may also be abundant. In coastal areas and in the most northern area, the Bothnian Bay, fresh-water species of the genera *Cyclops* and *Daphnia* are also important.

In coastal areas and in the Bothnian Bay, fresh-water species are also significant. In the Luleå archipelago a newly undertaken investigation has shown that the copepods *Diaptomus* sp., *Heterocope* sp. and *Cyclops* spp. are abundant although the dominant copepods are brackish-water species, viz. *Eurytemora* sp. and *Limnocalanus macrurus* (Wulff et al., 1977; Ackefors et al., 1978). The dominant cladoceran is the brackish-water species *Bosmina coregoni maritima*. The fresh-water species *Daphnia cristata* is also important, while the holoeuryhaline species *Pleopsis polyphemoides* and *Evadne nordmanni* are of less importance.

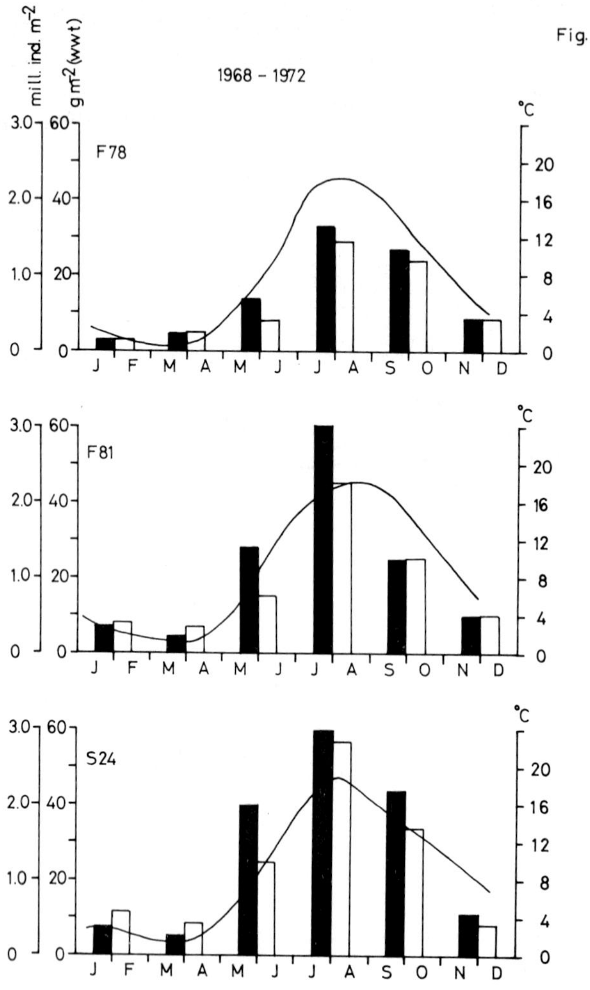

Fig. 5.9. Mean values of the total number of zooplankton individuals (black bars) and the biomass (open bars), 1968—1972, at the stations F 78, F 81 and S 24 in the Baltic Proper in relation to yearly temperature cycle. The values are calculated on two-monthly basis. (Modified according to Hernroth and Ackefors, 1979.)

Unfortunately, very little work has been done on microzooplankton in the Baltic Sea. The ciliates are probably as important in the Baltic Sea as in other seas. In coastal areas of the Bothnian Bay, Wulff et al. (1977) showed that they made 1/3 of the production of the whole plankton fauna.

Halme (1958) and Schwarz (1959) have published results on microzooplankton and larger zooplankton from inventories made in two different parts of the Baltic Sea.

Environment and fauna

Temperature

Temperature affects most processes in the plankton community, directly or indirectly. Temperature influences physiological processes such as mortality, survival, metabolic rate, feeding rate, embryonic development and ecological features such as community structure, diapause and energy flow.

Some species appear abundantly within a broad temperature range, e.g., *Aurelia aurita, Evadne nordmanni* or *Acartia bifilosa* (see Fig. 5.10). In contrast to eurytherm species, others are abundant only within a small temperature interval, e.g., the stenotherm species *Pleopsis polyphemoides*, which is frequent within the temperature range of 15—16° C. There are also some warm stenotherm species, which appear abundantly at temperatures higher than 15—16° C, viz., *Bosmina cor. maritima, Podon intermedius, Eurytemora* sp., *Centropages hamatus* and *Temora longicornis* (see Fig. 5.10) (Ackefors, 1969a; Hernroth and Ackefors, 1979).

Some species prefer cold water all the year round. They are never found above the thermocline in summer, e.g., *Pseudocalanus m. elongatus* and *Fritillaria borealis*. Other cold stenotherm species are *Limnocalanus macrurus, Mysis relicta* and *Mysis mixta* and the larvae of *Pleurobrachia pileus*.

The reproduction of some species is greatly affected by the seasonal variation in temperature. Rotifers, cladocerans and at least one copepod species overwinter mainly as "resting" eggs. While the presence of dormant eggs among rotifers and cladocerans has long been rather well known, it was only recently described for a marine copepod in laboratory and field studies (Zillioux and Gonzales, 1972). The existence of winter egg dormancy for *Acartia bifilosa* is conspicuous in the Baltic Proper. Very few adults appear in the winter, and in February—March there is an explosive development of nauplii due to the hatching of winter eggs.

In the Baltic Proper the first part of successive spawning periods of the various copepod species with significant number of nauplii is correlated with increasing temperature during spring and early summer; *Acartia* spp. in January—April (0—4° C), *Pseudocalanus m. elongatus* in April—May (4—8° C), *T. longicornis* in May—June (6—10° C), *C. hamatus* in May—June (8—10° C) *and Eurytemora* sp. in May—June (8—10° C).

The seasonal variation of species, biomass and production is to some extent regulated by temperature conditions. In general, the number of individuals of most species increases with increasing temperature in May—June. The maximum number of individuals of the total fauna appears in August in conjuction with maximum temperature (or slightly after the maximum) at the end of August or September (see Fig. 5.9). During maximum abundance, about 3 million individuals per m^2 may occur in the southern Baltic Proper while the corresponding figure for the northern Baltic Proper is 1.5 million.

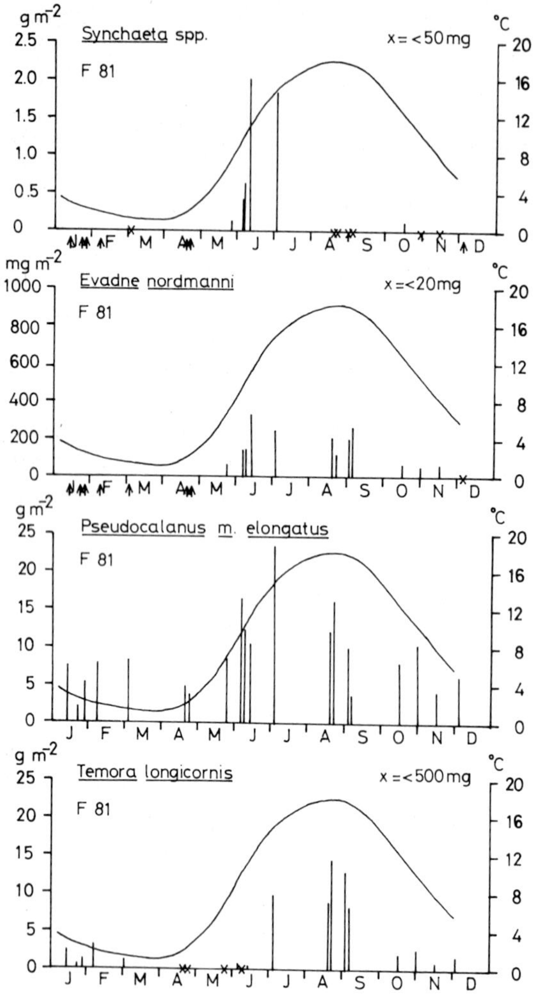

Fig. 5.10. Biomass of four different species, 1968—1972, at station F 81 on various sampling occasions in relation to the annual temperature cycle. Arrows indicate samplings with no specimens of the species concerned. (Modified according to Hernroth and Ackefors, 1979.)

However, the maximum abundance for an individual species may occur during a certain period of the summer season due to individual optimum temperature, competition, etc., which is obviously the case for the four different species given as examples in Fig. 5.10. Other species, which prefer cold water, may be abundant all the year round as, for example, *Pseudocalanus m. elongatus*. For such species the distribution is restricted to the waters below the thermocline in summer.

The amount of the total biomass in the Baltic Proper follows the yearly temperature cycle (see Fig. 5.9). During the first four months of the year the mean biomass in 1968—1972 at 7 different plankton stations was in the range 1.5--11.5 g m^{-2} (wet weight, wwt) when the surface temperature was below 4° C (Hemroth and Ackefors, 1979). Most of the biomass consisted of two species, viz., *P. m. elongatus* and *Fritillaria borealis*. In May—June, when the temperature rose from 4° C to 12° C, the mean values for the biomass ranged from 6 g m^{-2} to 25 g m^{-2} (wwt). In addition to the above-mentioned species, *Synchaeta* spp. and *Evadne nordmanni* were then also important. Beyond all comparison, *P.m. elongatus* was still the most important species. The biomass of this species alone exceeded 20 g m^{-2} on certain occasions (see Fig. 5.10). During the warm season, in July—September, the cladoceran *Bosmina cor. maritima*, and the copepods *Acartia* spp., *Eurytemora* sp., *Centropages hamatus* and *Temora longicornis* were very abundant and contributed greatly to the total biomass. At the end of the year when the temperature was in the range 4—12° C, all copepods except *Eurytemora* sp. and *Centropages hamatus* were important, and the biomass was in the range from 9 g m^{-2} to 12 g m^{-2}.

Zooplankton production has been studied thoroughly at a few plankton stations with dense sampling during the year (Lindahl, 1977a, b; Ackefors and Lindahl, 1979). East of Gotland (station 2, Fig. 5.7) production was low and stable during February until the beginning of May (4—7 mg C m^{-2} d^{-1}). With increasing temperature in May it rapidly increased and amounted to 60—70 mg C^{-2} d^{-1} during June—July and the first part of August. At the end of August and in September, 1976, there occurred an explosive development of *Bosmina coregoni maritima*, and production went up to about 190 mg C m^{-2} d^{-1}. From the end of September it decreased from about 60 mg to about 15 mg C m^{-2} d^{-1} at the end of November and the beginning of December.

According to our analyses of the zooplankton production in various parts of the Baltic Proper, the values may amount to about 20 gC m^{-2} a^{-1} in the southern Baltic Proper and 10 gC m^{-2} a^{-1} in the northern Baltic Proper (Hernroth and Ackefors, 1979). The values do not include the production of microzooplankton such as ciliates or macrozooplankton such as *Aurelia aurita*.

Salinity

The low salinity in the Baltic Sea affects the physiological processes of most species, and many species live under osmotic stress. The euryhaline marine species live in a hypo-osmotic milieu at the lower range of their tolerance limits and the fresh-water species in a hyper-osmotic milieu in the upper range of their tolerance limits. There are very few brackish-water species, and only a restricted number of marine species have adapted themselves to the brackish-water milieu. These organisms are either osmoregulators (homeoosmotic species) or osmo-conformers (poikilo-osmotic species) (cf., Kinne, 1964a, b). Certain euryhaline species have a greater ability to maintain an acceptable internal ion concentration or adapt themselves to the surrounding salinity. The energy used by an individual of marine origin for maintaining the internal ion concentration will increase with dilution. Certain organisms will be outnumbered more rapidly than others, which are better adapted to live in a low salinity medium.

The ability of various species to adapt themselves to decreasing salinity conditions will therefore greatly influence the horizontal as well as the vertical distribution of species in the Baltic Sea. A few marine species, e.g., *Oithona similis* and *Sagitta elegans baltica*, occur permanently and propagate only in the deep basins of the southern Baltic Proper, where the salinity exceeds 14—15‰. Other marine organisms, with a broader range of salinity tolerance, occur in water with a salinity down to 6‰, i.e., as far to the north as the border line between the Baltic Proper and the Gulf of Bothnia. *Pseudocalanus m. elongatus*, *Fritillaria borealis* and the larvae of *Pleurobrachia pileus* belong to species appearing at salinities as low as 6‰, but only in waters with a temperature lower than 10° C. The regulation of the body fluid in euryhaline marine organisms is often facilitated when the organisms live in cold water. Other species such as the cnidarian *Aurelia aurita*, the copepods *Temora longicornis*, *Centropages hamatus* and the cladoceran *Podon leuckarti*, for example, which also live and propagate in water with a salinity down to 6‰, appear also in warm water. A fourth group of marine species occur even in the Gulf of Bothnia at salinity as low as 2‰, the so-called holeurysaline species (Ackefors, 1969a). They are represented by the cladocerans *Evadne nordmanni*, *Pleopsis polyphemoides* and *Podon intermedius*.

The brackish-water cladoceran *Bosmina cor. maritima* may occur in the salinity range from close to zero up to at least 8‰, which means that this species occurs both in the Gulf of Bothnia and the Baltic proper. Other brackish-water species, such as *Keratella* spp. and *Limnocalanus macrurus*, appear mainly in the Gulf of Bothnia and only occasionally in the northern Baltic Proper, indicating a narrower salinity tolerance with the upper range around 6‰.

Oxygen

The oxygen consumption of euryhaline marine organisms increases with decreasing salinity due to the increasing matabolic rate. This might explain the low number of organisms other than *Pseudocalanus m. elongatus* below the halocline. The oxygen content is less than 3 g O_2 m^{-3} in that part of the water column. The alternating periods of adequate and deficient oxygen conditions with development of hydrogen sulphide in the bottom water of the deep basins affect the bottom fauna. At times there are zones from which the bottom fauna has died. During such anoxic periods, malformed *Harmothoë sarsi* larvae are found in water above the bottom.

Vertical distribution and diel migration

The vertical distribution is greatly affected by the abiotic factors temperature, salinity, oxygen, pressure, light etc. Access to food and other biotic factors may also be of great importance. During daytime, different species may accumulate above the thermocline in the warm surface water, within the thermocline, below the thermocline in the cold water, or be rather homogeneously distributed in certain parts of the water column (see Fig. 5.11 and Ackefors, 1969b). For instance, the cladoceran *Pleopsis polyphemoides* very often accumulates in the thermocline where the temperature is about 7—8° C. Laboratory experiments have shown that the preferred temperature is about 8° C (Ackefors and Rosén, 1970; Ackefors, 1971b). This

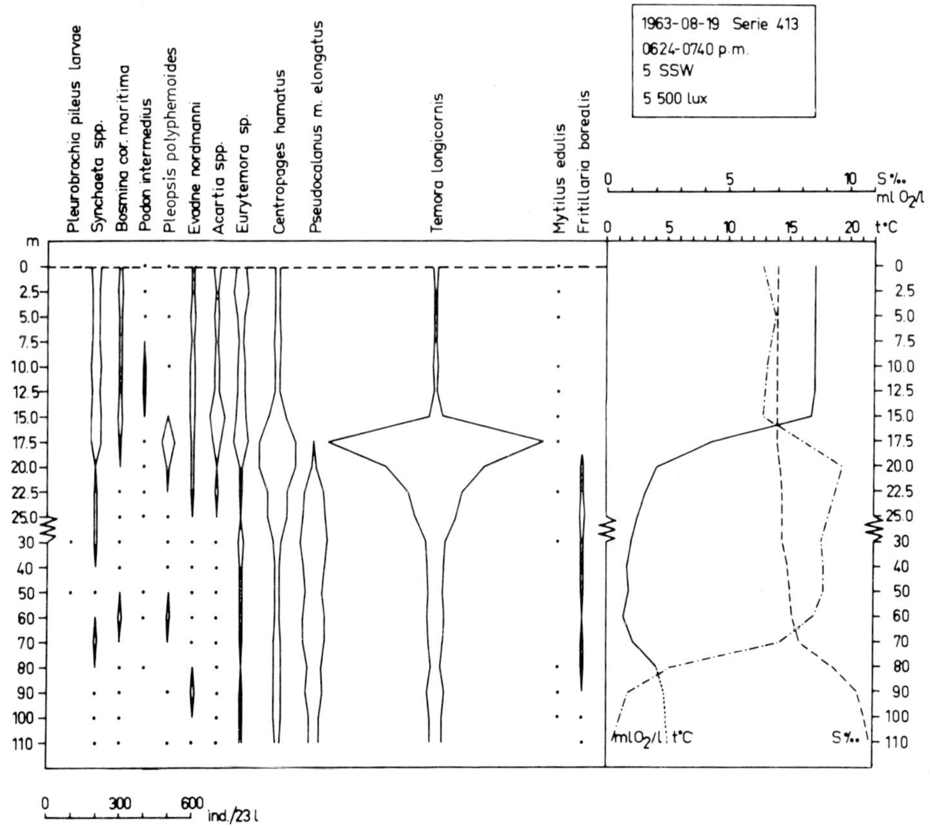

Fig. 5.11. Vertical distribution of zooplankton, August 1963, about one hour before sunset, at station F 79 (N58° 25′, E20° 25′). Samples were taken every 2.5 m from surface down to 25 m and every 10 m from 30 m to bottom with a 23 l sampler (Ackefors, unpubl.).

indicates that temperature may have a decisive influence upon the vertical distribution.

In Fig. 5.11 the vertical distribution from the surface down to a depth of 110 m in daytime is shown. Samples were taken every 2.5 m from the surface down to 25 m and then every 10 m from 30 m down to 110 m with a 23 l water sampler (see Ackefors, 1971a). In Fig. 5.12 the vertical distribution is shown at night. The differences are obvious. The euryhaline copepods *Centropages hamatus* and *Temora longicornis* accumulate within the thermocline or below the thermocline in daytime but occur mainly above the thermocline at night. *Pseudocalanus m. elongatus* occurs rather homogeneously from the thermocline down to the bottom in daytime, but accumulates just below the thermocline at night. These three examples indicate a diel migration. Other species such as the cladocerans *Pleopsis polyphemoides* and

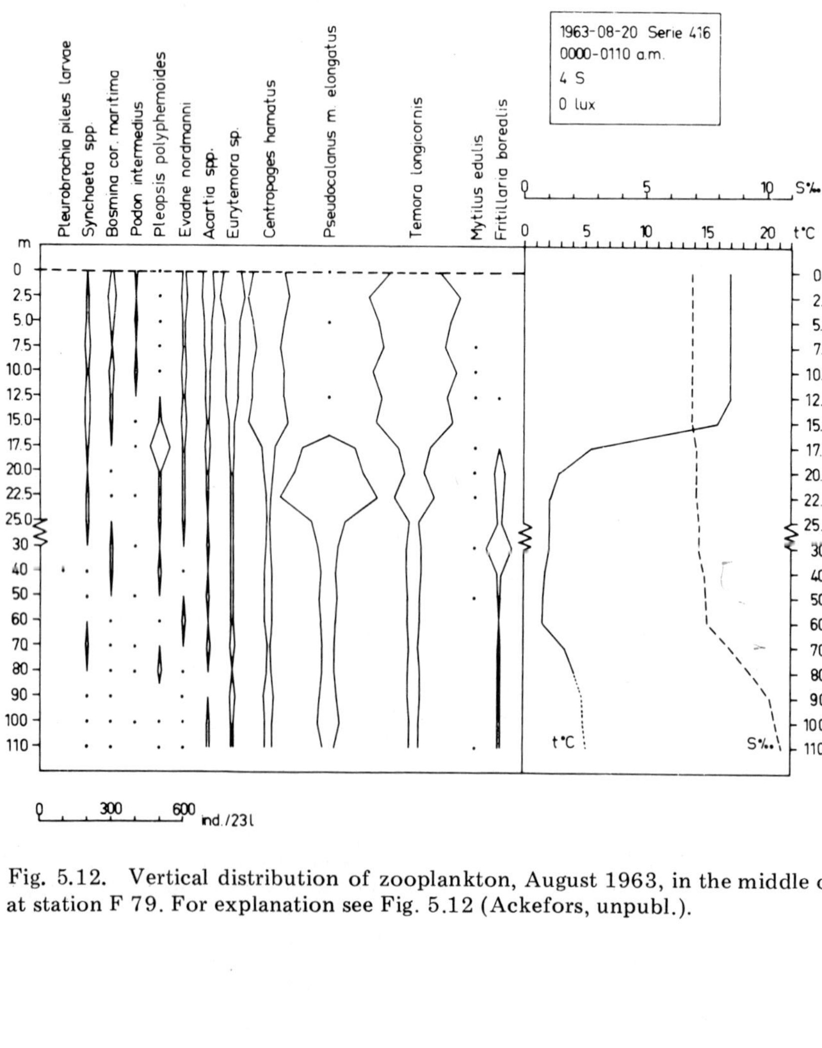

Fig. 5.12. Vertical distribution of zooplankton, August 1963, in the middle of the night at station F 79. For explanation see Fig. 5.12 (Ackefors, unpubl.).

Evadne nordmanni for example, do not seem to have a pronounced diel migration.

Our studies of the vertical distribution during various seasons show that there is no diel migration in winter. During that time the copepods may be distributed rather homogeneously from the surface down to certain levels in the water column, usually with the older copepodite stages occurring deeper than the younger ones. In May, when the surface temperature increases, the diel migration starts.

Food web

Recent studies of the energy flow of the pelagic ecosystem describe several possible prey—predator relationships within the plankton community of the Baltic Sea (Hillebrandt, 1972; Schnack, 1975; Lenz, 1977; Hobro et al., 1975; Å. Hagström, pers. commun., 1977). Including studies from other areas (see e.g., Beers and Stewart, 1970; Petipa et al., 1970; Strickland, 1972), the author has tried to outline a model of the pelagic ecosystem of the Baltic Sea using the energy circuit language of Odum (1972; see Fig. 5.13). The model indicates that the insolation (*I*) and the nutrient salts (*N*) in the water are the prerequisites for the production of phytoplankton (*P*), which is the basic food item for microzooplankton (*M*) and herbivorous zooplankton (*HZ*). An important food item for those two groups is also detritus consisting of inorganic and organic matter.

Particulate organic matter (*POM*) with absorbed dissolved organic matter and attached bacteria seems to be an important food item for herbivorous zooplankton, e.g., for the copepod *Pseudocalanus m. elongatus* and younger stages of other copepods, the cladoceran *Bosmina coregoni maritima* and microzooplankton (Strickland, 1972; Lenz, 1977). The omnivorous zooplankton (*OZ*), e.g., older copepodite stages of most copepods, consume microzooplankton, phytoplankton and probably also detritus. Carnivorous zooplankton (*CZ*) may feed at different trophic levels with *Aurelia aurita* as the top carnivore and at lower levels with the carnivorous copepod *Oithona similis*, the cladocerans of the genera *Podon*, *Pleopsis*, *Evadne* and the chaetognath *Sagitta elegans baltica*.

The leak of dissolved organic matter from phytoplankton (and from other sources in the ecosystem) is the basis for the production of pelagic bacteria (*B*). Bacteria in their turn are probably the main food item for microzooplankton such as ciliates, which serve as nourishment for some carnivorous species like the rotifers *Synchaeta* spp.

Dynamics of the plankton community

The annual cycle of primary production and phytoplankton biomass in offshore conditions in the Baltic Proper is described, e.g., by Ackefors and

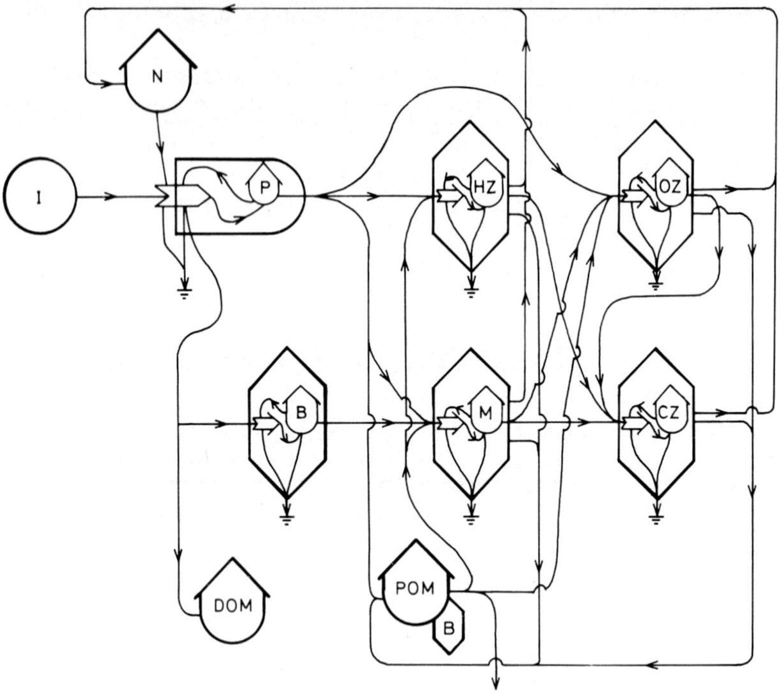

Fig. 5.13. Energy flow in the food web of the plankton community described with the energy circuit language according to Odum (1972). *I* = insolation; *N* = nutrients; *T* = temperature; *P* = phytoplankton; *M* = microzooplankton; *HZ* = herbivorous zooplankton; *OZ* = omnivorous zooplankton; *CZ* = carnivorous zooplankton; *B* = bacteria; *DOM* = dissolved organic matter; *POM/B* = particulate organic matter ("detritus") with bacteria. For further explanation, see text.

Lindahl (1975a, b), Lindahl (1977a, b), and Ackefors and Lindahl (1979). The relationship between primary and secondary production is described by Lindahl (1977a, b), Ackefors and Lindahl (1979). This relationship at one station east of Gotland is shown in Fig. 5.14.

During winter the primary production is in the order of 45 mg C m^{-2} d^{-1} and the biomass is about 0.1 g m^{-2} (wwt). The secondary production is about 5 mg C m^{-2} d^{-1}, and the biomass is in the order of 2—10 g m^{-2}. The flora is dominated by diatoms and the fauna of *Pseudocalanus m. elongatus* and to a lesser extent of *Temora longicornis* and *Fritillaria borealis*. Before the spring bloom starts, pelagic bacteria are abundant. From the end of February onwards, primary production increases, and a large phytoplankton biomass accumulates during March—May, which is only to a small extent utilized directly by the zooplankton community. The primary production reaches levels of 1.1—1.5 g C m^{-2} d^{-1} and the phytoplankton biomass exceeds 10 g m^{-2}. A large part of phytoplankton biomass sinks towards the

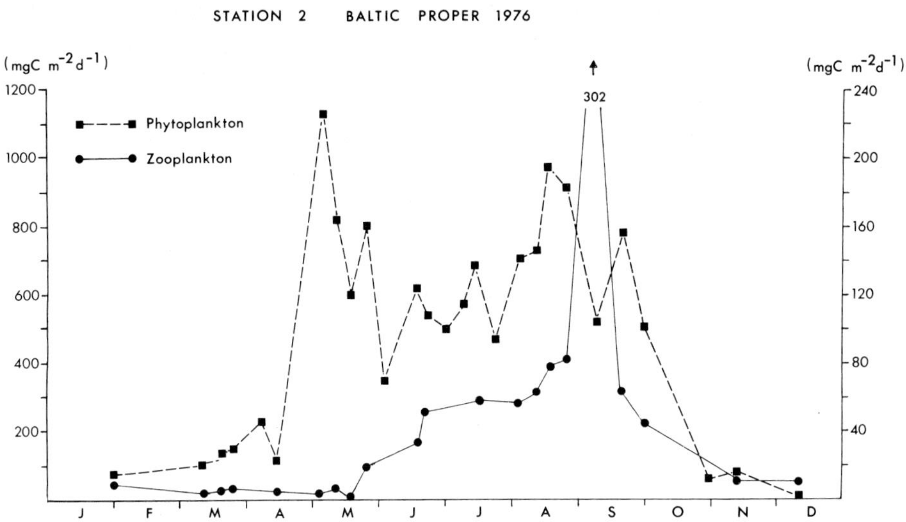

Fig. 5.14. Daily primary (hatched line, left scale) and secondary (solid line, right scale) production of station 2 in the Baltic Proper in 1976 (Ackefors and Lindahl, 1979).

bottom and supports the bottom fauna. The spawning of the polychaete *Harmothoë sarsi* is triggered by the spring bloom (Sarvala, 1971). At the beginning of the spring bloom, the nauplii of *Acartia bifilosa*, hatched from overwintering eggs, and ciliates are numerous. At the same time the bacteria reach their lowest abundance.

From March until June there is a successive spawning of the various copepod species, and the nauplii of the various species display separate peak abundances. In early summer the primary phytoplankton production drops to 300—700 mg C m^{-2} d^{-1} while secondary zooplankton production increases from 10 to 50 mg C m^{-2} d^{-1}. The phytoplankton biomass, consisting mainly of diatoms and monads, is small, or 1—2 g m^{-2}, and the zooplankton biomass is mostly of the order of 5—25 g m^{-2}. The percentage of carnivorous zooplankton species increases during this period of the year. The rotifers *Synchaeta* spp. and the cladocerans *Evadne nordmanni*, *Podon leuckarti* and *Pleopsis polyphemoides* attain their maximum abundance from late spring until the middle of summer. In late May the rotifers, *Synchaeta* spp., have an explosive development. This coincides with the decreasing populations of bacteria and ciliates, indicating a prey—predator relationship between these species. From midsummer and onwards the top carnivore *Aurelia aurita* is very abundant. Although no estimate of the biomass and the production of this medusa is available due to an unsatisfactory sampling technique for this species, it is obvious that the biomass and production are very high in certain years.

In late summer and at the beginning of autumn, primary production increases to 750—1000 mg C m^{-2} d^{-1}. Grazing by zooplankton is extensive and, consequently, the phytoplankton biomass is small. The secondary production and the zooplankton biomass reach maximum levels of 100—300 mg C m^{-2} d^{-1} and 60 g m^{-2}, respectively. The biomass consists largely of the herbivorous cladoceran *Bosmina coregoni maritima.* In the southern Baltic Proper, it may account for 60% of the production during a few weeks in August—September. The phytoplankton flora is dominated by the blue-green algae, the diatoms and the monads.

During summer the diversity of the zooplankton community reaches its maximum. In addition to an abundance of most species of cladocerans and copepods, the larvae of bivalves, gastropods and polychaetes also appear but do not play an important role.

During late autumn the importance of all the species, including the copepods, decreases. The production decreases from 50—60 mg C m^{-2} d^{-1} in October to 15—30 mg C m^{-2} d^{-1} in November. The populations of the warm stenotherm copepods *C. hamatus* and *Eurytemora* sp. decrease rapidly after September—October. *P.m. elongatus, T. longicornis* and *Acartia* spp. are still important in October—November. The primary production value decreases to less than 200—300 mg C m^{-2}. The relative importance of diatoms and monads increases. The phytoplankton biomass is of the order of 1—3 g m^{-2} while the corresponding zooplankton biomass is 10—15 g m^{-2}.

D. BENTHIC FAUNA OF THE BALTIC SEA*

Origin of the benthic fauna

Many of the special features of the animal life in the present-day Baltic Sea can be seen against the very peculiar history of this sea. Its short age as a habitat and the huge variations in its main brackish-water properties since the Pleistocene glaciation (from the limnetic Ancylus Lake to the Litorina Sea with salinities exceeding those of the recent Baltic Sea by about 5—6 ‰) and the restricted connection with the oceans through narrow channels have all left their marks on the existing fauna. The length of the Baltic Sea, about 1500 km in the S—N direction, involving a wide climatic range, the strong decrease in salinity towards the inner ends of the Gulfs, and the characteristic stratification of water masses create both horizontal and vertical gradients, the ecological effects of which are particularly well manifested by the bottom fauna. During the twentieth century, the uneven distribution of the effects of human activities has added a new aspect to the gradients predominating within the Baltic ecosystem.

The benthic animals of the Baltic Sea are of mixed zoogeographical origin.

* By Julius Lassig and Erkki Leppäkoski.

One of the best known special features is the presence of relicts, which have survived there or in neighbouring areas from earlier stages of the Baltic Sea (mainly the crustaceans *Mesidotea entomon*, *Mysis relicta*, *M. mixta*, *Pontoporeia affinis* and *P. femorata*, the bivalves *Astarte borealis* and the priapulid *Halicryptus spinulosus*). Some of the most euryhaline elements of the Atlantic-boreal and often more or less cosmopolitan species have succeeded in penetrating from the North Sea into the Baltic Sea. During the last decades some of these fairly opportunistic species (mainly polychaetous annelids) have been able to penetrate farther east and north in the Baltic Sea, benefited by a slight increase in salinity (see p. 144). Some of them are also capable of effectively occupying niches and habitats temporarily opened by influx of oxygen-rich water into the stagnant basins (Leppäkoski, 1975b; Andersin et al., 1978b).

Fresh-water animals are important in the coastal waters of the Baltic Sea, and some of them penetrate as far out as into the β-mesohaline (salinity range 5—10‰) water. For instance in the inner parts of the Archipelago Sea, they comprise one third of the total macrobenthic population on soft bottoms and their dominance tends to increase by increasing organic load (Leppäkoski, 1975a). Some of the fresh-water species avoid their original limnetic habitat, especially in the northern part of the Baltic Sea area and exist as "secondary" brackish-water species. Such a species is the fresh-water gastropod *Theodoxus fluviatilis* which is numerous in inshore waters but does not generally occur in Fennoscandian fresh-waters, obviously because of the low electrolyte content of those waters.

Potamopyrgus jenkinsi is a recent, successful brackish-water immigrant, which originated from New Zealand, and was first recorded in 1887 in the Baltic Sea. In 1926 this anthropochorous snail had reached the Åland archipelago, and today it occurs in the littoral zone of the whole Baltic Sea. This parthenogenetic gastropod coexists in parts of the Baltic Sea with three other species of Hydrobiidae (*Hydrobia ulvae*, *H. ventrosa* and *H. neglecta*). The immigration of *Potamopyrgus* has led to interesting theoretical considerations with respect to interspecific relationships, niche diversification and habitat selection (Muus, 1967; cf. Fenchel, 1977).

Other examples of anthropochorous distribution are the crabs *Eriocheir sinensis* from east Asia, and *Rhithropanopeus harrisi* from North America (Segerstråle, 1957). *E. sinensis* was recorded in the entrance area of the Baltic Sea in the early 1930s, and fullgrown specimens have now been observed in the whole Baltic Sea area (it is not capable of reproducing in brackish-water). *R. harrisi* is restricted to the bays and river mouths of the southern part of the Baltic Proper.

The strong decrease in the number of taxa of benthic animals when moving east- and northwards from the Arkona Basin has been known since long and described in classic papers of e.g. Välikangas (1933), Remane (1934), Segerstråle (1957), and Zenkevich (1963), but every year new faunistic data enhance the knowledge on the distribution of benthic animals, especially those belonging to less-studied groups. In this section we restrict ourselves to the general features of Baltic Sea benthos and the large-scale trends which are observed in the recent history of Baltic zoobenthos.

According to, e.g., Zenkevitch (1963) there are about 1500 macroscopic marine animal species in the Skagerrak but only 150 species in the Arkona area, 80 in the waters off Gotland, and about 50 in the Åland archipelago. In fact, for production estimates of the soft bottom subsystem in the Askö area (northern Baltic Proper), only about 6 species of macrofauna and 4 species of bottom feeding fish have to be considered (Elmgren, 1978). The marked decrease in the number of marine species towards the inner parts of the Baltic Sea is caused not only by decreasing salinity but also by unfavourable temperature conditions, especially in the northernmost parts. There are several entirely marine groups of benthos of which none, or only a few species are capable of penetrating into the Baltic Sea through the Danish Sounds, e.g., Porifera, Actiniaria, Madreporaria, Octocorallia, Echiuroidea, Scaphopoda, Crinoidea (Zenkevitch, 1963).

Littoral zone

The effects of the horizontal salinity and temperature gradients are best manifested by the animal assemblages of the littoral zones. The littoral zones of the Baltic Sea differ markedly from those of oceanic coasts. The low salinity decreases transiently even more during the time of snow melting. There are practically no tides in the Baltic Sea. However, variations in the sea level are caused by variations in the atmospheric pressure and prevailing winds. These irregular variations are especially pronounced at the heads of the Gulf of Finland and the Gulf of Bothnia. The northern location means a heavy seasonal pulse of illumination and temperature. The range of annual temperature variations in the littoral zone may be about 20° C. Ice cover lasts in general from less than one month in the coastal zone of the northern part of the Baltic Proper to five or six months in the innermost Bothnian Bay (see Section A of this chapter).

In the outer and middle archipelago zones as well as along open coasts, the vegetation is dominated by marine algae (see p. 230), whereas in the most diluted sheltered coastal bays limnetic plants, e.g., *Phragmites australis*, *Potamogeton* spp., and charophytes form dense macrophyte belts, especially hospitable for a variety of fresh-water fauna, e.g., Gastropoda and insect larvae of Odonata, Trichoptera and Chironomidae. The fauna of the uppermost vegetation belt, that of filamentous algae, mainly the fresh-water green algae *Cladophora glomerata*, has been comprehensively studied at the Askö and Tvärminne laboratories and on Åland (Jansson, 1974; Fagerholm, 1975; Hällfors et al., 1975). The fauna of the belt consists mainly of tiny gastropods (*Hydrobia* spp.), juvenile bivalves (*Mytilus edulis*, *Cardium* spp.), crustaceans and representatives of permanent meiofauna such as Rotatoria, Oligochaeta, Turbellaria and Nematoda. At Tvärminne a total of 30 macrofaunal taxa was recorded in this zone by Hällfors et al. (1975). The most important of these were the crustaceans *Gammarus* spp., *Idotea* spp. and *Iarea albifrons* coll., the gastropods *Theodoxus fluviatilis* and *Lymnaea* spp., and dipteran (*Chironomidae*) larvae. The *Cladophora* belt offers both protection and nourishment, and enlarges the hospitable area available for this variety of species. According to Jansson (1967), one dm^2 of rock surface may produce 3 km of algal filaments per month, i.e., an algal surface of 50 dm^2, colonized by approx. 50 macrofaunal and 12,000 microscopic individual animals.

For several species the Cladophora belt acts as a nursery from which the animals migrate towards the next belt, characterized by bladderwrack (*Fucus vesiculosus*). Many of the animals living here are highly vagile and, e.g., Mysidacea, Isopoda and Amphipoda are known to undertake diurnal migrations (Jansson and Källander, 1968). Also sessile and hemisessile forms (the bryozoan *Electra crustulenta*, the barnacle *Balanus improvisus*, the bivalve *Mytilus edulis*, and hydroid polyps) can live attached to *Fucus*. In the Askö area Haage (1975) recorded from 30 to about 1200 individuals of macrofauna per 100 g algae (dry weight). Of the 55 species found, as many as 24 were of limnetic origin, mainly gastropods and larvae of Trichoptera and Chironomidae.

Along the southern coasts of the Baltic Proper, some marine gastropods (*Littorina saxatilis, L. littorea*), amphipods (*Microdeutopus gryllotalpa*), and isopods are added to the *Fucus* belt fauna (v. Oertzen, 1968). In general, however, the resemblance of the fauna of the littoral zones in different parts of the Baltic Sea is great compared to the faunal assemblages of deeper waters. However, in the very diluted parts of the Gulfs the situation is quite different. *Fucus vesiculosus* disappears in the northern Bothnian Sea: north of its limit also the rocky-shore fauna is greatly impoverished.

Sublittoral zone

The sublittoral zone, here defined as the zone below the lower limit of benthic vegetation, is in the Baltic Sea a very extensive region, both geographically and as far as the amplitude of environmental factors is concerned. The main factors regulating the benthic populations are salinity, oxygen, temperature, substrate structure, and quality and quantity of food. The heterotrophic benthic subsystem is coupled to the pelagic and littoral producer systems through the seasonal pulse of sedimenting organic matter.

In the Baltic Proper the stability of the stratification is the basic phenomenon which rules, through its different physical and chemical consequences, the ecological conditions below the permanent halocline both for pelagic and benthic life. The special conditions prevailing below the permanent halo-

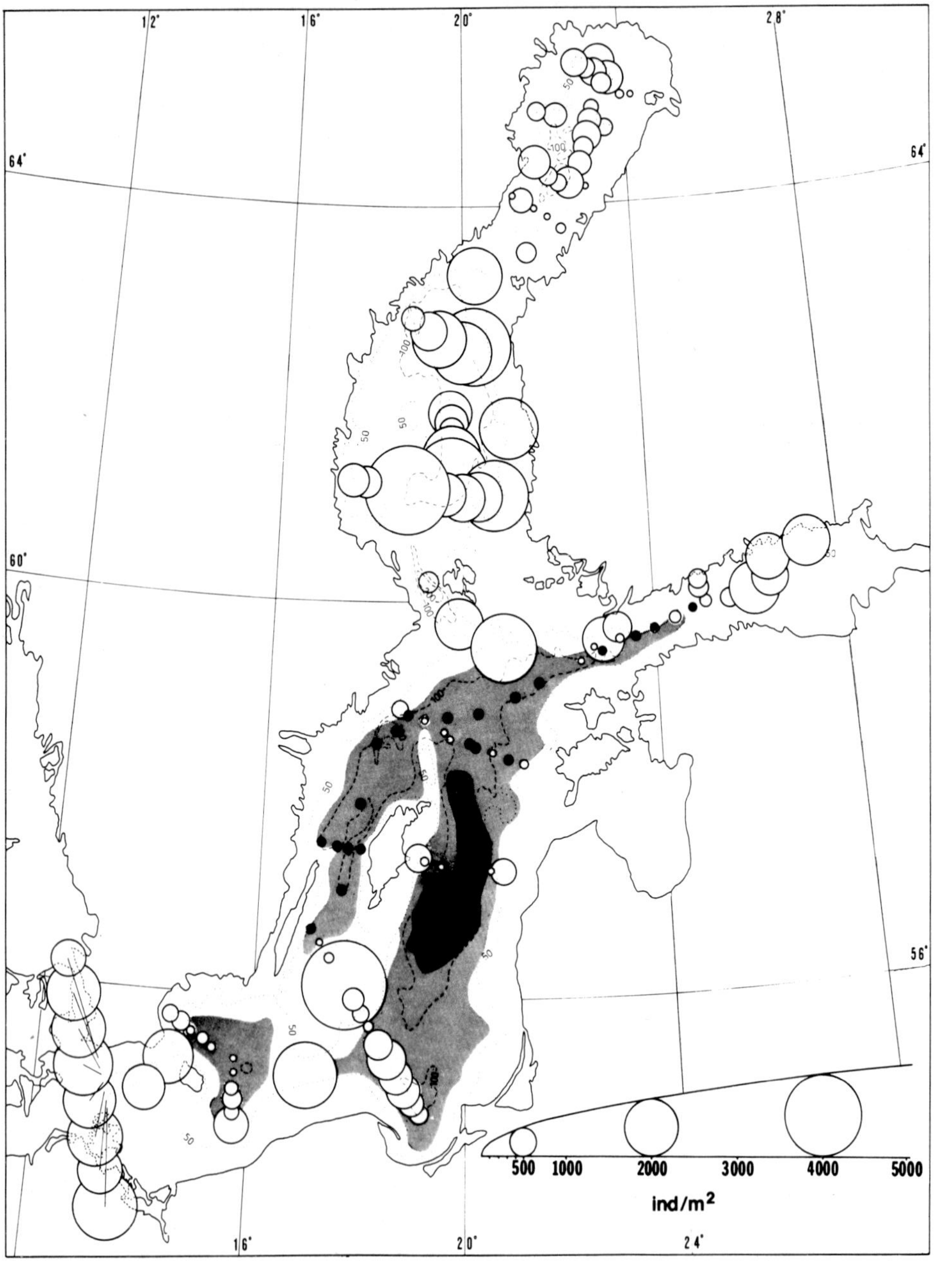

Fig. 5.15. Density of macrofauna in the Baltic Sea in June and July 1967. Filled circles indicate stations devoid of macrofauna. The light shading shows the areas in which oxygen contents below 2.86 g m^{-3} (2 ml/l) were recorded in the bottom water in 1967. The dark shading shows the area where hydrogen sulphide occurred (Andersin et al., 1977).

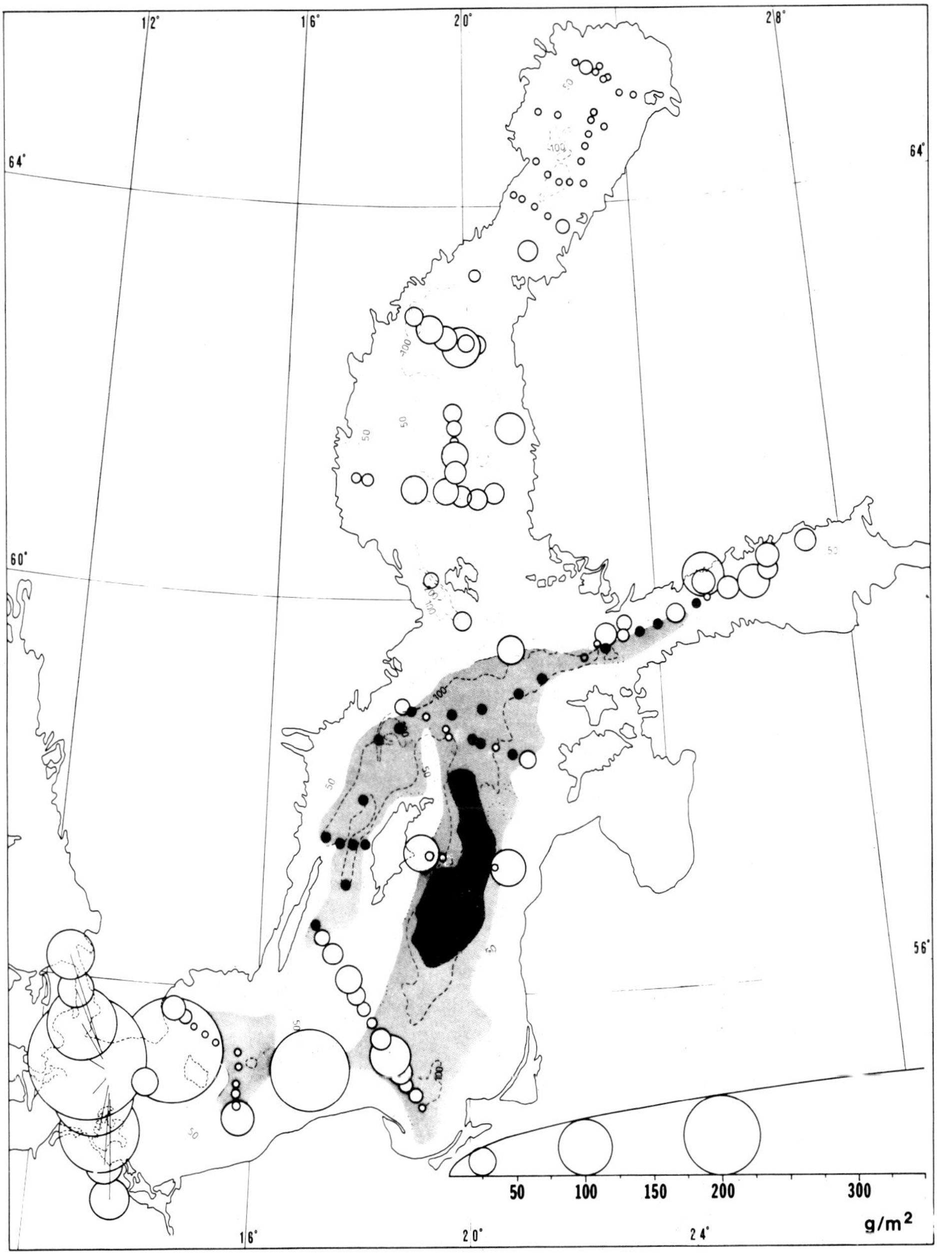

Fig. 5.16. Biomass (wet weight) of macrofauna in the Baltic Sea in June and July 1967. Shaded areas and filled circles as in Fig. 5.15 (Andersin et al., 1977).

cline, manifested mainly in the oxygen content, are described in Chapter 4 (Section on p. 188). The conditions in the Gulf of Bothnia differ in many respects from those of the Baltic Proper, including the Gulf of Finland. For reasons explained in Chapters 3 and 4, the oxygen conditions are good or satisfactory even in the deepest parts of the Gulf of Bothnia. The temperature amplitude in the sublittoral zone is considerable when the conditions in the southern part of the Baltic Sea are compared with those prevailing in the Bothnian Bay. However, the ecological effect of this difference is still poorly known. The occurrence of benthic macrofauna in a recent favourable year as regards the deep parts of the Baltic Sea is here exemplified by abundance and biomass values (Figs. 5.15 and 5.16).

In the Arkona Basin where the oxygen conditions are always satisfactory, the abundance values are generally fairly high without any extreme changes. Drastic changes are typical of the deep parts (below 80 m) of the Bornholm Basin where azoic periods alternate with periods of recolonization (Tulkki, 1965; Leppäkoski, 1975b; Zmudzinski, 1977; Andersin et al., 1978b) and this is also true of the deeper parts of the Gulf of Gdansk. These changes resemble in many respects the macrobenthic succession characteristic of polluted marine environments as well as faunal recovery subsequent to pollution abatement (Leppäkoski, 1975a; Pearson and Rosenberg, 1978). In the Central Basin, west, east and north of Gotland and in the deeper parts of the Gulf of Finland the macrofauna has been completely absent in vast areas for many years since the early 1960s (Andersin et al., 1978b). Where macrofauna exists it is strongly impoverished.

In the Gulf of Bothnia an extreme low-diversity community occurs. The amphipod *Pontoporeia affinis* predominates yielding in the southern part of the Gulf, the Bothnian Sea, the highest density values for macrofauna in the entire Baltic Sea. In the Bothnian Bay, the northernmost part of the Baltic Sea the benthic density is low, comprising normally some hundreds of individuals per m^2. Despite the high abundance values in the Bothnian Sea, the biomass values are not especially high. Like the abundance values the biomass values are strikingly small in the Bothnian Bay, being below 1 g m^{-2} in the greatest part of the area (Andersin et al., 1977). The reasons for the low biomass values in the Bothnian Bay are, in addition to the low abundance values, in general the relatively lower abundance of the large isopod *Mesidotea entomon* which forms 30—80% of the total biomass in the Bothnian Sea. Further, the average weight of *Pontoporeia* individuals is lower in the Bothnian Bay than in the Bothnian Sea. This is partly due to differences in the size distribution but partly also to differences in the weight of animals of the same length groups, the animals in the Bothnian Bay being both smaller and lighter than in the Bothnian Sea.

The highest biomass values of benthic macrofauna in this deep zone of the Baltic Sea are found in those parts of the southern Baltic Sea, where the oxygen conditions are good enough for the development of lamellibranch populations. In the Central Basin, where the oxygen content is the limiting factor for the occurrence of zoobenthos, the biomass values are low.

The highest diversity for soft-bottom macrofauna in the deeper parts of the Baltic Sea is, as expected, found in the Arkona Basin and in the transition area between the Bornholm Basin and the Central Basin, where both salinity, oxygen content and presumably food availability are favourable (Andersin et al., 1977). If the meiofauna species are included into diversity calculations, much higher values will be obtained. An illustrative example is given by Elmgren (1978): in the Askö area (at a depth of 44 m) a single meiofauna sample may contain 19 species per cm^2, whereas a macrofauna sample normally yields 5 or 6 species per m^2. Also in this respect the situation in the Gulf of Bothnia is an extreme situation with strikingly low diversity figures because of the strong dominance of *Pontoporeia affinis* which, particularly in the Bothnian Bay, forms a single species community in vast areas.

Concerning the dominance relations there is an obvious change when moving northwards in the Baltic Sea, the dominance being concentrated in a decreasing number of species. In the Arkona Basin the three most dominant species of macrofauna in 1967 accounted for about 70% of the total number of individuals, while the corresponding percentages for the Gulf of Finland and the Gulf of Bothnia were 97% and 99%, respectively (see Fig. 5.17). Provided the oxygen conditions are favourable, the major part of the macrofauna in the southern Baltic Sea is made up of various combinations of the polychaetes *Scoloplos armiger* and *Aricidea suecica*, the crustaceans *Diastylis rathkei* and *Pontoporeia femorata*, and the lamellibranchs *Astarte borealis* and *Macoma baltica*. As far as the biomass is concerned the lamellibranchs

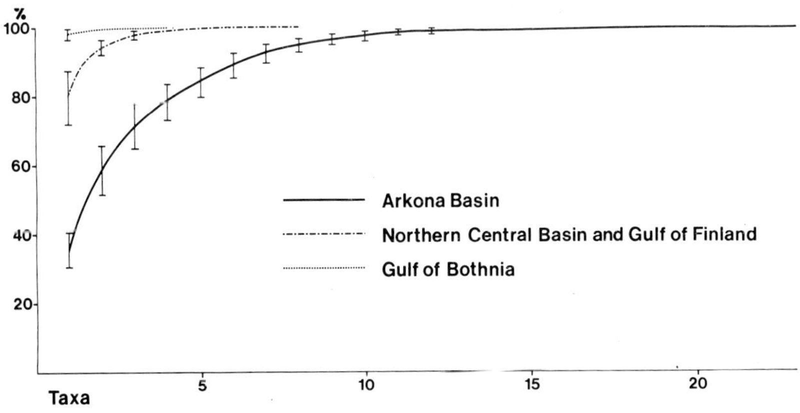

Fig. 5.17. Increase in percentage of total number of individuals with increasing number of taxa in different parts of the Baltic Sea in June and July 1967. Means and standard deviations (vertical lines) of all the stations sampled in each region (Andersin et al., 1977).

dominate in those regions (shell weight included). When going northwards, the increasing dominance of crustaceans is obvious as regards both abundance and biomass. In the Gulf of Finland *Pontoporeia affinis, P. femorata, Mesidotea entomon*, and *Macoma baltica* are important species. In the Gulf of Bothnia the dominance of crustaceans is striking. In addition to the predominant amphipod *Pontoporeia affinis* the isopod *Mesidotea entomon* occurs in low numbers. However, *Mesidotea* contributes strongly to the biomass values. At those depths of the Baltic Sea (Bornholm Basin, the Gulf of Gdansk, the Central Basin, and the Gulf of Finland) where the oxygen conditions are critical, but zoobenthos still exists, polychaetes predominate, mainly *Harmothoë sarsi, Scoloplos armiger*, and *Capitella capitata* in the southern part of the Baltic Sea, and *Harmothoë* in the Gulf of Finland (Andersin et al., 1978b).

Production and utilization of zoobenthos

The benthic animals of the Baltic Sea are normally not utilized for human consumption and thus have no direct commercial value. Big edible crustaceans do not exist. The brown shrimp (*Crangon crangon*) is very scarce or absent in most parts of the Baltic Sea, and the edible bivalves: the mussel (*Mytilus*), the cockle (*Cardium*) and the soft-shell clam (*Mya*) are reduced in size to an extent which makes commercial use unprofitable. However, the role of benthic animals is significant in the food web, and the estimation of their production thus very important.

In the estimation of benthic secondary production the complicated food web has to be considered. The present knowledge of the trophic pathways within the benthic community typical of the well-oxygenated bottom at intermediate depths in the northern Baltic Proper was summarized by Ankar (1977) including also the most important consumers which utilize macrobenthos as food (see Fig. 5.18). Elmgren (1978) who gives a review of structure and dynamics of benthic communities of the Baltic Sea, includes a discussion of production estimates.

For the Askö—Landsort area, Ankar and Elmgren (1975) reported a macrofauna-to-meiofauna biomass ratio of about 23:1 at depths between 3 m and 50 m. The ostracods made up 50% of the meiofauna biomass, whereas nematods made up 90% of the abundance values. Other groups recorded were Kinorhyncha, Turbellaria, Harpacticoida and copepod nauplii. With increasing depth, both the biomass and the number of taxa are strongly reduced, and in the oxygen-poor basins only Nematoda are found (see Fig. 5.19).

The first attempt to estimate the level of the benthic secondary production of these bottoms was made by Ankar and Elmgren (1976). They considered both the macro- and meiofaunal groups. The ratio of macrofauna to meiofauna was estimated as 2:1 for total benthic faunal production which

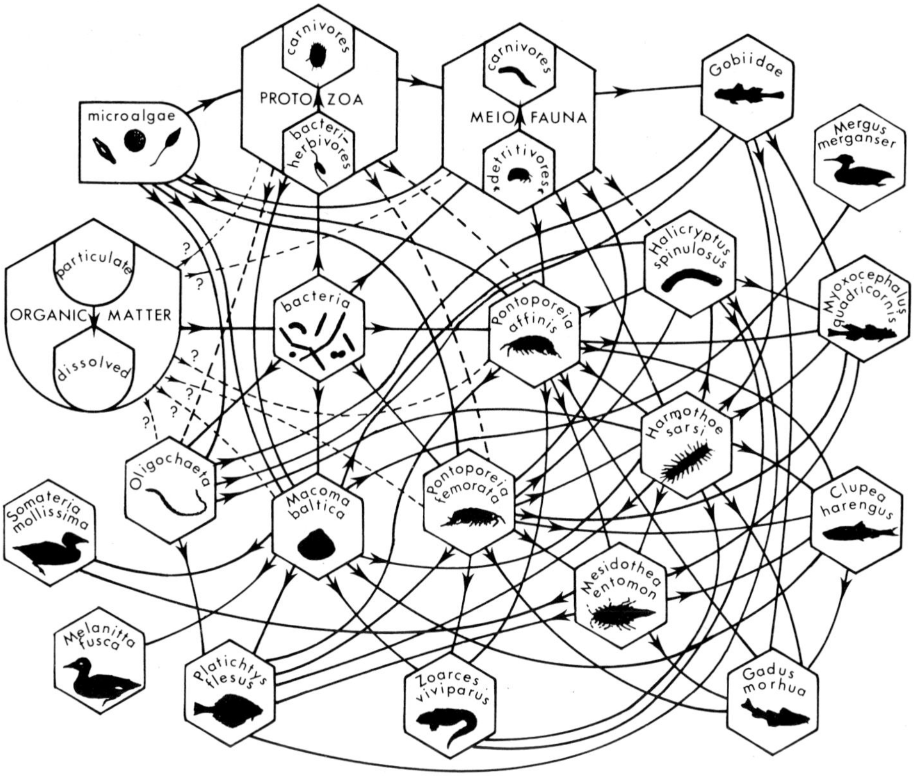

Fig. 5.18. Trophic relations of the dominating soft-bottom macrofauna taxa in the northern Baltic Proper. Dashed lines indicate possible but not yet proved food items and dashed lines with question marks questionable energy sources probably of minor importance (Ankar, 1977).

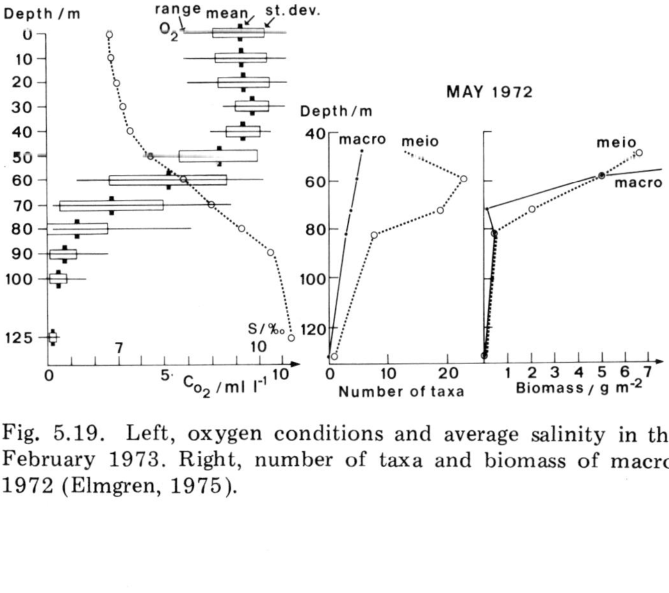

Fig. 5.19. Left, oxygen conditions and average salinity in the Landsort Deep 1969 in February 1973. Right, number of taxa and biomass of macro- and meiofauna in May 1972 (Elmgren, 1975).

equals about 340 kJ m^{-2} a^{-1}, whereas it was about 8:1 for energy in the benthic biomass and as high as 20:1 for wetweight biomass. Of this production, about one third was estimated to be left for the bottom-living fish. Very little is known of the sedimentation rates and the input of energy into the detritus-based, heterotrophic benthic ecosystem of the Baltic Sea. A total energy input corresponding to a carbon content of 40—60 gC m^{-2} a^{-1} is needed to fuel the benthos subsystem (including bacteria and microfauna) according to the estimates of Ankar and Elmgren (1976) and Ankar (1977). Ankar (1977) summarized the production estimates from two parts of the Baltic Sea as follows:

	Southern Baltic Proper Kiel Bay (Arntz and Brunswig, 1976) (kJ m^{-2} a^{-1})	Northern Baltic Proper Askö-Landsort (Ankar, 1977) (kJ m^{-2} a^{-1})
Pelagic primary production	9200	6000
Sedimentation	4900	2400
Micro-organisms		1000
Meiofauna		100
Macrofauna	440	200
Macrofauna, carnivores	20	20
Demersal fish	30	5

Benthos research

Since the 1960s, attention has been paid to the use of the benthic animals as indicators of pollution in different inshore areas of the Baltic Sea. In the 1970s, they have been used also in bioassays and in monitoring long-term changes caused by oil pollution, contamination by, and accumulation of, environmental chemical pollutants, such as chlorinated hydrocarbons and heavy metals (see Segerstråle, 1976, for a review). These new fields of benthic research are based not only on the aggravation of environmental problems and the development of analytical techniques but also on the scientific tradition of more than a hundred years. Benthos investigations made in, e.g., the Danish waters, the Kiel Bay, the Bornholm and Gotland Basins and the Gulf of Finland all cover a period of 50 years dating at least back to the classical work of Petersen (1913). By quantitative studies on benthic meiofauna, started in the 1970s (Elmgren, 1975, 1976, 1978), the knowledge of energy pathways within the benthos system and interactions between sediment and macrofauna have increased considerably.

The importance of benthic research has been realized in different forms of international cooperation in the Baltic Sea area. Since its start the organisation "Baltic Marine Biologists" (BMB), has worked on development, intercalibration and standardization of methods for studies in zoobenthos, covering both macro- and meiofauna of soft bottoms and littoral fauna as well (see Dybern et al., 1976). Studies on benthic fauna have been included in the programmes of different cooperative studies, based on bilateral agreements between different Baltic Sea states, and are also included in a joint monitoring programme under the auspices of the Convention on the Protection of the Marine Environment of the Baltic Sea Area (see Chapter 8, Section B).

REFERENCES

Abshagen, G., 1980. Das Phytoplankton des Greifswalder Boddens. Diss., Greifswald, 100 pp.

Ackefors, H., 1969a. Ecological zooplankton investigations in the Baltic Proper 1963—1965. Inst. Mar. Res., Lysekil, Ser. Biol. Rep., 18: 1—139.

Ackefors, H., 1969b. Seasonal and vertical distribution of the zooplankton in the Askö area (northern Baltic Proper) in relation to hydrographical conditions. Oikos, 20(2): 480—492.

Ackefors, H., 1971a. A quantitative plankton sampler. Oikos, 22(1): 114—118.

Ackefors, H., 1971b. Podon polyphemoides Leuckart and Bosmina coregoni maritima (P.E. Müller) in relation to temperature and salinity in field studies and laboratory experiments. J. Exp. Mar. Biol. Ecol., 7(1): 51—70.

Ackefors, H., 1972. The amount of zooplankton expressed as numbers, wet weight and carbon content in the Askö area (the northern Baltic Proper). Medd. Havsfiskelab., Lysekil, 129: 1—10, appendices.

Ackefors, H. and Hernroth, L., 1972. Djurplankton i Östersjöområdet. Zool. Revy, 34 (1—4): 6—31.

Ackefors, H. and Lindahl, O., 1975a. Investigations on primary phytoplankton production in the Baltic in 1974. Medd. Havsfiskelab. Lysekil, 195: 1—13, appendices.

Ackefors, H. and Lindahl, O., 1975b. Investigations on primary production in the Baltic in 1973. Medd. Havsfiskelab. Lysekil, 279: 1—14, appendices.

Ackefors, H. and Lindahl, O., 1979. Primary phytoplankton production in the Baltic Proper, 1973 1976, in relation to secondary zooplankton production. ICES C.M. 1979/L: p. 32 (mimeogr.).

Ackefors, H. and Rosén, C.-G, 1970. Temperature preference experiments with Podon polyphemoides Leuckart in a new type of alternative chamber. J. Exp. Mar. Biol. Ecol., 4(3): 221—228.

Ackefors, H., Hernroth, L., Lindahl, O. and Wulff, F., 1978. Ecological production studies of the phytoplankton and zooplankton in the Gulf of Bothnia. Finn. Mar. Res., 244: 116—126.

Alasaarela, E., 1979. Ecology of phytoplankton in the northern part of the Bothnian Bay. Acta Bot. Fenn., 110: 63—70.

Alasaarela, E. and Siira, J., 1976. Of the phytoplankton in the Liminganlahti Bay. Acta Univ. Oul. A, 42: 63—71.

Andersin, A.-B., Lassig, J. and Sandler, H., 1977. Community structures of soft-bottom macrofauna in different parts of the Baltic. In: B.F. Keegan, P. O'Céidigh and P.J.S. Boaden (Editors). Biology of Benthic Organisms. 11th European Symposium on Marine Biology, Galway, Oct. 1976, pp. 7—20.

Andersin, A.-B., Lassig, J., Parkkonen, L. and Sandler, H., 1978a. Long-term fluctuations in the deep areas of the Gulf of Bothnia 1954—1974; with special reference to Pontoporeia affinis Lindström (Amphipoda). Finn. Mar. Res., 244: 137—144.

Andersin, A.-B., Lassig, J., Parkkonen, L. and Sandler, H., 1978b. The decline of macrofauna in the deeper parts of the Baltic Proper and the Gulf of Finland. Kieler Meeresforsch., Sonderh., 4: 23—52.

Ankar, S., 1977. The soft bottom ecosystem of the northern Baltic Proper with special reference to the macrofauna. Contrib. Askö Lab., 19: 1—62.

Ankar, S. and Elmgren, R., 1975. A survey of the benthic macro- and meiofauna of the Askö—Landsort area. Merentutkimuslaitoksen Julk. Havsforskningsinst. Skr., 239: 257—264.

Ankar, S. and Elmgren, R., 1976. Benthic macro- and meiofauna of the Askö—Landsort area (northern Baltic Proper). A stratified random sampling survey. Contrib. Askö Lab., 11: 1—115.

Arntz, W.E. and Brunswig, D., 1976. Studies on structure and dynamics of macrobenthos in the western Baltic carried out by the joint research programme "Interaction sea—sea bottom" (SFB 95-Kiel) Proc. 10th European Symp. Mar. Biol., Ostend, Belgium, Sept. 17—23, 1975. Vol. 2: 17—42.

Bagge, P. and Lehmusluoto, P.O., 1971. Phytoplankton primary production in some Finnish coastal areas in relation to pollution. Merentutkimuslaitoksen Julk. Havsforskningsinst. Skr., 235: 3—18.

Bagge, P. and Niemi, Å., 1971. Dynamics of phytoplankton primary production and biomass in Loviisa archipelago (Gulf of Finland). Merentutkimuslaitoksen Julk. Havsforskningsinst. Skr., 233: 19—41.

Beers, J.R. and Stewart, G.L., 1970. Numerical abundance and estimated biomass of microzooplankton. In: J.D.H. Strickland (Editor), The Ecology of the Plankton off La Jolla, California, in the Period April through September, 1967. Bull, Scripps Inst. Oceanography.

Brattberg, G. 1977. Nitrogen fixation in a polluted brackish water archipelago. Ambio. Spec. Rep., 5: 27—42.

Buch, K., 1932. Untersuchungen über gelöste Phosphate und Stickstoffverbindungen in den nordbaltischen Meeresgebieten. Merentutkimuslaitoksen Julk. Havsforskningsinst. Skr., 86: 1—30.

Buch, K., 1948. Ämnesomsättningen i skärgårdsvattnen. In: Skärgårdsboken. Nordenskiöld-Samfundet, Helsingfors, pp. 134—146.

Buch, K., 1954. Physikalische und chemische Verhältnisse in Beziehung zur biologischen Aktivität im Wasser der südwestlichen Schärenmeeres von Finnland. Merentutkimuslaitoksen Julk. Havsforskningsinst. Skr., 159: 1—36.

Bursa, A., Wojtusiak, H. and Wojtusiak, R.J., 1939. Untersuchungen über die Bodenfauna und Bodenflora der Danziger Bucht unter Anwendung eines Taucherhelms. Bull. Acad. Pol. Sci. Lett. Cl. Sci. Math. Nat., Sér. B (II), 1939: 61—97.

Bursa, A., Wojtusiak, H. and Wojtusiak, R.J., 1948. Investigations of the bottom fauna and flora in the Gulf of Gdańsk made by using a diving helmet. II. Bull. Acad. Pol. Sci. Lettr., Cl. Sci. Math. Nat., Sér. B (II), 1947: 213—239.

Cooper, L.H.N.., 1937. On the ratio of nitrogen to phosphorus in the sea. J. Mar. Biol. Assoc, U.K., 22: 177—182.

Dahlin, H. 1978. Översikt av Bottniska vikens vattenkemi och materialbalans. Finnish-Swedish seminar of the Gulf of Bothnia, Vaasa, Finland, March 8th—9th, 1978, 18 pp., appendices (mimeogr.).

Du Rietz, G.E., 1950. Phytogeographical excursion to the maritime birch forest zone and the maritime forest limit in the outermost archipelago of Stockholm. 7. Int. Bot. Congr. Excursion Guide B 1, 11 pp. Uppsala.

Dybern, B.I., Ackefors, H. and Elmgren, R. (Editors), 1976. Recommendations on Methods for Marine Biological Studies in the Baltic Sea. Baltic Mar. Biol., Publ., 1: 1—98.

Edler, L., 1975. Qualitative analysis of phytoplankton. In: Investigations on primary phytoplankton production in the Baltic in 1973 (Appendix). Medd. Havsfiskelab. Lysekil, 179: A1—A7.

Edler, L., 1977. Phytoplankton and Primary Production in the Sound. Diss., Univ. Gothenberg, Dept. Mar. Bot. 82 pp.
Edler, L., 1979. Phytoplankton succession in the Baltic Sea. Acta Bot. Fenn., 110: 75—78.
Elmgren, R., 1975. Benthic meiofauna as indicator of oxygen conditions in the northern Baltic Proper. Merentutkimuslaitoksen Julk. Havsforskningsinst. Skr., 239: 265—271.
Elmgren, R., 1976. Baltic benthos communities and the role of the meiofauna. Contrib. Askö Lab., 14: 1—31.
Elmgren, R., 1978. Structure and dynamics of Baltic benthos communities, with particular reference to the relationship between macro- and meiofauna. Kieler Meeresforsch., Sonderh., 4: 1—22.
Elmgren, R. and Ganning, B., 1974. Ecological studies of two shallow, brackish water ecosystems. Contrib. Askö Lab., 6: 1—56.
Fagerholm, H.P., 1975. The effect of ferry traffic on the rocky shore macrofauna in the southern Åland archipelago: 1. The Cladophora zone. Merentutkimuslaitoksen Julk. Havsforskningsinst. Skr., 239: 331—337.
Fenchel, T., 1977. Competition, coexistence and character displacement in mud snails (Hydrobiidae). In: B.C. Coull (Editor), Ecology of Marine Benthos. Univ. South Carolina, pp. 229—243.
Fonselius, S., 1972. On biogenic elements and organic matter in the Baltic. Ambio, Spec. Rep., 1: 29—36.
Forsman, B., 1972. Evertebrater vid svenska östersjökusten. Zool. Revy, 34(1—4): 32—56.
Guterstam, B., 1977. An in situ study of the primary production and the metabolism of a Baltic Fucus vesiculosus L. community. In: B.F. Keegan, P. O'Céidigh and P.J.S. Boaden (Editors), Biology of Benthic Organisms. 11th European Symposium on Marine Biology, Galway, Oct. 1976, pp. 311—319.
Haage, P., 1975. Quantitative investigations of the Baltic Fucus belt macrofauna. 2. Quantitative seasonal fluctuations. Contrib. Askö Lab., 9: 1—88.
Hällfors, G., 1976. The plant cover of some littoral biotopes at Krunnit (NE Bothnian Bay). Acta Univ. Oul. A, 42: 87—95.
Hällfors, G., 1979. A preliminary check-list of the phytoplankton of the northern Baltic Sea. Vesientutkimuslaitoksen Julkaisuja. Publications of the Water Research Institute, 34: 3—24.
Hällfors, G. and Niemi, Å., 1974. A Chrysochromulina (Haptophyceae) bloom under the ice in the Tvärminne archipelago, southern coast of Finland. Memo. Soc. Fauna Flora Fenn., 50: 89—104.
Hällfors, G., Kangas, P. and Lappalainen, A., 1975. Littoral benthos of the northern Baltic Sea. III. Macrobenthos of the hydrolittoral belt of filamentous algae on rocky shores in Tvärminne. Int. Revue Ges. Hydrobiol., 60: 313—333.
Halme, E., 1958. Planktologische Untersuchungen in der Pojo-Bucht und angrenzenden Gewässern. IV Zooplankton. Ann. Zool. Soc. Zool. Bot. Fenn. Vanamo, 19(3): 1—62.
Hasle, G.R., 1978. Some freshwater and brackish water species of the diatom genus Thalassiosira cleve. Phycologia, 17: 263—292.
Hensen, V., 1887. Über die Bestimmung des Planktons oder des im Meere treibenden Materials an Pflanzen und Thieren. Ber. Komm. Wiss. Unters. Deutsch. Meere, 5: 1—107.
Hernroth, L. in press, a. The zooplankton of the Baltic Proper — Zooplankton dynamics of a spring bloom in the northern Baltic Proper. Thesis, Univ. of Stockholm.
Hernroth, L., in press, b. Östersjöns zooplankton — En jämförande studie av Nansenhåven och WP 2-håven med avseende på kvalitativa och kvantitativa fångstegenskaper. Thesis, Univ. of Stockholm.

Hernroth, L. and Ackefors, H., 1979. The zooplankton of the Baltic Proper. A long-term investigation of the fauna, its biology and ecology. Fish. Board. Swed. Inst. Mar. Res. Rep., 2: 1—60.

Hickel, W., 1969. Planktologische und hydrographisch-chemische Untersuchungen in der Eckernförder Bucht (westliche Ostsee) 1962/1963. Helgol. Wiss. Meeresunters., 19: 318—331.

Hillebrandt, M., 1972. Untersuchungen über die qualitative und quantitative Zusammensetzung des Zooplanktons in der Kieler Bucht während der Jahre 1966—68. Diss. Kiel, 138 pp.

Hobro, R. and Nyqvist, B., 1972. Pelagial studies in the Landsort area during 1970-1971. In: Askölaboratoriet. Ekologiska Undersökningar i Landsortsområdet 1970—1971. Rep. Swed. Environm. Prot. Board. Contrib. I 10/35 and 7-85/70. Subrep. 7 pp. (mimeogr.).

Hobro, R., Larsson, U. and Wulff, F., 1975. Dynamics of a phytoplankton spring bloom in a coastal area of the northern Baltic. 4th Baltic Marine Biologists Symposium Gdańsk 13—18 Oct. 1975, Abstr. p. 15. (To be published in Prace Morskiego Instytut Rybacki).

Horner, R.A., 1976. Sea ice organisms. Oceanogr. Mar. Biol. Annu. Rev., 14: 167—182.

Horstmann, U., 1975. Eutrophication and mass production of blue-green algae in the Baltic. Merentutkimuslaitoksen Julk. Havsforskningsinst. Skr., 239: 83—90.

Hübel, H., 1968. Die Bestimmung der Primärproduktion des Phytoplanktons der Nord-Rügenschen Boddengewässer unter Verwendung der Radiokohlenstoffmethode. Int. Revue Ges. Hydrobiol., 53: 601—633.

Hübel, H. and Hübel, M., 1974. Stickstoff-Fixierung in Küstengewässern der mittleren Ostsee. Z. Allg. Mikrobiol., 14(7): 617—619.

Ilus, E. (Editor), 1979. Loviisan Hästholmenia ympäröivän merialueen tarkkailu vuonna 1977 (Investigations on the sea area surrounding Loviisa nuclear power plant in 1977). Institute of Radiation Protection, Helsinki, Rep., STL-B23: 1—50. (in Finnish, with an English summary).

Jansson, A.-M., 1967. The food-web of the Cladophora-belt fauna. Helgol. Wiss. Meeresunters., 15: 574—588.

Jansson, A.-M., 1974. Community structure, modelling and simulation of the Cladophora ecosystem in the Baltic Sea. Contrib. Askö Lab., 5: 1—130.

Jansson, A.-M., 1975. Systems analysis and simulation of the green algal belt (Cladophora) in the Baltic. Merentutkimuslaitoksen Julk. Havsforskningsinst. Skr., 239: 240—247.

Jansson, A.-M. and Kautsky, N., 1977. Quantitative survey of hard bottom communities in a Baltic archipelago. In: B.F. Keegan, P. O'Céidigh and P.J.S. Boaden (Editors), Biology of Benthic Organisms. 11th European Symposium on Marine Biology, Galway, Oct. 1976, pp. 359—366.

Jansson, B.-O., 1972. Ecosystem approach to the Baltic problem. Bull. Ecol. Res. Comm., 16: 1—82.

Jansson, B.-O., 1978. The Baltic — a systems analysis of a semi-enclosed sea. In: H. Charnock and G. Deacon (Editors), Advances in Oceanography. New York, London, pp. 131—183.

Jansson, B.-O. and Källander, C., 1968. On the diurnal activity of some littoral peracarid crustaceans in the Baltic Sea. J. Exp. Mar. Biol. Ecol., 2: 24—36.

Jansson, B.-O. and Wulff, F. (Editors), 1977. Ecosystem analysis of a shallow sound in the northern Baltic — a joint study by the Askö group. Contrib. Askö Lab., 18: 1—160.

Jones, K. and Stewart, W.D.P., 1969. Nitrogen turnover in marine and brackish habitats. III. The production of extracellular nitrogen by Calothrix scopulorum. J. Mar. Biol. Assoc. U.K., 49: 475—488.

Kaiser, W. and Schulz, S., 1975. On primary production in the Baltic. Merentutkimuslaitoksen Julk. Havsforskningsinst. Skr., 239: 29—33.
Kell, V., 1973. Die Darsser Schwelle — eine biologische Grenze? Wiss. Z. Univ. Rostock, Math.-Naturwiss. Reihe, 22: 617—623.
King, R.J. and Schramm, W., 1976. Photosynthetic rates of benthic marine algae in relation to light intensity and seasonal variations. Mar. Biol., 37: 215—222.
Kinne, O., 1964a. Non-genetic adaptation to temperature and salinity. Helgol. Wiss. Meeresunters., 9(1—4): 433—458.
Kinne, O., 1964b. Physiologische und ökologische Aspekte des Lebens in Ästuarian. Helgol. Wiss. Meeresunters., 11(3—4): 131—156.
Kornaś, J., 1959. Sea bottom vegetation of the Bay of Gdańsk off Rewa. Bull. Acad. Pol. Sci. Cl. II., Sér. Sci. Biol., 7(1): 5—10.
Kornaś, J. and Medwecka-Kornaś, A., 1949. Les associations végétales sousmarines dans le Golfe du Gdańsk (Baltique Polonaise). Bull. Acad. Pol. Sci. Lettr., Cl. Sc. Math. Nat., Sér. B (I), 1948: 71—88.
Kornaś, J., Pancer, E. and Brzyski, B., 1960. Studies on sea-bottom vegetation in the Bay of Gdańsk of Rewa. Fragm. Florist. Geobot. (Krakow), 6(1): 1—92.
Lappalainen, A., Hällfors, G. and Kangas, P., 1977. Littoral benthos of the northern Baltic Sea. IV. Pattern and dynamics of macrobenthos in a sandy-bottom Zostera marine community in Tvärminne. Int. Revue Ges. Hydrobiol., 62: 465—503.
Lassig, J. and Niemi, Å., 1975. Parameters of production in the Baltic measured during cruises with R/V Aranda in June and July 1970 and 1971. Merentutkimuslaitoksen Julk. Havsforskningsinst. Skr., 239: 34—40.
Lassig, J., Leppänen, J.-M., Niemi, Å. and Tamelander, G., 1978. Phytoplankton primary production in the Gulf of Bothnia 1972—1975 as compared with other parts of the Baltic Sea. Finn. Mar. Res., 244: 101—115.
Lehmusluoto, P.O., 1968. Kasviplanktonin perustuotanto Helsingin edustan merialueella (Phytoplankton primary production off the coast of Helsinki). Limnologisymposion Helsinki, 1967, pp. 31—42 (in Finnish, with an English summary).
Lenz, G., 1977. On detritus as a food source for pelagic filter-feeders. Mar. Biol., 41(1): 39—48.
Leppäkoski, E., 1975a. Assessment of degree of pollution on the basis of macrozoobenthos in marine and brackish-water environments. Acta Acad. Abo. Ser. B, 35(2): 1—90, appendices.
Leppäkoski, E., 1975b: Macrobenthic fauna as indicator of oceanization in the southern Baltic. Merentutkimuslaitoksen Julk. Havsforskningsinst. Skr., 239: 280—288.
Levring, T., 1940. Studien über die Algenvegetation von Blekinge, Südschweden. Diss. Lund, 178 pp.
Lindahl, G., Wallström, K. and Brattberg, G., 1978. On nitrogen fixation in a coastal area in the northern Baltic. Kieler Meeresforsch. Sonderh., 4: 171—177.
Lindahl, O., 1977a. Studies on the production of phytoplankton and zooplankton in the Baltic in 1975. Medd. Havsfiskelab. Lysekil, 217: 1—23.
Lindahl, O., 1977b. Studies on the production of phytoplankton and zooplankton in the Baltic in 1976, and a summary of results from 1973—1976. Medd. Havsfiskelab. Lysekil, 220: 1—30.
Lindner, A., 1975. Katalog der submersen Makrophyten in der Boddenkette südlich des Darss und des Zingst unter Berücksichtigung der Autökologie produktionsbiologisch wichtiger Species. Wiss. Z. Univ. Rostock, Math.-Naturwiss. Reihe, 24: 735—742.
Lindquist, A., 1959. Studien über das Zooplankton der Bottensee. II. Zur Verbreitung und Zusammensetzung des Zooplanktons. Inst. Mar. Res., Lysekil, Ser. Biol. Rep., 11: 1—136, appendices.
Lohmann, H., 1908. Untersuchungen zur Feststellung des vollständigen Gehaltes des Meeres an Plankton. Wiss. Meeresunters. Abt. Kiel, N.F., 10: 131—370.

Luther, H., 1951a. Verbreitung und Ökologie der höheren Wasserpflanzen im Brackwasser der Ekenäs-Gegend in Südfinnland. I. Allgemeiner Teil. Acta Bot. Fenn., 49: 1—231.
Luther, H., 1951b. Verbreitung und Ökologie der höheren Wasserpflanzen im Brackwasser der Ekenäs-Gegend in Südfinnland. II. Spezieller Teil. Acta Bot. Fenn., 50: 1—370.
Luther, H., Hällfors, G., Lappalainen, A. and Kangas, P., 1975. Littoral benthos of the northern Baltic Sea. I. Introduction. Int. Revue Ges. Hydrobiol., 60: 289—296.
Melin, K.E.R. and Lindahl, P.E.B., 1973. Algal biotest of Stockholm archipelago waters — qualitative aspects. Oikos Suppl., 15: 189—194.
Melvasalo, T., 1971. Havaintoja Helsingin ja Espoon merialueiden kasviplanktonlajistosta ja -biomassoista vuosina 1966—1970 (Observations on phytoplankton species and biomass in the sea area of Helsinki and Espoo in 1966—1970). Vesiensuojelulaboratorion Tiedonantoja. Reports of the Water Conservation Laboratory, 3(10): 1—98 (in Finnish, with an English summary).
Melvasalo, T., Viljamaa, H. and Huttunen, M., 1973. Planktonanalyysimenetelmät vuosina 1966—1972 (Plankton-methods in the Water Conservation Laboratory in 1966—1972). Vesiensuojelulaboratorion Tiedonantoja. Reports of the Water Conservation Laboratory, 5(2): 1—21 (in Finnish, with an English summary).
Menge, J.L., 1975. Effect of Herbivores on Community Structure of the New England Rocky Intertidal Region: Distribution, Abundance and Diversity of Algae. Diss. Harvard Univ., Cambridge, Mass., 164 pp.
Meskus, E., 1976. The primary production of phytoplankton in the northeastern Bothnian Bay. Acta Univ. Oul. A, 42: 55—62.
Michanek, G., 1970. Data on primary production measured by the C-14 method at two lightships in the Baltic. Medd. Havsfiskelab. Lysekil, 86: 1—15.
Muus, B.J., 1967. The fauna of Danish estuaries and lagoons. Medd. Danm. Fisk. Havunders. N.S., 5: 7—314.
Niemi, Å., 1972. Observations on phytoplankton in eutrophied and non-eutrophied archipelago waters of the southern coast of Finland. Memo. Soc. Fauna Flora Fenn., 48: 63—74.
Niemi, Å., 1973. Ecology of phytoplankton in the Tvärminne area, SW coast of Finland. I. Dynamics of hydrography, nutrients, chlorophyll a and phytoplankton. Acta Bot. Fenn., 100: 1—68.
Niemi, Å., 1975. Ecology of phytoplankton in the Tvärminne area. SW coast of Finland. II. Primary production and environmental conditions in the archipelago and the sea zone. Acta Bot. Fenn., 105: 1—68.
Niemi, Å., 1978. Ecology of phytoplankton in the Tvärminne area. SW coast of Finland. III. Environmental conditions and primary production in Pojoviken in the 1970s. Acta Bot. Fenn., 106: 1—28.
Niemi, Å., 1979. Blue-green algal blooms and N:P ratio in the Baltic Sea. Acta Bot. Fenn., 110: 57—61.
Niemi, Å. and Ray, I.-L., 1977. Phytoplankton production in Finnish coastal waters. Report 2: Phytoplankton biomass and composition in 1973. Meri, 4: 6—22.
Nikolaev, I.I., 1957. Biologičeskie sezony Baltijskogo morja. Trudy Latv. Otdelenija VNIRO, 2: 115—140.
Odum, H.T., 1972. An energy circuit language for ecological and social systems: its physical basis. In: B.C. Patten (Editor), Systems Analyses and Simulation in Ecology. New York, London, 2: 140—211.
v. Oertzen, J.A., 1968. Untersuchungen über die Besiedlung der Fucusvegetation der Gewässer um Hiddensee. Z. Fisch. NF, 16: 253—277.
Olsson, B., 1971. A statistical study of depth distribution in the Baltic Sea. Inst. Meteorol., Univ. of Stockholm, Rep., GH-3: 1—18.
Öström, B., 1976. Fertilization of the Baltic by nitrogen fixation in the blue-green alga Nodularia spumigena. Remote Sensing Environm., 4: 305—310.

Overbeck, J., 1965. Die Meeresalgen und ihre Gesellschaften an den Küsten der Insel Hiddensee (Ostsee). Bot. Mar., 3: 218—233.

Palosuo, E., 1966. Ice in the Baltic. Oceanogr. Mar. Biol. Annu. Rev., 4: 79—90.

Pankow, H., 1976. Algenflora der Ostsee. II. Plankton (einschl. benthischer Kieselalgen). Jena, 493 pp.

Parsons, T.R., Takahashi, M. and Hargrave, B., 1977. Biological Oceanographic Processes. Oxford, 2nd ed., 332 pp.

Pearson, T.H. and Rosenberg, R., 1978. Macrobenthic succession in relation to organic enrichment and pollution of the marine environment. Oceanogr. Mar. Biol. Annu. Rev., 16: 229—311.

Petersen, C.G.J., 1913. Havets bonitering II. Om havbundens dyresamfund og om disses betydning for den marine zoogeografi. Beretn. Danske Biol. Stat., 21: 3—42.

Petipa, T.S., Pavlova, E.V. and Mironov, C.N., 1970. The food web structure, utilization and transport of energy by trophic levels in the planktonic communities. In: J.H. Steele (Editor), Marine Food Chains, pp. 142—167.

Pietikäinen, S., Niemi, Å., Tulkki, P. and Aurimaa, K., 1978. Mean values and trends of physical and chemical properties in the Gulf of Bothnia 1962—1975. Finnish Mar. Res., 244: 64—75.

Rasmus, J., Lemons, F. and Zimmerman, C., 1977. Adaptation of light-harvesting pigments to downwelling light and the consequent photosynthetic performance of the eulittoral rockweeds Ascophyllum nodosum and Fucus vesiculosus. Mar. Biol., 42: 293—303.

Rautiainen, H. and Ravanko, O., 1972. The epiphytic diatom flora of the benthic macrophyte communities on rocky shores in the southwestern archipelago of Finland, Seili Islands. Nova Hedwigia Z. Kryptogamenkd., 23: 827—842.

Ravanko, O., 1968. Macroscopic green, brown, and red algae in the southwestern archipelago of Finland. Acta Bot. Fenn., 79: 1—50.

Ravanko, O., 1972. The physiognomy and structure of the benthic macrophyte communities on rocky shores in the southwestern archipelago of Finland (Seili Islands). Nova Hedwigia Z. Kryptogamenkd., 23: 363—403.

Remane, A., 1934. Die Brackwasserfauna. (Mit besonderer Berücksichtigung der Ostsee). Verh. Dtsch. Zool. Ges. E.V., 36: 34—74. Zool. Anz. Suppl., 7: 34—74.

Remane, A., 1940. Einführung in die zoologische Ökologie der Nord- und Ostsee. Tierwelt Nord-Ostsee, 1a: 1—238.

Renk, H., 1973. Primary production and chlorophyll content in the Baltic Sea. Part II. Chlorophyll-a distribution. Pol. Arch. Hydrobiol., 20: 237—255.

Renk, H., Torbicki, H. and Ochocki, S., 1975. Primary production in the Gdańsk Deep. Merentutkimuslaitoksen Julk. Havsforskningsinst. Skr., 239: 45—48.

Ringer, Z., 1973. Phytoplankton of the southern Baltic Sea. Pol. Arch. Hydrobiol., 20: 371—378.

Rinne, I. and Tarkiainen, E., 1978. Algal tests used to study the chemical factors regulating the growth of planktonic algae in the Helsinki sea area. Mitt. Int. Verein. Limnol., 21: 527—546.

Rinne, I., Melvasalo, T., Niemi, Å. and Niemistö, L., 1978. Nitrogen fixation by blue-green algae in the Baltic Sea. Kieler Meeresforsch., Sonderh., 4: 178—187.

Rothe, F., 1941. Quantitative Untersuchungen über die Planktonverteilung in der östlichen Ostsee. Ber. Dtsch. Wiss. Komm. Meeresforsch. N.F., 10: 291—368.

Rönnberg, O., 1975. The effects of ferry traffic on rocky shore vegetation in the southern Åland archipelago. Merentutkimuslaitoksen Julk. Havsforskningsinst. Skr., 239: 325—330.

Sarvala, J., 1971. Ecology of Harmothoë sarsi (Malmgren) (Polychaeta, Polynoidae) in the northern Baltic area. Ann. Zool. Fenn., 8(2): 231—309.

Schnack, S., 1975. Untersuchungen zur Nahrungsbiologie der Copepoden (Crustacea) in der Kieler Bucht. Diss., Kiel.
Schnese, W., 1973. Untersuchungen zur Produktionsbiologie des Greifswalder Boddens (südliche Ostsee). III. Abundanzen und Biomasseverteilung des Phytoplanktons im Jahrescyklus (1962—1965). Wiss. Z. Univ. Rostock, Math.-Naturwiss. Reihe, 22: 657—673.
Schramm, W., 1973. Langfristige in situ -Messungen zum Sauerstoffgaswechsel benthischer Meerespflanzen in einem kontinuerlich registrierenden Durchflussystem. Mar. Biol. 22: 335—339.
Schramm, W. and Guterstam, B., in press. Studies of the metabolic activity in Fucus communities from the Baltic. IV. Baltic Marine Biologists Symposium, Gdańsk, Poland, 13—18 October 1975.
Schwarz, S., 1959. Vergleichende Studien an tierischem Netz - und Vollplankton aus Brackwassergebieten den Ostsee. Z. Fisch. N.F., 8(4-6): 351—370.
Schwenke, H., 1964. Vegetation und Vegetationsbedingungen in der westlichen Ostsee (Kieler Bucht). Kieler Meeresforsch., 20: 157—168.
Schwenke, H., 1966. Untersuchungen zur marinen Vegetationskunde. I. Über den Aufbau der marinen Benthosvegetation im Westteil der Kieler Bucht (westliche Ostsee). Kieler Meeresforsch., 22: 163—170.
Schwenke, H., 1969. Meeresbotanische Untersuchungen in der westlichen Ostsee als Beitrag zu einer marinen Vegetationskunde. Int. Revue Ges. Hydrobiol., 54: 35—94.
Schwenke, H., 1974. Die Benthosvegetation. In: L. Magaard and G. Rheinheimer (Editors), Meereskunde der Ostsee. Springer, Berlin, pp. 131—146.
Schwenke, H., Schramm, W. and Black, H.J., 1975. Investigations on the phytobenthos in the "Hausgarten-Project SFB 95" in the Kiel Bight, western Baltic. Merentutkimuslaitoksen Julk. Havsforskningsinst. Skr., 239: 167—172.
Segerstråle, S.G., 1944. Weitere Studien über die Tierwelt der Fucus-Vegetation an der Südküste Finnlands. Commentat. Biol., 9(4): 1—28.
Segerstråle, S.G., 1957. Baltic Sea. In: J.W. Hedgpeth (Editor), Treatise on Marine Ecology and Paleoecology I. Ecology. Geol. Soc. Am. Mem., 67: 751—800.
Segerstråle, S.G., 1962. The immigration and prehistory of the glacial relicts of Eurasia and North America. A survey and discussion of modern views. Int. Revue Ges. Hydrobiol., 47(1): 1—25.
Segerstråle, S.G., 1976. Recent trends in Baltic marine biological research. Oceanogr. Mar. Biol. Annu. Rev., 14: 347—369.
Sen Gupta, R., 1972. Photosynthetic production and its regulating factors in the Baltic Sea. Mar. Biol., 17: 82—92.
Sjöblom, V. and Parmanne, R., 1978. Itämeren kalastus (Baltic fisheries). Suomen Luonto, 37: 120—125, 197 (in Finnish, with an English summary).
Smetacek, V., 1975. Die Sukzession des Phytoplanktons in der westlichen Kieler Bucht. Diss. Univ. Kiel, 151 pp.
Steemann Nielsen, E., 1952. The use of radioactive carbon (C^{14}) for measuring organic production in the sea. J. Cons. Int. Explor. Mer, 18: 117—140.
Stewart, W.D.P. and Alexander, G., 1971. Phosphorus availability and nitrogenase activity in aquatic blue-green algae. Freshwater Biol., 1: 389—404.
Strickland, J.D.H., 1972. Research on the marine planktonic food web at the Institute of Marine Resources: a review of the past seven years of work. Oceanogr. Mar. Biol. Annu. Rev., 10: 349—414.
Svedelius, N., 1901. Studier öfver Östersjöns hafsalgflora. Diss., Uppsala, 140 pp.
Sverdrup, H.U., 1953. On conditions for the vernal blooming of phytoplankton. J. Cons. Int. Explor. Mer, 18: 287—295.
Tarkiainen, E., Rinne, I. and Niemistö, L., 1974. On the chemical factors regulating the

primary production of phytoplankton in the Baltic Proper. Merentutkimuslaitoksen Julk. Havsforskningsinst. Skr., 238: 39—52.

Thomsen, H.A., 1977. Chrysochromulina pyramidosa sp. nov. (Prymnesiophyceae) from Danish coastal waters. Bot. Not., 130: 147—154.

Thomsen, H.A., 1979. Electron microscopical observations on brackish water nannoplankton from the Tvärminne area. SW coast of Finland. Acta Bot. Fenn., 110: 11—37.

Thorson, G., 1957. Bottom communities (Sublittoral or shallow shelf). In: J.W. Hedgpeth (Editor), Treatise on Marine Ecology and Paleoecology. I. Ecology Ged. Soc. Am. Mem., 461—534.

Trahms, O.-K., 1939. Beiträge zur Ökologie küstennaher Brackwässer. 1. Das Plankton des Grossen Jasmunder Boddens. Pol. Arch. Hydrobiol., 35: 529—551.

Trei, T., 1975. Flora and vegetation in the coastal waters of western Estonia. Merentutkimuslaitoksen Julk. Havsforskningsinst. Skr., 239: 348—351.

Tulkki, P., 1965. Disappearance of the benthic fauna from the basin of Bornholm (Southern Baltic), due to oxygen deficiency. Cah. Biol. Mar., 6: 455—463.

Unesco, 1968. Zooplankton sampling. Monographs on Oceanographic Methodology, 2: 1—174.

Utermöhl, H., 1958. Zur Vervollkommnung der quantitativen Phytoplankton-Methodik. Mitt. Int. Ver. Theor. Angew. Limnol., 9: 1—38.

Välikangas, I., 1933. Über die Biologie der Ostsee als Brackwassergebiet. Verh. Int. Ver. Theor. Angew. Limnol., 6: 62—112.

Voipio, A., 1961. The silicate in the Baltic Sea. Ann. Acad. Sci. Fenn. Ser. A II, 106: 1—15.

Voipio, A., 1976. Variations in nutrient content in the Bothnian Bay in 1966—1974. Acta Univ. Oul. A, 42: 73—78.

Wærn, M., 1949. Remarks on Swedish Lithoderma. Svensk Bot. Tidskr., 43: 633—670.

Wærn, M., 1952. Rocky shore algae in the Öregrund archipelago. Acta Phytogeogr. Suecica, 30: 1—298.

Wallentinus, I., 1974. Fluctuations of Scytosiphon lomentaria in the northern Baltic Proper. Memo. Soc. Fauna Flora Fenn., 50: 81—88.

Wallentinus, I., 1975. Primary production of macroalgae measured by the ^{14}C method. Merentutkimuslaitoksen Julk. Havsforskningsinst. Skr., 239: 72—77.

Wallentinus, I., 1976a. Environmental influences on benthic macrovegetation in the Trosa—Askö area, northern Baltic Proper. I. Hydrographical and chemical parameters, and the macrophytic communities. Contrib. Askö Lab., 15: 1—138.

Wallentinus, I., 1976b. Productivity studies on Cladophora glomerata (L.) Kützing in the northern Baltic Proper. Proc. 10th European Symposium on Marine Biology, Ostend, Belgium, Sept. 17—23, 1975, Vol. 2: 631—651.

Wallentinus, I., 1978. Productivity studies on Baltic macroalgae. Bot. Mar., 21: 365—380.

Wallentinus, I., 1979. Environmental influences on benthic macrovegetation in the Trosa—Askö area, northern Baltic Proper. II. The ecology of macroalgae and submersal phanerogams. Contrib. Askö Lab., 25: 1—210.

Whitton, B.A., 1973. Freshwater plankton. In: N.G. Carr and B.A. Whitton (Editors), The Biology of Blue-green Algae. Bot. Monogr., Blackwell, Oxford, 9: 353—367.

Wulff, F., 1977. En metod för beräkningar av primärproduktion under dygnet och för längre tidsperioder. Bilaga till rapport till Statens Naturvårdsverk, Kontr. 5860401-8. Askölaboratoriet, Trosa. 8 pp. (mimeogr.).

Wulff, F., Flyg, C., Foberg, M., Hansson, S., Johansson, S., Kautsky, H., Klintberg, T., Samberg, H., Skärlund, K., Sörlin, T. and Widbom, B., 1977. Ekologiska undersökningar i Luleå skärgård 1976. Slutrapport till Statens Naturvårdsverk, Kontr. 5860401-8, 30.6.1977 (mimeogr.).

Zenkevitch, L., 1963. Biology of the seas of the USSR. London, 955 pp.
Zillioux, E.J. and Gonzales, J.G., 1972. Egg dormancy in a neritic calanoid copepod and its implications to overwintering in boreal waters. In: B. Battaglia (Editor), Fifth European Marine Biology Symposium, Padova, pp. 217—230.
Zmudzinski, L., 1977. The Baltic deserts. Ann. Biol., 32: 50—51.

Chapter 6

FISHES AND FISHERIES

EVALD OJAVEER, ARNE LINDROTH, OLE BAGGE,
HANNU LEHTONEN and JORMA TOIVONEN

A. FISH FAUNA OF THE BALTIC SEA*

The evolution of the contemporary fish fauna of the Baltic Sea has been mostly influenced by: (1) changeability of environmental conditions during the short history of this sea (see Section 1.2); (2) horizontal and vertical configuration of the Baltic Sea determining its brackish-water character with a dominating two-layered water mass, as well as the character of currents (see Chapter 3, Section on p. 149); (3) climate.

The main components of the contemporary fish fauna of the Baltic Sea — the arctic, marine-boreal, anadromous, katadromous and fresh-water fishes — have immigrated into this sea at different times by various ways.

Probably one of the first fish immigrants into the post-glacial Yoldia Sea was the very eurihaline *Myoxocephalus quadricornis* from the White Sea basin. It has probably lived in the Baltic Sea throughout the later stages of development.

From marine species in the Baltic Sea could have adapted mainly eurihaline species being capable to endure a relatively low temperature. In the areas near the Danish Sounds, the number of marine fishes is higher; their number diminishes eastwards and northwards: the number of marine fishes is in the North Sea 120, in the Kiel Bay and Mecklenburg Bay 69, in the southern and middle parts of the Baltic Sea 41, in the Åland Sea, Gulf of Finland and Gulf of Bothnia 20 (Remane, 1958). In the Bothnian Bay, there live 6—10 stationary marine fishes (Haahtela, 1974). The spreading of marine fishes is limited mainly by wintering and reproduction conditions, the last affecting especially the fishes having floating eggs.

In each period of evolution of the Baltic Sea, the species composition of its fish fauna was probably different. It is probable that especially during the Litorina time with the warmest climate and the highest salinity in the history of the Baltic Sea, a number of marine fishes immigrated into the Baltic Basin (for instance, the autumn herring — Munthe, 1956). The great majority of marine fishes immigrated into the Baltic Sea from the southwest. The similarity

* By Evald Ojaveer.

of the Baltic spring herring with the White Sea herring has caused the assumption that the spring herring could have immigrated into the Baltic Sea also from the White Sea (Hempel and Nellen, 1974). By its number and structure of chromosomes the Gulf of Finland spring herring is similar both to the Atlantic herring and to the Onega Bay, Dvina and Ivanovsk herrings of the Kandalaksha Bay in the White Sea, but it differs from the Pacific herring and the Yegoryevsk herring of the Kandalaksha Bay (Skvorcova, 1975). This might show the common origin of the Baltic spring herring and the three groups of the White Sea herring mentioned but does not support, however, the hypothesis of the evolution of the Baltic herring from the Pacific herring.

Freshwater, anadromous and catadromous fishes. Compared with the fishes living in the North Sea and in fresh water, the fishes of the same species adapted to the conditions in the Baltic Sea have a number of differences. In marine fishes the most important of them are the following:

(1) In the Baltic Sea the body size of the fishes is smaller than in the North Sea. This is partly the result of a generally earlier maturation. Furthermore, in the Baltic Sea a general decrease of body size and growth rate of marine fish species occurs towards the North and East reflecting probably the influence of climate and decrease of salinity.

(2) In connection with their adaptation to a new environment, the morphological characters of fishes have changed.

(3) In a comparatively small area there is a great number of groups (populations), especially in fish species closely connected with coast.

(4) Adaptation to the environmental conditions has brought about an increase in fecundity in marine fishes and an increase of egg diameter of the marine species having floating eggs, corresponding to the decrease of salinity.

As a result of environmental conditions, in the Baltic Sea a characteristic distribution of fishes has become established. Due to a low content or absence of oxygen in the bottom layers of the deep areas, the fishes are periodically or permanently absent. In the northern part of the Baltic Sea, mainly plankton- and nektobenthos-eating herring and sprat are constantly abundant and to a smaller degree also the flounder. In the southern and southwestern Baltic Sea, more marine fishes including benthos-eaters and predatory fishes are relatively abundant. Therefore, in this area, the composition of fish fauna and their trophic relations are more complicated.

B. MARINE PELAGIC FISHES*

Geographical distribution and groups

The marine pelagic fishes playing the most important role in the communities of the Baltic Sea are the spring and autumn herring and sprat.

* By Evald Ojaveer.

Herring and sprat groups form a continuous chain extending from the North Sea to the northernmost parts of the Baltic Sea. Neighbouring groups of the same species mix not only on feeding and wintering grounds but also on spawning grounds. Therefore there is no sharp biological difference between any neighbouring groups of spring and autumn herring or sprat, and no clear borderline separating *Clupea harengus harengus* from *Clupea harengus membras* or *Sprattus sprattus sprattus* from *Sprattus sprattus balticus.*

On the other hand, the farther east and north one moves in the Baltic Sea, the more the Baltic subspecies of herring and sprat differ from the Atlantic ones. The Baltic herring differs from the Atlantic and North Sea herring in the lower number of vertebrae and keeled scales, body proportions, smaller size at the same age, reproduction at low salinity and earlier maturation of the spring herring. The Baltic sprat differs from the Atlantic subspecies mainly in the lower growth rate, distribution and age composition. Although some migration of pelagic fishes occurs in both directions through the Danish sounds (Jensen, 1955), it seems that the steep environmental gradients existing in this area limit the mass mixing of pelagic fishes of the Baltic Sea with their western neighbours.

Owing to their probably heterogeneous origin, different times of immigration and adaptation to geographically differing environmental conditions, the herring at present inhabiting the Baltic Sea belong to a number of groups, which differ on the following three levels:

(1) The greatest differences exist between the spring-spawning and the autumn-spawning herring. Similarly to other Atlantic spring and autumn herrings they can be individually differentiated by their otoliths. In the northern areas of the Baltic Sea some overlapping of the spawning periods and spawning grounds of the spring and autumn herring can occur, especially in the years when the spring-spawning period is prolonged by unusual environmental conditions. In that case hybrids of the spring and autumn herring can originate. The normal development and survival of such hybrids and also of the offspring of one seasonal group hatched in the spawning period of another are strongly limited by: (a) the low quality of the sexual products of parent fishes spawning outside their normal spawning period; and (b) the very different and specific conditions prevailing during the larval development of the two groups, in which only the larvae of the group adapted to these conditions can survive. Therefore Baltic spring and autumn herring are reproductively practically isolated from one another and could be regarded as belonging to different sibling species.

(2) Both the spring and the autumn herring can be divided into two sets of populations — the sea and the gulf (fjord) herring, which differ from one another in morphological and otolith features, horizontal and vertical distribution, maturation cycle, reproduction conditions, fecundity, feeding, growth, abundance dynamics, etc. (Rannak, 1954; Suškina, 1954; Popiel, 1958a; Sjöblom, 1961; Ojaveer, 1969; Ojaveer and Simm, 1975; etc.). Although the re-

production of the bulk of the sea and gulf herring is separated spatially and temporally, they are not reproductively isolated and a numerous transitional group exists.

(3) Both the sea and the gulf herring are usually divided into different stocks or populations, inhabiting different areas (Fig. 6.1) and differing to a certain extent in morphological features, growth rate, fecundity etc. However, the different populations are not sharply delimited and there are transitional areas in which gradual changes occur in their characteristics.

The stability and survival of these populations are probably due to a complex of factors, the most important being: (a) temporal differences in the pro-

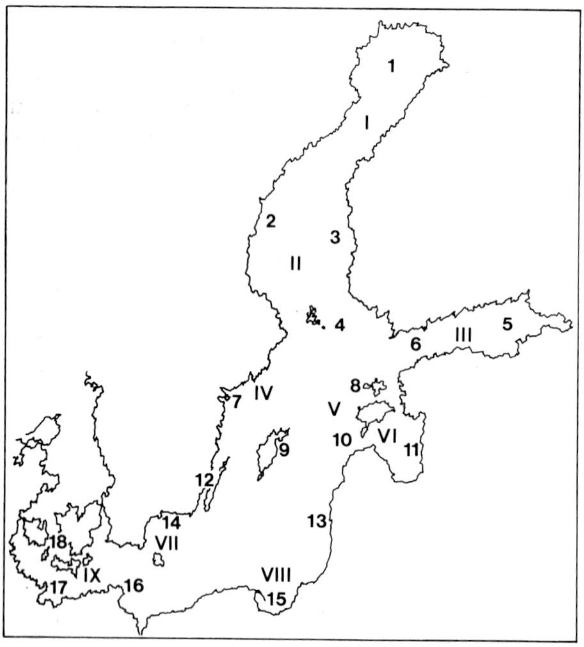

Fig. 6.1. The main populations (groups of populations) of the spring and autumn herring in the Baltic Sea. Spring herring populations: *1* = Bothnian Bay herring; *2* = Western Bothnian Sea herring; *3* = Eastern Bothnian Sea herring; *4* = Archipelago Sea herring; *5* = Eastern Gulf of Finland herring; *6* = Western Gulf Finland herring; *7* = herring of the Stockholm—Västervik Archipelagos; *8* = Hiiumaa herring; *9* = Gotland herring; *10* = Saaremaa — Ventspils sea herring; *11* = Gulf of Riga herring; *12* = Kalmarsund herring; *13* = Liepaja — Klaipeda sea herring; *14* = herring of the south coast of Sweden; *15* = Gulf of Gdansk herring; *16* = Rügen herring; *17* = southwestern Baltic herring; *18* = Belt Sea herring. Autumn herring populations: *I* = Bothnian Bay herring; *II* = Bothnian Sea herring; *III* = Gulf of Finland herring; *IV* = herring of the east coast of Sweden; *V* = sea herring of the northern and middle part of the east coast; *VI* = Gulf of Riga herring; *VII* = Gulf of Hanö — Bornholm herring; *VIII* = Gdansk Bay herring; *IX* = herring of the Belt Sea, Sound and southwestern Baltic Sea (consisting possibly of three separate units). After Hessle, 1925; Jensen, 1950; Popiel, 1958a; Otterlind, 1961; Ojaveer, 1969; etc.

duction cycles in different parts of the Baltic Sea, which affect the time of the highest survival of herring larvae at their transition to exogenous feeding (Ojaveer and Simm, 1975); (b) spatial differences in the conditions determining the growth, development and wintering of the herring, which cause differences in its sexual cycle; (c) a certain "homing instinct" in the herring (Otterlind, 1962a).

The living conditions of the sea herring are fairly uniform over large areas but those of the gulf herring groups, which inhabit gulfs and fjords and are more closely connected with the coasts, can differ significantly between areas situated relatively close to one another. Therefore the number of groups in the gulf herring is greater than in the sea herring.

Compared with the herring, the Baltic sprat clearly has a smaller tendency to form separate populations (Hessle, 1927). It is much less closely connected with the coasts than the herring and therefore sprat originating from different areas mix more easily, especially in the wintering period and during spawning time in the open sea. Also, the eggs and larvae can drift with currents to other areas. However, it is possible to divide the Baltic sprat into several populations (Poulsen, 1950; Elwertowski, 1960; Seleckaja, 1970; Veldre, 1974; etc.) that inhabit large areas corresponding roughly to the basins formed by the bottom relief and to the system of currents: (1) the sprat of the northern part of the Gotland Basin, the northern part of the Baltic Proper and the Gulf of Finland; (2) the sprat of the Gulf of Gdansk and the southern part of the Gotland Basin; (3) the Bornholm sprat; (4) the sprat of the southwestern Baltic Sea and the Belt Sea.

Seasonal distribution patterns and migrations

The distribution of the pelagic fishes in the Baltic Sea shows clear seasonal variation and depends on the species and physiological condition of the fish, the distribution of food, the temperature and oxygen content in different water layers and the light conditions. In the greater part of the Baltic Proper and the western Gulf of Finland, the herring and sprat are mainly found at depths of 70—100 (120) m in winter, concentrating in warmer (3—5° C) water at the lower limit of the homohaline layer or in the heterohaline layer (Fig. 6.2). The lower limit of their distribution coincides with an oxygen content of 1.0—1.5 g m^{-3}.

The concentrations are generally largest in the zone of intense mixing of the homo- and heterohaline layers. In severe winters, there is usually one fish layer which is rather compressed in the vertical direction. In the lower part are the older herring and in the upper part sprat and younger herring. In milder conditions, the sprat and younger herring keep markedly higher than the older herring, giving rise to a two-layered fish distribution. The young fish keep mainly on the coastward edge of the layer. The wintering concentrations of herring generally keep close to the bottom, and herring are usually avoiding

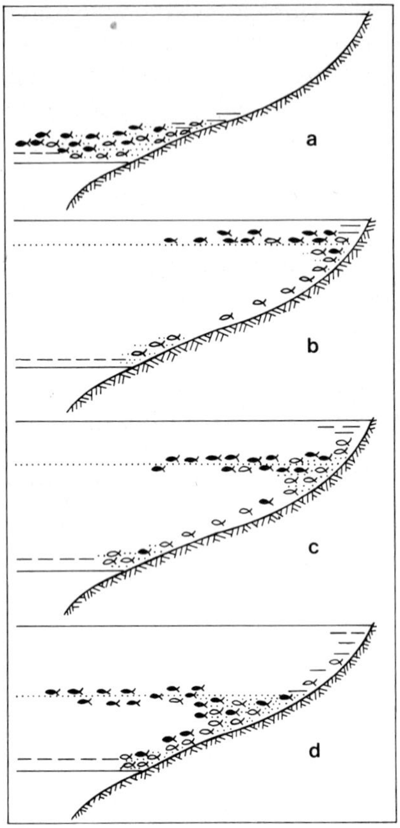

Fig. 6.2. Distribution of herring and sprat in the middle and northern Baltic Sea in winter (a), spring (b), summer (c) and autumn (d). — — halocline; . . . . thermocline; —— oxygen content 1 ml/l; ::: intense mixing zone; ∝ herring; ⤙ sprat; — young fish (age-group 0 +).

the deeps. The distribution of the sprat depends more on temperature and as the temperature falls in winter it shifts over to the central parts of the deeps where the water is warmer; in severe winters it concentrates in the central parts of the deeps. In shallow areas without a heterohaline layer (the Gulf of Riga, the eastern Gulf of Finland, the Archipelago Sea) the wintering fish concentrations keep at depths of 30—50 m or even less, mainly at 0.5—2° but in good oxygen conditions.

In spring sprat and young herring rise to the warmer surface water or the thermocline, where their food — the warm-water zooplankton — is concentrated. The older herring feed near the bottom, in the open sea commonly at depths of 50—80 m, in the eastern Gulf of Finland and in the Gulf of Riga at 30—50 m. In both layers the age of the fish usually increases with the increase of distance from the coast. In summer a third fish layer, which consists of whitebaits and is most extensive in gulfs, can be found near the coast in the warm surface layer. From spring to autumn, the pelagic fishes concentrated near the thermocline move together with it deeper and farther from the coast (Fig. 6.2) eating plankton in the water through which they migrate and, as far as the temperature conditions will allow, shifting towards deeps or areas richer in food organisms. Similar movement also occurs in the bottom fish layer.

In both the pelagic and bottom layer the distribution of pelagic fish is patchy. The largest concentrations are met with in the productive zones with strong mixing of homo- and heterohaline water layers on the coastal slopes of deeps, especially in the vicinity of sills, banks, isles and peninsulas. In the Baltic Proper, the best feeding grounds of herring and sprat are situated on the slopes of the Bornholm and Gdansk Deeps, in Slupsk Furrow, in the area of Middle Bank and Hoburgs Bank, on the eastern slope of the Gotland Deep, west of Irben Sound and the Estonian archipelago, on the slopes of the northern part of the Baltic Proper (especially in the mouth of the Gulf of Finland) and in the sill area between Gotska Sandön and Saaremaa.

The distribution and abundance of the herring and sprat concentrations vary considerably between the different parts of the sea and the different years. They depend on the abiotic conditions determining the distribution and abundance of their food organisms, as well as on the location of their spawning and wintering areas.

The diurnal vertical migration of sprat and herring is most intense in spring and in the first half of summer and less intense in winter.

The horizontal migrations of the herring and sprat, especially of their young stages, are little known. It seems that the migrations of the gulf (fjord) herring populations are limited because their spawning grounds are often situated near relatively rich feeding grounds, separated from the grounds of neighbouring populations by areas poorer in food. Nevertheless, in the herring populations of the Gulf of Riga and the Gulf of Finland the tendency to migrate towards the open sea increases with increasing age. The migrations of the sea herring are longer. The spring and autumn sea herring migrates for spawning into gulfs and sounds. A clear migration pattern has been observed in the spring herring populations off the east coast of Sweden, which migrate after spawning to feeding grounds in the southern Baltic and back (Otterlind, 1962a). The intensity of this migration depends on the feeding conditions. In the years rich in plankton, fewer herring migrate to the southern Baltic Sea than in the years poor in this respect (Popiel and Strzyżewska, 1971).

Spawning, larval and adolescent phase

In the Baltic Sea the spawning of the spring and the autumn herring takes place at different times, depths and distances from the coast, depending on the latitude.

This is probably connected with the adaptation of embryonic development to a certain temperature range and to the timing of the transition of the larvae to exogenous nutrition to coincide with a certain point in the production cycle (to the maximum of the copepod young stages).

The spawning of the spring herring starts and finishes earlier in the south than in the north, but the spawning of the autumn herring commences and ends earlier in the north than in the south. The spring herring of the Belt Sea, the southwestern part of the Baltic Sea, Rügen and the Gulf of Gdansk spawn from March to May, but those in the Quark spawn from the second half of May to midsummer (Jensen, 1950; Popiel, 1958a; Sjöblom, 1961; etc.) In the Gulf of Bothnia the autumn herring spawn from the end of July to early September, but those in the southwestern Baltic Sea from September to November (Hessle, 1925; Rechlin, 1964).

The spawning grounds of the spring herring occur patchily near the coast, commonly in the places of intense mixing of water layers and of higher productivity — in the vicinity of river mouths, in gulfs and sounds. The spawning begins in shallow water (1—5 m), which warms earliest in spring, at a temperature of 2—4° C, and shifts constantly towards the sea, finishing at depths of 6—20 m and at 17—20° C (Jensen, 1950; Rannak, 1954; Sjöblom, 1961). The range of salinity on the spawning grounds from the Gulf of Bothnia to the Gulf of Gdansk is 3—4‰ to 5—7‰. In the middle and northern parts of the Baltic Sea the sea herring spawns closer to the coast and at lower temperatures in the first half of the spawning period, and the gulf herring reproduces in deeper water and at higher temperatures in the second half of the period. Older and larger fish spawn earlier; at the end of the spawning time the recruitment spawns.

Depending on the temperature, the embryonic development lasts 1500—3300 degree-hours. At hatching the length of the sea herring averages 5.5—8 mm and that of the Gulf of Riga herring 5—7 mm; at the yolk resorption stage, the corresponding lengths are 8—8.5 and 7.5—8 mm. The larvae live in the surface layer. Metamorphosis takes place 2—2.5 months after hatching, the Gulf of Riga herring then being 25—30 mm long. The whitebait live in the coastal zone, in or above the thermocline.

The autumn herring spawns at depths of 3—25 m on offshore banks or shoals on the seaward slopes of archipelagoes. The spawning grounds extend from the Danish Sounds along the east and west coast of the open sea as well as into the Gulfs of Riga, Finland and Bothnia. Like the spring herring, the autumn herring spawns on sand, gravel or stony bottoms, on vegetation or other substrates during the embryonic period, the eggs being fastened to the substrate. Off the east shores of the Baltic Sea, mass spawning occurs at 12—16° C, off the Swedish coast at 11—14° C (Hessle, 1925; Ojaveer, 1969). The mass spawning of both the spring and the autumn herring starts after vigorous mixing of the water on the spawning grounds. The spawning of the autumn herring starts on deeper grounds and in the second half of the spawning period it shifts to more shallow grounds. The sea herring spawns before the gulf herring. The percentage of older fish rises towards the end of spawning (Ojaveer, 1969; Ojaveer and Simm, 1975). The embryonic development lasts 1600—3400 degree-hours. The length of the autumn herring of the middle parts of the Baltic Sea at hatching is 5.5—8 mm and at yolk resorption 7.5—8.5 mm. The eggs of the autumn herring can develop parthenogenetically resulting in viable hatch. The main enemies of both the spring and autumn herring embryos are eel-pout, whitefish and, in the Danish Sounds, some invertebrates that feed on herring spawn (Jensen, 1950; Ojaveer, 1969) as well as moulds.

In the autumn herring the larvae develop at greater depths than in the spring herring — in winter they have been caught at depths down to 62 m. The majority metamorphose in shallow water near the coast the following spring, in the middle parts of the Baltic Sea 8—9 months after hatching at a length of 40—44 mm (Ojaveer, 1969).

The majority of the herring populations attain sexual maturity at the age of 3—4 years but some spring-spawning gulf herring (Gulf of Riga, eastern Gulf of Finland) populations mainly mature at the age of 2 years, a small part maturing at 1 year.

The spawning area of the sprat is determined by salinity and temperature. Its pelagic eggs are spawned at salinities of 5‰ and more at 4—14° C (optimum range 6—12° C) in the area of the coastal slopes of deeps in the open sea, but also in the deeps, less often in fjords or archipelagoes (Hessle, 1927; Elwertowski, 1957; Grauman, 1969; Veldre, 1974). The spawning area extends from the western shores to the Åland Sea, and includes the Gulf of Finland and the Gulf of Riga. No sprat eggs have been found in the Gulf of Bothnia (Hessle, 1927; Poulsen, 1950). Sprat eggs are spawned in 8—10 batches in the south and in 6—9 in the north, probably at intervals of 8—10 days (Veldre, 1974). In the Belts, and the southwestern and southern Baltic Sea spawning starts in March—April and lasts till July—August, in the northern parts of the sea spawning occurs in June—August. If the temperature of the surface layers at the beginning of spawning is too low, the sprat may spawn in deeper layers, at 4—7° C, but the bulk of the sprat spawns in the surface layers at 7—14° C (Poulsen, 1950; Elwertowski, 1957; Grauman, 1969). The embryonic period commonly lasts 3—4 days. The length at hatching is 2.0—3.6 mm. Metamorphosis takes place at (25)30—40 mm, 6—8 weeks after hatching. In the first year the sprat hatched from the earlier egg batches reach a length of 9—10 cm in the southern Baltic Sea and 8.5 cm in the northern areas. The Baltic sprat attains sexual maturity at the age of (1)2—3(4) years, usually earlier in the south than in the north (Elwertowski, 1957; Veldre, 1974).

Fecundity

The individual fecundity (the number of eggs per individual) of both the spring and the autumn herring varies considerably. Compared with that of autumn herring of the same area, the fecundity of the spring herring is lower and increases more slowly with the growth of body size (Fig. 6.3). The fecundity/weight (in grams) regression lines for the autumn herring populations are much closer to one another than are those of the spring herring populations. Even in a relatively limited area, in different parts of the Gulf of Finland, the regression lines of the spring herring vary considerably (Fig. 6.3). In the Baltic Sea autumn herring fecundity is higher and increases more rapidly with the growth in body weight than in the North Sea herring (Kändler and Dutt, 1958). This can be the result of adaptation to the less favourable conditions for survival of the young stages in the Baltic Sea.

In both the spring and the autumn herring reproduction is optimal in the medium age-groups, the relative fecundity and the increase of fecundity with body size being then highest and the quality of the sexual products best (Rannak, 1970; Ojaveer, 1974; etc.).

The fecundity of the sprat increases towards the southwestern Baltic Sea. In the Gdansk area its fecundity in the length groups 106—115 mm and 126—135 mm is 8,150 and 13,700, whereas in the Kiel Bay area at the lengths of 105 and 136 mm its fecundity is 11,000 and 36,000. The number of eggs in the batches (500—5000) depends on the size and age of the fish (Elwertowski, 1957, 1960).

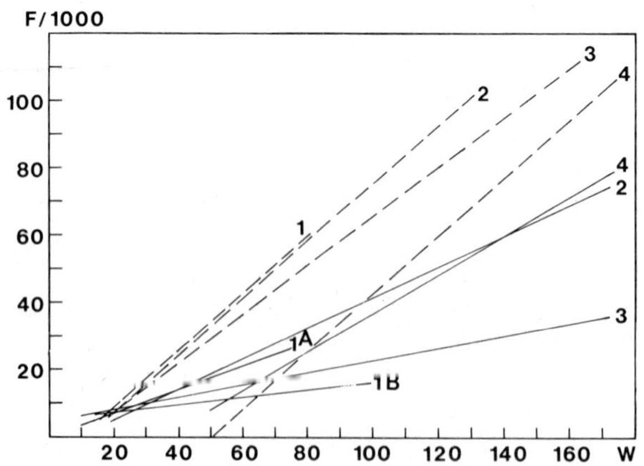

Fig. 6.3. Fecundity (F)/weight (W) regression lines for the Baltic spring (——) and autumn (— — —) herring populations according to equation $F = a + bW$. *1* – Gulf of Finland herring (*1A* – eastern, *1B* = western population); *2* = spring herring of Northern Öland, autumn herring of the Saaremaa-Ventspils area; *3* = Gulf of Gdansk herring; *4* = southwestern Baltic herring. According to Kändler and Dutt (1958), Strzyżewska (1960), Rannak (1970), Ojaveer (1974).

Feeding

The first exogenous food of the herring is copepod nauplii. The diversity of its food increases but up to the end of the first year of life the herring feeds only or mainly on plankton (Popiel, 1951; Suškina, 1954; etc). With the increase in age and length the importance of plankton decreases and the proportion of larger crustaceans (Mysidae, Amphipoda), young stages of fishes, etc. increases. However, the food varies considerably between populations and individuals. For instance, some gulf herrings feed on plankton all their life and their growth rate decreases rapidly with age. In contrast, in the central and northern parts of the Baltic Proper a certain percentage of the spring herrings pass to a fish diet (sprats, sticklebacks etc.), and the growth rate of these so-called giant herring considerably exceeds that of the other herrings, their total length reaching up to 50 cm (Ojaveer, 1976).

In general, the most stable component of the food of adult herring is copepods (*Pseudocalanus*, *Limnocalanus*, *Temora*, *Eurytemora*, *Acartia*, *Centropages*). Of the Cladocera, *Evadne* and *Podon* are found in greater quantity in summer and Bosmina in autumn. Mysidae occur mainly in the second half of summer, in autumn and winter, and Amphipoda in autumn and winter (Jespersen, 1936; Popiel, 1951; Suškina, 1954; Bitjukov, 1961).

Sprat larvae feed on eggs and young stages of copepods, diatoms and flagellates. Whitebait live mainly on young stages of copepods and molluscs (Veldre, 1974). The all-year-round component of sprat food is copepods. The most important species is *Temora longicornis*, which occurs throughout the year in great quantity and dominates in sprat food together with *Acartia* and *Eurytemora* in spring and in autumn. In winter *Pseudocalanus* is dominant in sprat food and in summer Cladocera (mainly *Bosmina* and *Podon*). Mysidae occur mainly in spring and summer, the larvae of Mollusca and Cirripedia in summer (Mankowski, 1947; etc.).

Depending on the composition and abundance of plankton and benthos, the sprat and herring diets can differ between years and areas. In the large gulfs the importance of mysids, amphipods and occasional food components in the herring diet is considerably higher than in the open sea, whereas the proportion of cladocerans is significantly less. The importance, time of occurrence and species composition of fish larvae in herring food differ markedly between various parts of the sea (Jespersen, 1936; Popiel, 1951; Suškina, 1954; Bitjukov, 1961; Kostrichkina and Starodub, 1976). The feeding intensity and species composition of the diet also differ in different water layers. In the eastern Baltic Proper in July, herring feed on warm-water plankton in the thermocline twice as intensely as on the cold-water plankton and larger crustaceans below the thermocline (Kostrichkina and Starodub, 1976).

For all groups in winter and for the spring herring and sprat throughout the year, the feeding intensity generally accords with the water temperature in the layer in which the fish are concentrated (Fig. 6.4). In the southwestern and southern parts of the sea the sinking of temperature in winter is accompanied by a decrease in feeding intensity (Jespersen, 1936; Mankowski, 1947; Popiel, 1951), in the northern parts feeding finally ceases. The length of the period of starvation depends on the severity of the winter and in the eastern Gulf of Finland it lasts from the second half of October to the middle of May (Bitjukov, 1961). In the vegetation period the influence of temperature is modified by the sexual cycle: the feeding intensity of the herring decreases sharply after its gonads have reached stage III—IV of maturity; in the spawning time feeding practically stops in herring and considerably decreases in sprat. Feeding intensity is highest after spawning, in the period of regeneration and rapid development of the gonads. Under the combined influence of internal and external factors, the annual courses of feeding intensity in the spring and autumn herring are completely different (Fig. 6.4).

The spring-spawning gulf herring spawns in June—July and the rapid development of its gonads takes place in spring (especially in younger fish). There-

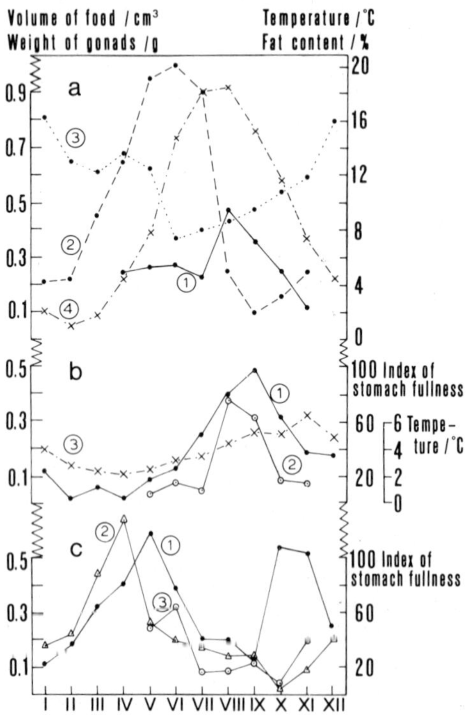

Fig. 6.4. a. Sprat (southern Baltic Sea): *1* = mean volume of food; *2* = weight of gonads; *3* = fat content; *4* = water temperature. b. Spring herring: *1* = mean volume of food (southern Baltic Sea); *2* = index of stomach fullness (Gulf of Riga and eastern Baltic Sea); *3* = water temperature in the Gdansk Deep at a depth between 50 and 60 m. c. Autumn herring: *1* = mean volume of food (southern Baltic Sea); *2* = mean volume of food (Bornholm region); *3* = index of stomach fullness (Gulf of Riga and eastern Baltic Sea). According to Popiel (1951), Suškina (1954) and Elwertowski (1960).

fore, unlike the sea herring, the spring-spawning gulf herring has a feeding period in spring before spawning (Sjöblom), 1961).

Both herring and sprat have two to three periods of maximum stomach fullness a day. Feeding is clearly selective — the sprat commonly prefers Temora, the herring larger food animals or dense concentrations of plankters. The feeding relations of the spring and autumn herring and sprat differ with the year and season. In young herring the diet is more similar to that of sprat than in older herring. The herring has a greater ability to change the composition of its food than the sprat (Popiel, 1951; Kostrichkina and Starodub, 1976; etc.).

Growth and age

In the Baltic Sea, the theoretical asymptotic lengths (L_∞) and growth rates of the sea herring and sprat decrease towards the north (Otterlind, 1962a; Table 6.I). The growth rate of the gulf herring is slower. In general, the L_∞ is higher in the spring herring than in the autumn herring of the same area. *k* (coefficient of catabolism), varies in the spring herring between 0.25 and 0.54 (commonly higher in the south than in the north) and in the autumn herring between 0.33 and 0.42 (Stryżewska, 1975; Ojaveer, 1976). During the last two decades the growth rate of herring has had an increasing trend both in the southern Baltic Sea and in the Gulf of Riga. A considerable rise in growth rate

TABLE 6.I

Growth parameters (theoretical asymptotic length (L_∞), in centimeters, and mean length of three-year-old fish) of the spring and autumn herring and sprat in the various parts of the Baltic Sea (after Elwertowski, 1957; Ojaveer, 1976; Popiel, 1958b; Strzyżewska and Popiel, 1974, Rechlin and Friess, 1975; Sjöblom and Parmanne, 1976; Strzyźewska, 1975; Veldre, 1974; Weber, 1974)

Area	Spring herring		Autumn herring		Sprat
	L_∞	L_3	L_∞	L_3	L_3
Bothnian Sea	20.0—20.6	15.9			
Western Gulf of Finland	19.1	15.9	18.4	16.2	11.9
Gulf of Riga	20.6	15.2	19.8	16.1	
Hiiumaa	21.8	16.9	19.7	16.6	11.8
Saaremaa	21.8	18.2	21.0	18.4	
Gulf of Gdansk	25.2	19.6	24.4	21.6	12.7
Bornholm			27.9	23.2	13.3
Rügen	28.2—28.4	22.0			
Southwestern Baltic Sea		23.9		24.8	

occurred in the Bornholm autumn herring during 1939—1945 (Strzyżewska and Popiel, 1974; Ojaveer, 1976).

In both the herring and sprat the maximum age is highest in the northern part of the Baltic Proper, decreasing towards the south and in gulfs. The maximum age of the spring herring in the Gulf of Finland is 16 years, in the Hiiumaa—Saaremaa area 16—20 years and in the southern and southwestern Baltic Sea 9—12 years; the maximum age of the autumn herring in the northern and middle parts of the Baltic Sea is 12—15 years and in the southern and southwestern Baltic Sea 9—12 years. In catches the herring of the northern and middle part of the Baltic Proper, and of the Gulfs of Bothnia and Finland is important up to the age of 8—10 years, the herring of the southern and western Baltic Sea up to 5—7 years (Rechlin, 1964; Ojaveer, 1969; Popiel and Strzyżewska, 1974; Sjöblom and Parmanne, 1976).

The maximum age of sprat in the north east part of the sea is 18 years, in the Gulf of Gdansk area 7—8 years and in the Bornholm Basin 8—9 years. In fisheries the sprat age-groups are of importance in the northern Baltic Proper up to 6—10 years, in the southern and southwestern Baltic Sea up to 4—5 years (Elwertowski, 1957, 1960; Veldre, 1974).

Year-class abundance

In the various parts of the Baltic Sea the abundance dynamics of the year-classes of pelagic fishes is markedly different (Table 6.II). The greatest geographical differences occur in the abundance of the spring herring year-classes, for its populations and their spawning grounds are best isolated from one another. The year-class abundance of this fish in different years and parts of the sea is probably determined by different factors (the quantity and quality of eggs, success of the embryonic and larval development, etc.) the most important general factor being the food available for the larvae (Rannak, 1974; etc.). The abundance of the autumn herring year-classes seems to be most limited by the survival rate of the larvae in the first winter of their life. It is probable that the general decrease in the abundance of this fish towards the north and east (in the southwestern Baltic Sea and Bornholm Basin the autumn herring commonly constitutes the bulk of the herring catches whereas in the eastern Gulf of Finland its percentage in the total catches of herring does not exceed 1—2) is caused by the increase in the severity of the winter, which decreases the survival rate of its larvae. In the northeastern Baltic Sea, the abundance of the autumn herring year-classes correlates well with the winter temperatures. This correlation is much poorer in the southern and southwestern Baltic Sea — in spite of mild winters in 1970—1974, no good year-classes developed in these areas in the period 1965—1974 (Table 6.II). In the sprat, year-class abundance depends mainly on the quantity of eggs, the food supply of the larvae and whitebait and the temperature regime, and also on the oxygen conditions in the deeps at the beginning of spawning and

TABLE 6.II

Abundance of year-classes of the spring- and autumn-spawning herring and sprat in various parts of the Baltic Sea (according to Elwertowski, 1960, 1976; Elwertowski and Popiel, 1961; Ojaveer et al., 1975; Popiel and Strzyżewska, 1974; Rannak, 1974; Rechlin, 1967; Rechlin and Friess, 1975; Veldre and Polivajko, 1975)

Year class	Spring herring					Autumn herring		Sprat		
	Eastern Gulf of Finland	Gulf of Riga	NW part of the open sea	Gulf of Gdansk	Rügen	NW part of the sea	Southern Baltic Sea	NW part of the sea	Gulf of Gdansk and southern Gotland	Bornholm
1948		‡	+	‡		+	+			
1949	+	+	‡	+		—				
1950	±	±	±	+						
1951	+	±	+	+		—				
1952	±	—	±	‡	‡	±	±			
1953	±	+	—	+	±	—	—	=	‡	—
1954	—	—	±	+	±	+	‡	=	±	—
1955	‡	±	±	±	±	=	—	‡	‡	‡
1956	=	=	—	=	±	‡	—	=	=	=
1957	+	+	±	=	±	—	+	‡	‡	±
1958	±	—	—	=	±	+	+	=	—	—
1959	+	+	+	—	+	—	+	‡	+	+
1960	—	=	=	—	±	±	+	=	—	+
1961	+	‡	+	—	+	+	+	=	+	+
1962	±	+	±	±	+	=	—	=	—	
1963	—	±	—	+	±	—	—	—	+	+
1964	+	+	+	‡	±	‡	‡	±	+	+
1965	—	—	—	±	+	±	—	=	—	
1966	±	—	—	+	+	±	—	—	+	+
1967	‡	+	+	‡	+	—	—	‡	+	+
1968	—	—		—	+	=	—	—	—	
1969	—	—	±	‡	+	=	—	+	±	
1970	+	+	+	+	+	+	—	—	±	
1971	—	—	±	+	+	+	—	±	±	
1972	±	±	±	+	±	+	—	+	+	+
1973	±	±	±	+		+	—	—	+	+
1974	—	—	—	+		±		—	±	

= very poor; — poor; ± moderate; + rich; ‡ very rich

the strength of winds later on when the sprat is spawning in the surface layers. Both the latter factors influence embryonic mortality, which may reach 50—96% (Hessle, 1927; Elwertowski, 1960; Grauman, 1969; etc.).

Catches and mortality

Long-term decreases or increases in the size of one or several herring or sprat populations brought about by changes in year-class abundance have always been of great economic importance, for they have seriously affected catches.

During the last 50 years, advanced fishing gear has almost completely replaced the gear traditionally used in the past — beach seines, fixed nets, etc., with which pelagic fish were caught by human power in the coastal zone only. With the development of fish-searching devices, mechanization of the catching process and introduction of pelagic trawls, which enable fishing for pelagic fishes in all seasons and water layers, the herring and especially the sprat catches have markedly increased (Fig. 6.5). Together with the increase in catches and intensity of exploitation the total instantaneous mortality of the herring and sprat stocks has risen: in the Gulf of Finland spring herring from 0.39 in 1959—1965 to 0.66—0.70 in 1969—1973, in the Rügen spring herring from 0.76 in 1964 to 1.10 in 1972 and in the Gulf of Finland sprat from 0.14 in 1960/1961—1964/1965 to 0.40 in 1969/1970—1973/1974.

The total mortality of the herring and sprat populations varies greatly: in the Bothnian Bay spring herring in 1970—1975 it was 0.27, in the Saaremaa—Ventspils spring herring in 1971—1973 0.89, in the Gulf of Pomerania and Odra Bank spring herring in 1970—1974 1.09, in the Bornholm spring and

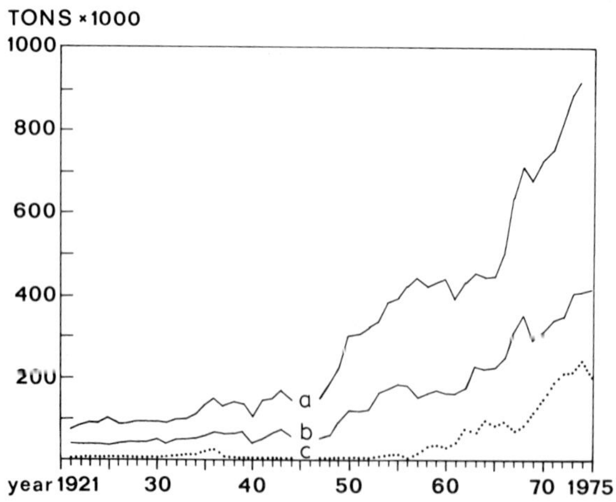

Fig. 6.5. Total catches (*a*) and the catches of herring (*b*) and sprat (*c*) in the Baltic Sea in thousands of tons. According to ICES (1953—1973), International Baltic Sea Fishery Commission (1976), Lomniewski et al. (1975).

autumn herring in 1967—1974, respectively, 0.41 and 0.47, in the sprat of the Gulf of Gdansk and Bornholm Basin in 1954—1958, respectively 1.13 and 0.93 (Elwertowski, 1960; Ojaveer et al., 1975; Rechlin and Friess, 1975; Stryżewska, 1975; Veldre and Polivajko, 1975; Sjöblom and Parmanne, 1975; 1976). Analysis of the catch curves shows that at present most of the Baltic herring and sprat stocks are rather heavily exploited; generally their exploitation rate should not be increased and in some populations it must be decreased to introduce the regime of rational utilization of the stocks.

Both constituents of total mortality — the natural and the fishing mortality — are probably selective. The coefficient of natural mortality has been directly determined only in the Hiiumaa autumn herring in 1963—1964 (average for age-groups 3—8 0.21—0.26, increasing with age from 0.08 for 3-year-old fish to 0.49 for 8-year-olds). The causes of natural mortality have been little studied. Herring larvae can be eaten by stickleback, herring and other fishes, and planktonic invertebrates; adolescent and adult herring and sprat — by cod, salmon, garpike, giant herring, lamprey, birds, seals, etc. Mortality can also be caused or promoted by parasites and epizootics (Jensen, 1950; Poulsen, 1950; Veldre and Polivajko, 1975; Ojaveer et al., 1975). Consideration of the number of enemies suggests that in both the herring and sprat the natural mortality is markedly lower in the northern part of the sea than in the southern and southwestern Baltic Sea.

The character of fishing mortality depends not only on the catch intensity but also on the gear used. In net and pound-net fishery on the herring spawning grounds and in net fishery for sprat, mainly older age-groups are affected, whereas in trawl fishery, especially with pelagic trawls in the coastal zone, the mortality of younger age-groups is increased.

Other fishes

In addition to the spring and autumn herring and sprat, the Baltic Sea contains a number of less abundant marine pelagic fishes that live only in certain habitats in the coastal zone. The abundance of all these species decreases from the Danish sounds and the southwestern part of the Baltic Sea towards the north and east.

The garpike, *Belone belone* (L.), is also relatively abundant and important for fishery in some areas in the middle parts of the Baltic Sea (on the Estonian west coast). The species has been found in the Gulf of Bothnia up the Quark and in the Gulf of Finland up to Kronstadt (Valle, 1934; Berg, 1961). The spawn is produced in batches off the Estonian west coast and in the Gulf of Finland in May—July. It feeds on fishes (sticklebacks, sprat, herring), crustaceans and insects.

The fifteen-spined stickleback, *Spinachia spinachia* (L.), occurs in the Gulf of Finland up to its eastern part (Luga Bay) and in the Gulf of Bothnia up to the middle of the Bothnian Sea (Hudiksvall). It lives mainly in the zone of bottom vegetation singly or in small shoals; spawns in May—June; feeds on Amphipoda, Isopoda, Vermes, and young stages and eggs of other fishes (Berg, 1961; Järvekülg, 1963).

Syngnathus typhle L. can be found in the Gulf of Bothnia up to the northern Bothnian Sea (Maalahti) and in the Gulf of Finland up to Helsinki. It spawns from April to August; feeds on mysids, gastropod larvae and fish larvae.

Nerophis ophidion (L.) lives in the Gulf of Bothnia up to the middle parts of the Bothnian Bay (Raahe) and in the Gulf of Finland up to Kronstadt, inhabiting the belt of bottom vegetation like the foregoing fish; spawns May—June to August; feeds mainly on planktonic crustaceans (Valle, 1934; Berg, 1961; Järvekülg, 1963).

A number of marine pelagic fishes encountered in the Baltic Sea do not commonly reproduce there but migrate periodically into this sea, especially with inflowing North Sea water. The most important of them is the mackerel (*Scomber scombrus* L.) which has some periods of importance for fishery in the southwestern Baltic Sea, and the anchovy, *Engraulis encrasicholus* (L.). Both fishes have also been found in other parts of the Baltic Sea, including the Gulf of Finland (Berg, 1961).

C. ANADROMOUS AND CATADROMOUS FISHES*

All fishes eat. All fishes spawn. Few fishes spawn where they eat. Consequently, most fishes undertake migrations from feeding areas to spawning areas and, usually, vice versa.

Anadromous fishes migrate from the sea upstream to spawn. Even though this movement applies to many populations of various fish species, the most noteworthy in the Baltic Sea are the following:

Atlantic salmon	*Salmo salar* L.
Brown (sea) trout	*S. trutta* L.
Grayling	*Thymallus thymallus* (L.)
Whitefish	*Coregonus lavaretus* (L.)
Vimba	*Vimba vimba* (L.)
Lamprey	*Lampetra fluviatilis* (L.)

Catadromous fishes make their spawning migration downriver to the sea. Here one species is to be mentioned:

European eel	*Anguilla anguilla* (L.)

This chapter will be devoted to the seven species listed. Other fishes, with more or less anadromous populations along parts of the Baltic Sea coasts, will

* By Arne Lindroth.

not be treated in this section. The reader is referred to Sections D and E of this chapter.

Salmon

Distribution in space and time

The Atlantic salmon is distributed along and off the coastal areas of the North Atlantic. On the American side of the ocean, it is found from New York State to Labrador and Hudson Bay (and in one river on West Greenland). On the European side, it lives in the area from Portugal to Petchora River in the USSR and in Iceland. The brackish water of the Baltic Sea contains populations practically isolated from the Atlantic since the last phase of the Late Cenozoic (Pleistocene) Glaciation, about 10,000 a — 15,000 a ago.

Some 50 Baltic Sea rivers have originally accommodated spawning salmon populations. Now many of them are ruined by water pollution and hydro-electric power plants. Less than one half are left to naturally reproducing stocks, often seriously reduced. More than ten rivers are maintained by release of artificially reared smolt-sized parr, solely or as complement (Fig. 6.6).

Beyond the Baltic Sea rivers the Baltic Sea area contains landlocked salmon in a few lakes, e.g., Lake Ladoga.

The salmon spawns in fresh-water streams and remains there as egg, fry and parr until metamorphosis to smolt at an age of from one to several years after hatching and after reaching a length of 10—20 cm. As smolt it migrates in spring and early summer downstream to the sea. Most salmon move south to the Baltic Proper where the sea fishery mainly takes place. The mortality caused by fishing is so heavy that now few salmon attain an age of more than 3 years in the sea. Many males make their return journey to the rivers as grilse late in their second sea-summer, but females seldom before their third summer. The fish, surviving from fishing on the coast and in the rivers, and from the spawning, return to the sea usually in spring. Atlantic salmon may spawn several times.

Spawning migration

One of the most conspicuous features of the salmon is its spawning migration and its ability to climb rapids and jump falls; vertical heights of 4—5 m are reported. Before reaching the rivers, however, the salmon has left the feeding grounds early in spring for the older fish, later in summer for the grilse. They travel at an apparent speed of 100 km/day (Alm, 1942) and head remarkably straight for the home river which they left as smolts. They run in the rivers, the older in summer, the grilse in late summer—autumn. Even if they are released as smolt-sized parr from a hatchery, situated anywhere, they return to the river of liberation. Strayers amount only to about 2% (reared fish; Carlin, 1969). If and how far the different river populations are genetically segregated is not known for the Baltic Sea (but data exist for Canadian populations; Møller, 1970).

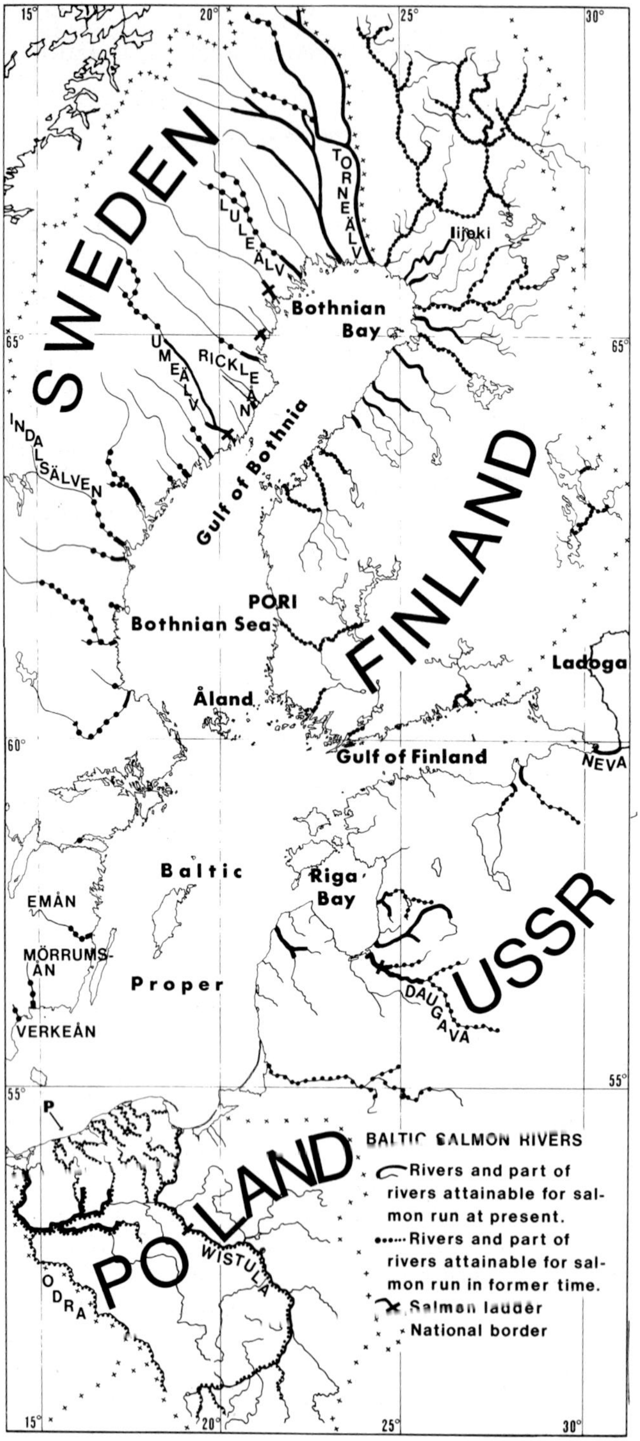

Fig. 6.6 Salmon rivers in the Baltic Sea area. According to Christensen and Johansson (1975).

What guides the salmon migrating in the sea is obscure. The sense of smell, however, is a necessary condition for the recognition of the proper estuary and ascending the home river. Experimentally deprived of smell the fish do not enter fresh water but remain roaming in the sea (Toft, 1975).

Spawning and river life

The spawning takes place in autumn (October to November) over gravelly bottom indicating fast currents, and preferably in places where water is seeping down into the bottom. The female, after selecting a suitable site and attended by the male, digs by vigorous flapping of body and tail a deep depression in the gravel. At a certain signal he pulls up alongside and in a sort of spasm both sexes give their sex products off, a portion of eggs sinking to the bottom through a cloud of milt in the relatively quiet water of the depression. Immediately afterwards the female starts the digging activity again upstream whereby the just deposited eggs become covered. The fertilization may be accomplished also by male parr of which many are ripe at this time and present as wedding guests.

The fecundity of the female salmon amounts to some 1000—1200 eggs per kg ripe fish (Carlin 1951; Carlin and Johansson, 1971; Lindroth, 1952a). During winter, the deposited eggs remain in the holes and crevices of the bottom, and even after hatching, as fry. Not until the yolk-sac is consumed does the young parr rise to the bottom surface. Provided that spawning is not repeated by several pairs on the same spot, the period from egg to fry should be rather safe. Fertilization and survival rates are high.

The salmon parr is a typically territorial fish normally occupying stations on the bottom from which it undertakes excursions to catch drifting food particles, mostly insect larvae and other small invertebrates, and to defend its territory (Kalleberg, 1958). This behaviour is instrumental in the dispersal of the parr over the appropriate kinds of bottom which shifts with the size of the parr. Significantly coarser gravel and stony bottoms are preferred by the older parr (Karlström, 1977a). Density of "living drift" beside coarseness of the bottom ought to affect the size of territories (see Fig. 6.7).

In the southernmost part of the Baltic Sea and in the fertile rivers the parr may largely run to the sea after one year in the river (e.g., Mörrumsån River; Lindroth, 1977). In Daugava River the mean smolt age is about 2 years (Lishev and Rims, 1961) and in central and northern Sweden 2—4 years (Alm, 1934; Järvi, 1938, 1948). The winters are spent partly in crevices between stones and boulders (Lindroth, 1955a; Karlström, 1977a). (Fig. 6.8).

The river mortality of salmon parr is caused by several species of fish and birds, e.g., pike, burbot and mergansers. The goosander and redbreasted merganser are calculated to have taken in a single year (1953) 150,000 salmon parr in Indalsälven River in Sweden, the estimated smolt production having been about 300,000 (Lindroth, 1955b).

In all, adequate spawning being provided, territorial behaviour and pred-

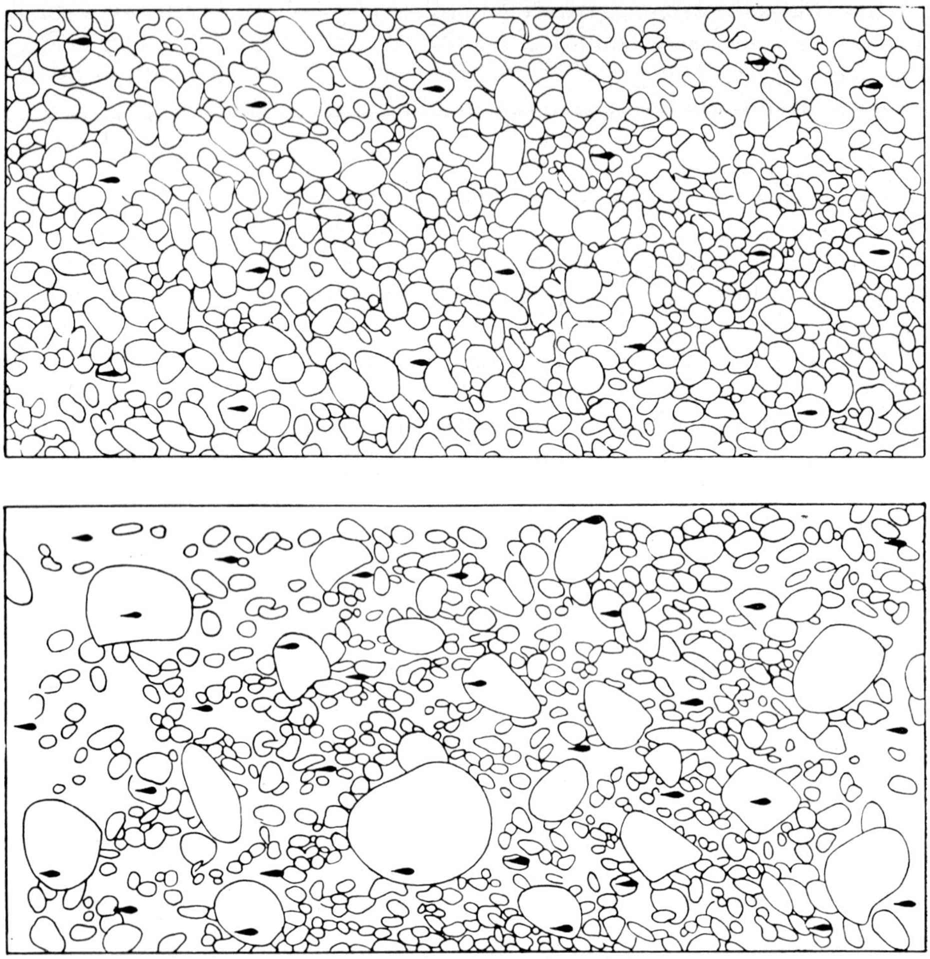

Fig. 6.7. Distribution of underyearling salmon parr on two bottoms of different coarseness (Kalleberg, 1958).

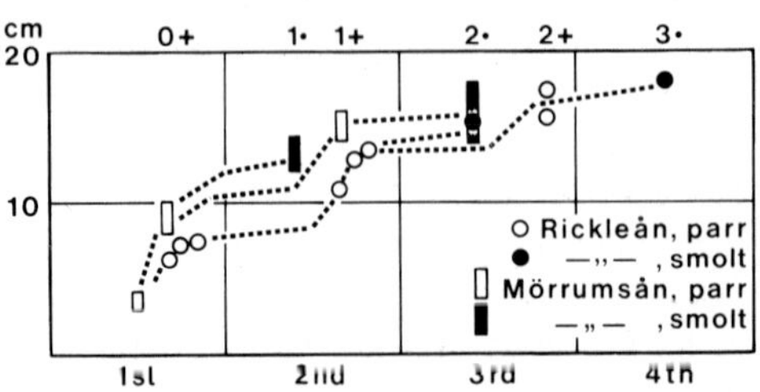

Fig. 6.8. Schematic growth of salmon parr in two rivers, Rickleån and Mörrumsån. See map in Fig. 6.6 (Karlström, 1966; and Lindroth, unpubl. material).

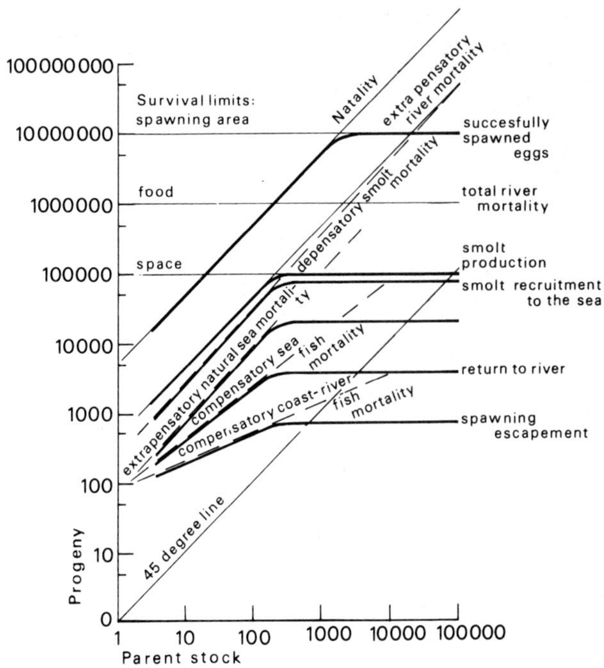

Fig. 6.9. Hypothetical reproduction diagram for Atlantic (Baltic) salmon. The numerical values of survival limits and mortality lines may fluctuate (Lindroth, 1965).

ation that first and foremost remove surplus parr, tend to produce such survival limits during river life that correspond to the carrying capacity of the river (see the reproduction diagram Fig. 6.9). There are instances of a remarkable constancy in smolt production within a river (Rickleån River in Sweden; Österdahl, 1969). Among different rivers, on the contrary, a very great span in parr density obtains, in autumn from some 5 parr/100 m^2 of all (mostly three) age classes in the north up to about 75 parr 100 m^2 in Daugava River (Lishev and Rims, 1961) and Mörrumsån River (dominating one-year-old smolts; Karlström, 1977a).

Older male parr often turn sexually ripe in autumn and may take part in the spawning but the primary sex ratio in young parr does not differ significantly from 1:1 (Rickleån River in Sweden; Karlström, 1966).

Smolt run and postsmolts

Under the influence very likely of photoperiod, the larger parr prepare in spring for the physiological and morphological changes resulting in the smolt stage. The parr coloration is coated by guanine in the cutis, and the body of the fish changes into a thinner shape as do the scales (Lindroth, 1963). The tie to the territory is loosened, and the smolts disappear, partly by active swimming, from the rivers (Kalleberg, 1954).

The smolt run may be triggered by temperature — at the least there seems to be a causal connection with its start. In the southern part of the area (Mörrumsån River in Sweden; Lindroth, 1977) the run starts in the last week of April to the first week of May and lasts for 4 or 5 weeks. In northern Sweden, the duration is about the same but the run takes place 3 or 4 weeks later (Österdahl, 1969). No autumn migration of smolts is reported for the Baltic Sea. The migration starts out with the older and larger smolts running at night shifting to younger fish running at day as the migration period proceeds (Österdahl, 1969). The size of the smolts is important for the rate of return (Lishev and Rims, 1961); released reared smolt-sized parr have increased their return rate to the fishery by 2.3% for every cm increase in size (Carlin, 1969), wild smolts even by 5% (Lindroth, 1977). The sex ratio in the smolts is lower than 1:1 (males:females), probably associated with the sexual maturation of the males (Österdahl, 1969).

When migrating downriver the smolts feed mainly on the living drift (Södergren and Österdahl, 1966). Their feeding habits are maintained in the sea, inasmuch as insects from the water surface, often found drifting in thousands in foam lines with algae and other debris, constitute the bulk of food at least for a considerable part of the first summer (Lindroth, 1961).

A special study of predation by burbot (*Lota vulgaris* Jenyns) on released reared smolts in Lule älv River in 1975 has given extremely high figures. About 175,000 smolts, or 1/3 of the number released, had fallen victim to burbot. Smolts released only 5 km further downstream from the ordinary release site, a power station, showed an essentially better survival rate (return to the fishery 11% as against 6.3%; Larsson and Larsson, 1975).

Sea life

In the sea the salmon displays its remarkable capacity for growth and attains a catchable size in its second summer-autumn at sea (see Fig. 6.10). Not less remarkable is a shift in growth observed from the year classes 1938—1939 onwards. Before those a salmon, 3 years of sea age, attained a weight of about 12 kg, after it was 7—8, at least not 10 kg. This is true at least for the northern rivers (Järvi, 1948; Lindroth, 1965). A food limit in the sea, affecting the larger year classes apparent in the catches from around 1945, seems not plausible.

Whether the salmon from the diverse reproduction areas — Gulf of Bothnia, Gulf of Finland, Riga Bay — show preferences for different feeding grounds in the Baltic Proper is not evident, even though probable (Toivonen, 1973). During its sea life, the salmon is roaming about apparently in aggregates if not true schools in search for food in the superficial water strata. First of all sprat and Baltic herring and stickleback dominate as food. Thurow (1968) has estimated the consumption of sprat as 16,500 metric tons annually or 40% of the fishing yield.

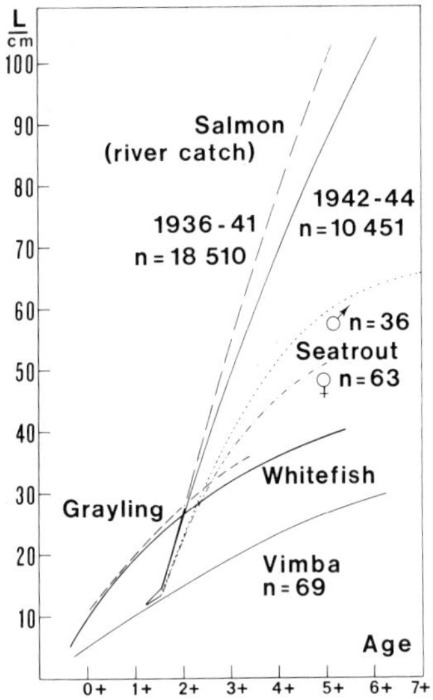

Fig. 6.10. Schematical examples of growth in anadromous fishes in the Baltic Sea. Salmon (Järvi, 1948), sea trout (Hessle, 1935), grayling (Peterson, 1964), whitefish (Lindroth, unpubl. material), and vimba (Papadopol et al., 1970).

Fluctuation of abundance

From time immemorial it has been known that good salmon years have alternated with bad years and that this phenomenon has been common to the Baltic Sea rivers at least in the north. This fact is corroborated by statistical records (see Fig. 6.11). There seems to be no counterpart of this behaviour on the Atlantic coasts. If this conclusion is valid, it is tempting to assume that the common life on the feeding grounds in this enclosed brackish sea is instrumental, if not through food limits so through predation. Much has been speculated on these matters (e.g., Alm, 1924; Lindroth, 1950, 1957b, 1965; Svärdson, 1955, 1957b), but no definite conclusions have been reached. It is somewhat disturbing that the rivers whose populations are maintained solely through release of hatchery smolt-sized parr do not seem to follow the few remaining virgin rivers fluctuation phase downward. This would mean that the fluctuations were generated in the river.

Population regulation

Survival limits, operating during the fresh-water phase, are suggested above: spawning area, food density, number of satisfactory territories. These factors are indicated in Fig. 6.9. They are of an absolute nature and tend to stabilize the populations with respect to the carrying capacity of the actual rivers. It is thought that generally the interaction of food and space is the operating factor provided that spawning is adequate.

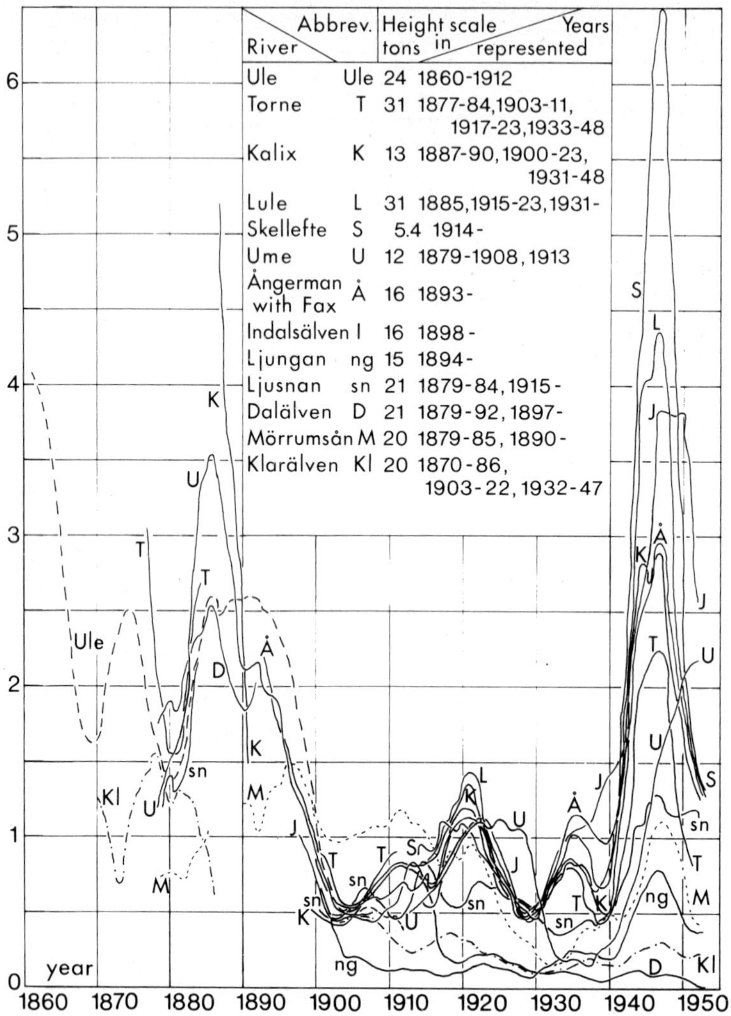

Fig. 6.11. Salmon stock fluctuation in the Baltic Sea illustrated by salmon catch figures in Swedish rivers (Lindroth, 1965).

On the resulting smolt population the sea factors work. As there seems to be little reason to believe in processes limiting survival in the sea, these factors should be density dependent. They could be depensatory smolt mortality (predation type A; Ricker, 1954) through burbot etc., extrapensatory natural mortality (predation?) and compensatory fishing mortality (see Fig. 6.9). However, for the time being, specific predators or diseases cannot be blamed for the natural mortality even though conjectures exist, primarily referring to mortality during the first year at sea: lamprey, trout, salmon, porpoise, seals. After the recruitment of salmon to the catchable stock, man appears as a predator of compensatory nature, i.e., he increases his fishing activities with increasing salmon abundance.

Based on the Swedish taggings, Carlin (1962) and P.-O. Larsson (1975) have quantified a survival diagram for Baltic Sea salmon. Both assume an estuarial and first sea-year mortality of about 85% for the reared and released smolt-sized parr, this figure varying inversely to the size and quality of fish (see Fig. 6.12). Wild smolts have given about twice as large returns. After the first sea-year, the natural mortality decreases to very low values, in fact so low that, considering the growth capacity of the fish and the efficiency of the coastal and river fishery, the sea fishery is an economically unsound undertaking (Fig. 6.13) (see Lindroth, 1964; P.-O. Larsson, 1975).

Artificial stocking

In many countries legislative prescriptions exist for the conservation of the salmon stocks. Beside steps to enforce pertinent industries to control pollution, to erect fish-ways etc., stocking of rivers has in many cases been the verdict of water courts or comparable authorities. After a private start in the middle of the 19th century with release of fry, around 1930 summerlings were used, but, as a rule, positive results failed to appear. It is now apparent that the artificial fry and parr released were destroyed in the competition with

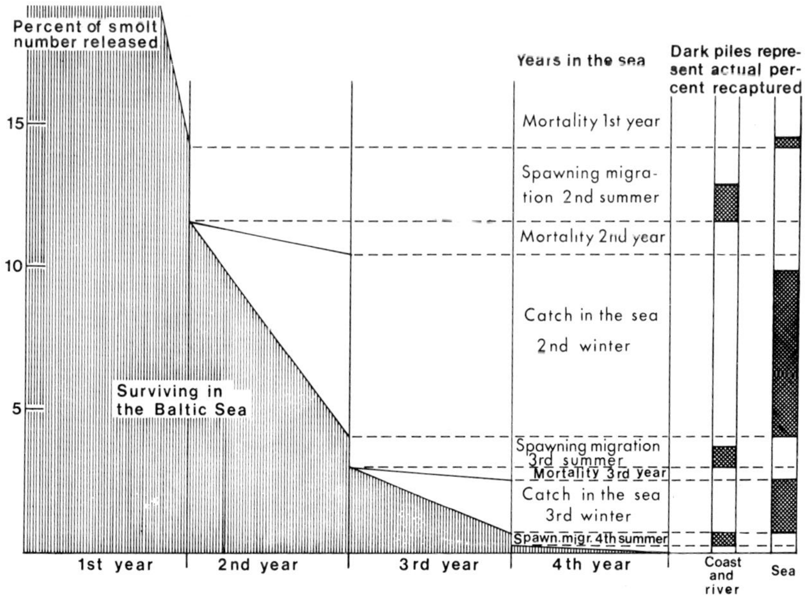

Fig. 6.12. Model of the Baltic salmon population in the sea according to tagging results of the hatchery smolt releases 1958—1967. 10% natural mortality assumed (Larsson, 1975).

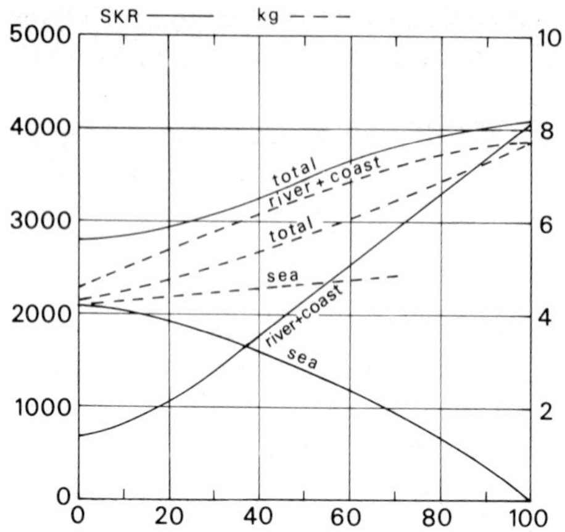

Fig. 6.13. Change in fishing yield and in mean weight of salmon in the Baltic Sea area at hypothetical reduction of the sea fishery. On the abscissa "100" stands for total cessation of the offshore fishery (Lindroth, 1965).

wild conspecies in an already parr-saturated biotope. Not until the stocking was performed during the 1950s through the Swedish Salmon Research Institute (Carlin, 1963) with large parr, substituting a missing wild smolt recruitment to the sea, did the operation turn to success (Fig. 6.14). At present, a high percentage of the salmon taken by fishermen of many countries around the Baltic Sea have been raised in hatcheries and released as smolts; estimates indicate 30% of the catch or more. Sweden alone produces about 1.7 million smolt-sized parr annually, Finland is expected to substantially increase its present 0.1 million, the USSR takes part in the release activities, and Denmark and the Federal Republic of Germany have paid Sweden for the liberation of quite a number in addition.

The overall return from the stockings is of the order of 10% (Carlin, 1969), but the figure varies among rivers, years and even family groups.

Exploitation

The salmon resources have been exploited as far as memory goes. In old times the fishing was concentrated to running salmon in the rivers and along the coasts — biologically, the right manner of procedure. Around the turn of the century, fishermen were more reluctant to fish in the open sea, but with the installation of engines in the ever-increasing number of larger vessels there was a remarkable change evident in the statistically recorded shift in yield from running salmon to feeding fish in the sea. For how long this kind of fishing can continue without affecting the supply of an adequate number of spawning fish for the rivers and for the hatcheries remains to be seen (Lishev and Rims, 1961). There are voices warning that it has already gone too far (Karlström, 1977b).

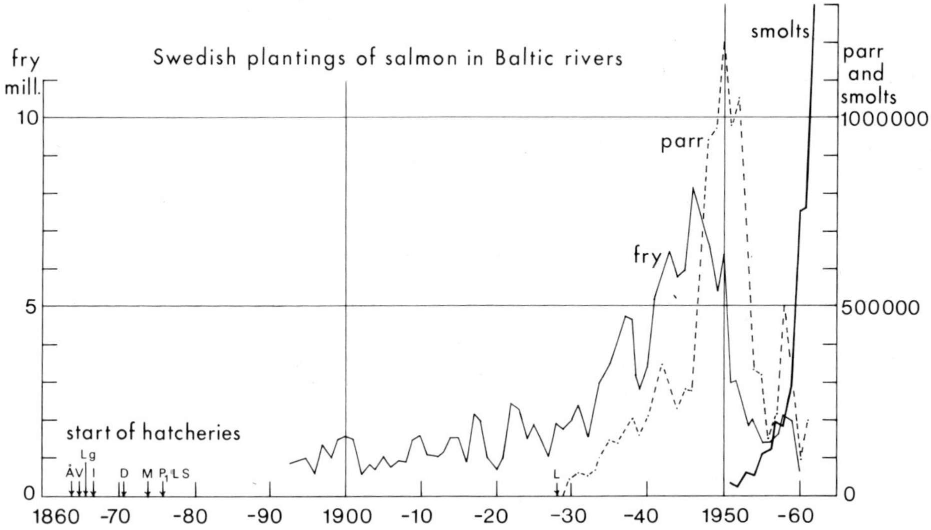

Fig. 6.14. Stocking of salmon in Swedish rivers (Lindroth, 1965).

Salmon has been caught by the most diverse methods. In the rivers, different weirs, traps or seines, standing and drifting nets and dip nets have been used according to the topography and hydrology of the salmon river. On the coasts nets and fixed engines are operated (fyke nets, bag nets, etc.). The running salmon is not likely to take the hook in the north, because running fish do not eat, but in the off-shore fishery long lines with hooks are in extensive use along with drift nets. Hooks are most effective in winter when food is more scarce (Thurow, 1968); see Table 6.III.

Sea-running brown trout

Many forms of brown trout exist, the systematic status of which is not clear. One of these forms is ecologically very similar to salmon. It uses rivers for spawning and as a habitat as fry and parr and passes most of its adult life in the sea. In some respects it deviates from the salmon, however.

The distribution of the sea trout covers not only the salmon rivers, but the fish lives in most large brooks emptying into the sea all along the northeastern coast of the Atlantic Ocean including the Baltic Sea. In the upper tributaries of the large rivers one may, however, find the lake or brook trouts, which do not migrate to the sea; the trout is more likely to form resident populations in lakes and running water.

During its spawning migration, trout navigates about as well as salmon; of river-caught fish only a few per cent have strayed to a foreign river (1 of 62 from the northern Rickleån River and 1 of 54 from the Mörrumsån River; all

TABLE 6.III

Some catch figures (tons a^{-1}). (Sources: Christensen, 1976, 1977; J. Toivonen, pers. commun, 1977; ICES, 1961-1973)

Species and country	1961—1965	1966—1970	1971—1975	thereof 1971—1975	
				Gulf of Bothnia	Gulf of Finland
Salmon					
Sweden	497	455	529	185	
Finland	392[a]	433	549	213	141
USSR	0	0	162[b]		
Poland	44	12	14		
Federal Republic of Germany	278	172	89	+	
Denmark	1460	1585	1129	+	
Total	2671	2657	2472	398	141
Sea trout					
Sweden	64	80	88		
Finland			94[c]		
USSR			108		
Poland			76		
Federal Republic of Germany			+		
Denmark			3		
Total			369		
Whitefish (2 species)					
Sweden	546	395	476		
Finland	1392	1628	1955		
Total	1938	2023	2431		
Eel					
Sweden	1643	1376	1005		
USSR			326		
Poland			294		
German Democratic Republic			308		
Federal Republic of Germany			2		
Denmark			618		
Total			2553		

1) [a] Mean of 1963—1965.
2) [b] Off-shore fishery only 1972—1975.
3) [c] Mean of 1974—1975.

TABLE 6.III (continued)

Catch distribution in the Baltic Proper (tons month^{-1})

Month	I	II	III	IV	V	VI	VII	VIII	IX	X	XI	XII
Denmark 1971—1975, Salmon, trout	92	60	66	40	46	23	0	4	183	226	141	144
Sweden 1971—1975, Salmon	12	15	26	20	42	8	1	17	74	58	28	13
Sweden 1965—1969, Eel	4	3	3	27	76	56	98	172	351	458	163	46

wild fish). They run in the rivers during summer and autumn, the time varying after the size of river, hydrological regime, and other criteria. In the Gulf of Bothnia, the large fish arrive in May and June, the small later, even in late autumn; the latecomers are often silvery with red flesh and do not intend to spawn (e.g., Rosén, 1918). Wistula River was noted for having two populations of sea trout, one running in summer and spawning in the lower reaches of the river, another running late and not spawning until the next year in tributaries, Dunajec and others, far in the uplands of the Carpathian Mountains (Zarnecki, 1963).

Spawning and hatching takes place a little earlier for trout than for salmon, and in most, but not all, rivers the parr is a little larger. The trout parr prefers shallower water, slower current and coarser bottom than does salmon (Karlström, 1977a), and the two species may exert interactive segregation in these respects (Lindroth, 1955a). Trout parr may be found in brackish estuarine water, and spawning is reported to have occurred on the coast. Trout parr remain in the river a little longer and migrate to the sea a little earlier in the year as older and larger smolts. In spite of their large size, wild trout smolts give a little less return (as reported) to the fishery where the two species have been tested side by side (Mörrumsån River; Lindroth, 1977).

The behaviour during sea life of the Baltic Sea trout is different in the northern populations and in the populations from at least the larger rivers emptying into the Baltic Proper. The northern sea trout is comparatively stationary, only a few moving more than 200 km away from its estuary (at most about 5% of those caught in the sea). The large-growing fish from some rivers in the south behave differently, scattering in the Baltic Sea as salmon does. Sea trout from the Emån and Mörrumsån Rivers are taken in great numbers (25—50% of the sea catch) more than 200 km from the estuary (even in the Bothnian Sea), and the same widely roaming habits characterize other river populations, e.g., in the Verkeån River (Svärdson and Anheden, 1963) and the Vistula River (sometimes called Weichsellachs). This tendency is not so pronounced for the older fish after their first return to the river (Svärdson, 1964).

The fishing yield fluctuates also for the sea trout. As it occurs at the same time as for salmon, it is plausible that there exist common biological causes.

However, as fishing activities follow the abundance of salmon and the gear is common to the two species, other interpretations are possible.

Also sea trout parr is stocked to compensate for a loss in wild smolt production. The number of parr released is about 10% of the number of salmon, and their exploitation is, on account of their more limited dispersal from the home estuary, of a greater benefit to local fishermen (some data in Table 6.III).

Grayling

The popular sports fish grayling is originally a fresh-water fish. In the brackish water of the Bothnian Sea and the Bothnian Bay (salinity about 5‰) it lives on the coast and north of Umeå in Sweden and of Pori in Finland it even spawns in the sea. The grayling justifies a position in this chapter because of the habit of its coastal populations to migrate upriver in spring to spawn. Taggings in the River Indalsälven in Sweden have shown that there exist two populations, one more stationary in the river and the other anadromous (Peterson, 1964). The last mentioned consists of fish sometimes attaining a weight of more than 1 kg and spawning in the lower reaches. There may be no genetic barrier between the two populations.

The grayling differs ecologically from salmon and trout. It spawns in spring over shallow water and gravelly bottom, and during the fertilization act the female bores her anal part into the substratum in order to get the eggs somewhat covered (Fabricius and Gustafson, 1955). A 0.5 kg female may produce 4000 eggs (Alm, 1942). The young are more vagile than those of the *Salmo* species, not so bound to a territory, and the adults of the coastal population seem to leave the river after spawning. The food, of course, is different on the coast: young fish, gastropods, *Mysis* and *Gammarus* and, during the summer half year, flying insects from the surface (Peterson, 1964).

Commercial exploitation of the grayling is virtually nonexistent.

Whitefish

Among the many palearctic *Coregonus* species there is one truly anadromous whitefish in the Baltic Sea, *Coregonus lavaretus* (L); see Svärdson (1957a). Its populations spawn in most large rivers, the fry or the young are swept out in spring or in summer, and the adults return to the river to spawn in autumn. At least those tagged as adults in a river are homing to the same river, probably annually (Lindroth, 1957a). The migrations may cover up to 500 km (from the River Iijoki in Finland to the Archipelago of Åland; J. Toivonen pers. commun., 1977).

The spawning of this whitefish occurs in late autumn in rivers over fine gravel. The males may spawn at an age of 2 years or older, females a year later. Sex ratio in the population is 1:1, but seine nets operated on the spawning grounds in late autumn (November) give a majority of males; these seem to hold on to the grounds awaiting brief visits from the females (Lindroth,

1957a). The fecundity of this species is 18,000—20,000 eggs per kg of mature female (Lindroth, 1952b; Johansson, 1956). Both sexes show at dusk some kind of a circular swimming display close over the bottom followed by a suden coincident rush to the surface by a couple giving the sex products off, sperm and a portion of 50—100 eggs (Fabricius and Lindroth, 1954). The eggs sink to the bottom. Some are hidden between small stones and pebbles, but many are immediately eaten by the parents or other fish or roll downstream along with the current.

Hatching occurs in connection with the vernal rise of water temperature. The fry swim upwards and are swept downstream. In dark nights this happens at very weak currents, 0.5 cm s^{-1} ; in light the fry may head against a current with a speed not far from 10 cm s^{-1}. Some fry are detained in eddies for weeks and do not leave the river until having reached a size of about 6—8 cm. The fry start eating copepods, chironomids and blackfly larvae in the river and later also pupae and diverse items of the living drift (Lindroth, 1957a).

The whitefish treated here, *C. lavaretus*, seems to occur beyond spawning time on the coast mixed with a coast-spawning species (the nomenclature of which is under reconsideration by Svärdson). So far their only distinguishing criteria are gillraker counts: *C. lavaretus* about 31, the other species about 27—28, with a slight clinal depression to the north from the Bothnian Sea to which the figures refer (see Svärdson, 1957a; Lindroth, 1957a).

The food of both species on the coast consists of benthic gastropods, such as *Lymnaea* and *Theodoxus* and, north of the Quark, *Valvata. Mysis* and *Gammarus* are common items as are small crustaceans; planktonic forms are eaten especially by the younger fish. In summer, flying insects are often found in the stomachs. If food segregation occurs between the two coastliving species it has not been studied (Hansson and Sandström, 1968; Karlsson and Larsson, 1969).

Both whitefishes are heavily exploited in the coastal fishery by professional fishermen and others; nets and fixed gear are used. Table 6.III gives some figures for Sweden and Finland. How much of the catch refers to the river-spawning species cannot be ascertained.

Vimba

Till now this chapter has dealt with salmoniform fishes, and it may be questioned if the next one to be treated, the lamprey, is a fish at all. However, many Baltic Sea cypriniforms contain populations which migrate upriver to spawn from their coastal habitats. Among them the most pronounced anadromous species to be treated here is the vimba.

The vimba has an easterly distribution, the main areas of its many forms of uncertain systematic rank (Banarescu et al., 1970) being the Black Sea and the Caspian Sea and the southeastern Baltic Sea with its tributary streams, and a few inland lakes. From the brackish water the mature fish migrate into the rivers, if possible far up, e.g., in the Wistula River in a distance of 865 km.

The migration occurs at high water in autumn or in early spring. Power dams may delay the migration and cause resorption of eggs as is the case at the Kegum dam in Daugava (Erm et al., 1970).

The vimba spawns in May on stony-gravelly bottom after having cleaned stones at a moderate depth in preparation for egg deposition at a current velocity of 0.3—0.7 m s^{-1} (Volskis et al., 1970a). The newly hatched fry remain in the bottom material for 5 or 7 days. Then they spread upstream, partly in pools, feeding on small invertebrates (crustaceans and insects). They migrate to the sea in schools from late summer to winter, sometimes after a year or more, at a size of (24)35—44(57) mm (Sukhanova et al., 1970; Kublickas et al., 1970).

In the sea the vimba keeps to the coast. Molluscs and larger crustaceans are their preferred food items (Kublickas et al., 1970). It is the subject of a not insignificant fishery in the USSR and in Poland in spring and autumn when it rises into bays and rivers (Volskis et al., 1970b). Its commercial importance as an esteemed food fish has justified conservation measures on artificial spawning grounds and its artificial rearing (Volskis, 1970; Troitsky, 1970).

Lamprey

Though not a fish in the true sense but a cyclostome, the lamprey is nevertheless fished and deserves to be mentioned here. The lamprey inhabits the European coasts and large lakes and migrates upstream in running water to spawn. In the Baltic Sea, it spawns probably in all rivers not too small. The brook lamprey, *Lampetra fluviatilis planeri* (Bloch), is intraspecific. Its local populations live upstream of insurmountable falls and do not migrate to the sea (Enequist, 1937).

The spawning occurs on sandy bottom covered with small stones which are moved away by the male by stirring movements cleaning a patch of sand and afterwards defending the site. A watching female now adheres to a stone in front of the depression, the male moves to either side of the female, glides along her, and attaches his oral disc to her head. After a few similar courting performances he winds his tail around the female. The loop glides backwards and the female shakes vigorously, stirring up sand at the same time as a portion of eggs and milt are ejected and the eggs sink to the bottom (Hagelin and Steffner, 1958; Hagelin, 1959).

The spent adults drift downstream, wounded and exhausted, only to die. At the spawning time the lampreys attract lots of fish-eating birds, especially the goosander (*Mergus merganser* L.) but also gulls (*Larus argentatus* Pontoppidan and *L. fuscus* L.). The predatory birds adapt their visits to the stream (Rickleån River in Sweden) according to the diel (nocturnal) activity rhythm of the lamprey, but they also prey upon dead and dying lampreys (Sjöberg, 1973).

The eggs develop in the river, and the ammocoetes larvae burrow in the sandy or muddy bottom and lie there concealed probably for about 4 years (Hardisty and Potter, 1971). During this time, spontaneous nocturnal swimming activity is reported (Enequist, 1937, 1963). The larvae filtrate microorganisms for food.

After having reached a size of 8—15 cm, the ammocoetes larvae metamorphose into young lampreys and migrate to the sea in March or May (in northern Sweden), mainly at night (Lindroth, 1957a; Fogelin, 1972).

In the sea, the river lamprey leads a parasitic life also rasping particulate food off its host (A.G. Johnels, pers. commun., 1955). It has been observed feeding on salmon parr in a rearing station already before encountering salt water (Lindroth, 1957b). It is suspected that the lamprey could be a predator of some importance for smolts migrating to the sea a few weeks after the transformed lampreys. Lamprey scars have been observed on large salmon in the Baltic Sea and are reported from Canada and Scotland (P.E. Elson and W.H. Shearer, pers. commun.).

Having grown to maturation in the sea, probably within 1.5 year, the lamprey starts its spawning migration to the rivers. There is no food intake from this time on. The localities in the sea where the different populations reside are unknown. If they return to a home river is also not known. In the Rickleån River in Sweden the migration occurs from August to the end of the year. Full moon and low run-off are avoided. The opinion has been expressed that in summer the lamprey runs high upstream, that the catch of the autumn fishery consists of lampreys dropping downstream who later turn to swim upstream again to spawn (Enequist, 1937, 1963).

The size of the running lampreys decreases, at least partly due to a shrinkage of the body as well in length as in weight (8—19% in length, 25—29% in weight). The migrating population of the Rickleån River in Sweden has been estimated by a mark-and-recapture technique as about 200,000 individuals in 1973 (the accessible area of the river is 65 hectares). Of this run 4.7 tons were caught (Sjöberg, 1974; Asplund and Södergren, 1976).

Before spawning in the first half of June the lamprey has a nocturnal activity rhythm (Wikgren, 1953; Sjöberg, 1974), during the spawning, it is active all day with a maximum activity at night (Sjöberg, 1974).

European eel

It is a well known fact that the European eel is not reproduced in Europe. This catadromous species spawns in the Atlantic Southeast of Bermuda. The fry drift with the Gulf Stream for three years and eventually reach as metamorphosed elvers the coasts of western Europe and northwestern Africa including the Baltic Sea and the Mediterranean Sea plus the Black Sea. A few American eels (*Anguilla rostrata* Le Sueur) have been caught after having gone astray to the west coast of Denmark (Boëtius, 1976).

In Skagerak and Kattegatt the elvers are observed in the period from January to April, carried with the Jutland current at night in the surface layers of water (Lindquist, 1976). They proceed to the coast and through the Sounds to the Baltic Sea where they continue more slowly, growing into yellow eels. The sexes behave differently. Whereas the males remain on the coast the females pursue their dispersal even though with decreasing density all over the Baltic Sea and its rivers, brooks and lakes until they are stopped by insurmountable waterfalls.

The growing yellow eel is relatively stationary and nocturnal (Westin and Nyman, 1976). In daytime it lies mostly hidden under cover and undertakes its feeding excursions during the dark. It feeds on mussels, snails, crustaceans, worms and small fish, even eels, and fish eggs, and the fat content of its flesh may attain 25% (Hessle, 1942). In fresh water it is a serious predator on the crayfish (*Astacus astacus* (L.); Svärdson, 1972).

At maturation the eels undergo a second transformation to silver eels. The males are now about 40 cm long at an age of from 7 to 10 years. The size of the female varies depending on time and the temperature regime. Few female eels are longer than 100 cm at an age of about 15 years. Captive eels with an age about 100 years have been reported. The fecundity of the female can reach 1.5 million eggs (Hinrichsen, 1977). The silver eels have stopped eating and change the yellowish colour of their bellies into a silvery white. A few eels who may have been delayed in the Baltic Sea show a beginning nuptial dress. This is characterized by a metallic lustre of the skin, grossly enlarged eyes and pointed pectorals. Apparently this stage is normally not attained until the eels have reached the Atlantic Ocean. The eels die after spawning.

The migration of the silver eel out of the Baltic Sea has always intrigued biologists and professional fishermen. To judge from the catch, the migration starts, or is resumed, in spring—summer and goes on, with a peak in September—October, all autumn long (Table 6.III). The eels in the Gulf of Finland keep a west-southwest direction across the Baltic Proper (e.g., Määr, 1947), and if it is assumed that this direction corresponds to some internal compass leading to the Sargasso Sea, it would take the eels south and west along the coast of Sweden, and west along the coasts of Poland and Germany. Denmark, then, would constitute a problem. Along with this hypothesis, many others have been forwarded, but as yet no one, or no combination of hypotheses, has won general acceptance if not possibly the last that at least has the advantage of simplicity. After experimental studies of diel activity, level of activity, reaction to temperature, swimming direction and speed of yellow and silver eels, the authors (Westin and Nyman, 1976, 1977) conclude that their findings all suit the hypothesis that "avoidance of colder water and guidance by the position of the isotherms" (1976 abstract) could take the eel, in some cases after hibernation, out of the Baltic Sea and to the Sargasso Sea. The reaction of other fisheries biologists has to be awaited before position can be taken to this unifying hypothesis (which is here not discussed in detail).

Many tagging experiments have been performed on eel, especially silver eel. Of these many have had practical purposes such as study of the influence of chemical or heat pollution on migration routes. Other results have appeared: apparent maximum swimming speed is about 50—60 km/day (e.g., Määr, 1947; Lindroth, 1976); the full moon delays the eel; consequently, the smaller catches then seem not to be caused by a more successful avoidance of the nets (Lindroth, 1976).

The sex ratio of the eel in the Baltic Sea is very low as most of the eels are females growing bigger and staying for a longer time in the Baltic waters. Svärdson (1976) sees this as an evolutionary result and combines it with a hypothesis that females, contrary to the males, avoid crowding and, anxious to arrive at favourable feeding areas, disperse more widely the more abundant the elver population has been. Thus, an increasing share of females remaining on the coast should imply a decrease of the eel population. Now the relative share of females has increased along the southern Swedish south and east coasts which would explain a "decline of the Baltic Sea eel population" (Svärdson, 1976) manifest in a decreased yield of the eel fisheries in the Baltic Sea (see Table 6.III). Records from the Swedish catch stations for elvers point in the same direction, and the cause may be found in a successively cooler climate (Svärdson, 1976).

Table 6.III shows that the exploitation of the eel in the Baltic Sea is concentrated to the autumn. Yellow eel is taken by hook, single line or long-line, or by small fyke nets almost all year round. The large engines, however, with fyke nets of varying design with leaders of up to 500 m long, are not brought in position until summer. They are sometimes so densely placed along the coast that it is a wonder that enough eel slip in between to make these expensive gear profitable. In older days fish-gigs were used from boat or ice aiming at eels aggregated in the mud at this time of the year.

Table 6.III contains also statistical catch data.

D. DEMERSAL FISHES*

Möbius and Heincke (1883) bound the Baltic Sea on the west by the southern entrance of the Danish straits, which means that the Kiel Bay and the southern Little Belt are included and, consequently, areas west of the thresholds at Darss. Bearing in mind that the threshold at Darss and Drogden (in the Sound) are very important limiting factors for the inflow of saline water to the Baltic Sea from the North Sea, only the area to the east of Darss and to the south of Drogden are considered here the Baltic Sea and the areas to the west and north thereof the Transition Area.

Demersal fishes include fishes which ordinarily live on or near the bottom.

* By Ole Bagge.

In this section only demersal fish species which are able to reproduce within the area mentioned above are dealt with and only among them species which are of direct or indirect importance for the fishery will be discussed in greater detail. The species are not listed from a systematic point of view but according to their importance in the ecosystem.

Cod

Distribution

In the Baltic Sea there are two stocks of cod. The Baltic cod (*Gadus morhua callarias*) is found frequently east of 14° 30′E (just west of Bornholm) up to the northern part of the Gulf of Bothnia (the Quark), and occasionally further north. The Atlantic cod (*Gadus morhua morhua*) is distributed west of 14° 30′E.

The stock which occurs in the Baltic Sea in the Kiel Bay, the Danish Straits and in the Kattegat should be called the Transition Area cod rather than the Atlantic cod because the Atlantic cod includes a lot of different stocks (Jamieson, 1967).

The difference between the two cod stocks in the Baltic Sea was shown by Schmidt (1930), who found that the cod east of Bornholm had low mean numbers of rays in their second dorsal fins and high mean numbers of vertebrae. Schmidt's results were confirmed by Kändler (1944b). Sick (1965) who worked with cod haemoglobin types using a zone electrophoresis method showed a distinct and sudden difference in the frequency of the HbI^1 allele just west of Bornholm. Jamieson and Otterlind (1971) who repeated the work of Sick and in addition used cod transferrin types, which is another protein found in cod serum, found a more diffuse border between the two stocks suggesting a fluctuation in the relative proportions of the cod stocks so that the relative proportions of the innermost cod stock (*Gadus morhua callarias*) would contribute relatively less to the fishery in 1967—1968 than in 1961—1962.

Migrations

Thanks to many tagging experiments carried out in the Baltic Sea and in the Transition Area the migrations of the two cod stocks are well known especially since 1955.

From the results of these experiments it appears that the cod east of 15° 30′E migrate only little to the west of that longitude (less than 1—2%), emigrators from the west being more common and possibly related to conditions with a strong influx of North Sea water during westerly gales.

A part of the cod stock west of 15° E which presumably to a certain extent depends on an influx of eggs, larvae and 0-group cod* from the south-

* The age group 0 corresponds to ages younger than one year, age groups I and II correspond to one year, two years, etc.

ern Kattegat and the Belts migrates to the Belts and the Kattegat during winter and spring to spawn (Berner, 1962, 1967, 1968, 1969, 1971a, b, 1973; Bagge, 1961, 1969) but no migration of adult cod in the opposite direction has been observed so far. The remaining part of the stock performs only short migrations from shallow water where the cod stay during summer and autumn to the spawning areas which are found in the Arkona Basin, Fehmarn Belt, Kiel Bay and the Danish Straits and back to shallow water after spawning.

The cod east of 14° 30′E may perform rather long migrations within its area of distribution, especially in connection with spawning and/or unfavourable content of oxygen in the deep water. There are four main spawning areas in this region, viz., the Bornholm Deep, the Slupsk Furrow, the Gdansk Deep and the Gotland Deep.

After spawning the adult cod perform feeding migrations. In the southern Baltic Sea mostly in easterly and westerly directions, the biggest individuals seem to be the most active. The following spring the cod migrate back to their respective spawning grounds, or maybe to the nearest one.

The young cod age groups 0 and I live along the coasts most frequently at a depth of 40—70 m, but are especially abundant along the Polish and Soviet coasts from Oder Bank to Ventspils. A part of the young cod presumably drift slowly with the current eastwards and northwards, which should explain the sometimes abundant occurrence of I and II group cods in the Åland Sea, where no cod eggs, larvae or 0 group cods are found (Otterlind, 1959). It would also explain the long migrations of the Åland Sea cod shown by Otterlind (1961, 1962b) and Sjöblom and Parmanne (1977) see Fig. 6.15, and further the frequent migrations in the southern Baltic Sea from east to west (Rutkowicz, 1959; Biriukov, 1969; Bagge et al., 1974; Netzel, 1963, 1974).

Reproduction

Spawning area. In order to develop cod eggs successfully in the Baltic Sea the salinity must be minimum 10‰. At a lower salinity the eggs cannot float and will sink to the bottom where they presumably cannot survive. This means that to the east of Bornholm the spawning is limited to areas where the water is at least 60—80 m deep. Accordingly, the main spawning areas east of 14° 30′E are the Bornholm Deep (minimum spawning depth 60 m), the Gotland Deep (minimum spawning depth 80 m), the Gdansk Deep and the Slupsk Furrow. In the Baltic Sea west of 14° 30′E the main spawning area is the Arkona Deep just north of Rügen at a depth of 40—50 m.

In Table 6.IV the frequency of cod eggs according to depth in the Gotland Deep, the Gdansk Deep (Grauman, 1970; Bagge, unpubl.) and the Bornholm Deep (Bagge, unpubl.) are shown as percentages.

It is seen that in 1960 the maximum number of eggs were found at a

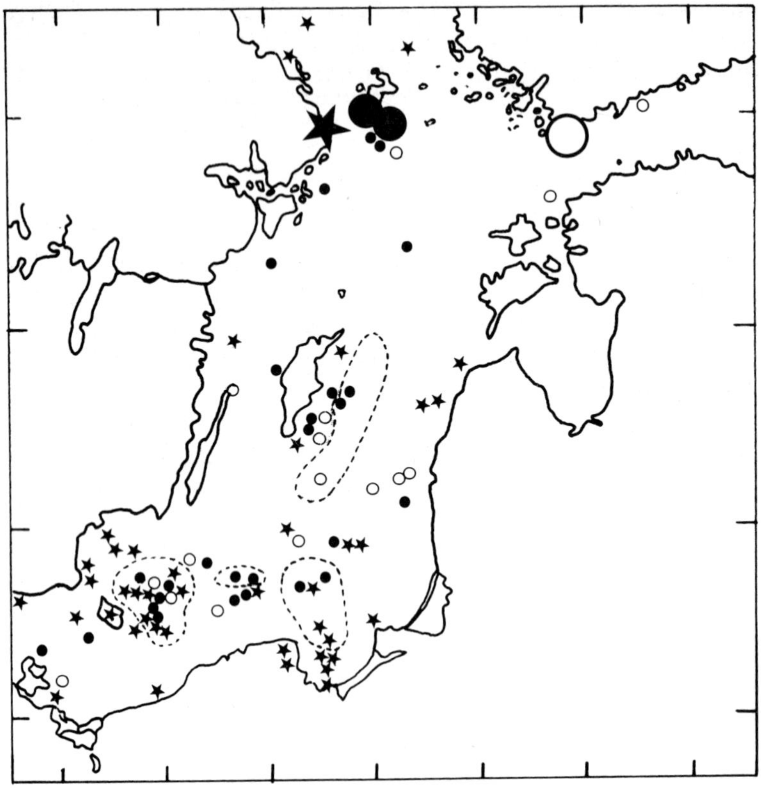

Fig. 6.15. Recaptures of cod tagged at the SW-Finnish coast and in the Åland archipelago. Otterlind (1961, 1962), Sjöblom and Parmanne (1977).

TABLE 6.IV

Number of cod eggs below 1 m^2 according to depth (Grauman, 1970; Bagge, unpubl.)

Depth (m)	Gdansk Deep		South Gotland Deep		Bornholm Deep	
	1968	1969	1968	1969	1976	1977
41— 50	—	—	—	—	2	—
51— 60	—	—	—	—	55	16
61— 70	—	—	—	—	24	46
71— 80	—	—	—	—	11	19
81— 90	10	1	—	1	7	19
91—100	62	8	57	25	—	—
101—130	22	91	43	74	—	—

greater depth than in 1968, possibly because of changes in hydrographical conditions.

For the successful development of the cod eggs the oxygen content of the water is another limiting factor in addition to the salinity. 1 g O_2 m^{-3} is the lower limit. This implies that the success of the spawning in the eastern spawning areas, the Gdansk Deep and the Gotland Deep is more variable and usually less than in the Bornholm Deep and the Arkona Deep. This is illustrated in Table 6.V, where the mean number of eggs per square meter, the mean rate of egg survival, the range of salinity and oxygen content are shown for the period 1954—1970 (Grauman, 1974).

The survival of eggs and the number of eggs and larvae decrease from the Bornholm area towards east and northeast. The survival of cod eggs estimated by Grauman (1974) may be an overestimate by a factor 5—10 (Bagge and Müller, 1976; Müller and Bagge, 1977).

Spawning time. To compensate for the sometimes critical abiotic conditions, the spawning season in the Baltic Sea east of Bornholm is very long. It begins in February—March and ends in August, beginning earliest in the Bornholm area. In the Arkona Deep the spawning starts in February and ends in April. The most intensive spawning period is usually May—June, but varies from year to year and from area to area probably depending on temperature, salinity and oxygen contents. This variation is illustrated in Table 6.VI (Grauman, 1978).

TABLE 6.V

Abiotic conditions and spawning intensity of cod in different areas during the years 1954—1970 (Grauman, 1974)

Spawning areas	Salinity range (0/00)	Oxygen range (ml/l)	Egg survival (%)	Mean catches as numbers per m^2 of	
				eggs	larvae
Bornholm	13—16	0.5 4.7	8.9	52	4.5
Stolpen	10—14	1.0—5.0	7.5	15	4.5
Gdansk	9.8—12.8	0.8—4.0	7.2	21	3.0
Southern Gotland	10—13	0.6—3.5	6.5	25	3.5
Middle Gotland*	10—13	0.2—1.7	4.8	32	3.0
Northern Gotland*	9.5—11.9	0.2—1.0	0.8	3.5	0
Öland*	10—11.2	0.1—1.1	0.9	4.0	0

* Data for 1968—1970.

TABLE 6.VI

Number of cod eggs below 1 m^2 in the Bornholm Deep and the Gotland Deep according to month in 1969, 1971 and 1973 (Grauman, 1974)

	February	March	April	May	June	July	August
Bornholm Deep:							
1969	—	29	59	71	82	87	8
1971	—	10	57	50	13	23	6
1973	2	60	129	48	37	54	—
Southern Gotland Deep:							
1969	—	0	23	18	24	20	—
1971	2	0	4.5	5	4	6	3
1973	—	4	16	10	13	0	—

The eggs hatch after about 18 days at +5° C.

When the larvae have grown to a length of about 10—15 mm, they leave the spawning area more or less actively approaching the coasts close to the spawning areas, of which the German, Polish and Soviet coasts seem to be preferred, possibly because of a higher production there, possibly because the main current in the Baltic Sea is anticlockwise and the Coreolis effect.

The young cod spend their first two years of life in rather shallow water at a depth of 10—70 m before they enter the spawning stock as described on p. 257.

Fecundity. Marine fish species in the Baltic Sea are slowly growing but have a higher fecundity. According to Schopka (1971) in the North Sea, a female cod 75 cm long produces an average of 1,830,000 eggs, but 3,660,000 eggs in the Baltic Sea, i.e., 100% more than in the North Sea. A part of the Baltic cod mature at an age of 2 years and all when 3 years old. In the northern Gotland area and the Aland area they mature later.

The size of cod eggs increases with decreasing salinity. The osmotic conditions increase the uptake of fluid by the eggs from the ovary. The decreasing osmotic pressure further reduces the density of the eggs and thus facilitates the floating. The mean diameter of cod eggs in the Baltic Sea is between 1.7 mm and 1.8 mm and in the North Sea 1.45 mm.

Food and feeding

Food consumed by cod in the Gotland area has been investigated by Uzars (1975) who found that cod less than 30 cm feed mainly on crustaceans like *Mysis mixta, Pontoporeia femorata, Pontoporeia affinis, Mesidotea entomon* and the polychaete *Harmothoë sarsi.* Fishes like herring and sprat play a minor role as food except in the spring when small cod between 20 cm and 30 cm long eat 14—29% sprat by weight.

For cod longer than 30 cm fish are more important as food, especially the sprat which during the first 6 months of the year constitute 30—40% of the cod food, herring 9—25% and other fishes 2—10%. The corresponding figures for the last 6 months of the year are 6—9%, 3—25% and 2—7%, respectively. The invertebrates are also important as food. Thus the first 6 months *Mesidotea* constitute 19—21%, Mysidacea 2—8%, other crustaceans 2—6% and *Harmothoë* 12—23% of cod food. The corresponding figures for the last 6 months of the year are 28—44%, 11—37%, 2—7% and 6—13%, respectively.

The cod is feeding all the year long, except in very severe winters. The feeding of mature female cod is especially intensive under maturity stage IV*. During stages V and VI the feeding does not stop, but decreases. For males the feeding decreases during stages IV—VI.

The annual food consumption of cod in the eastern Baltic Sea was estimated to be 2,995,000 tons, of which 276,000 tons consist of herring and 408,000 tons of sprat (Uzars, 1975).

The food of cod in the Gdansk and the Bornholm areas has been studied by Zalachowski et al. (1976).

Almost the same difference between the food of fish below and above 30 cm as found by Uzars (1975) in the Gotland area was shown in the Gdansk and the Bornholm areas, but in these areas cannibalism especially on cod of 0 and I groups was important in late autumn; further feeding on gobies and sandeels in the autumn and the winter. Danish investigations not published have shown that *Lumpenus lampretaeformis* and *Onos cimbrius* during some years are frequently eaten in early spring in the southern Baltic Sea and in the Arkona area. The annual food consumption in the southern and central parts of the Baltic Sea was estimated to be 1,112,000 metric tons (315,000 metric tons of herring and 266,000 metric tons of sprat) and thus amount to only 37% of the total amount of food estimated by Uzars (1975).

Age and growth

In marine fishes which tolerate a wide range of salinity (euryhaline species) a reduced growth and a shorter life span are ordinarily observed for that stock of a given species living in the lowest salinities.

Before 1935, the exploitation of cod in the Baltic Sea was very limited and the stock was presumably large, consisting of many very old individuals and was also growing very slowly. During the World War II the fishing intensity increased rather much and except in the years 1945—1946 the increase continued causing a reduction in the stock of cod. The reduction has caused an increase in the rate of growth and a decrease of the mean age.

Today only very few cod older than 10 years are caught.

* To describe the stage of maturity: A scale with seven steps is used (I—VII), of which I and II represent a weak development of the gonads, VI is the spawning stage and VII the spent stage.

In Table 6.VII the variation of the mean age and mean length during the years 1929—1977 has been shown.

The increase in mean age in 1974—1977 is partly caused by the introduction of a minimum landing size of 30 cm in 1974 and probably partly due to differences in age readings, which are very difficult to make with reference to the eastern Baltic cod.

The von Bertalanffy growth parameters which are considered relevant for the Baltic cod are: $L_\infty = 105.0$ cm, $k = 0.15$, $t_0 = 0.5$. The length at a given age can be calculated from the equation $L_t = L_\infty(1 - \exp[-k(t - t_0)])$.

Exploitation

The cod has no natural enemies in the Baltic Sea except the species itself (cannibalism). The most important limiting factors are fishing, the food supply, and the oxygen and salinity of the water.

In Table 6.VIII the total catch and the catch by countries are shown. Since 1935, the annual catch has increased from about 11,000 metric tons to 216,000 metric tons. This increase is probably not caused by an increased stock of cod but by an increased effort (fishery), which the increased length at age and the decreased mean age shown above support (see Table 6.VII).

The increase in fishing effort has been especially heavy during the last 4 to 5 years, partly due to transfer of fishing effort from the North Sea as a result of limitations of the fishery by quotas, partly due to development of pelagic fishery on the spawning stock in years with bad oxygen conditions.

The state of the stock was evaluated in 1973, 1974 and 1977 (ICES,

TABLE 6.VII

Mean age and mean length of cod at age according to year (1929—1977) (According to Thurow, 1974). Data from 1974—1977 have been added. (Antonovicha et al., 1974). Danish data 1974—1977. Polish data 1975, 1976 presented to the ICES Working Group on Assessment of Demersal Stocks in the Baltic Sea.)

Period	Kiel Bay	Bornholm Deep	Gdansk Deep	Gotland Deep
Mean age				
1929—1938	4.77	4.51	4.74	5.95
1939—1944	2.77	4.22	3.48	4.48
1946—1957	2.02	3.58	3.71	4.33
1960—1967	1.59	3.60	3.68	3.78
1965—1977	2.34	3.57	3.94	4.00
Mean length — age group III				
1931—1938	38	35	35	34
1950—1956	44	42	41	41
1962—1965	44	38	38	—
1974—1977	44	41	36	—

TABLE 6.VIII

Total catch of cod in the Baltic Sea as defined in the introduction (in metric tons)[b]. (Thurow, 1974; ICES, 1978)

Year	Denmark	Finland	German Dem. Rep.	Fed. Rep. Germany[a]	Poland	Sweden	U.S.S.R.	Total
1935	1 700	—		3 000	500	5 000	1 000	11,200
1943	9 900	—		53 100	—	11,900	6 100	81,000
1951	11,700	—	8 000	3 200	51,200	20,500	46,900	141,500
1965	21,450	23	8 186	5 203	41,498	19,523	22,420	118,303
1966	22,658	26	6 671	3 297	56,007	20,415	38,270	147,344
1967	25,839	27	16,839	2 520	56,003	21,367	42,980	165,575
1968	28,320	70	19,381	4 808	63,245	21,895	43,610	181,329
1969	29,371	58	21,802	7 217	60,749	20,888	41,580	181,665
1970	28,014	70	13,604	6 954	68,440	16,467	32,250	165,799
1971	30,000	53	7 375	4 562	54,151	14,251	20,910	131,302
1972	42,000	76	9 160	4 985	57,093	15,194	30,140	158,648
1973	44,650	95	10,404	15,873	49,790	16,734	20,083	157,629
1974	39,510	160	7 948	12,226	48,650	14,498	38,131	161,123
1975	46,543	298	11,271	12,465	69,318	16,033	49,289	205,217
1976	57,094	287	7 260	14,313	70,466	18,388	49,047	216,854
1977[c]	55,749	310	9 990	19,899	47,702	16,061	23,767	173,478

[a] Gutted weight.
[b] East of Gedser-Dars, south of Stevns-Falsterbo.
[c] Provisional data.

1973, 1974, 1977). The yield per recruit curve[1] in the 1977 report shows for the stocks east of Bornholm that fishing ($F = 0.7$)[2] has exceeded the maximum yield per recruit ($F = 0.35$)[3]. It was further shown that an increase in the average length of cod at first capture from 30 cm to 35 cm would result in a gain of 12% in total catch in the course of five years.

Since 1975 quotation of cod has been introduced by the International Baltic Sea Fishery Commission (see Chapter 8, Section A), together with a minimum landing size of 30 cm and a minimum mesh size of 90 mm in cod trawls.

[1] The yield per recruit is the expected yield in weight from a single recruit. It is a function of the total amount of fishing (effort).
[2] F — fishing mortality, $e^{-0.7} = 50\%$ per year.
[3] F — fishing mortality, $e^{-0.35} = 30\%$ per year.

Flounder

Distribution

The flounder (*Platichthys flesus*) lives in all parts of the Baltic Sea except in the deeper parts of the Gotland Deep, the easternmost part of the Gulf of Finland and the Bothnian Bay. It is frequently caught in the Bothnian Sea up to the Quark. North of the Quark it exists only occasionally and sparsely. The flounder shows a wide tolerance to changes in salinity and sometimes enters fresh water.

Within its total European area of distribution the flounder is divided into six geographical subspecies and presumably the flounder in the Baltic Sea and in the Transition Area might be divided into additional subspecies like the cod, but so far no work has been done on the subdivision.

Migration

Tagging experiments (Otterlind, 1965, 1967; Vitinsh, 1972) have shown that there is a rather distinct boundary between mature flounder in the southern and central parts of the Baltic Sea to be drawn from the southern part of Öland to Rosewie (Rixhöft) in Poland. North of this boundary the flounder seem to be strongly confined to the different coastal areas and they perform mainly short migrations from shallow water to deeper water and migrate along the coasts mainly towards the north. The deep water around Gotland and between Gotland and Soviet coasts with a very low oxygen content acts as barrier which only to a small extent is forced pelagically. According to Vitinsh (1972) the pelagic crossings should be more frequent when the oxygen content in the bottom water is very low.

In the southern Baltic Sea south and west of the boundary, the flounder is distributed all over the area with the exception of the Bornholm Deep during summer and autumn when the oxygen content is low in the bottom water. There are distinct spawning migrations in winter from shallow water along the coasts to the spawning grounds at a depth of 40—80 m in the Gdansk Deep, the Slupsk Furrow, the Bornholm Deep and the Arkona Deep. After spawning, the fish migrate to very shallow water with a depth of only 0.5—10 m, the females leaving the spawning grounds first.

Tagging experiments show that migrations are rather short and that a lot of local populations exist. A few migrations out of the southern parts of the Baltic Sea to the Sound and the southern Kattegat were demonstrated by Otterlind (1967) and very few migrations from the Transition Area and the Sound to the Baltic Sea were observed by Bagge (1966).

Reproduction

Spawning grounds. In order to keep eggs of flounder floating the salinity of water must be not less than 10‰. It means that the flounder is able to spawn in all the deeps of the Baltic Sea, but the success of the spawning depends on the oxygen content of water. Values of about 1 g O_2 m^{-3} and less are critical for the survival of the eggs. In the eastern parts of the Baltic Sea, a special flounder population which is able to spawn in shallow water has developed. This population is called bank flounder. Sandman (1906) found flounder eggs off southwestern Finland on the bottom at 20 m; the salinity of water was 6 ‰. The eggs were alive and in different stages of development. Sandman stated that the bank flounder spawns at a depth of from 3 m to 20 m. Eggs of flounder also are found on the bottom on Oder Bank. The eggs of the flounder with pelagic eggs have like cod eggs an increasing diameter with decreasing salinity as is shown in the following tabulation (Hempel and Nellen, 1974):

Mean diameter (mm) of eggs of flounder

North Sea	Kiel Bay	Central Baltic Sea	Eastern Baltic Sea
0.95	1.1	1.25	1.3

Accordingly, the eggs of the bank flounder should be expected to be the largest, but, surprisingly enough, the eggs are as small as the eggs of the North Sea flounders.

Spawning time. In the Arkona Deep the flounder spawn in March and April, in the Bornholm Deep from March to May and in the Gdansk and Gotland Deeps from March to June.

The bank flounder of the southwestern Finnish coast spawn from May to July. The eggs hatch in 5 to 6 days at 10° C and in 10 days at 5° C. On hatching the larvae are about 3 mm long and live a pelagic life until 7—10 mm long. They move to the bottom in shallow coastal waters before they are metamorphosed.

At the end of their first year of life the flounder off Bornholm and to the east thereof are about 4—5 cm long, to the west larger. In the Transition Area they are 11—12 cm long.

Fecundity. According to Kändler and Pirwitz (1957) the highest fecundity of flounder is observed in the Akona—Bornholm area, decreasing towards the west. The mean fecundity of a female flounder 35 cm long is:

Number of eggs

North Sea	Kattegat	Kiel Bay	Arkona—Bornholm
920,000	1,190,000	1,120,000	1,310,000

Food and feeding

As regards the 0-group, crustaceans are the most important food. According to Blegvad (1932), 83—87% of the food in the Arkona—Bornholm area consisted of *Bathyporeia* sp. and *Gammarus* sp.

The I-group flounder still prefer crustaceans, but polychaetes are also important. The older fish prefer bivalves like *Mya arenaria* and *Cardium* sp. in the Oder Bank and *Macoma baltica* and *Mytilus edulis* in the Bornholm area. Polychaetes like *Nephthys ciliata*, *Terbellides strömi* and *Harmothoë sarsi* are also important together with crustaceans such as *Mysis mixta*, *Diastylis rathkei*, *Pontoporeia affinis* and *Crangon crangon*.

Age and growth

Only a small part of the flounder in the Baltic Sea gets older than 8 or 9 years. The mean length at age of both males and females in different parts of the Baltic Sea is shown in Table 6.IX.

It appears that the growth decreases from the Kiel Bay towards the east. In the Gdansk Bay it seems to be better than in the Bornholm area, which may be due to a different representation of males and females in the samples.

TABLE 6.IX

The mean length at age of flounder in different parts of the Baltic Sea. In the Bertalanffy equation: $L_t = L_\infty [1-e^{-k(t-t_0)}]$ L_∞ is the length at which the increment per unit of time is zero, k is the growth rate, L_t being the length at time t and t_0 is time zero.

Age group	I	II	III	IV	V	VI	VII	L_∞	k	t_0
Kiel Bay[1]	—	24.6	29.3	33.5	38.8	44.1	46.0	—	—	—
Bornholm Area[2]	—	21.4	23.0	25.5	28.5	31.8	34.2	—	—	—
Gdansk Bay[3]	—	21.3	23.7	26.4	29.5	32.3	35.5	35.2	0.49	0.86
Latvian waters[4]	13.0	18.5	22.2	24.6	27.1	28.2	30.5	35.1	0.25	—0.19

[1] Kändler and Thurow (1959);
[2] Mulicki (1959);
[3] Mulicki (1959);
[4] Zemskaya (1959).

Exploitation

Flounder is the species of major importance among flatfish in the Baltic Sea. The catch of flounder in the Baltic Sea by countries and the total catch is shown in Table 6.X. The catch is probably somewhat higher as flounder caught by cod trawlers often are thrown out dead or moribund. Yield-recruit curves were given in the ICES (1974). The fishery was shown to be near the maximum yield in the Bornholm and Gotland stocks, but it had exceeded the maximum yield in the Gdansk area.

It was recommended to increase the length at first capture from 21 cm to 27 cm, which should produce a gain of 10% and 17%, respectively. The fisheries rules of the International Baltic Sea Fishery Commission specify a minimum landing size of 19—25 cm, protection periods of mature fish, and minimum mesh sizes in trawls.

Plaice

Distribution

The regular area of occurrence of plaice (*Pleuronectes platessa*) extends eastwards to the Gulf of Gdansk. Towards the north it is found sporadically in the Gotland area.

TABLE 6.X

Catch of flounder in the Baltic Sea in metric tons (ICES, 1974 and 1976)

Year	Denmark	Fed. Rep. Germany	German Dem. Rep.	Poland	Sweden	USSR	Total
1960	575	—	—	1523	642	2160	4900
1961	537	—	1288	1621	688	2540	6674
1962	581	—	1092	2467	631	2180	6951
1963	514	14	1119	1937	679	2550	6813
1964	458	50	1180	1257	576	4440	7961
1965	475	23	1050	1797	586	5580	9511
1966	785	28	1378	2886	632	5660	11,369
1967	1262	43	1734	2036	642	4060	9777
1968	502	9	2105	3058	616	3240	9530
1969	314	11	1865	2987	559	3300	9036
1970	457	11	1587	3464	484	3680	9683
1971	327	7	1745	2409	444	4080	9012
1972	324	3	2438	4171	466	3930	11,332
1973	967	6	2885	4278	502	2660	11,298
1974	286	5	1804	4668	470	2510	9743
1975	275	—	1682	5139	400	3238	10,734

Migration

The plaice is more inclined to migrate than is the flounder, possibly due to its lesser tolerance of low salinity and oxygen content of water and, possibly, due to some recruitment from the Transition Area by drifting eggs and larvae.

Otterlind (1967), who tagged plaice in the Arkona area and east of Bornholm, has shown that the plaice east of Bornholm are more stationary than those in the Arkona Basin and in the area between Bornholm and Scania, which to a rather great extent migrate to the Sound, the Belts, the Kattegat and even to the Skagerrak. The most common migrations performed by true Baltic stocks of plaice are from the spawning areas to shallow water along the coasts (feeding migration) in late spring and in the summer and back to deep water in the autumn and the winter (wintering and spawning migration).

Reproduction

The plaice spawns, 3 years old, in the Bornholm Deep and the Arkona Deep at a depth of 50—90 m from November to May, most likely in the Slupsk Furrow, but not in the Gdansk Deep (Mulicki, 1959). February and March seem to be most important. The eggs which are pelagic with a diameter of 1.7—2.2 mm need a salinity of 12—13‰ to float and hatch successfully. The hatching time is 20—22 days at 6—7° C. The newly hatched larvae are 6 mm long; they are pelagic and have a normal symmetric fish shape. Having reached a length of about 10 mm, they start to metamorphose, and at a length of 12—14 mm they start to live on the bottom in shallow coastal water where they stay until they are one year old.

Fecundity. A plaice, 35 cm long, has, according to Kändler and Pirwitz (1957), the highest fecundity in the Transition Area, the lowest in the North Sea and an intermediate one in the Baltic Sea:

Number of eggs

North Sea	Kiel Bay	Arkona—Bornholm
64,000	308,000	250,000

Food and feeding

According to Blegvad (1916) the most important food for adult plaice consists of thin-shelled bivalves such as *Macoma baltica, Mya arenaria, Cardium edule* and small *Mytilus*. Important are also *Hydrobia sp.* and polychaetes, e.g., *Arenicola marina, Scoloplos armiger* and *Harmothoë sarsi.* Crustaceans are not eaten. The young bottom stages feed largely on crustaceans, e.g., *Corophium*.

Age and growth

Plaice older than ten years are not common. The growth of plaice decreases towards the east. Length at age is shown in Table 6.XI (males and females together).

The shape of the plaice is related to its condition. In the Baltic Sea and in the Transition Area, where the abundance of plaice in relation to food supply is small, the shape of the plaice is more circular than in the Kattegat and in the North Sea.

Exploitation

The highest catch of plaice in the Baltic Sea was 5201 metric tons in 1928. After the World War II the catches have been less than 2000 metric tons, since 1969 less than 1000 metric tons, and thus the commercial importance of plaice is small. The stock is not evaluated but no doubt is heavily overexploited.

The fisheries rules of the International Baltic Sea Fishery Commission specify minimum landing sizes (21 cm and 25 cm) and a protection period during spawning time.

Turbot

Distribution

In the Baltic Sea the turbot (*Scophthalmus maximus*) is frequently found up to the northern part of the Archipelago Sea and in the Gulf of Finland to the area west of Porvoo.

Migration

The turbot only performs short migrations from deeper water to shallow water in the spring in order to eat and spawn, and back to deeper water in autumn to a more stable temperature.

TABLE 6.XI

The mean length at age of plaice in different parts of the Baltic Sea. According to Cieglewicz (1964), Kändler (1964) and Munch-Petersen (1974). For notations see legend for Table 6.IX

Age group	II	III	IV	V	VI	VII	L_∞	k	t_0
Kiel Bay	24.7	30.7	33.6	37.1	38.4	40.0	47.0	0.33	−3.0
Oder Bank	22.4	24.7	27.4	30.3	32.0	—	—	—	—
Bornholm Deep	—	23.8	26.2	29.5	32.4	35.4	—	—	—
Gdansk Bay	—	22.4	26.5	29.9	32.5	35.5	—	—	—

Reproduction

Successful spawning of turbot is possible in water with a salinity of 6—7‰. The turbot spawns from May or June to August in shallow water at a depth of 5—40 m on the banks (Oder Bank, Stolpe Bank, Rønne Bank, Great Middelbank) and along the coast in its Baltic area of distribution (Kändler, 1944a). Like the eggs of the so-called bank flounder, the eggs of turbot are not bigger than the eggs of the North Sea turbot (0.90—1.20). The eggs are often covered with diatomeans. Due to the high temperature in shallow water during the summer the eggs and larvae develop rapidly, and already in August the young bottom stages are found in very shallow waters at a depth of 0.5—1 m.

Fecundity. The fecundity of turbot in the Baltic Sea and in the North Sea is almost equal, but is somewhat higher in the Transition Area (Kändler and Pirwitz, 1957). The fecundity of a turbot with a weight of 1 kg is.:

Number of eggs

North Sea	Kiel Bay	Arkona—Bornholm
846,000	983,000	834,000

Food and feeding

The turbot feed mainly on bottom-living fish like sandeels and gobies, but bivalves, e.g., *Macoma baltica* and crustaceans, e.g., *Crangon crangon* are also important as food.

Age and growth

The growth of turbot in the Baltic Sea is variable, being smallest in the northern and eastern parts.

Cięglewicz et al. (1969) have estimated the growth in the Gdansk area which may represent a medium growth:

Age (years)	1	2	3	4	5	0	7	8
Male length (cm)	5.9	14.2	20.7	24.1	26.7	26.7	28.6	—
Female length (cm)	6.6	14.5	22.2	27.1	31.8	31.8	35.8	40.5

According to these results, the L infinite (L_∞), length at which the growth is zero, should be 32.9 cm and 51.3 cm for males and females, respectively, k being 0.38 and 0.35, t_0 0.45 and 0.22.

Exploitation

The catch of turbot in the Baltic Sea before the World War II was from 250 tons to 450 tons annually. After the war the catch has not exceeded 30 tons and is thus of minor importance. The fisheries rules of the International Baltic Sea Fishery Commission specify a minimum landing size of 30 cm and protection periods.

Brill

Brill (*Scophthalmus rhombus*) is found very sparsely in the waters west of Bornholm.

It is doubtful whether brill spawns in the Baltic Sea, but as it is common in the Sound and in the Transition Area west of Darss, the brill found in the Baltic Sea may derive from these sources.

Its spawning, fecundity, food and feeding, age and growth are similar to what is described for turbot.

Brill is of no commercial importance. The quality of the fish is poorer than that of the turbot. The brill is covered by the same protection as the turbot.

Dab

A salinity of not less than 15‰ is necessary for the successful spawning of the dab (*Limanda limanda*), which means that its spawning area is in the Arkona Deep. The spawning takes place from April to August at a depth of 35—40 m. The eggs are pelagic and about 1 mm in diameter. The eggs hatch into larvae in 7—14 days. When the larvae are about 14 mm long they move to the bottom. The small bottom stages remain in rather deep water. It is unknown how much the Baltic stock of dab depends on eggs and larvae from the Kattegat and the Belts drifting in with the current.

According to Kändler and Pirwitz (1957) the fecundity of the dab in the Baltic Sea is higher than in the North Sea, but the highest fecundity is found in the Transition Area. Kändler and Pirwitz found the following fecundity for dab 35 cm long:

Number of eggs

North Sea	Kattegat	Kiel Bay	Arkona—Bornholm
420,000	735,000	765,000	675,000

Dabs feed principally on brittle stars, sandeels, polychaetes and molluscs, e.g., *Macoma baltica* and *Cyprina islandica.*

No data on growth of dab in the Baltic Sea, east of the Gedser-Dars are available, but Kändler and Thurow (1959) have estimated the mean growth in Kiel Bay for males and females as follows:

Age (years)	2	3	4	5	6
Length (cm)	16.98	22.27	28.02	31.52	32.83

The females have a higher growth rate than males. The growth rate in the Baltic Sea is presumably less.

The catch of the dab in the Baltic Sea is a by-catch which does not exceed 30 metric tons per year. It is thus of small importance.

Sandeel

The sandeel (*Ammodytes tobianus*, syn. *A. lancea*) lives in the whole Baltic Sea and is more abundant than the greater sandeel.

The sandeel spawns in small piles on sandy bottom at a depth of about 20 m. The eggs are oval and 0.8—1.0 mm in diameter. According to Kändler (1941), there are spring spawners and autumn spawners. The larvae are pelagic from September to June.

The maximum age of the sandeel is 3 years. It obtains a length of 17—18 cm at the end of the third year and a maximum length of 20 cm.

Like the greater sandeel, this species has no direct commercial importance, but forms an important part of the diet of cod and turbot.

Greater sandeel

The greater sandeel (*Hyperoplus lanceolatus*) is distributed in the whole Baltic Sea. The fish spawns from May to August on the bottom at a depth of 20 m. The eggs are oval; the larvae are pelagic. The fish feeds on zooplankton such as larvae of young sandeel (*Ammodytes tobianus*). It reaches an age of four years. The length at various ages is (Kändler, 1941):

Age (years)	1	2	3	4
Length (cm)	9.7	15.1	20.2	23.8

Its maximum length is about 25 cm. In the Baltic Sea, the species has no direct commercial importance but is very important as food for the cod and the turbot.

The biology and identification of the different species need further investigations.

Snake blenny

In the Baltic Sea the snake blenny (*Lumpenus lampretaeformis*) is frequently found at depths exceeding 90 m. Its northern boundary of distribution is just north of the Quark. In the Transition Area, with the exception of the southern Little Belt, it is not common but is more abundant in the Skagerrak. The distribution has caused that the Baltic population is a relict from the last glaciation. Hansen (1975), who compared the number of vertebrae of samples from the Baltic Sea and the Skagerrak, did not find a significant difference, the numbers being 79.24 and 80.73, respectively. The larvae of the snake blenny are abundant in the Transition Area in February and in March.

The fish spawns on the bottom in December and in January. Its larvae are pelagic. Details about the spawning localities are not known.

According to Hansen (1975), the snake blenny feeds mainly on *Diastylis rathkei*, amphipods and polychaetes (*Harmothoë sarsi*), but also on *Mycella bidentata*, ostracods, isopods and small sea cucumbers.

Hansen (1975) found an age range from 2 years to 16 years and that males grow faster than females. The calculated infinite lengths (L_∞) from length at age are 35.3 and 30.7, respectively, with k as 0.29 and 0.30 and t_0 as —0.40 and —0.79. Observed maximum length was 42 cm.

The species has no direct commercial importance but forms a part of the diet of the cod.

Four-bearded rockling

In the Baltic Sea the four-bearded rockling (*Rhinonemus cimbrius*) lives at depths exceeding 20 m, on a soft bottom. The Archipelago Sea and the western Gulf of Finland form its northern boundary of distribution.

Like the snake blenny this fish is not common in the Transition Area with the exception of Kiel Bay, but is rather abundant in the northern Kattegat and in Skagerrak.

Fester (1974) found a significant difference between the mean number of vertebrae of samples from the Gdansk Deep and the Skagerrak, viz., 53.63 and 54.56, respectively. Hoffmeister (1957) found a lower number in the Kiel Bay (53.13).

The fish spawns in the Arkona, the Bornholm, the Gdansk and the Gotland Deeps, and in the Slupsk Furrow from March to August. The eggs are pelagic and have a diameter of 0.81—1.10 mm with oil globule.

According to Fester (1974), the most important food item is *Diastylis rathkei*, followed by *Mysis mixta*, polychaetes, amphipods (*Caprella*, *Corophium*), *Idothea granulosa* and *Crangon crangon*.

Fester (1974) found a range of age from 1 to 8 years. The growth of males and females is not different. The calculated infinite length (L_∞) was 41.6 cm and 43.1 cm, respectively, with k as 0.22 and 0.16 and t_0 as 0.07 and —0.87. The maximum length observed was 36 cm.

The species has no direct commercial importance but forms a part of the cod diet.

Father lasher

The father lasher (*Myoxocephalus scorpius*) is distributed in the whole Baltic Sea in coastal waters in the algal zone.

The eggs are large (2.5 mm in diameter) and yellow and are spawned in clumps on the bottom in February and March and guarded by the male. The larvae are pelagic until they are about 15 mm long. Before spawning pairing takes place. The male has at that time a deep red belly with white spots.

The fish obtains maturity in its second year of life.

In the southern Sound, Nielsen (1971) has estimated the following length and fecundity (thousands of eggs) at age:

Age (years)	1	2	3	4	5	6	7	⩾8	L_∞	k	t_0
Female length (cm)	15.6	18.9	20.9	22.9	22.1	25.0	22.5	23.5	25.5	0.43	—1.29
Male length (cm)	15.9	17.4	18.4	19.0	20.3	20.3	21.0	—	21.3	0.33	—3.29
Fecundity	5.5	8.5	10.6	13.0	—	15.9	12.6	13.8	—	—	—

The fish has no commercial importance but is eaten by cod and used as bait on long lines.

Sea scorpion

The northern boundary of the sea scorpion (*Tarulus bubalis*) is the south-western Finnish and in the Stockholm Archipelago.

The species is smaller (14—18 cm long) and less abundant than the father lasher. It spawns in February and March.

Four-horned cottus

The four-horned cottus (*Oncocottus quadricornis*) has a circumpolar distribution. Its distribution in the Baltic Sea is peculiar, as it lives only in the eastern and northern parts. In addition it lives in fresh water. It is assumed to be a relict from the last glaciation when there was an open connection between the Baltic Sea and the White Sea. In fresh water its characteristic horns are much smaller.

Its food in the Baltic Sea consists mainly of crustaceans and worms. It is found down to 80 m but spawns in shallow water and on the bottom from November to February after pairing.

According to Ojaveer (1971), the length, weight and fecundity (thousands of eggs) at age for females are:

Age	0	1	2	3	4	5	6	7	8	9	10	11	12
Length (cm)	5.2	10.6	14.4	16.8	20.3	23.0	24.3	24.8	25.1	25.3	27.9	27.9	30.6
Weight (g)	—	12	44	66	137	195	224	235	285	295	315	393	436
Fecundity	—	—	—	—	—	9.2	9.0	12.7	13.2	—	—	26.1	—

The species has no commercial importance even though it is considered a titbit.

Eelpout

The eelpout (*Zoarces viviparus*) is distributed in coastal waters (the algal zone) in the whole Baltic Sea. Its maximum length is 35 cm, and it is viviparus. Internal fertilisation of the eggs takes place after copulation in August and September. It feeds on sandhoppers and worms.

The fish is of no commercial importance but is caught for human consumption and for use as live bait.

Black goby

The black goby (*Gobius niger*) occurs in coastal waters as far north as the southwestern Finnish and the Stockholm Archipelagos. Its maximum length is 15 cm.

It spawns in the summer and its eggs adhere to algae and stones guarded by the male.

It is important for cod and turbot. It is used as a live bait for catching cod and eel.

Sand goby

The sand goby (*Gobius minutus* (Pallas)) occurs mainly in coastal waters as far north as the Quark. It migrates to deeper water in the winter down to a depth of 100 m.

Its maximum length is 10—11 cm. It spawns in the summer in shallow water. The eggs adhere to the inner surface of empty shells of bivalves. Its larvae are 3 mm long at hatching and move to the bottom at a length of 17—18 mm.

The fish has no commercial role but is important as food for cod, turbot, eel and sandeel.

Lumpsucker

The lumpsucker (*Cyclopterus lumpus*) lives up to the northern part of the Archipelago Sea and in the western part of the Gulf of Finland.

There are two stocks of lumpsucker in the Baltic Sea, of which one occurs in the eastern part along the Soviet coast and in the Gotland area, while the other lives mainly west of Bornholm. The latter is the same migrating stock that lives in the Kattegat and in the western North Sea, while the former is an isolated Baltic stock, which spends its whole life in the Baltic Sea.

The spawning takes place from February to April on the bottom in shallow waters. The eggs are guarded by the male.

The fecundity of the Baltic stock is higher than that of the North Sea stock. According to Bagge (1964) the mean number of eggs per gram of fish is 70.3 and 37.8, respectively.

The length at age in the Baltic Sea and in the North Sea for males and females is shown below (Bagge, 1964).

	Mean length (cm)											
Age group	III		IV		V		VI		VII		VIII	
	♂	♀	♂	♀	♂	♀	♂	♀	♂	♀	♂	♀
North Sea	—	—	31.7	—	36.1	44.9	37.2	45.6	38.2	47.0	40.0	53.0
Baltic Sea	13.5	15.0	14.1	15.8	15.5	16.8	—	—	—	—	—	—

The Baltic lumpsucker has no commercial importance. In spite of its small size it is often caught on salmon hooks during the winter.

Sea snail

The sea snail (*Liparis liparis*) lives along the coasts in the eastern Baltic Sea in the algal zone. It is not observed in the southwestern Baltic Sea (possibly an Arctic relict).

It spawns from November to February, and the eggs are adhered on algae or polyp colonies.

The fish is called sea snail because the ventral fins form an adhesion disc. Its maximum length is from 6 cm to 8 cm. It feeds on small crustaceans.

Gold sinny

The gold sinny (*Ctenolabrus rupestris*) lives in the algal zone in coastal waters as far east as Bornholm. It spawns in the summer at an age of two years and seems to die after spawning. Its eggs and larvae are pelagic. It feeds on small crustaceans (Blegvad, 1932).

Butterfish

The butterfish (*Pholis gunellis*) lives in the algal zone in coastal waters at a depth down to 20 m as far north as the coast of Helsingland. It spawns at an age of three years from November to January. The eggs are hidden under stones or empty shells and are guarded by the parents. Its maximum length is 20 cm. The young fish remain pelagic until they are 3 cm long.

E. FRESH-WATER FISHES*

Species composition and distribution

Thirty-eight species of fresh-water fish are regularly found in the Baltic Sea, five of which have been introduced. The original species are the following:

Salmonidae
- Vendace (*Coregonus albula* (L.))
- Whitefish (*Coregonus lavaretus* (L.) s.l.)
  - — Migratory whitefish (*Coregonus lavaretus* (L.) s.str.)
  - — Sea-spawning whitefish (*Coregonus nasus* (Pallas) sensu Svärdson)
- Arctic char (*Salvelinus alpinus* (L.))

Thymallidae
- Grayling (*Thymallus thymallus* (L.))

Osmeridae
- Smelt (*Osmerus eperlanus* (L.))

Esocidae
- Pike (*Esox lucius* L.)

Cyprinidae
- Bream (*Abramis brama* (L.))
- Silver bream (*Abramis ballerus* (L.))
- Bleak (*Alburnus alburnus* (L.))
- Asp (*Aspius aspius* (L.))
- White bream (*Blicca bjoerkna* (L.))

* By Hannu Lehtonen and Jorma Toivonen.

Crusian carp (*Carassius carassius* (L.))
Gudgeon (*Gobio gobio* (L.))
Dace (*Leuciscus leuciscus* (L.))
Ide (*Leuciscus idus* (L.))
Chub (*Leuciscus cephalus* (L.))
Chekhon (*Pelecus cultratus* (L.))
Minnow (*Phoxinus phoxinus* (L.))
Roach (*Rutilus rutilus* (L.))
Rudd (*Scardinius erythrophthalmus* (L.))
Tench (*Tinca tinca* (L.))

Cobitidae
Spined loach (*Cobitis taenia* L.)
Stoneloach (*Nemacheilus barbatulus* (L.))

Siluridae
Sheatfish (*Silurus glanis* L.)

Gadidae
Burbot (*Lota lota* (L.))

Gasterosteidae
Three-spined stickleback (*Gasterosteus aculeatus* L.)
Ten-spined stickleback (*Pungitius pungitius* (L.))

Percidae
Perch (*Perca fluviatilis* L.)
Ruff (*Gymnocephalus cernua* (L.))
Pike-perch (*Stizostedion lucioperca* (L.))

Cottidae
Miller's thumb (*Cottus gobio* L.)
Mottlefoot sculpin (*Cottus poecilopus* Heckel)

The following introduced species are occasionally found in the Baltic Sea:

Salmonidae
Rainbow trout (*Salmo gairdneri* Richardson)
Lake trout (*Salvelinus namaycush* (Walbaum))
Peled whitefish (*Coregonus peled* (Gmelin) sensu Berg)

Cyprinidae
Carp (*Cyprinus carpio* L.)

Ictaluridae
Brown bullhead (*Ictalurus nebulosus* (Le Sueur))

This list does not include the following anadromous and catadromous species which are dealt with in section C, this chapter: Atlantic salmon (*Salmo salar* L.), sea trout (*Salmo trutta* L.), vimba (*Vimba vimba* (L.)), eel (*Anguilla anguilla* (L.)) and lamprey (*Lampetra fluviatilis* (L.)).

In addition, almost all the species of fresh-water fish inhabiting the waters of the adjoining land areas occur sporadically in the Baltic Sea.

The habitats of the different species vary considerably. The number of fresh-water species is largest in inlets and areas sheltered by the archipelago. Similarly, the proportion of fresh-water species is largest in the northern and eastern parts of the Baltic Sea. According to Hempel and Nellen (1974), fresh-water fish are practically absent from the southwestern parts of the Baltic Sea. In the central parts, half of the 36 constant species are fresh-water fish. In the eastern and northern parts fresh-water species predominate.

Fresh-water fish never form stocks comparable to those of marine fish and do not occur in the open sea. Their movements are usually rather limited and it is possible that even small bays and inlets have their own stocks of fish, which intermix very little with other stocks.

In the bays and inlets on the southwestern coast of Finland, the fish fauna typically consists of: roach, perch, bream, white bream, rudd, pike, bleak, pike-perch and ruff. Some marine fish are also present, particularly the Baltic herring. The proportion of fresh-water species decreases towards the seaward limit of the archipelago, especially that of Cyprinids and the amount of marine species increases. Pike and perch, however, occur abundantly throughout the archipelago. Fresh-water fish are practically absent from the open sea; in addition to truly marine fish, only sticklebacks and migratory fish occur, such as the Atlantic salmon and the sea trout. The change in the composition of the fish fauna towards the open sea corresponds approximately to the change from the southern waters, dominated by Cyprinids, to the northern waters, where Salmonids are dominant.

In the inlets and archipelago of the Bothnian Bay, the fish fauna consists mainly of perch, roach, pike, vendace and whitefish. Cyprinids other than roach are of little importance, while Salmonids play a more important role. The Bothnian Bay is the main area in the Baltic Sea for such species as whitefish, vendace and grayling.

The following species occur in all the coastal water of the Baltic Sea: smelt, whitefish, pike, roach, ide, ten-spined stickleback, three-spined stickleback, ruff and perch.

Species absent only from the southwesternmost parts of the Baltic Sea but occurring elsewhere in coastal waters are minnow, bream and burbot.

The distribution of nearly all the species is patchy, owing to the general nature of the coast. For instance the occurrence of the pike-perch depends on the presence of an archipelago or sheltered bays and inlets. There may thus be large stretches along the coast where it is found only occasionally (Toivonen, 1968). The distribution of some species is shown in Figs. 6.16—6.19.

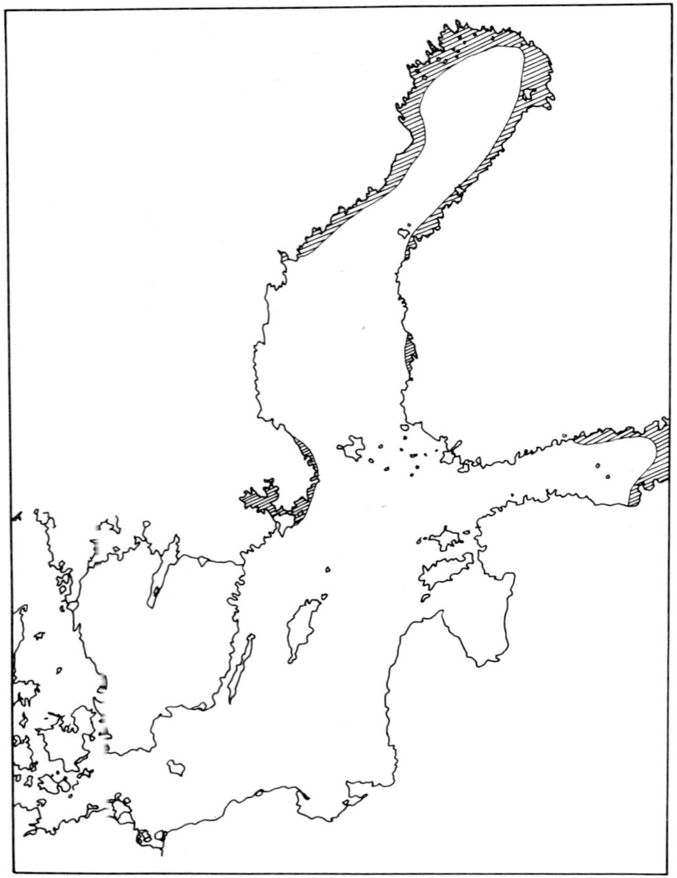

Fig. 6.16. Distribution of vendace (*Coregonus albula*).

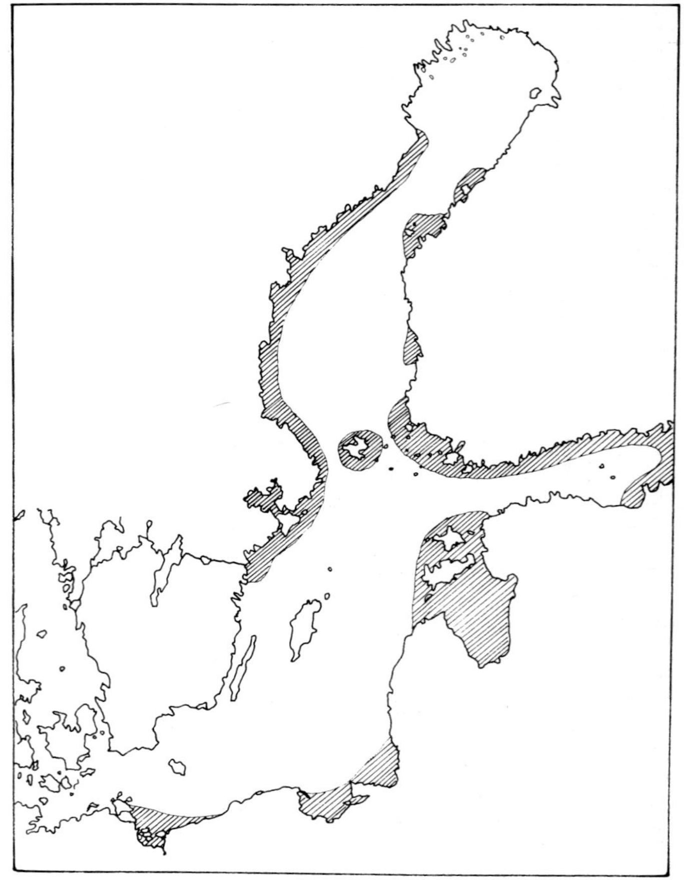

Fig. 6.17. Distribution of pike-perch (*Stizostedion lucioperca*).

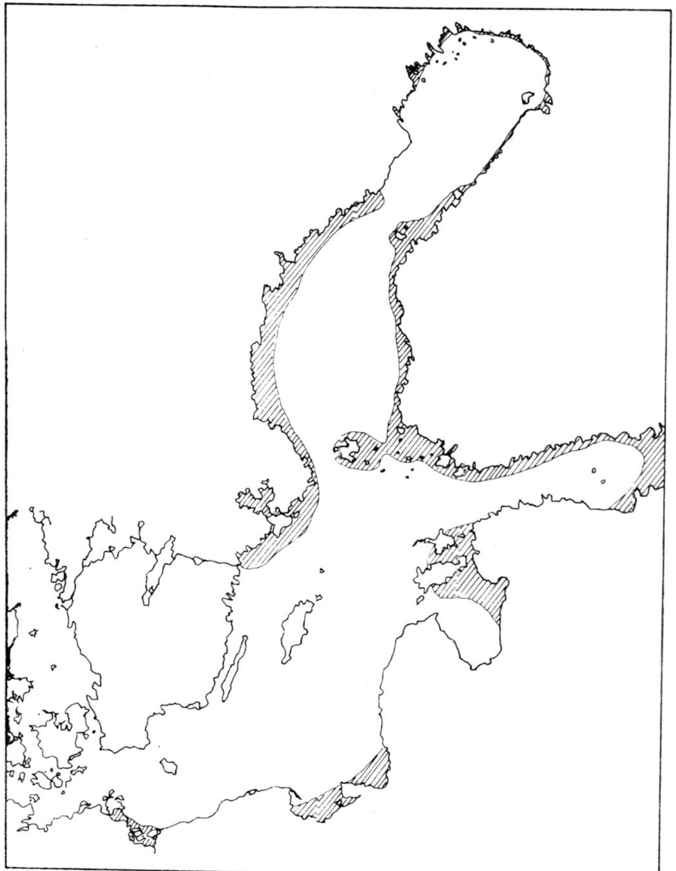

Fig. 6.18. Distribution of bream (*Abramis brama*).

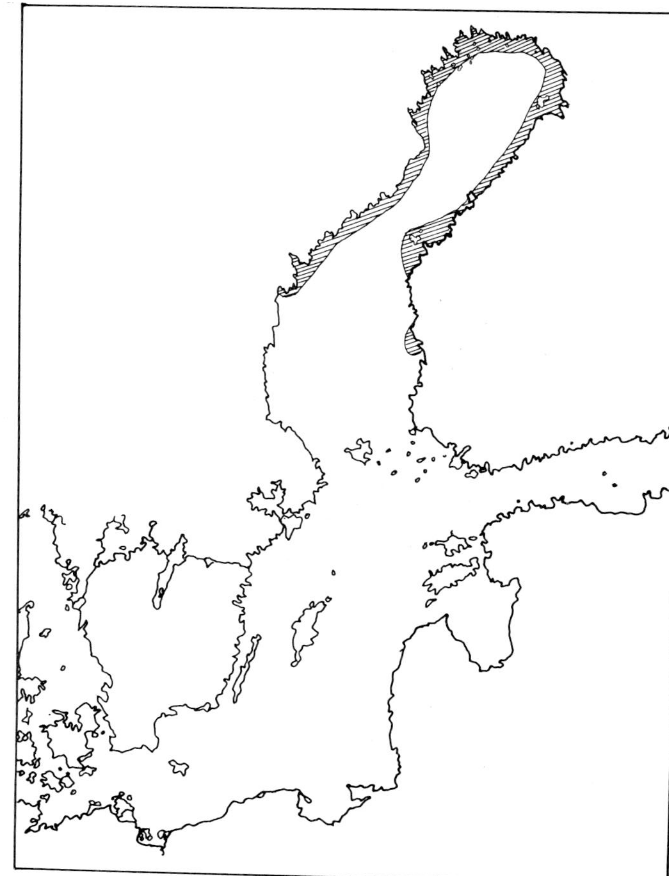

Fig. 6.19. Distribution of grayling (*Thymallus thymallus*).

Migrations

The movements of the Baltic Sea fresh-water fish do not stand comparison with the migrations of marine fish. The fresh-water fish do not travel across large areas of open sea; in most cases their migrations are short and take place between the spawning, wintering and feeding areas.

In the Gulf of Finland the pike-perch winters in deep places seaward of the archipelago and migrates to inlets to spawn in April and May. After spawning, feeding migration takes place in the same areas as the spawning. Migration to the wintering places begins in August or September, and migration parallel to the coastline also occurs at the same time (Toivonen, 1968). In the Bay of Szczecin and in Zalew Wislany feeding migration takes place in the open Baltic Sea and wintering migration in inlets (Henking, 1923; Filuk, 1962).

The bream has a similar movement pattern but does not necessarily winter in deep places (Lehtonen, 1977). The burbot also migrates perpendicularly to the coastline but the seasonal pattern of its movements differs from that of the other fresh-water fish. It migrates into shallow coastal waters to spawn in winter and moves seaward to deeper places when the water is becoming warmer in the spring.

The migrations of fresh-water fish along the coast are usually rather short, but the river-spawning whitefish has been observed to migrate from the Bothnian Bay to Åland Sea and back (Wikgren, 1962). The movements of the sea-spawning whitefish are confined to a much smaller area.

There are also some fresh-water fish whose migrations apparently lack regular patterns or are very short. For example, the pike has been shown to spend all its life within a few square kilometres (Ekman, 1915), and when displaced attempts to return to its original home range (Halme and Korhonen, 1960). A similar though weaker homing instinct has been demonstrated in pike-perch when displaced during their spawning period (Lehtonen, 1979).

Spawning grounds and reproduction

The spawning grounds of fresh-water fish show considerable interspecific variation. Some species spawn in fresh water at the mouths of rivers, some by reefs outside the archipelago, but the majority spawn in inlets in the coast or in the archipelago.

As is normal in inland waters, the pike spawns throughout the Baltic Sea on aquatic vegetation in shallow waters, though at depths somewhat greater than inland, but it has also adapted to a spawning substrate of marine type and usually chooses bladderwrack (Fucus) at a depth of about 0.5—2 m. However, it does not reproduce successfully in salinities exceeding 7‰. Thus in the southern Baltic Sea, successful spawning is limited to the mouths of rivers and to inlets where the salinity is low. The same applies to most of the other fresh-water fish in the Baltic Sea. Gosteeva (1957) has demonstrated experimentally that the eggs of bream can develop in salinities exceeding 10‰, so the bream may be able to reproduce fairly far out in the south. Fresh-water fish often seek their spawning grounds in river mouths and inlets, where salinity and temperature are more favourable. The vimba and the river-spawning whitefish, classified as migratory fish, ascend rivers to spawn, and several species spawn mainly in running water at the mouths of rivers, the most numerous being the smelt.

The fish that spawn on reefs in the open sea are usually marine. The only fresh-water fish using these areas as their spawning grounds are vendace, grayling and sea-spawning whitefish. These species occur mainly in the northernmost parts of the Baltic Sea, where the salinity is low.

Growth

The growth of fresh-water fish shows considerable regional variation, depending on environmental factors. The growth rate is usually higher in brackish than in fresh water, as has been demonstrated experimentally, for example, in rainbow trout (Smith and Thorpe, 1976). There is a general tendency for the growth rate to increase from north to south. Exceptions are found, however: the growth of perch showed no marked differences between various parts of the Finnish and Swedish coasts. The most important factor affecting the length of the growing period is the temperature of the water. Figures 6.20 and 6.21 show the growth rates of burbot and pike-perch in dif-

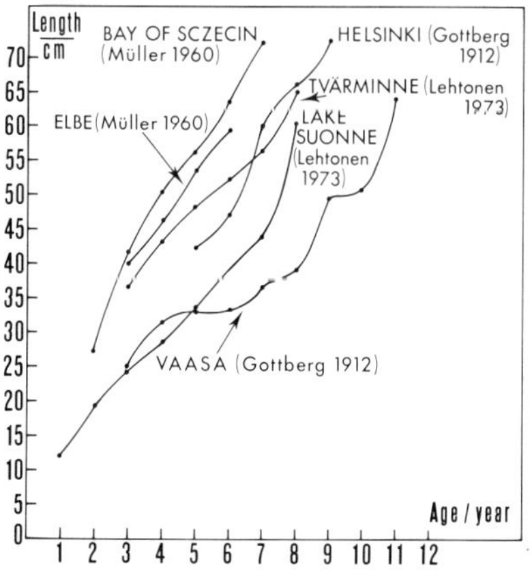

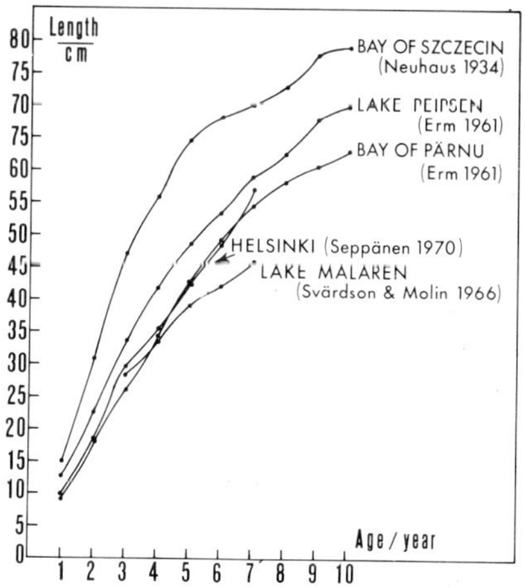

Fig. 6.20. Growth curves of burbot (*Lota lota*) in the Baltic Sea and some lakes and rivers.

Fig. 6.21. Growth curves of pike-perch (*Stizostedion lucioperca*) in the Baltic Sea and some lakes.

ferent parts of the Baltic Sea and in some inland lakes. The farther north the two species live, the lower is their growth rate, but the difference between brackish and fresh water is small.

The fresh-water fish of the Baltic Sea have usually only a short growing period in the summer. In pike-perch in the Helsinki sea area, growth is usually restricted to the time between the end of July and September (Toivonen, 1968). In some other species, e.g., the pike, growth starts earlier and lasts a little longer. The growth of spring-spawning species can be considered to start only after spawning and to continue until the waters cool in September and October.

Fishery and catches

Fresh-water fish are especially important in the catch from the northern and eastern parts of the Baltic Sea. In 1976 they accounted for 38% of the value of the Finnish catch from the Baltic Sea. The importance of fresh-water fish is greatest in areas where the archipelago is extensive, or in certain bays or haffs, particularly the Bay of Szczecin, the Bay of Pärnu and Kurskij Saliv.

The significance of the whitefish fishery is greatest in the Gulf of Bothnia. Whitefish are caught with gill nets and fyke nets and, in the Bothnian Bay, also with trawls. The whitefish fishery is also of great importance in the eastern parts of the Gulf of Finland, in Åland, in areas north of Stockholm and in the waters off western Esthonia and Latvia. In the central and southern parts of the Baltic Sea, whitefish are caught mainly with gill nets.

The vendace fishery is almost entirely limited to the Bothnian Bay, where vendace are caught mainly with trawls and fykes. Less extensive vendace fishery occurs in the eastern parts of the Gulf of Finland and here and there along the coasts of Finland and Sweden.

Smelt are caught mainly in the eastern parts of the Gulf of Finland and the Gulf of Riga, with trawls and fykes. Fyke-fishery is also important in the Quark in the Gulf of Bothnia.

Pike are fished intensively, particularly by amateur fishermen, who catch them mainly with bait-casting or spinning equipment. They are also commonly caught with fykes and gill nets and hooks in winter. The fishery is most important in Åland, the Finnish Archipelago Sea and the archipelago of Stockholm.

The pike-perch fishery is especially important in the Gulf of Finland, the Finnish Archipelago Sea, the Gulf of Riga, certain areas off the Swedish coast and the bays of the southern Baltic Sea. Pike-perch are caught mainly with gill nets.

Cyprinids (bream, roach, ide, etc.) are mainly fished in archipelagos and inlets, i.e., off southern and central Sweden, off the northern shores of the Gulf of Finland, in the Finnish Archipelago Sea, the Gulf of Riga and bays in the southern Baltic Sea. They are caught with gill nets, fykes and hooks.

TABLE 6.XII

The catches of the fresh-water fish in the Baltic Sea in 1977 in Finland and Sweden in metric tons (Anonymous, 1978a, b).

	Finland	Sweden
Vendace	565	1092
Smelt	388	1)
Whitefish	1967	475
Pike	2048	321
Bream	1036	4
Pike-perch	517	1)
Perch	3003	151
Burbot	603	19
Ide	295	1)
Others	1046	100
	11,468	2162

1) Included in "others".

Perch are fished in all the coastal waters. The most important fishing areas are situated in the archipelagos of southwestern Finland and Stockholm. Perch are caught with gill nets and fykes, and fished by amateur fishermen with hooks and lines.

Fresh-water fish abound in catches from the Gulf of Bothnia, the Finnish Archipelago Sea, the Gulf of Finland, the Gulf of Riga, some bays in the southern Baltic Sea and in the Swedish eastcoast. Table 6.XII shows the catches in metric tons made by Finland and Sweden in 1977.

REFERENCES

Alm, G., 1924. Laxen och laxfiskets växlingar i Mörrumsån och andra Östersjö-älvar. Meddn K. LantbrStyr., 244: 1—136.

Alm, G., 1934. Salmon in the Baltic precincts. ICES Rapport, 92: 1—63.

Alm, G., 1942. Laxfamiljen, Salmonidae. In: K.A. Andersson (Editor), Fiskar och Fiske i Norden. Vol. 2, pp. 594—641.

Anonymous, 1978a. Fiskeristatistisk årsbok 1978 (Yearbook of fishery statistics 1978). Stockholm, 107 pp.

Anonymous, 1978b. Kalastus vuonna 1977 (Fishing in 1977). Suomen Kalatalous, 48: 60—61.

Antonovicha, L., Kondratovich, E. and Uzars, D., 1974. Soviet investigations on the eastern stock of Baltic cod 1973—1974. Ann. Biol., 31: 92—93.

Asplund, C. and Södergren, S., 1976. Flodnejonögats (Lampetra fluvialitis L.) lekvandring i Rickleån. Zool. Revy, 36(4): 111—119.

Bagge, O., 1961. Mærkning af torsk i indre danske farvande. Fiskeriundersøgelser, 1960: 10—22. Skr. Danm. Fisk.-Havunders., 21: 1—12.

Bagge, O., 1964. Some observations on the biology of the lumpsucker (Cyclopterus lumpus). ICES C.M. 1964. Baltic Belt Seas Committee 150, 7 pp. (mimeogr.).

Bagge, O., 1966. Tagging of flounder in the western Baltic, the Sound and the Belt Sea. ICES C.M. 1966/D:1, 16 pp. (mimeogr.).

Bagge, O., 1969. Mærkning af torsk i den vestlige Østersø. — Fiskeriundersøgelser 1969. Skr. Danm. Fisk.-Havunders., 30: 1—8.

Bagge, O. and Müller, A., 1976. The spawning of cod in the Bornholm Basin and the size of the spawning stock. Meeresforschung. Ber. Dtsch. Wiss. Komm. Meeresforsch., 25: 172—185.

Bagge, O., Tiews, K., Lamp, F. and Otterlind, G., 1974. German, Swedish and Danish cod tagging experiments in the Baltic 1968—69. Rapp. P.-V. Réun. Cons. Int. Explor. Mer, 166: 22—39.

Banarescu, P., Papadopol, M. and Mikhailova, L., 1970: Sistematika (Systematics). In: Biologija i Promyslovoe Značenie Rybcov (Vimba) Evropy (Biology and Fisheries of Vimba in Europe). Vilnius, pp. 23—70 (in Russian, with an English summary).

Berg, L.S., 1961. Ryby Finskogo zaliva. Izbr. Tr., 4: 304—335. Moskva, Leningrad.

Berner, M., 1962. Ergebnisse der Markierungsexperiment an Ostseedorsch (Gadus morhua L.) in der Bornholm- und Arkonasee, 1959. Z. Fisch. N.F., 11(516): 453—462.

Berner, M., 1967. Results of cod tagging in the western and central Baltic 1962—65. ICES C.M. 1967/F: 5, 5 pp., appendices (mimeogr.).

Berner, M., 1968. Dorschmarkierungen in der Mecklenburger Bucht im Jahre 1965. Fischereiforschung, 6(1): 73—76.

Berner, M., 1969. Ergebnisse der Dorschmarkierungen des Jahres 1967 in der Arkonasee und der Mecklenburger Bucht. Fischereiforschung, 7(1): 39—43.

Berner, M., 1971a. Ergebnisse der Dorschmarkierungen des Jahres 1968 in der Bornholm- und Arkonasee. Fischereiforschung, 9(1): 15—20.

Berner, M., 1971b. Ergebnisse der Dorschmarkierungen des Jahres 1969 in der Mecklenburger Bucht und der Arkonasee. Fischereiforschung, 9(1): 21—27.

Berner, M., 1973. Ergebnisse der Dorschmarkierungen des Jahres 1970 in der Mecklenburger Bucht. Fischereiforschung, 11(1): 19—25.

Biriukov, N.P., 1969. Spawning communities of Baltic cod and the extent of their mixing. ICES C.M. 1969/F: 7, (mimeogr.).

Bitjukov, E.P., 1961. Pitanie salaki (Clupea harengus membras L.) vostočnoj časti Finskogo zaliva. Vopr. Ikhtiol. 1(4)(21): 725—736.

Blegvad, H., 1916. Om fiskenes føde i de danske farvande inden for Skagen. Beret. Dan. Biol. Station, 24: 17—72.

Blegvad, H., 1932. On the flounder (Pleuronectes flesus L.) and the Danish flounder fishery in the Baltic. Rapp. P.-V. Réun. Cons. Int. Explor. Mer, 78(1): 1—28.

Boëtius, J., 1976. Elvers, Anguilla anguilla and Anguilla rostrata, from two Danish localities. Medd. Danm. Fisk.-Havunders. N.S., 7: 199—220.

Carlin, B., 1951. Undersökning över laxens rommängd. Vandringsfiskutredningens Medd., 6: 1—5, appendices (mimeogr.).

Carlin, B., 1962. Synpunkter på Östersjöns laxbestånd i belysning av de svenska märkningsförsöken. Laxforskningsinst. Medd., 5: 1—10 (mimeogr.).

Carlin, B., 1963. Laxforskningsinstitutet och dess hittillsvarande verksamhet. Sven. Vattenkraftfören. Publ., 502: 223—247.

Carlin, B., 1969. Salmon tagging experiments. Atl. Salmon Assoc., Ser. Lectures by B. Carlin, pp. 8—13.

Carlin, B. and Johansson, N., 1971. Undersökningar över rommängd hos lax i Ljusnan 1960 och 1970 — en jämförelse. Laxforskningsinst. Medd., 8: 1—6 (mimeogr.).

Christensen, O., 1976. Notes on the catches of Baltic salmon 1971—1975. ICES C.M. 1976/M: 12, 2 pp., appendices (mimeogr.).

Christensen, O., 1977. Catches of Baltic salmon in 1976. ICES C.M. 1977/M: 25 (mimeogr.).
Christensen, O. and Johansson, N. (Editors), 1975. Reference report on Baltic salmon with additional information on Baltic Sea trout compiled by the Working Group on Baltic Salmon. ICES, Coop. Res. Rep., 45: 1—160.
Cięglewicz, W., 1964. Composition of Polish flatfish catches in the Baltic in 1962. Ann. Biol., 19: 84.
Cięglewicz, W., Draganik, B. and Zukowski, C., 1969. Chrakterystyka wzrostu storni (Platichthys flesus L.) i skampia (Scophthalmus maximum L.) za pomocą równania v. Bertalanffy'ego (Characteristic of growth of flounder (Platichthys flesus L.) and turbot (Scophthalmus maximus L.) by means of the v. Bertalanffy's equation). Pr. Morsk. Inst. Ryb. Gdyni A, 15: 121—132 (in Polish with English and Russian summaries).
Ekman, T., 1915. Meddelanden rörande utförda märkningar af gädda m. fl. fiskar i Södermanlands och Östergötlands skärgård. Sven. Fisk. Tidskr., 1914—1916: 62—68.
Elwertowski, J., 1957. Biologiczma charakterystyka polskich polowów szprota w Baltyku poludniowym w latach 1950—1954 (Biological characteristic of Polish sprat catches in southern Baltic in 1950—1954). Pr. Morsk. Inst. Ryb. Gdyni, 9: 175—219 (in Polish, with English and Russian summaries).
Elwertowski, J., 1960. Biologische Grundlagen der Sprottenfischerei in der östlichen und mittleren Ostsee. Fischerei Forsch. Inf. Prax., 3(4): 1—19.
Elwertowski, J., 1976. The recruitment of Baltic sprat stocks in relation to the environmental conditions. ICES C.M. 1976/P: 17, 8 pp., appendices (mimeogr.).
Elwertowski, J. and Popiel, J., 1961. Recruitment to the herring stock in the southern Baltic. Rapp. P.-V. Réun. Cons. Int. Explor. Mer, 154: 243—245.
Enequist, P., 1937. Das Bachneunauge als ökologische Modifikation des Flussneunauges. Über die Fluss- und Bachneunaugen Schwedens. Ark. Zool., 29A20: 1—22.
Enequist, P., 1963. Angående "nättingfisket" och nejonögonens vandringar i älvarna. Svensk Fisk. Tidskr., 72: 177—178.
Erm, V., 1961. Eesti riim — ja magevete kohade bioloogiliistest ja morfoloogilistest erinuvustest. Hiidrobiol. Uurim., 2: 289—342.
Erm, V., Bontemps, S., Volskis, R. and Spirina, L.I., 1970. Migracii proizvoditelej (Spawning migrations). In: Biologija i Promyslovoe Značenie Rybcov (Vimba) Evropy (Biology and Fisheries of Vimba in Europe), Vilnius, pp. 71—104 (in Russian, with an English summary).
Fabricius, E. and Gustafson, K.-J., 1955. Observations on the spawning behaviour of the grayling, Thymallus thymallus (L.). Rep. Inst. Freshwater Res. Drottningholm, 36: 75—103.
Fabricius, E. and Lindroth, A., 1954. Experimental observations on the spawning of whitefish. Rep. Inst. Freshwater Res. Drottningholm, 35: 105—112.
Fester, U., 1974. Parameterbestemmelse for Enchelyopus (Onos) Cimbrius. Internal Report, Danish Fisheries Investigations.
Filuk, J., 1962. Nachkriegsstudium über Biologie und Fang des Zanders des Frischen Haffs. Z. Fisch. N.F., 10: 705—709.
Fogelin, P., 1972. Studier över drift av larver och nymetamorfoserade ungar av flodnejonöga (Lampetra fluviatilis L.) i Indalsälven och Rickleån. Rep. Rickleå Fältstation 20: 1—27, (mimeogr.).
Gosteeva, M.N., 1957. Ekologo-morfologičeskaja harakteristika razvitija aralskogo lešča. Trudy Inst. Morf. Životnyh, 20: 121—147.
Gottberg, G., 1912. Studier öfver lakens (Lota lota L.) ålder, lek och födoämnen hos oss. Fisk- Tidskr. Finl., 21: 114—121.
Grauman, G.N., 1969. Nekotorye dannye po razmnoženiju kilki v južnoj časti Baltijskogo morja (1958—1964 gg.). In: Zapasy i biologija prom. ryb južnoj časti Baltijskogo morja i ego limanov. Kaliningrad.
Grauman, G.N., 1970. Peculiarities of cod spawning in the Baltic Sea in the spring—sum-

mer period of 1969. Ann. Biol., 26: 136—137.
Grauman, G.N., 1973. The spawning of Baltic cod in 1973. Ann. Biol., 30: 96.
Grauman, G.N., 1974. Investigations on the spawning of Baltic cod in 1968—1970. Rapp. P.-V. Réun. Cons. Int. Explor. Mer, 166: 7—9.
Haahtela, I., 1974. The marine element in the fauna of the Bothnian Bay. Hydrobiol. Bull., 8: 232—241.
Hagelin, L.-O., 1959. Further aquarium observations on the spawning habits of the river lamprey (Petromyzon fluviatilis). Oikos, 10: 50—64.
Hagelin, L.-O. and Steffner, N., 1958. Notes on the spawning habits of the river lamprey (Petromyzon fluviatilis). Oikos, 9: 221—238.
Halme, E. and Korhonen, M., 1960. Haukien vaelluksista rannikoillamme. Kalamies, 4: 1—12.
Hansen, U.J., 1975. Vækst, dødelighed, meristiske karakterer hos Lumpenus lampretasformis (Wolb.) i Østersøen, Kattegat og Nordsøen. Internal Report, Danish Fisheries Investigations.
Hansson, R. and Sandström, O., 1968. Siken vid södra Västerbottens kust. Inst. Ecol. Zool., Univ. Umeå, 55 pp. (mimeogr.).
Hardisty, M.W. and Potter, I.C. (Editors), 1971. The biology of Lampreys 1. Academic Press, New York, London, 423 pp.
Hempel, G. and Nellen, W., 1974. Fische der Ostsee. In: L. Magaard and G. Rheinheimer (Editors), Meereskunde der Ostsee. Springer, Berlin, pp. 215—232.
Henking, H., 1923. Die Fischwanderungen zwischen Stettiner Haff und Ostsee. Z. Fisch., 22: 1—92.
Hessle, Chr., 1925. The herrings along the Baltic coast of Sweden. Publ. Circonst. Conseil Int. Explor. Mer, 89: 1—55.
Hessle, Chr., 1927. Sprat and sprat-fishery on the Baltic coast of Sweden. Medd. K. LantbrStyr., 262: 1—29.
Hessle, Chr., 1935. Gotlands havslaxöringar. Medd. Unders.-Försöksanst. Sötvattensfisket, 7: 1: 1—12.
Hessle, Chr., 1942. Ålen Anguilla vulgaris Turton. In: K.A. Andersson (Editor), Fiskar och Fiske i Norden. Vol. 1, pp. 210—221.
Hinrichsen, D., 1977. All about eels. Scanorama June—July 1977: 70—72 (Interview with I. and J. Boëtius).
Hoffmeister, H., 1957. Beiträge zur Biologie des Whitlings (Gadus merlangus L.) unter der Seequabbe (Onos cimbrius L.) in der Ostsee. Internal Report, Kiel University.
ICES, 1953—1973. Bulletin statistique des pêches maritimes. Vols. 26—57.
ICES, 1973. Report of the Working Group on Assessment of Demersal Stocks in the Baltic. ICES C.M. 1973/F:6, (mimeogr.).
ICES, 1974. Report of the Working Group on Assessment of Demersal Stocks in the Baltic. ICES C.M. 1974/F:4, 21 pp. (mimeogr.).
ICES, 1976. Report of the Working Group on Assessment of Demersal Stocks in the Baltic. ICES C.M. 1976/P:2, 11 pp. (mimeogr.).
ICES, 1977. Report of the Working Group on Assessment of Demersal Stocks in the Baltic. ICES C.M. 1977/P:2, 17 pp. (mimeogr.).
ICES, 1978. Report of the Working Group on Assessment of Demersal Stocks in the Baltic. ICES C.M. 1978/J:3, 24 pp. (mimeogr.).
International Baltic Sea Fishery Commission, 1976. Proceedings of the second session, Warsaw, Sept. 27—Oct. 1, 1976. Warsaw.
Jamieson, A., 1967. Genetic diversity in cod. ICES C.M. 1967/F 20, 6 pp. (mimeogr.).
Jamieson, A. and Otterlind, G., 1971. The use of cod blood protein polymorphism in the Belt Sea, the Sound and the Baltic Sea. Rapp. P.-V. Réun. Cons. Int. Explor. Mer, 161: 55—59.
Järvekülg, A., 1963. Huvitavaid kalu Väinameres. Eesti Loodus, 4: 210—217.

Järvi, T.H., 1938. Fluctuations in the Baltic stock of salmon 1921—1935. Rapp. P.-V. Réun. Cons. Int. Explor. Mer, 106: 1—114.

Järvi, T.H., 1948. On the periodicity of salmon reproduction in the Northern Baltic area and its causes. Rapp. P.-V. Réun. Cons. Int. Explor. Mer, 119: 1—131.

Jensen, A.J.C., 1950. Amount and growth of herring fry in the Danish waters. Dan. Biol. Stat., Rep., 51: 19—48.

Jensen, A.J.C., 1955. Danish herring tagging inside the Scaw, 1949—1952. Rapp. P.-V. Réun. Cons. Int. Explor. Mer, 140(2): 30—32.

Jespersen, P., 1936. The food of herring in the waters round Bornholm. Medd. Komm. Havunders., Ser. Plankton, 3(2): 1—40.

Johansson, B., 1956. Rommängd och romsvällning hos Indalsälvens sik. Sven. Fisk-Tidskr., 65: 144—145.

Kalleberg, H., 1954. Redogörelse för undersökningar rörande smolttransformationen hos lax. Vandringsfiskutredningens Medd., 14: 1—5 (mimeogr.).

Kalleberg, H., 1958. Observations in a stream tank of territoriality and competition in juvenile salmon and trout. Rep. Inst. Freshwater Res. Drottningholm, 39: 55—98.

Kändler, R., 1941. Untersuchungen über Fortpflanzung, Wachstum und Variabilität der Arten des Sandaals in Ost- und Nordsee, mit besonderer Berücksichtigung der Saisonrassen von Ammodytes tobianus L. Kieler Meeresforsch., 5(1): 45—145.

Kändler, R., 1944a. Über den Steinbutt der Ostsee. Ber. Dtsch. Wiss. Komm. Meeresforsch. N.F., 11(2): 73—136.

Kändler, R., 1944b. Untersuchungen über den Ostseedorsch während der Forschungsfahrten mit dem R.F.D. "Poseidon" in den Jahren 1925—1938. Ber. Dtsch. Wiss. Komm. Meeresforsch., 11(2): 137—245.

Kändler, R., 1964. Quantitative fishing of flatfish with the standard trawl. Ann. Biol., 19: 83—84.

Kändler, R. and Dutt, S., 1958. Fecundity of Baltic herring. Rapp. P.-V. Réun. Cons. Int. Explor. Mer, 143(2): 99—108.

Kändler, R. and Pirwitz, W., 1957. Über die Fruchtbarkeit der Plattfische im Nordsee—Ostsee Raum. Kieler Meeresforsch., 13(1): 11—34.

Kändler, R. and Thurow, F., 1959. On the stock of flatfish and cod and the yields of the German fishery in the Baltic. Rapp. P.-V. Réun. Cons. Int. Explor. Mer, 147: 24—38.

Karlsson, S. and Larsson, C., 1969. Siken vid Norrbottenskusten. Inst. Ecol. Zool., Univ. Umeå. 69 pp. (mimeogr.).

Karlström, Ö., 1966. Redogörelse över elfisken efter laxungar i Rickleån 1963 och 1964. Laxforskningsinst. Medd., 6: 1—14.

Karlström, Ö., 1977a. Biotopval och besättningstäthet hos lax- och öringungar i svenska vattendrag. Inform. Sötvattenslab., 6: 1—71 (mimeogr.).

Karlström, Ö., 1977b. Effects of Human Activities on the Population Densities of Salmon and Trout Parr in Swedish Rivers. Diss., Univ. Uppsala, Inst. Zool., 7 pp.

Kostrichkina, E.M. and Starodub, M.L., 1976. Feeding and forage reserve use by herring and sprat. ICES C.M. 1976/H:4, 13 pp. (mimeogr.).

Kublickas, A., Zheltenkova, M.V., Erm, V. and Murzabekova, N.M., 1970. Pitanie (Feeding). In: Biologija i promyslovoe Značenie Rybcov (Vimba) Evropy (Biology and Fisheries of Vimba in Europe) Vilnius, pp. 371—400 (in Russian, with an English summary).

Larsson, H.-O. and Larsson, P.-O. 1975. Predation på nyutsatt odlad smolt i Luleälven 1974. Laxforskningsinst. Medd., 9: 1—13 (mimeogr.).

Larsson, P.-O., 1975. En modell av Östersjöns population av odlad lax. Laxforskningsinst. Medd., 4: 1—12 (mimeogr.).

Lehtonen, H., 1973. Mateen biologiasta Suonteenjärvessä ja Tvärminnessä. Luonnon Tutkija, 77(3—4): 91—100.

Lehtonen, H., 1977. Kuhan (Stizostedion lucioperca L.), hauen (Esox lucius L.) ja lahnan

(Abramis brama L.) vaelluksista, kuolevuudesta ja kasvusta Helsingin edustan merialueella. Manuscript, Univ. Helsinki, Dept. Zool., 73 pp.

Lehtonen, H., 1979. Stock assessment of the pike-perch (Stizostedion lucioperca L.) in the Helsinki sea area. Finn. Fish. Res., 3: 1—12.

Lindquist, A., 1976. Observations on eels in the Skagerak and Kattegat. ICES/EIFAC Symp. on Eel, 48: 1—17 (mimeogr.).

Lindroth, A., 1950. Fluctuations of the salmon stock in the rivers of northern Sweden. Sven. Vattenkraftfören. Publ., 415: 99—224.

Lindroth, A., 1952a. Rommängden hos Indalsälvens lax. Vandringsfiskutredningens Medd., 6: 1—2, appendices (mimeogr.).

Lindroth, A., 1952b. Rommängden hos Indalsälvens sik. Vandringsfiskutredningens Medd., 7: 1—2, appendices (mimeogr.).

Lindroth, A., 1955a. Distribution, territorial behaviour and movement of sea trout fry in the river Indalsälven. Rep. Inst. Freshwater Res. Drottningholm, 36: 104—119.

Lindroth, A., 1955b. Mergansers as salmon and trout predators in the river Indalsälven. Rep. Inst. Freshwater Res. Drottningholm, 36: 126—132.

Lindroth, A., 1957a. A study of the whitefish (Coregonus) of the Sundsvall Bay district. Rep. Inst. Freshwater Res. Drottningholm, 38: 70—108.

Lindroth, A., 1957b. Baltic salmon fluctuations. A reply. Rep. Inst. Freshwater Res. Drottningholm, 38: 109—130.

Lindroth, A., 1961. Sea food of Baltic smolt. ICES C.M. 1961/8, 2 pp., appendices (mimeogr.).

Lindroth, A., 1963. Body/scale relationship in Atlantic salmon. ICES C.M. 1963/J:28, pp. 137—152 (mimeogr.).

Lindroth, A., 1964. Östersjölaxens storlek. Svensk Fisk. Tidskr., 73: 29—34.

Lindroth, A., 1965. The Baltic salmon stock. Mitt. Int. Verein. Theor. Angew. Limnol., 13: 163—192.

Lindroth, A., 1976. Eel catch and lunar cycle on the Swedish east coast. ICES/EIFAC Symp. on Eel, 10: 1—5 (mimeogr.).

Lindroth, A., 1977. Smolt migration in the river Mörrumsån. ICES C.M. 1977/M:8, 11 pp., appendices (mimeogr.).

Lishev, M.N. and Rims, E.Ja., 1961. Some regularities of the population dynamics of Baltic salmon. Trudy Naučno-Issled. Inst. Ryb. Hozjajastvo Sovnark Latv. SSR 3.

Lomniewski, K., Mankowski, W. and Zaleski, J., 1975. Morze Baltyckie Warszawa, 507 pp.

Määr, A., 1947. Über die Aalwanderung im Baltischen Meer auf Grund der Wanderaalmarkierungsversuche im Finnischen und Livischen Meerbusen i.d.J. 1937—1939. Mitt. Anst. Binnenfisch. Drottningholm, 27: 1—56.

Mańkowski, W., 1947. Odzywianie sie i pokarm szprota (Clupea sprattus L.) Bałtyku Środkowego (Feeding and food of sprat (Clupea sprattus L.) in the middle Baltic). Arch. Hydrobiol. Ryb., 13: 37—90 (in Polish, with an English summary).

Möbius, K. and Heincke, F.R., 1883. Die Fische der Ostsee. Berlin, 206 pp.

Møller, D., 1970. Genetic diversity in Atlantic salmon and salmon management in relation to genetic factors. Int. Atl. Salmon Foundation, Spec. Bull. Ser., 1(1): 5—29.

Mulicki, Z., 1959. The state of the South Baltic flatfish stock. Rapp. P.-V. Réun. Cons. Int. Explor. Mer, 147: 39—47.

Müller, A. and Bagge, O., 1977. The mortality of cod eggs, the spawning of cod and the biomass of the spawning stock in the Bornholm Basin. ICES C.M. 1977/P:14 (mimeogr.).

Müller, W., 1960. Beiträge zur Biologie der Quappe (Lota lota L.) nach Untersuchungen in den Gewässern zwischen Elbe und Oder. Z. Fisch., 9(1—2): 1—72.

Munch-Petersen, S., 1974. The flatfish in the Baltic. Data for consideration. Internal Report, Danish Fisheries Investigations.

Munthe, H., 1956. On the development of the Baltic herring in the light of the late quaternary history of the Baltic. Ark. Zool., 9(10): 333—341.
Netzel, J., 1963. Polish cod tagging experiments in the Gdansk Area 1957—62. ICES C.M. 1963. Cadoid Fish Committee No 96.
Netzel, J., 1974. Polish cod tagging experiments in the Baltic in 1969 and 1970. Rapp. P.-V. Réun. Cons. Int. Explor. Mer, 166: 40—46.
Neuhaus, E., 1934. Studien über das Stettiner Haff und seine Nebengewässer. III. Untersuchungen über den Zander. Z. Fisch., 32: 599—634.
Nielsen, E., 1971. On growth and fertility of Myoxocephalus scorpius in the Danish waters. ICES C.M. 1971/F:19, 6 pp., appendices (mimeogr.).
Ojaveer, E., 1969. Räim—Läänemere heeringas. Eesti Loodus, 4: 230—235; 5: 296—301.
Ojaveer, E., 1971. Merihärjast ja mõnest teisest meie mere põhjakalast. Eesti Loodus, 12: 734—737.
Ojaveer, E., 1974. Plodovitost populjacij osennej salaki (Clupea harengus membras L.) severovostočnoj časti Baltijskogo morja. Vopr. Ikhtiol., 14(4)(87): 645—654.
Ojaveer, E., 1976. O roste populjacij salaki vostočnogo poberežja Baltijskogo morja. Rybohoz. Issled. v Bassejne Baltijskogo Morja, 12: 89—113.
Ojaveer, E., Evtjuchova, B.K. and Naglis, A.K., 1975. Stand der Vorkommen und Nutzungsgrad des Herings ander Ostküste der Ostsee. Fischereiforschung, 13(1. Sonderh.) 11—20.
Ojaveer, E. and Simm, M., 1975. Effect of zooplankton abundance and temperature on time and place of reproduction of Baltic herring groups. Merentutkimuslaitoksen Julk. Havsforskningsinst. Skr., 239: 139—145.
Österdahl, L., 1969. The smolt run of a small Swedish river. In: T.G. Northcote (Editor), McMillan Lectures in Fisheries. Univ. B.C. 1969, pp. 205—215.
Otterlind, G., 1959. Swedish cod investigations in the Baltic. Rapp. P.-V. Réun. Cons. Int. Explor. Mer, 147: 65—67.
Otterlind, G., 1961. Swedish cod taggings in the Baltic. ICES C.M. 1961. Cadoid Fish Committee No. 122 (mimeogr.).
Otterlind, G., 1962a. Sillens/strömmingens vandringsvanor vid svenska syd- och ostkusten. Ostkusten, 34(1): 15—21.
Otterlind, G., 1962b. Zoogeographical aspects of the southern Baltic. ICES C.M. 1962. Baltic-Belt Seas Committee No. 103, 6 pp., appendices (mimeogr.).
Otterlind, G., 1965. Migration of flounder in the central Baltic. ICES C.M. 1965. Baltic-Belt Seas Committee No. 162 (mimeogr.).
Otterlind, G., 1967. Migration of plaice and flounder in the southern Baltic. ICES C.M. 1967/F:34, Medd. Havsfiskelab. Lysekil, 40: 1—10, appendices.
Papadopol, M., Volskis, R., Moroz, V.N. and Vladimirov, M.Z., 1970. Rost i vozrast (Growth and age). In: Biologija i Promyslovoe Značenie Rybcov (Vimba) Evropy (Biology and Fisheries of Vimba in Europe), Vilnius, pp. 343—370 (in Russian, with an English summary).
Peterson, H., 1964. Om harrens biologi. In: O.W. Nilsson and H.H. Peterson, Harren. Gebers, pp. 7—48.
Popiel, J., 1951. Pokarm i odżywianie się śledzia (Clupea harengus L.) na terenie zatoki Gdańskiej i wód przyległych (Feeding and food of the herring (Clupea harengus L.) in the Gulf of Gdansk and in the adjoining waters). Pr. Morsk. Inst. Ryb. Gdyni, 6: 29—56 (in Polish, with an English summary).
Popiel, J., 1958a. Differentation of the biological groups of herring in the southern Baltic. Rapp. P.-V. Réun. Cons. Int. Explor. Mer, 143(2): 114—121.
Popiel, J., 1958b. The growth rate of the southern Baltic herring. ICES C.M. 1958. Herring Committee No. 48, 8 pp. (mimeogr.)
Popiel, J. and Strzyżewska, K., 1971. Zmiany stanu zasobów śledzi południowego Bałtyku w latach 1957—1967 (The changes in the state of herring resources in the southern

Baltic in the years 1957—1967). Pr. Morsk. Inst. Ryb. Gdyni A, 16: 85—108 (in Polish, with an English summary).
Popiel, J. and Strzyżewska, K., 1974. Recent changes in the herring stocks of the southern Baltic. Rapp. P.-V. Réun. Cons. Int. Explor. Mer, 166: 133—139.
Poulsen, E., 1950. Brislingefiskeriet og brislingebestanden i de danske farvande. Beretn. Dan. Biol. Station, 52: 3—26.
Rannak, L.A., 1954. Nerestovye arealy, nerest i ocenka moščnosti pokolenij salaki v vodah Estonskoj SSR. Trudy VNIRO, 26: 24—48.
Rannak, L.A., 1970. Plodovitost salaki Finskogo zaliva. Trudy Balt NIIRH, 4: 228—255.
Rannak, L.A., 1974. On the factors determining the abundance of recruitment in the spring spawning herring of the eastern Baltic. Rapp. P.-V. Réun. Cons. Int. Explor. Mer, 166: 145—149.
Rechlin, O., 1964. Untersuchungen zum Verlauf der Laichperiode des Herbstherings in der Mecklenburger Bucht im Jahre 1963. Fischereiforschung, 2(1): 25—29.
Rechlin, O., 1967. Heringsfischerei und Heringsbestand in der Ostsee während der Jahre 1963 bis 1965. Fischereiforschung, 5(2): 23—31.
Rechlin, O. and Friess, C.-C., 1975. Beitrag zur Bestandsbiologie und zur Fangenentwicklung beim Frühjahrshering in den Gewässern um Rügen. Fischereiforschung, 13(1. Sonderh.): 81—84.
Remane, A., 1958. Ökologie des Brackwassers. In: A. Remane and C. Schlieper (Editors), Die Biologie des Brackwassers. Die Binnengewässer, 22: 1—216. Stuttgart.
Ricker, W.E., 1954. Stock and recruitment. J. Fish. Res. Board Can., 11: 559—623.
Rosén, N., 1918. Om havslaxöringen i övre Norrland. Meddn K. LantbrStyr., 212: 1—23.
Rutkowicz, S., 1959. The state of the stock of cod in the southern Baltic as indicated by Polish investigations. Rapp. P.-V. Réun. Cons. Int. Explor. Mer, 147: 48—55.
Sandman, J.A., 1906. Kurzer Bericht über in Finnland ausgeführte Untersuchungen über den Flunder, den Steinbutt und den Kabeljau. Rapp. P.-V. Réun. Cons. Int. Explor. Mer, 5: 37—44, appendices.
Schmidt, J., 1930. Racial investigations. X. The Atlantic cod (Gadus callarias L.) and local races of the same. C.-R. Trav. Lab. Carlsberg, 18(6): 1—72, charts.
Schopka, S.A., 1971. Vergleichende Untersuchungen zur Fortpflanzungsrate bei Herings- und Kabeljaupopulationen (Clupea harengus L. und Gadus morhua L.). Ber. Dtsch. Wiss. Komm. Meeresforsch. N.F., 22(1): 31—79.
Seleckaja, A.V., 1970. Morfologičeskaja harakteristika kilki Baltijskogo morja. Rybokhoz. Issled. Basseine Balt. Morja, 5: 76—92.
Sick, K., 1965. Haemoglobin polymorphism of cod in the Baltic and the Danish Belt Sea. Hereditas, 54(1): 19—48.
Sjöberg, K., 1973. Fiskätandefåglar i ett strömekosystem. Zool. Revy, 35(3): 131—134.
Sjöberg, K., 1974. Lektid, rörelseaktivitet och längd/vikt hos flodnejonöga. Zool. Revy, 36: 41—47.
Sjöblom, V., 1961. Wanderungen des Strömlings (Clupea harengus L.) in einigen Schären- und Hochseegebieten der nördlichen Ostsee. Ann. Zool. Soc. Zool. Bot. Fenn. Vanamo, 23(1): 1—193. Merentutkimuslaitoksen Julk. Havsforskningsinst. Skr., 199: 1—193.
Sjöblom, V. and Parmanne, R., 1976. Assessment of the state of exploitation of Baltic herring in Finnish fisheries in 1970—1975. ICES C.M. 1976/P:11, 17 pp. (mimeogr.).
Sjöblom, V. and Parmanne, R., 1977. Cod off the coast of Finland in 1974 and 1975. Ann. Biol., 32: 84—85.
Skvorcova, T.A., 1975. Hromosomnye kompleksy belomorskoj seldi (Clupea harengus pallasi n. maris-albi Berg) i salaki (Clupea harengus harengus n. membras L.). Issl. Fauny Morej, 16(24): 104—108.
Smith, M.A.K. and Thorpe, A., 1976. Nitrogen metabolism and trophic input in relation to growth in freshwater and saltwater Salmo gairdneri. Biol. Bull., 150: 139—151.

Södergren, S. and Österdahl, L., 1966. Laxungarnas föda under utvandringen. Sven. Fisk. Tidskr. 75: 33—37.
Strzyżewska, K., 1960. Fecundity of the Baltic herring in the Gulf of Gdansk region. ICES C.M. 1960, Herring Committee No. 69 (mimeogr.).
Strżyzewska, K., 1975. Growth and mortality coefficients in Baltic herring local populations. ICES C.M. 1975/P:15, 10 pp. (mimeogr.).
Strżyzewska, K. and Popiel, J., 1974. Changes in the growth of herring in the southern Baltic. Ber. Dtsch. Wiss. Komm. Meeresforsch., 23(3): 268—272.
Sukhanova, E.R., Volskis, R., Moroz, V.N. and Erm, V., 1970. Rečnoj period žizni molodi (River period of the young life). In: Biologija i Promyslovoe Značenie Rybcov (Vimba) Evropy (Biology and Fisheries of Vimba in Europe). Vilnius, pp. 291—342 (in Russian, with an English summary).
Suškina, A.P., 1954. Pitanie salaki Baltijskogo morja i Rižskogo zaliva. Trudy VNIRO, 26: 118—136.
Svärdson, G., 1955. Salmon stock fluctuations in the Baltic Sea. Rep. Inst. Freshwater Res. Drottningholm, 36: 226—262.
Svärdson, G., 1957a. The coregonid problem 6. The palearctic species and their integrades. Rep. Inst. Freshwater Res. Drottningholm, 38: 257—356.
Svärdson, G., 1957b. Laxen och klimatet. Rep. Inst. Freshwater Res. Drottningholm, 38: 357—384.
Svärdson, G., 1964. Verkeåns öring. In: G. Svärdson and N.-A. Nilsson (Editors), Fiskebiologi. LTs Förslag Tema, Halmstad, pp. 103—110.
Svärdson, G., 1972. The predatory impact of eel (Anguilla anguilla L.) on populations of crayfish (Astacus astacus L.). Rep. Inst. Freshwater Res. Drottningholm, 52: 149—191.
Svärdson, G., 1976. The decline of the Baltic eel population. Rep. Inst. Freshwater Res. Drottningholm, 55: 136—143.
Svärdson, G. and Anheden, H., 1963. Långvandrande skånsk havsöring. Sven. Fisk. Tidskr. 72: 109—113.
Svärdson, G. and Molin, G., 1966. Gösen i Hjälmaren och Mälaren. Information från Sötvattenslaboratoriet, Drottningholm, 1: 1—25.
Thurow, F., 1968. On food, behaviour and population mechanism of Baltic salmon. Laxforskningsinst. Medd., 4: 1—16. (mimeogr.)
Thurow, F., 1974. Changes in population parameters of cod in the Baltic. Rapp. P.-V. Réun. Cons. Int. Explor. Mer, 166: 85—93.
Toft, R., 1975. Lukt- och synsinnets roll för lekvandringsbeteendet hos Östersjö-lax. Laxforskningsinst. Medd., 10: 1—40 (mimeogr.).
Toivonen, J., 1968. Kuhan (Lucioperca lucioperca L.) vaelluksista, kasvusta ja kuolleisuudesta Suomenlahden saaristossa, Saaristomeressä ja Ahvenanmaalla. Manuscript, Univ. Helsinki, Dept. Zool., 203 pp.
Toivonen, J., 1973. The stock of salmon in the Gulf of Finland. ICES C.M. 1973/M:17, 5 pp. (mimeogr.).
Troitskij, S.K., 1970. Iskusstvennoe razvedenie (Artificial reproduction). In: Biologija i Promyslovoe Značenie Rybcov (Vimba) Evropy (Biology and Fisheries of Vimba in Europe). Vilnius, pp. 485—508 (in Russian, with an English summary).
Uzars, A., 1975. Peculiarities of feeding and quantitative food consumption of eastern Baltic cod. ICES C.M. 1975/P:4, 9 pp. (mimeogr.).
Valle, K.J., 1934. Suomen kalat. Otava, Helsinki, 228 pp.
Veldre, I., 1974. Kilu. Eesti loodus, 8: 475—479.
Veldre, I. and Polivajko, A.G., 1975. Über die Sprottvorkommen und ihre Nutzung in der nördlichen und östlichen Ostsee. Fischereiforschung, 13(1. Sonderh.): 21—29.
Vitinsh, M., 1972. Migration of flounder in the eastern Baltic. ICES C.M. 1972/F:4 (mimeogr.).

Volskis, R., 1970. Ulučšenie uslovij neresta i nagula molodi (The improvement of the conditions of spawning and fattening of the young). In: Biologija i Promyslovoe Značenie Rybcov (Vimba) Evropy (Biology and Fisheries of Vimba in Europe). Vilnius, pp. 467—483 (in Russian, with an English summary).

Volskis, R., Erm, V., Vladimirov, M.Z., Moroz, V.N., Bontemps, S. and Sukhanova, E.R., 1970a. Ylovy i regulirovanie promysla (Catches and regulation of the fishery). In: Biologija i Promyslovoe Značenie Rybcov (Vimba) Evropy (Biology and Fisheries of Vimba in Europe). Vilnius, pp. 419—466 (in Russian, with an English summary).

Volskis, R., Moroz, V.N. and Sukhanova, E.R. 1970b. Nerest (Spawning). In: Biologija i Promyslovoe Značenie Rybcov (Vimba) Evropy (Biology and Fisheries of Vimba in Europe). Vilnius, pp. 105—133 (in Russian, with an English summary).

Weber, W. 1974. Prediction of feeding catches in the western Baltic. Rapp. P.-V. Réun. Cons. Int. Explor. Mer, 166: 150—151.

Westin, L. and Nyman, L., 1976. Activity, orientation and migration of Baltic eel (Anguilla anguilla L.). ICES/EIFAC Symposium on Eel, 31: 1—22 (mimeogr.).

Westin, L. and Nyman, L., 1977. Temperature as orientation cue in migrating silver eels, Anguilla anguilla (L.). Contrib. Askö Lab., Univ. Stockholm, 17: 3—16.

Wikgren, B.J.P., 1953. Dygnsrytmiken hos nejonöga (Petromyzon fluviatilis L.). Memo. Soc. Fauna Flora Fenn., 29: 24—27.

Wikgren, B.J.P., 1962. Resultaten av sikmärkningar inom Åland och vid Luvia. Husö Biol. Stat. Medd., 3: 1—34.

Zalachowski, W., Szypula, J., Krzykawski, S. and Krzykawska, I., 1976. Composition and amount of food consumed by sprat, herring and cod in the southern Baltic in the years 1971—1974. ICES C.M. 1976/P:23 (mimeogr.).

Zarnecki, S., 1963. Times of entering into the Vistula River of summer and winter population of sea trout and Atlantic salmon in the 1952 year cycle. ICES C.M. 1963/29: 1—8 (mimeogr.).

Zemskaya, K.A., 1959. Characteristics of the fishing stock of flounder in the central Baltic. Rapp. P.-V. Réun. Cons. Int. Explor. Mer, 147: 73—74.

Chapter 7

POLLUTION

BERNT I. DYBERN and STIG H. FONSELIUS

INTRODUCTION

Global aspects of pollution

GESAMP, which is a permanent Working Group composed of members from several specialized agencies of the United Nations, has formulated the generally accepted definition of marine pollution as follows:
"Introduction by man, directly or indirectly, of substances or energy into the marine environment (including estuaries), resulting in such deleterious effects as harm to living resources, hazard to human health, hindrance to marine activities, including fishing, impairing the quality for use of sea water and reduction of amenities". (GESAMP I/11, para. 12, 1974)

Polluting substances may cause changes in the physical-chemical environment, e.g., in temperature, salinity, pH, gas content, chemistry, colour or turbidity of the water, which in turn may influence the marine life in different ways. They may also cause direct damage to organisms, e.g., on physiological processes, including alimentation and reproduction, after having reached a certain threshold concentration.

Pollutants have spread enormously, especially since the early 1950s, and no part of the seas, even at great depths, seems now to be completely free from them. Nobody knows how many pollutants there are, but already the pertinent new chemical compounds can be counted in tens of thousands.

Fig. 7.1 gives a schematic picture of the present-day pollution of the oceans. The most disturbed areas are generally located outside industrialized land areas, from which pollutants are carried out directly through runoff and through sewers or indirectly through rivers and the atmosphere. However, there is also a certain contamination of the remaining sea areas, due, among other things, to the long-distance transport of pollutants by water and air currents.

Semi-enclosed sea areas, which, like the Mediterranean and the Baltic Sea are surrounded by industrialized countries, are among the most polluted in the world.

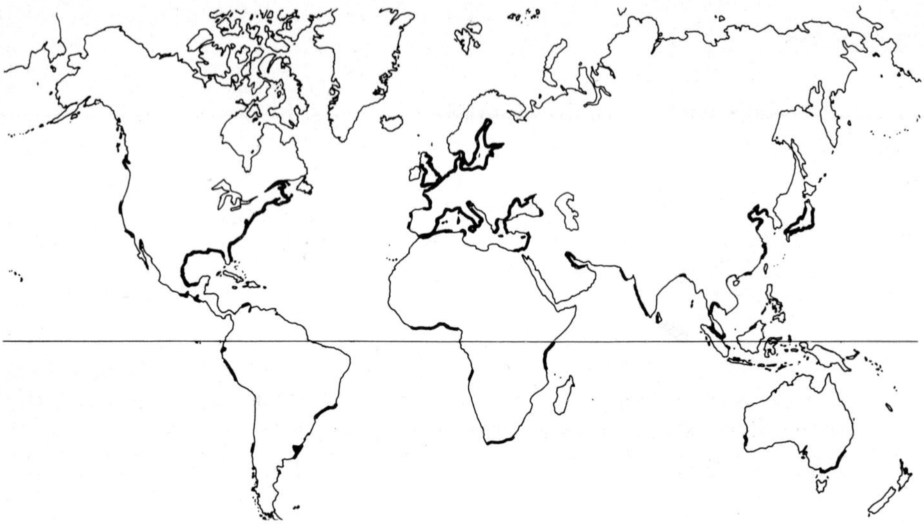

Fig. 7.1. Coastal areas with pronounced pollution problems. Simplified after Dybern (1974a).

Spread and accumulation of pollutants in the Baltic Sea

The Baltic Sea has a positive water balance with a supply of fresh water of 433 km^3 a^{-1} (Falkenmark and Mikulski, 1975), less evaporation (see also Chapter 2). The annual fresh-water excess leaves the Baltic Sea together with an equal amount of water which has flowed in through the entrance sounds as bottom currents. Only one-third of the inflowing water is really Kattegat water, which has a high salinity (34.8‰). The rest, two thirds, is Baltic surface water, which in the Danish Sounds is mixed with the inflowing bottom water. In order to illustrate the magnitude of the residence time of the water, the following calculation, with several oversimplifications, is presented.

We may very roughly compute the residence time of the water in the Baltic Sea by dividing the volume, which is about 21,000 km^3 (Ehlin et al., 1974), with the amount of water which really leaves the Baltic Sea [$\frac{2}{3}$(433 + 433) km^3 a^{-1}]. Thus we get the residence time as approximately 36 years. This means theoretically that only 577.3:21,000 = 2.75% of a "conservative" pollutant annually leaves the Baltic Sea. A "conservative" pollutant is a pollutant which does not react chemically with the water, is biologically inactive, and is not involved in sedimentation processes. By multiplying the yearly input with the residence time, we get the amount of the pollutant which will accumulate in the water when steady state is reached in the system. If, e.g., 10,000 tons of a conservative pollutant is discharged yearly,

362,000 tons will have accumulated in the Baltic water at steady-state conditions (Svansson, 1975). However, there are hardly any really conservative pollutants, and the above calculations only give the possible maximum rate of accumulation in the Baltic Sea. In reality, the amount of sewage from communities and other organic industrial wastes will be considerably reduced through oxidation and sedimentation. The amount of radioactive matter will be reduced through decay, biological uptake and sedimentation, metal compounds, organochlorine compounds and other chemicals through break-down, transformation, accumulation in organisms, or through sedimentation, thus delaying the steady state.

The sensitivity of the Baltic Sea ecosystems

According to their distribution, there are three main categories of organisms in the Baltic Sea: species belonging to ecosystems of the North Sea that penetrate into the Baltic Sea through the entrance sounds, species belonging to the ecosystems of neighbouring fresh waters and invading the sea from the coastal areas, and finally, a small number of species surviving from earlier stages of evolution of the Baltic Sea (see Chapter 5).

Since the hydrographical conditions are very peculiar and even extreme with a series of gradients (salinity, temperature, gas content, turbidity, etc.), many marine species are entirely excluded from the Baltic Sea, and most of the others can penetrate only to certain zones, the locations of which depend on the capacity of the species to adjust to the prevailing conditions. A change in the environmental conditions may cause a species to withdraw from an area or to spread to new areas, depending on the character of the change. Since the ecosystems are simple with few species and few links between them, even a relatively small environmental change may cause severe imbalance in a whole ecosystem. No doubt additional stress by pollutants is of great importance in such an environment. Many pollutants which are carried out into the Baltic Sea accumulate in organisms into concentrations that may become deleterious, for a single species, ecosystems or even man, if this development should not be slowed down.

EUTROPHICATION

Waste discharge from communities and industries

The total discharge of domestic sewage into the Baltic Sea has been roughly estimated by an ICES/SCOR Working Group on Pollution of the Baltic Sea to be about 2.3×10^6 m^3 d^{-1} as direct input and 1.2×10^6 m^3 d^{-1} as indirect (through rivers, etc.) input (ICES/SCOR, 1977). Some of the main inputs occur in the Öresund region, around the Gulf of Finland and in the

Stockholm area. About 40% of the direct and 20% of the indirect input are discharged without any kind of treatment. About 20% of the direct and 30% of the indirect sewage discharge get only mechanical treatment, and about 30% of all discharge receives mechanical and biological treatment. Additional treatment (e.g., phosphate precipitation) is given only to 3% of the direct and about 15% of the indirect discharge. This kind of chemical treatment is used mainly in Finland and Sweden.

The total number of people contributing to the direct and indirect input of domestic sewage is estimated to be around 17.5 millions (ICES/SCOR, 1977). In some countries an additional temporary population of tourists during the summer season is included. About two thirds of the population contribute directly to the discharge to the sea, and one third indirectly via estuaries and rivers.

There are no reliable figures available for the total industrial discharge. Considerable quantities of waste waters are, however, emitted by the paper and pulp industry and other forestry industries along the coasts of the Gulf of Bothnia. The degree of treatment of industrial waste waters varies very much from industry to industry and from country to country. All countries often treat and discharge industrial wastes together with domestic sewage.

Table 7.I (compiled from the report of the ICES/SCOR Working Group 1977) shows the input to the Baltic Sea of BOD load, total phosphorus and total nitrogen from communities and industries in the surrounding countries. Lignin values from Finland and Sweden are also included. Detailed information is often missing, especially on the input of nitrogen. The total amount of phosphorus is estimated as about 34,000 t a^{-1} and of sewage nitrogen as about 77,000 t a^{-1}, after extrapolation of the missing figures in accordance with the calculations for Finland, Sweden and Denmark.

Fig. 7.2 shows the distribution of BOD load in the different sub-areas of the Baltic area according to Dybern (1974b). It is evident that most organic pollution in the Belt Sea derives from domestic sewage, while in the Bothnian Sea and the Bothnian Bay most of it derives from the forestry-based industries. In the Baltic Proper and the Gulf of Finland, population and industries share more equally.

Oxygen utilization

The discharge of readily oxidized organic matter, e.g., from sewage systems or industries, consumes much oxygen, because one oxygen molecule is needed to produce one carbon dioxide molecule during the oxidation of one atom of carbon in organic matter. The oxygen utilization in the water is generally measured as biochemical oxygen demand (BOD) under a certain time, e.g., 5 or 7 days. In such a short time the organic matter cannot become completely oxidized, and therefore, BOD_5 or BOD_7 only measures a part of the oxidability. The real oxygen utilization may be much larger, and

TABLE 7.I

Input of BOD, total N, total P and lignin into the Baltic Sea in t a^{-1}. The BOD values contain partly BOD_5 and partly BOD_7. For calculations of BOD 25 kg O_2 per year and person was mostly used as factor, but 20 kg and 22 kg were used in some cases. For calculations of total N 4.4 kg per person and year was used as a factor, and for total P generally 1.0 kg per person and year, but in one case 1.4 kg per person and year. After ICES/SCOR, 1977.

Country	BOD			Total nitrogen			Total phosporus			Lignin	
	Direct	Indirect	Industry	Direct	Indirect	Industry	Direct	Indirect	Industry	Direct	Indirect
Denmark	76,500	14,300		6400	6600		2860	1520			
F.R.G.	9000						400	1100			
Finland	16,500	640	144,000	5500	175	1383	950	45	213	116,000	2800
G.D.R.	3450	18,000					190	1040			
Poland	35,820	30,400			97,650			7970			
Sweden	15,232	17,900	381,000	6252	6753	4009	1819	1671	804		386,100
U.S.S.R.	140,000	19,600	107,000				5600	790			

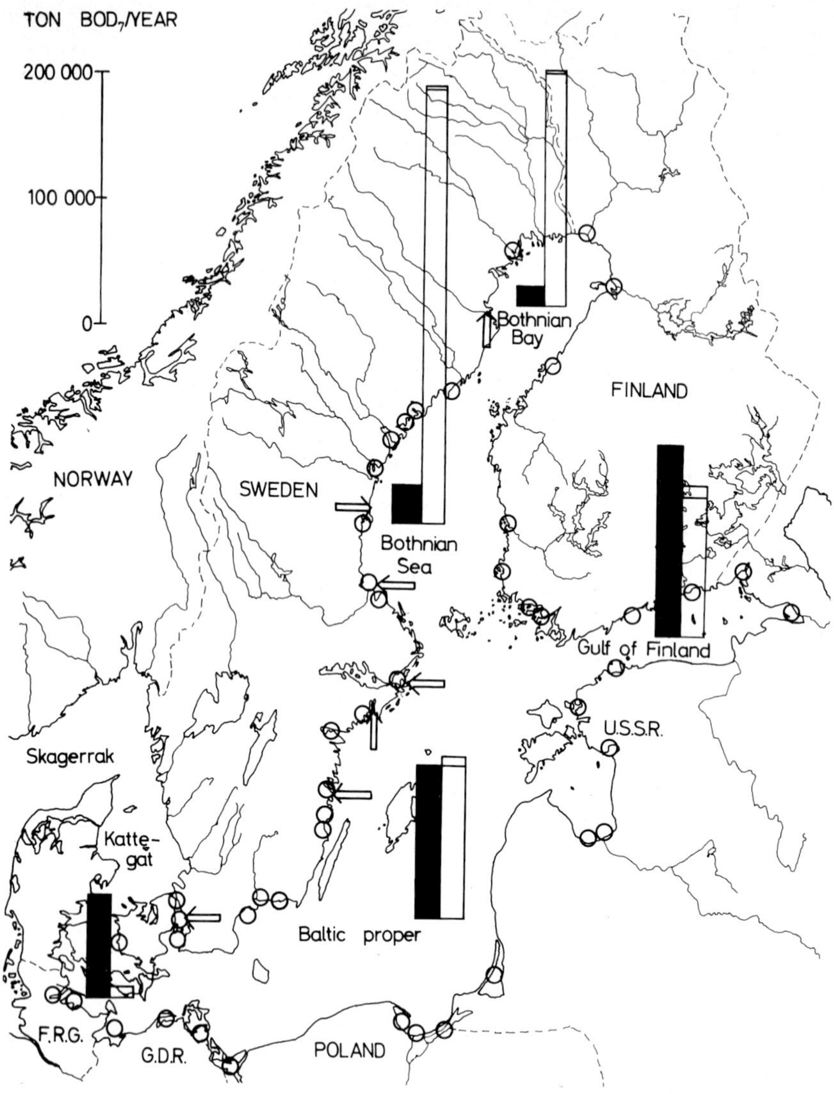

Fig. 7.2. Baltic pollution. Black piles show tons BOD_7 per year from sewage, agriculture, etc. White piles show tons BOD_7 per year from industries (the lower, longer, part forestry-based industries and the upper, smaller, part other industries). After Dybern (1974b). Circles show main areas mentioned as eutrophicated or otherwise influenced by organic pollution. Arrows show Swedish areas where fishing was entirely or partly black-listed in 1977.

estimation of BOD_{21} (3 weeks) or the demand during longer periods would certainly give a more reliable value but is not much used for practical reasons. Swedish investigations indicate that BOD_5 is only about one third of the total BOD, being lower for industrial wastes and higher for, e.g., ensilage juice from silos in agriculture (S.O.U. 1967). When the total BOD value for the Baltic Sea from Table 7.I (1.1 million tons) is multiplied by 3 and the volume of the Baltic Sea is assumed to be 21,000 km^3, a "total average BOD" value of 0.43 mg O_2 m^{-3} day^{-1} is obtained.

Through the oxidation of the organic matter nutrients are converted into inorganic salts causing a "fertilization" of the water and growth of algae. When the algae break down, a "secondary oxygen utilization" results, which may be from two to five times as great as the "primary oxygen utilization". Therefore, the actual average oxygen demand caused by the discharge of organic substances is rather 1—2 mg O_2 m^{-3} day^{-1}. This is, of course, true only if the waste would be evenly distributed into the whole water mass of the sea. All important waste discharges into the Baltic Sea, however, occur along the coasts, and often large discharges are concentrated in single spots, e.g., outside a large city. Therefore, the actual oxygen demand in the neighbourhood of densely populated areas may be 10 to 100 times larger than the figures mentioned above.

In shallow coastal areas with restricted water exchange, e.g., in archipelagos, gulfs and bays, severe eutrophication can be caused in this way. Mass occurrence of algae may be found along the shores and in the surface water. Where the water is deeper, with a thermocline during summer, the deeper layers can be more or less depleted of oxygen due to break-down of sinking organic matter. Generally, oxygen is replenished during the autumn turnover, unless special circumstances prevail, as in some fjord-like bays on the Swedish coast. During the turnover, additional nutrients are carried upwards from the bottom layers to the surface layers, and during next production period a "vicious cycle" of more pronounced fertilization of the surface layers may take place and continue during several years until the discharge of nutrients from land decreases. The conditions described are somewhat similar to what may happen in a polluted inland lake in temperate regions.

In more open areas of the Baltic Proper, with constant halocline, summer thermocline, and a lower natural nutrient content, release of organic wastes and nutrients causes also a more or less pronounced fertilization of the surface waters. The increased biological production may be beneficial by increasing the fish yield but can have serious effects on the conditions in the deep waters under the discontinuity layer. Breakdown of increased quantities of decaying organic matter there may lead to total loss of oxygen and, due to the great depth and the hindrance formed by the discontinuity layers, new oxygen is only very slowly brought down. Hydrogen sulphide will form, and the area is transformed into a toxic oceanic desert, bottom fauna being completely destroyed. When hydrogen sulphide is formed, the environment changes from an oxidizing to a reducing one. In a reducing environment, nutrient salts, especially phosphates, are dissolved from the bottom sediments and may slowly be brought up to the surface layers and increase the nutrient load there followed by increased production and increased breakdown of sinking organic matter in the deeper layers. Oxygen may again become depleted, hydrogen sulphide may be formed, and so on. It seems to be very difficult for nature to restore oxidizing conditions when a fertilization cycle of this kind has begun, especially in a partly stagnant basin as the Baltic Proper (Fonselius, 1969).

As long as oxygen measurements have been carried out, the deep water has had a low oxygen concentration in the Baltic Proper. Temporary formation of hydrogen sulphide has taken place in several deep areas. The long series of oxygen measurements made since the 1890s at the international hydrographic stations show that the oxygen concentration of the deep water during this century has decreased from about 3 ml l^{-1} to values close to zero (Fig. 7.3, see also Chapter 4, Section p. 188). Hydrogen sulphide has several times appeared in the bottom water of the central basin. Increasingly larger areas have been affected (Engström and Fonselius, 1974) (Fig. 7.4).

It is known from sediment cores that stagnation periods have occurred previously during the geological evolution of the Baltic basin (Hallberg, 1978). It is also known that natural factors, e.g., increased salinity and temperature (Fonselius, 1977), may have diminished the water exchange and increased the oxygen utilization, but it is obvious that the enormous load of urban and industrial waste discharged into the Baltic Sea has increased the oxygen utilization.

The discharge of organic substances and dissolved nutrient salts into the

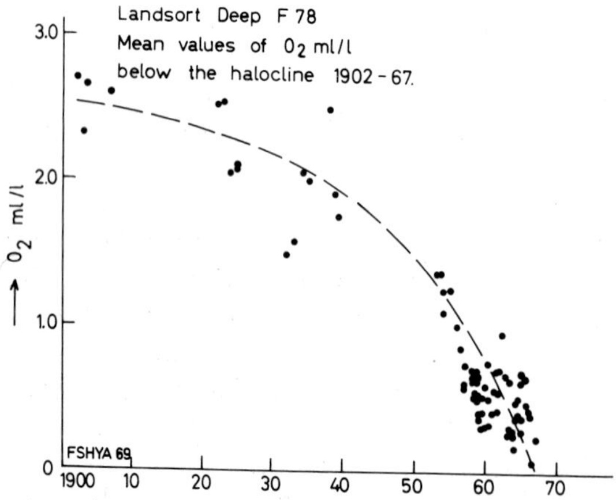

Fig. 7.3. Mean values of dissolved oxygen in ml/l below the halocline in the Landsort Deep from 1902—1967. (From Fonselius, 1969.) See also the legend to Fig. 4.1.

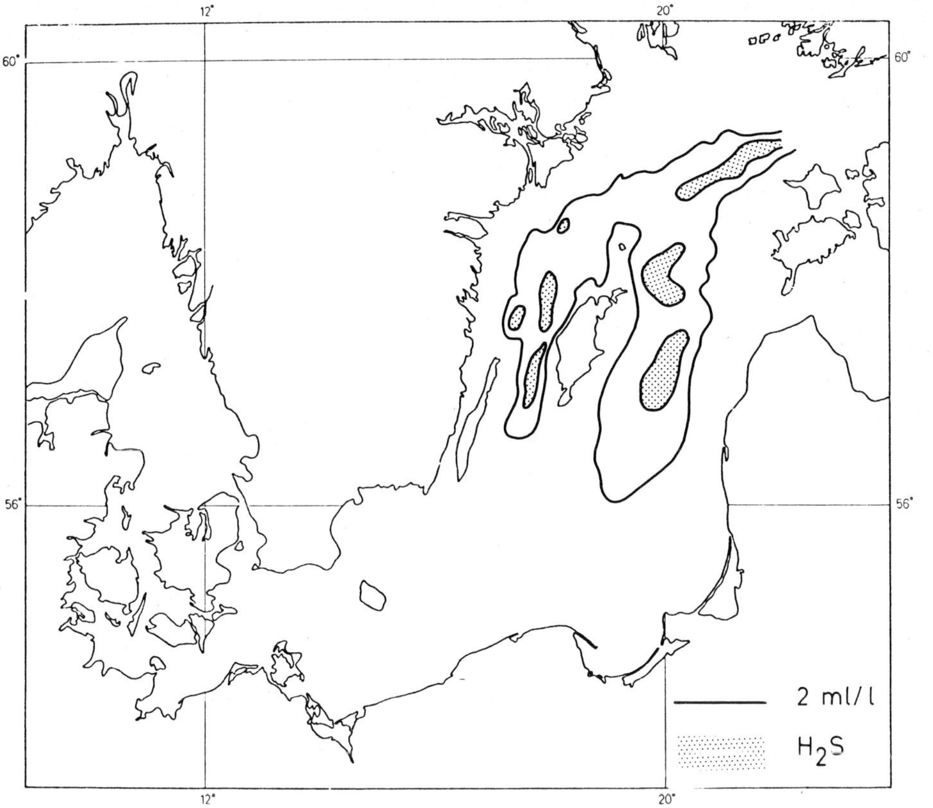

Fig. 7.4. Distribution of H_2S and low O_2 concentrations in the deep water of the Baltic Sea in May 1972.

Baltic Sea by man is now believed to have reached its peak, and thanks to abatement measures it has started to decrease slowly in somc parts.

Effects on the ecosystems

Waste waters rich in organic substances or nutrient salts, or both, caused very small changes in the conditions during previous centuries. The trivial changes that occurred were confined to the immediate vicinity of urban areas. However, from the turn of the last century, the addition of such substances to the Baltic environment has successively increased, the main reason being urbanization, industrialization (especially forestry-based industries such as pulp mills and paper mills), increasing use of fertilizers in agriculture, and introduction of synthetic detergents. Industries around the Baltic Sea increased with at least 100% between 1950 and 1970. Already in 1915 pollution from sulphite pulp mills was so evident that a

Swedish committee delivered a report with proposals for an air and water pollution abatement bill (Vasseur, 1966). However, the bill was not approved. At the same time increasing attention was also paid to the effects of sewage in Finland (Witting, 1922).

In Fig. 7.2 sites are shown at which considerable changes or damage to the natural ecosystem due to eutrophication or overfertilization have been reported. The Baltic Sea is one of the best investigated sea areas in the world, and a few hundred papers, mainly dealing with the coastal areas, deal with the pollution problem. The following "spot samples" give a rough picture of the prevailing conditions.

Occurrence of coliform and other fecal bacteria have been demonstrated outside almost all cities and towns around the Baltic Sea. Mellgren (1973) reported that coliform bacteria occur more than 50 km from the city of Stockholm extending through the archipelago and out to the open sea. Even at a distance of about 40 km from the city the numbers could temporarily rise to 3000/100 ml (in winter 1970). Bonde (1967a, b) found *Clostridium perfringens* to be very common in the sediments of the Öresund and also observed the species in the guts of plaice, flounder and mackerel.

The presence of *Salmonella* bacteria was demonstrated on bathing beaches in Poland by Buczowska and Nowicka (1960), and Krongård Kristensen (1971) found 40 serotypes of the species in the Öresund, the concentration being highest during the warmest months. Some of the serotypes were of exotic origin and probably introduced by tourists and foreign workers (so-called guest workers). Adenovirus 7 spread to a bathing beach of a suburb of Stockholm from a small sewage-treatment plant in 1959 and caused an epidemic (Kallings, 1961). Lund (1971) demonstrated the occurrence of enteroviruses ECHO 11 and Coxsachie B in the Öresund in 1967.

Increased production of planktonic and benthic algae, due to the increased content of nutrients in water (cf. Section on p. 354), has been reported from numerous sites along the Baltic Sea coast, especially from areas with limited water exchange, such as bays, fjords, archipelagos, etc., where even "blooms" and accumulations of drifting benthic algae frequently occur. Lehmusluoto and Pesonen (1973) showed, for instance, that phytoplankton primary production in the inner part of the polluted Helsinki archipelago was about 5—10 times as high as in the "unpolluted" area outside the archipelago, and Lindgren (1975) showed deviations in the zonation of the benthic algae in the same area.

Several algal species have during recent years disappeared from, or their distribution has been reduced in the Öresund due to increased pollution (v. Wachenfeldt, 1971). Results of investigations carried out in 1964—1970 showed that the littoral of the Gulf of Riga may be divided into three main parts according to the distribution of planktonic and sessile algae indicating various degrees of pollution (Andrushaitis, 1972). The green alga *Cladophora glomerata* has increased considerably along the Baltic Sea coast in Sweden, especially at river mouths (Jansson, 1978), Norin and Waern (1973) showed how differences in the growth of *C. glomerata* were closely related to the degree of pollution in the archipelago of Stockholm. In the Kiel Bight, it was demonstrated that the growth of the pollution-resistant *Fucus vesiculosus* was up to 100% better in polluted areas (Hentse; ref. in Caspers, 1975).

The question whether phosphorus or nitrogen is the limiting factor in algal production in Baltic coastal waters has been discussed by several authors, e.g., by Wallentinus (1976) and Karlgren (1978). The answer seems to depend on the kind of plant, the composition of the nutrient supply, the season and other variables. In the Bothnian Bay, phosphorus generally seems to be the limiting factor (Fonselius, 1978).

Changes in the composition of coastal zooplankton, due to pollution have been reported from, e.g., the Bay of Szczecin (Żmudziński, 1976), the Gulf of Riga (Andrushaitis, 1972), the Helsinki archipelago (Melvasalo and Viljamaa, 1975) and other sites along the Finnish coast (Bagge and Lehmusluoto, 1971).

Changes in the benthic fauna have been described from at least 50 coastal sites along the Baltic Sea, most of them caused by sewage discharges, e.g., the Öresund (Henriksson, 1967), the Kiel Bight (Kändler, 1953, Anger, 1975), the Bight of Lübeck (Schulz, 1968), Greifswalder Bodden (Arldt, 1975), the Bay of Gdansk (Żmudziński, 1975), the Gulf of Riga (Andrushaitis, 1972), the Tallinn area (Järvekülg, 1970), the archipelago of Helsinki (Viitanen, 1971), and the archipelago of Turku (Leppäkoski, 1975). Rosenberg et al. (1975) and Landner et al. (1977) reported effects on benthos in a number of areas along the east coast of Sweden, polluted by organic wastes, especially from paper and pulp industries.

Fish populations have been affected outside many paper mills and pulp mills along the coasts of the Gulf of Bothnia, and by sewage pollution in, e.g., the Helsinki archipelago (Anttila, 1972a) and the Öresund (Bagge, 1971). In many cases, oxygen depletion has been the immediate reason for the changes. Anttila (1972b) found differences in the taste of fishes from different parts of the Helsinki archipelago.

Nutrients carried out into the open sea have certainly in many cases stimulated the production of plankton and pelagic fish in the upper water layers where the oxygen content is adequate (Otterlind, 1978). But the increased supply of organic matter from the surface layers to the deeper water layers, where they are broken down, has certainly contributed to the increasing depletion of oxygen there, and thus the water volumes where bottom-living fish can exist have decreased (Dementieva, 1972). In the deeper waters of the Gulf of Bothnia this development is not so evident as in the Baltic Proper.

TOXIC MATTER

Organochlorine compounds

Most organochlorine compounds in the Baltic area are used in agriculture, horticulture and forestry (e.g., DDT, lindane, HCB) or in industry (e.g., PCBs).

There is so little information on the input of these chemicals to the Baltic Sea that it is impossible to make any reliable estimations about the quantities introduced. After the publication of reports on their toxic effects on living organisms in the 1960s their use has been restricted and the total amounts discharged into the Baltic Sea are probably decreasing. Thus, the use of DDT has been forbidden in most Baltic countries, but before that considerable quantities were certainly carried through runoff and also in atmospheric precipitation. The precipitation is probably still an important supply route to the Baltic area for organochlorine compounds, also from other parts of the world.

After the detection of PCBs in the Baltic Sea, their use has been more strictly controlled. However, there seems still to be a considerable supply to the sea for which the routes are partly unknown.

Most organochlorine compounds have a relatively long half-life and tend to accumulate in the Baltic Sea due to the long residence time of its water (see Section on p. 352). The low mean temperature of the water may also retard their breakdown. In the water they are taken up in different ways by organisms and accumulate in the food chains, the highest concentrations being generally found in the highest trophic levels. Particle-bound organochlorine compounds reach the bottom sediments where some of them again may enter the food chains.

Sampling for investigations of organochlorine compounds in Baltic fish was started in 1965, and the results from 180 analyses were published by Jensen et al. (1969). They stated that in fish from the Baltic Proper the average content of DDT and PCB was from 5 to 10 times as high as in fish from localities off the west coast of Sweden. In 1969—1971 more than 2000 fishes from different parts of the Baltic Sea, including the Gulf of Bothnia, were analyzed for DDT and PCB compounds by Jensen et al. (1972a). Table 7.II is based on the results of these investigations and on the results of another study carried out by an international working group in the North Sea (ICES, 1974).

It appears that the concentrations in Baltic fish are higher than those in fish from the North Sea, and also that different parts of the Baltic Sea are differently affected by the pollutants discussed.

The concentrations of organochlorine compounds are higher in fat than in muscular tissues. The liver is the organ mainly responsible for detoxification. Investigations by Luckas et al. (1977) show very high average values for DDT

TABLE 7.II

Average concentrations of DDT and PCB in muscle tissues of cod and herring from the Baltic Sea and the North Sea. In mg kg^{-1} for wet weight tissue. Sources, see text.

Area	Cod		Herring	
	Σ DDT	PCB	Σ DDT	PCB
Central North Sea	< 0.01	< 0.05	0.10	0.25
Skagerrak	0.01	0.03	0.12	0.30
Kattegat	0.01	0.04	0.18	0.25
Southern Baltic Proper	0.11	0.12	1.50	0.60
Central Baltic Proper	0.06	0.04	1.20	0.45
Gulf of Bothnia	0.08	0.05	0.49	0.39

and PCB in cod liver from the southern Baltic Sea. Denmark and Sweden have forbidden entry and sale of cod liver from the Baltic Sea.

The results of a recent investigation (ICES/SCOR, 1977) indicate that the high DDT contents, especially from the central and southern Baltic Sea proper, are now slowly decreasing. However, the PCB values from the same areas may be increasing. Linko et al. (1974) have shown a similar trend locally in the archipelago of Turku in Finland.

Regional variations do not seem to be evident in the case of coastal fish along the coast of Sweden, pike and eel having about the same contents of DDT and PCB in different parts of the coastal waters (Olsson, 1978).

Table 7.III shows the results of some investigations of the content of organochlorine compounds in Baltic Sea animals other than fish. Since the results are scattered, it is impossible to draw far-reaching conclusions, but it is evident that in animals at the end of the food chain, like fish-eating birds and seals, the content may be very high. Dead white-tailed eagles have been found in, e.g., the archipelago of Stockholm, and it is believed that their death was caused by a high content of organochlorine compounds in their bodies (Jensen et al., 1972b). The reproduction of birds, especially of fish-eating predators, may be hindered by the breaking of too thin egg shells (Odsjö, 1971). The decreasing population of seals in the Baltic Sea has been attributed to the effect of organochlorine compounds (Helle et al., 1976). Results in other sea areas indicate that fish reproduction and fish stock, and the entire ecosystem may be negatively affected when certain concentrations of organochlorine compounds are reached. Up to now there has, however, been very little research on this problem in the Baltic area. There is also very little research going on regarding other halogenated hydrocarbons, but, at least, it has been verified that Baltic fish may contain varying amounts of dieldrine, lindane, HCB and α-BHC (ICES/SCOR, 1977).

TABLE 7.III

Concentrations of DDT and PCB in Baltic organisms. Simplified from Jensen et al. (1972a, b), Olsson et al. (1973), and Helle et al. (1976). Values in ppm., h = homogenate.

Area	Organisms	Number in samples	DDT Fresh tissue	DDT Fat	PCB Dry tissue	PCB Fresh tissue	PCB Fat
Outer Stockholm archipelago	*Invertebrates:*						
	Gammarus sp.	h			0.16		3.3
	Idothea sp.	h			0.12		1.1
	Mysis sp.	h			0.12		1.6
	Plankton	h			0.38		4.9
	Aurelia aurita	h			1.60		33.0
	Mytilus edulis	h			0.22		3.4
Baltic Proper east of Bornholm, incl. Stockholm archipelago	*Birds, not fish-eating:*						
	Golden eye	5				0.37	13.0
	Long-tailed duck	3				0.35	14.0
	Birds, fish-eating:						
	Red-breasted merganser	3				2.3	87
	Goosander	2				5.0	170
	Cormorant	3	23.0	580		17.0	420
	Black guillemot	2	3.2	140		3.9	180
	Guillemot	10	14.0	610		3.4	160
	eggs 30.5.1968	11	40.0	590		17.0	250
	eggs 26.5.1969	10	56.0	590		19.0	200
	Herring gull	4				18.0	650
	Birds, birds-of-prey:						
	White-tailed eagle	7				180.0	10,000
	Eagle owl	3				96.0	9700
	Seals:						
Northern Bothnian Bay	Ringed seal	40		110			69
Gulf of Bothnia	Ringed seal	33		200			110
	Grey seal	15		210			100
Baltic Proper	Grey seal	18		420			140

Metals

Discharge of metals takes place from almost all kinds of industries, and the Baltic area is rather heavily industrialized with paper mills and pulp mills, tanneries, chemical and graphic industries, metal refineries, steelworks, plating works, electrical industries, and textile factories among others. Additional discharge of metals takes place through rivers, sewage systems and air. Deliberate dumping of metal-containing waste and accidental discharge from ships have taken place, especially at points outside the coast of Sweden. The river discharge also contains metal salts produced in rock weathering. Their quantities are unknown, but the total amount is presumably smaller than the quantities derived from man's industrial activities. Thus, Ahl (1977) calculated the total amount of zinc carried into the Baltic Sea by rivers in Sweden as 3140 t a^{-1} (from both natural and industrial sources), while Engwall (1972) calculated the total discharge into the Baltic Sea from only the plating industry of Sweden as about 5000 t a^{-1}.

Brügmann (1976) found that the average contents of zinc, cadmium and lead in the Baltic Sea water often exceeded their content in unpolluted water (Table 7.IV, see also Chapter 4, Section on p. 203). Somer (1977), using the information from various field measurements and calculations based on the theoretical discharge per person, attempted to calculate a mass balance for some metals (Table 7.V). Even though the results of his calculations are very preliminary, they probably indicate the order of magnitude.

TABLE 7.IV

Average values of zinc, cadmium and lead in water from different Baltic areas. In mg m^{-3} (after Brügmann, 1976)

Area	Number of samples	Zn	Cd	Pb
Great Belt, Öresund, Kattegat	20	9.8	1.99	2.1
Fehmarn Belt	54	8.1	0.26	0.8
Lübeck Bight	47	8.8	0.41	0.8
Mecklenburg Bight	111	10.6	0.25	0.8
Arkona Basin, Oder Bight	117	7.2	0.17	0.7
Bornholmsgat	32	7.7	0.38	1.0
Bornholm Basin	83	8.6	0.16	0.6
Southern Baltic	21	6.1	0.13	0.5
Gdansk Deep	27	8.0	0.51	0.8
Gotland Basin	73	6.9	0.11	0.8
East of Öland	6	7.6	0.67	1.3
West of Gotland	7	3.9	0.09	0.4
Landsort Deep	7	6.2	0.41	0.9
Northern Baltic Proper	2	5.0	0.06	0.4
Gulf of Finland	33	7.0	0.25	0.8

TABLE 7.V

Attempt to construct a mass-balance scheme for 4 metals in the Baltic Sea. In t a^{-1} (after Somer, 1977)

Metal	Input				Loss					Left
	Rivers	Sewage	Industry	Precip-itation	Inflow	Outflow	Sedimentation		Fishery	Stored in Baltic water
							coastal	open sea		
Hg	6	4	20	4	?	3	29	2	0.05	140
Pb	10^3		?	10^3–10^4	?	10–10^2	$> 10^3$	10^3–10^4	?	10^3
Cu	4×10^3	2×10^2	?	6×10^2	?	2×10^3	?	2×10^3	?	10^5
Zn	4×10^3	10^3	?	4×10^3	?	4×10^3	?	5×10^3	?	2×10^5

Many metals discharged from land certainly remain near the discharge sites. Thus, the highest contents should be found in coastal areas, especially near river mouths, towns and industrial sites. There is reason to believe that they are largely and rapidly removed from water to sediments. The discharge of metals from the refinery and smelter at Rönnskär on the Bothnian Bay is listed in Table 7.VI. Lithner and Samberg (1976) found high contents of several metals in sediments and animals within 50—100 km from the industry. Henriksson (1971) showed how the mercury deposition and water currents together caused a special distribution pattern for mercury in the Öresund.

The concern about metal compounds as dangerous pollutants in the Baltic Sea area started with the discovery in Sweden toward the end of the 1950s that bird populations decreased as a result of mercury poisoning (e.g., Borg et al., 1966). Further investigations soon showed that mercury contents could be high also in aquatic animals, especially outside paper mills, pulp mills and chlorine-alkali factories where this metal was used for different purposes. An extensive Swedish programme for continuous investigation of fish in fresh-water and sea water was started in the 1960s. High concentrations of mercury were found in fish from many waters and — in view of the alarming reports from Japan — a number of areas where the average values were more than 1 ppm (of fresh weight of muscle tissue in fish) were black-listed entirely or regarding certain species, some of them along the Baltic coast (cf. Fig. 7.2). The black-listing implies that fish from such areas must not be sold or given away as gifts; however, the fisherman is allowed to eat it at his own risk. From 1967 onwards, the use of mercury in Swedish paper mills and pulp mills was banned, and other uses were restricted. Since then the concentrations have often decreased (Westöö and Rydälv, 1971), and the

TABLE 7.VI

Discharge in t a^{-1} from the metal works at Rönnskär in 1973 (after Lithner, 1974)

As	1500	Bi	1.2
Zn	144	Sn	1.0
Pb	25	Te	0.7
Cu	50	Ag	0.5
Cd	9.7	Fe	3—500
Hg	1.1	Ga	0.3
Se	4.0	In	0.2
Cr	2.4	Au	0.02
Ni	2.0	F	86
Mn	5.6	SO_2	1200
Co	2.4	H_2SO_4	600

ban has been lifted for some areas, e.g., in the archipelago of Stockholm for Baltic herring.

Surveys of Finnish waters (Häsänen and Sjöblom, 1968; Miettinen, 1971) revealed similar problems as in Sweden. Engberg (1976) showed that the mercury content in flounder fillets from the Copenhagen area in Denmark reached average values of up to 0.8—0.9 mg kg^{-1} wet weight, while the corresponding content along the Zealand coast of the Great Belt rarely exceeded 0.2 mg kg^{-1}. Results of preliminary investigations carried out under the auspices of the ICES/SCOR Working Group on Study of the Pollution of the Baltic in 1974—1975 (ICES/SCOR, 1977) showed that average mercury content in fish from different parts of the open waters of the Baltic Sea generally were low, but that they sometimes reached values of more than 1 mg kg^{-1} as shown in Table 7.VII.

The mercury content in fish-eating birds and seals is often high. Jensen et al. (1972b) found a mercury content of 1.3 mg kg^{-1} and 1.8 mg kg^{-1} fresh weight in muscle tissue in two black guillemots and between 2.9 and 3.7 mg kg^{-1} in three cormorants from different parts of the Baltic Sea area. In four white-tailed eagles from the Stockholm archipelago the content in the pectoral muscle ranged from 7.3 to 26 mg kg^{-1}. The birds were dead and contained also much DDT and PCB. Häsänen (1973) reported high mercury contents in seals from Finland (up to 61 mg kg^{-1} in the liver). In some cases signs of decreasing trends in the mercury content are, however, observed (Somer and Appelqvist, 1974; Häsänen, 1975).

The occurrence of other metals in different organisms from the Baltic Sea is less known. Some data, mainly from the open Baltic Proper, are listed in Table 7.VII.

Apart from death of birds there exists very little information on the influence of metal pollution on the biota of the Baltic Sea. However, the results of Lithner and Samberg (1976), who demonstrated that a large area outside the metal refinery and smelter at Rönnskär on the shore of the Bay

TABLE 7.VII

Concentration ranges of some heavy metals in Baltic Sea fish species. Sampling sites irregularly scattered, most of them in the open sea. mg/kg wet weight muscle tissue. After ICES/SCOR 1977.

Metal	Cod	Herring	Flounder
Zn	1.2—9.2	3.4—32.0	3.5—11.3
Cu	0.08—2.4	0.3—1.9	0.10—0.89
Pb	0.03—1.3	0.01—1.4	0.02—0.26
Cd	0.002—0.05	0.002—0.07	0.002—0.04
Hg	0.02—0.88	0.004—0.3	0.01—1.50

of Bothnia in Sweden almost entirely lacks macrofauna, may presumably indicate similar influence outside other industrial areas. Some laboratory experiments, e.g., by Bengtsson (1974) have been made with Baltic animals, the results showing death, growth changes or changes in the reproduction or behaviour in minnow at different zinc contents in the water. Theede et al. (1979) showed how low salinity together with temperatures corresponding to Baltic Sea surface summer temperatures increased the toxicity of cadmium to the hydroid *Laomedea loveni*. In these and experiments from other parts of the world with similar results the content of metals in the water has often been low enough to correspond to those which are observed near industrial discharges.

OIL POLLUTION

According to rough calculations 30,000—60,000 metric tons of oil is annually discharged into the Baltic Sea from tankers, cargo ships, ferry boats, trawlers, pleasure boats, refineries, power plants, sewage outlets, run-off water from streets, and atmospheric precipitation. About 20—25 million tons of oil per year are transported into the Baltic Sea by tankers. An enclosed sea with often very complicated shorelines and archipelagos like the Baltic Sea, of course, is very sensitive to oil pollution. However, the 17 m deep entrance sill between Darsser Ort and Gedser puts a maximum cargo limit of about 120,000 tons for tankers entering the Baltic Sea. In the Baltic Sea it is entirely forbidden to discharge oil according to the Oil Pollution Convention (1954 and later revisions). In spite of this and in spite of the movements of tankers being very well observed, big tankers sometimes deliberately discharge oil in the open sea when cleaning their tanks. Accidents have also occurred, and in 1968—1977 there were 5 major and several minor tanker accidents in Finnish and Swedish coastal waters, due to navigation problems in archipelagos and — at least in one case — to thick fog. The total oil leakage from these accidents was thousands of tons. Mines laid during wars may still be found in the entrance sounds, in the southern parts of the Baltic Sea and even in the Gulf of Finland constitute a potential danger to navigation.

There are also a series of minor discharges of different kinds every year, especially in the coastal areas and harbours. Thus in 1976 about 160 cases were reported along the coast of Sweden from Öresund to the Bay of Bothnia, some of them causing the death of several hundred sea birds (Anonymous, 1977).

Oil prospecting is carried out in several parts of the Baltic Sea but no accidents have so far been reported. Some leakage into the sea has occurred outside places where oil shales are quarried in eastern Esthonia (ICES, 1970).

Due to the low average temperature of the Baltic Sea water bacterial and other breakdown of oil is slow, but the content of oil in the water of the open sea is not yet directly alarming. During the joint Soviet—Swedish "Musson" expedition in June 1976, Ahnoff and Johnsson (1977) reported contents of oil dissolved in the water generally between 2 mg m^{-3} and 4 mg m^{-3} for some 300 samples. The lowest value was 0.2 mg m^{-3} and the highest one 8 mg m^{-3}. According to another investigation, the average content of petroleum hydrocarbons in marine sediments was 10 mg/kg (dry weight; 0—5 cm sediment depth), but outside urban areas it rose up to 40—400 mg/kg dry weight (IVL, 1976).

When drifting ashore the oil can cause comprehensive damage, both to marine life and, e.g., to recreation facilities. As an example of the often very expensive cleaning operations even for a small discharge, the cost for restoring the shore and smaller boats after a discharge of only 25 tons fuel oil drifting ashore at Lidingö in the Stockholm archipelago was 250,000—300,000 SEK (60,000—70,000 USD), the damage to marine life not included (Newspaper Svenska Dagbladet, 19 June 1976). At the "Tsesis" accident in the Asköfjärd in the southern Stockholm archipelago in October 1977, about 2000 tons of fuel oil leaked out and were removed mechanically from the surface and from the shores. The cleaning operations were estimated to cost at least 5 million SEK (about 1.1 million USD) (Funck, 1978).

There is a certain cooperation between the Baltic Sea countries in the efforts to detect and abate oil spills in time. Commercial air and shipping lines report observed oil slicks. Receiving terminals for oil and oily water have been established in some of the larger harbours. In most countries the rule is that discharged oil should in the first place be removed mechanically from the water. Only in emergencies dispersants should be used.

Damage to sea birds in the Baltic Sea was reported already by Lemmetyinen (1966), and since then many other reports have dealt with bird deaths caused by oil. After the "Palva" oil tanker disaster in the southwestern Finnish archipelago on May 1st, 1969, about 150 tons oil escaped and polluted the shores of islands and skerries within the area of more than 200 km^2 (Heino, 1972). It was estimated that 3000—4000 eider ducks were killed in the Kökar and Föglö areas (Soikkeli and Virtanen, 1972). Numerous dead fish were observed after the oil spill and the subsequent cleaning operations with dispersants. Analyses of the littoral fauna, however, did not show any evident detrimental effects, except the absence of three littoral crustaceans a couple of weeks afterwards (Pelkonen and Tulkki, 1972). Nor was there any definite harm to the sessile algae, probably due to the fact that the annual species had not yet developed at the time of the accident (Ravanko, 1972). Ganning (1970) reported that still half a year after the "London Harmony" tanker accident in the Stockholm archipelago, also in 1969, sunken oil came to the surface, polluting the shores and their life forms. After this accident great quantities of dispersants were sprayed on the shores and the water, which contributed to increasing the damage.

Schramm (1972) demonstrated that the gas exchange in algal cell walls was inhibited after the covering of the algae with an oil film. Pautsch and collaborators (1978) stated that the reproduction of the shrimp, *Crangon crangon*, is retarded when exposed to certain concentrations of fuel oil. Kühnhold (1972) showed that oil films on the water surface release toxic substances harmful to fish eggs and larvae, and Lindén (1975) showed that application of dispersants strongly increased the toxic effect on Baltic herring larvae.

There are a few oil refineries along the coasts of the Baltic Sea. No severe pollution problems have been reported from them, except some cases of bad taste in fish caused by phenol observed in Denmark (ICES, 1970).

RADIOACTIVE POLLUTION

Sea water has a certain background radioactivity, mainly originating from the radioactive potassium isotope, ^{40}K. The radioactivity has measurably increased, especially by the detonations of nuclear bombs.

Previously, radioactive substances were discharged or dumped into the Baltic Sea without special control, but this practice has now ceased, and the control is rigorous. The nuclear power plants, however, still discharge small quantities of radioactive waste, mainly tritium, ^{3}H, but also ^{90}Sr, ^{137}Cs and other nuclides. Nuclear-powered ships pay occasional visits to the Baltic Sea and may also release small amounts of radionuclides.

Voipio and Salo (1971) found an increase in the ^{90}Sr and ^{137}Cs contents from 7 Bq m^{-3} to 40 Bq m^{-3} (0.2—1.0 pCi l^{-1}) during the period 1960—1967. They also stated that 1% of the annual ^{90}Sr supply and 10% of the annual ^{137}Cs supply were deposited in the sediments. Agnedal (1966) reported 1.9—4.1 Bq kg^{-1} (50—110 pCi kg^{-1}) ^{90}Sr in the bone of pike from the narrow bay of Tvären on the east coast of Sweden, the water of the bay having a ^{90}Sr content of 13—41 Bq m^{-3} (0.35—1.1 pCi l^{-1}). Simola et al. (1977) found about 40—110 Bq km^{-2} (1—3 mCi km^{-2}) plutonium in the Baltic sediments.

The content of radioactive matter in the Baltic Sea seems to be higher than in the Atlantic Ocean. Thus, Aarkrog (1974) reported 25 Bq ^{90}Sr m^{-3} (0.7 pCi l^{-1}) in the Belt Sea and 4 Bq m^{-3} (0.1 pCi l^{-1}) at the Faroe Islands. In cod tissues he found about 4 Bq ^{90}Sr kg^{-1} (100 pCi kg^{-1}) in the Belt Sea and about 0.6 Bq kg^{-1} (15 pCi kg^{-1}) at the Faroes. The $^{137}Cs/^{90}Sr$ ratio is lower in the Baltic water than in the more saline Kattegat water. This is mainly caused by a greater retention of ^{137}Cs from river water in comparison to ^{90}Sr. The ratio is around 1 in the Baltic Sea and 1.5 in ocean water (Voipio and Salo, 1971).

PHYSICAL POLLUTION

Warm-water effects

There is some discharge of warm cooling water from nuclear power plants, oil-powered plants and industries along the Baltic coasts. The discharge has up to now been relatively small, and to our knowledge it does not exceed 50 m^3 s^{-1} in any instance.

From an oil-powered plant at Karlshamn in southern Sweden, the quantity of cooling water discharged into the archipelago in 1970 was about 10 m^3 s^{-1}, and the temperature was about 8—9° C higher than that of the sea water. The distance at which surface temperature of the sea water was increased by 2° C or more varied between a few hundred meters and 3 km, directions varying according to the currents, and the area affected could be up to a few km^2 (Ehlin, 1973). The nuclear power plant Simpevarp, on the east coast of Sweden on the Baltic Proper, operated in 1974 at a maximum power level of 460 MW with a constant outflow rate of cooling water of about 20 m^3 s^{-1}, temperature increase being about 8—11° C. The distance from the power plant with an increased surface temperature of at least 2° C did rarely exceed 2 km (Weil, 1974). Entering the sea in the narrow bay of Hamnefjärden, which receives the discharge, the cooling water under certain ambient temperature conditions formed a sinking zone, increasing the temperature with at least 2° C down to the bottom at a depth of 10—15 m (Ehlin, 1974).

It is planned to build at least 10 big nuclear power plants in the Baltic area in the 1980s or 1990s, probably with an average cooling water discharge of about 100—200 m^3 s^{-1} for each. The areas affected by the cooling water will be considerably larger than the ones affected today, but they will still be small compared with the whole Baltic Sea, and the temperature of the sea as a whole would hardly be increased. However, there may be local negative influence on both the abiotic and biotic environment. Discharged cooling water from a small power plant at Värtan in the Stockholm area in Sweden in 1967—1968 caused an increased production of organic matter in the already polluted water. This caused increased production of H_2S in the deeper water layers due to increased break-down of the organic matter (Karlgren and Ljungström, 1969). Neuman (1974) found that several "cold-water" fish species avoided the warm effluent from the nuclear power plant at Simpevarp, while "warm-water" species increased their movement towards the discharge. Nyman (1975), using ultrasonic tracking, showed that, e.g., yellow eels were attracted by warm water in the summer but seemed to be repelled in late summer and autumn which is their main migration period. Sea-run brown trout were attracted by warm water when the ambient sea temperature did not exceed 15—16° C. The number of eye parasites in perch and roach increased in specimens affected by the discharge from the Simpevarp plant as compared with specimens from an area with normal temperature conditions (Grimås, 1974). Experimental work has indicated that the uptake rate of DDT and PCB increases in fish when the temperature increases (Grimås, 1974).

In 1953 the polychaete, *Mercierella enigmatica*, was found in the warm effluent of a power plant in the harbour of Copenhagen. It is a warm-water species and seems to originate from India. It has been brought, e.g., to France and southwestern England by ships and has probably spread to Copenhagen in the same manner and survived thanks to the warm water (Rasmussen, 1955).

Solid waste

Dumping of solids

Dumping of various solid matter from urban areas, industries, etc., into the sea has been common in the past in most coastal and some open areas of the Baltic Sea. Scrap-iron, containers, wires, ropes and even nets and other fishing gear have caused considerable trouble for the fishery. Once it was suggested to use the Landsort Deep as a dumping place for old cars, and the first "test-dumping" was carried out there in 1964 (Hult, 1968). The increasing use of plastics has caused the formation of what is sometimes called the "plastic zone" on many shores.

The frequent shipping has in some instances given rise to "tracks" of dumped ship garbage on the bottom, e.g., along the ferry routes between Finland and Sweden. A calculation in 1968 showed that 250—400 m^3 garbage day^{-1} was dumped from the ferries plying the Öresund between Denmark and Sweden (Rosén, 1969).

Recently dumpings of this kind have, in principle, been forbidden, according to the Helsinki Convention on the Protection of the Marine Environment of the Baltic Sea Area (see Chapter 8, Section on p. 386).

Sewage sludge and mud from harbour dredgings are dumped in many places around the Baltic Sea, but mainly in territorial waters. Local damage to the bottom fauna may occur. Fibres from the cellulose industry are often a nuisance in the Bothnian Sea (Landner et al., 1977).

Military waste

Many deep areas of the Baltic Sea have been used for dumping of ammunition and military waste material. Such places are, e.g., the Bornholm Basin, the Gdansk Basin and the southern part of the East Gotland Basin. These dumping grounds are marked on sea charts, and fishermen have been instructed to avoid them in order not to get ammunition and bombs in their trawls.

The accidental loss of bombs and torpedoes has to our knowledge not caused serious damage. Such accidents happen very seldom. Torpedoes are very expensive, and the ships responsible for their loss do everything possible to recover them. Airplane bombs are sometimes lost, but they are heavy and sink mostly deep down into the bottom sediments.

More serious is the problem with mines, both anchored and drifting. Large areas, especially at the entrances to the Baltic Sea and outside important ports, were during the World Wars I and II mined by both the defenders and the enemy. Lots of these mines have never been recovered and mine accidents still happen in the Baltic Sea. Thus a Danish fishing vessel was destroyed, most probably by a mine almost in sight of the ferries between Rödby and Puttgarden in August 1969 (Fonselius, 1970).

Wrecks are also a kind of pollution of the sea bottom. During the World Wars I and II many transport ships, cargo ships and men of war were lost. These wrecks often form big obstacles on the bottom to fishermen's trawls. Wrecks of aircraft have also caused similar damage. Some ships and airplanes are, of course, also lost during peacetime. Sites of known wrecks are marked on sea charts, but many unknown wrecks certainly exist in the Baltic Sea. Wrecks may also cause oil pollution when their fuel tanks are corroded and oil escapes.

After World War II German phosphorus bombs and war gases (some 20,000 tons of mustard gas and in some cases cyanide) were dumped in the Bornholm Basin at a depth of 110 m under the supervision of the allied occupation forces. The last dumpings were carried out in 1947—1948. Several types of gas containers were used, e.g., gas shells, airplane bombs and gas pots. Also fog or smoke developers (cylindrical sheet-iron barrels) have later been found in the area. Crates with containers were sometimes carried away from the dumping places by the currents, and finds have been reported from the coasts of Bornholm, southern Sweden, Öland and Gotland.

On several occasions gas containers have been caught in the trawls of fishing vessels. Mustard gas (actually an oily liquid) brought on board has severely hurt fishermen. The effects are generally felt only after several hours, and therefore the victim may have contaminated parts of his body and may also have spread the liquid to other persons before he has become aware of the danger (Fonselius, 1976).

Extraction of sand and gravel

Extraction of marine sand and gravel on a large scale has up to now mainly been carried out in territorial waters, and the quantities recovered have been relatively modest, e.g., in Finland less than 400,000 $m^3\ a^{-1}$, in Sweden (mainly in the Öresund) about 500,000 $m^3\ a^{-1}$, and in the Federal Republic of Germany (in the Kiel Bight) about 215,000 $m^3\ a^{-1}$. Denmark extracts about 6,000 000 $m^3\ a^{-1}$, but that includes also extraction in the Kattegat and Limfjord areas (ICES, 1975). Prospecting for new extraction sites is carried out in some places, also outside the territorial waters.

In the southwestern Baltic Sea the extraction is carried out by special ships which suck up sand from very shallow water down to a depth of about 15—20 m. In the holes formed the water may be stagnant or semi-stagnant with a very low oxygen content (Ackefors and Fonselius, 1968). Nordenberg (1971) showed that the bottom fauna may be locally affected. Investigations in other areas show that the increased turbidity arising from the operations may cause interference with visual orientation and feeding of fish species and irritate and even clog fish gills. The increased silting may change the bottom structure, and fish-spawning grounds may be destroyed (ICES, 1975).

Other kinds of physical pollution

Acoustic pollution, caused by the noise of machines and screws of ships and motorboats and from underwater operations such as oil prospecting, sand suction and certain military operations, is certainly felt by many fish. The ability to hear in most fish lies below 1000 Hz, but some species can perceive sounds of up to 5000 Hz or even more (Busnel, 1963). Motor sounds from fishing boats have been found to disturb herring schools which often try to escape by diving deeper (Mohr, 1967). Westerberg (1975) reported that when he was studying the migration of the Baltic eel by means of small radio transistors attached to the fish, they reacted to ships passing over them by diving deeper.

Observations by Ungsgård (1976) showed that when a mine containing 400 kg of explosive was detonated in shallow water, herring and perch were killed within a radius of 1500 m. In this case the main reason for most of the kills was the pressure waves generated. During reflection-seismic investigations in the Kiel Bight great numbers of fish were killed after having been damaged in different ways, e.g., by internal bleeding and torn kidneys. Some species, e.g., codfish, even showed displacement of the otoliths (Anonymous, 1966).

The "washing force" of waves produced by ferry boats in the archipelago of Åland was found to decrease the algal biomass on exposed sites in the hydrolittoral zone by 30% and in the sublittoral zone by 50% (Rönnberg, 1975).

Electric direct current is transmitted, e.g., from the mainland in Sweden to the island of Gotland using a single cable. Electrodes at the coasts lead the current back through the water. Höglund and Koczy (1971) showed that there was no influence on the behaviour of fish, except in the very close neighbourhood of the electrodes where fish were attracted to the anode and repelled from the cathode. At the anode the fish became unconscious and floated up to the surface. Schools of fish swimming over the cable or in the area between the electrodes did not react in any visible way when the current was opened or closed or when it was reversed.

REFERENCES

Aarkrog, A., 1974. Radioaktiv forurening af havet. Fisken Havet, 74: 3—11.

Ackefors, H. and Fonselius, S.H., 1968. Preliminary investigations on the effect of sand suction work on the bottom in the Öresund. Medd. Havsfiskelab. Lysekil, 58: 1—10.

Agnedal, P.O., 1966. Calcium and strontium in Swedish waters and fish and accumulation of ^{90}Sr. Proc. Int. Symp. Radioecol. Conc., Stockholm, April 1966, pp. 879—896.

Ahl, T., 1977. River discharges of Fe, Mn, Zn and Pb into the Baltic Sea from Sweden. Ambio, Spec. Rep., 5: 219—228.

Ahnoff, M. and Johnsson, L., 1977. Quantitation and characterization of petroleum hydrocarbons in Baltic Sea water. Rep. to the Soviet-Swedish expert meeting on evaluation of the results from the joint MUSSON-expedition and planning of future cooperation. Göteborg, April, 24—30. 34 pp. (mimeogr.)

Andrushaitis, G., 1972. Hydrobiological investigations of the Baltic Sea in the U.S.S.R. Ambio, Spec. Rep. 1: 67—70.

Anger, K., 1975. On the influence of sewage pollution on inshore benthic communities in the south of Kiel Bay. Part I. Qualitative studies on indicator species and communities. Merentutkimuslaitoksen Julk. Havsforskningsinst. Skr., 239: 116—122.

Anonymous, 1966. Beobachtungen und Versuche an Fischen, die durch reflexionsseismische Messungen in der westlichen Ostsee in der Zeit vom 26.9 bis 5.10.1965 verletzt bzw. getötet wurden. Veröff. Inst. Küst.-u. Binnenfisch., 37: 1—24.

Anonymous, 1977. Kronologisk förteckning över iakttagna oljeutsläpp under 1976 jämte andra anmärkningsvärda händelser. Mimeogr. Rep. till Nordiska Oljeskyddsunionen, 25 pp.

Anttila, R., 1972a. Helsingin merialueen kalasto. Fish fauna in the Helsinki Sea territory. Vesiensuojelulaboratorion Tiedonantoja. Reports of the Water Conservation Laboratory, 4(5): 1—157 (in Finnish. with an English summary).

Anttila, R., 1972b. Makuvirheet Helsingin merialueen kaloissa. Taste defects in fish in the sea area of Helsinki. Vesiensuojelulaboratorion Tiedonantoja. Reports of the Water Conservation Laboratory, 4(4): 1—52 (in Finnish, with an English summary).

Arldt, G., 1975. Remarks on indicator organisms (meiofauna) in the coastal waters of the G.D.R. Merentutkimuslaitoksen Julk. Havsforskningsinst. Skr., 239: 272—279.

Bagge, O., 1971. Fiskeriet i Øresund 1953—1969. Öresundsvattenkomm. Unders. 1965—1970, pp. 207—222.

Bagge, P. and Lehmusluoto, P.O., 1971. Phytoplankton primary production in some Finnish coastal areas in relation to pollution. Merentutkimuslaitoksen Julk. Havsforskningsinst. Skr., 235: 3—18.

Bengtsson, B.-E., 1974. Effect of zinc on growth of the minnow Phoxinus phoxinus. Oikos, 25: 370—373.

Bonde, G., 1967a. Bakteriologi. Sedimenter. Bacteriology of sediments. Øresunds-vandkomiteens Undersøgelser 1959—1964. Report on the Investigations of the Swedo-Danish Committee on Pollution of the Sound 1959—1964, København. pp. 156—195 (in Danish, with an English summary).

Bonde, G., 1967b. Bakteriologi. Fisk. Bacteriological examination of fish. Øresunds-vandkomiteens Undersøgelser 1959—1964. Report on the Investigations of the Swedo-Danish Committee on Pollution of the Sound 1959—1964, København, pp. 196—198 (in Danish, with an English summary).

Borg, K., Wanntorp, H., Erne, K. and Hanko, E., 1966. Mercury poisoning in Swedish wildlife. J. Appl. Ecol. Suppl., 3: 171—172.

Brügmann, L., 1976. On the distribution of some heavy metals in the Baltic. Proc. 10th Conf. Baltic Oceanogr., Göteborg, 2—4 June 1976, Paper 8, 10 pp. (mimeogr.).

Buczowska, Z. and Nowicka, B., 1960. Locating Salmonella infection sources of river and

bathing beaches by means of sewage examination. Bull. Inst. Mar. Med. Gdansk, 11: 139—146.
Busnel, R.G. (Editor), 1963. Acoustic Behaviour in Animals. Elsevier, Amsterdam, London, New York, 933 pp.
Caspers, H. (Editor), 1975. Pollution in Coastal Waters. An interim report on results of a priority programme of the German Research Society (1966—1974), Boppard, 142 pp.
Dementieva, T.F., 1972. Changes in abundance of the Baltic cod and sprat. Ambio, Spec. Rep., 1: 63—66.
Dybern, B.I., 1974a. Water pollution — a problem with global dimensions. Ambio, 3(3—4): 139—145.
Dybern, B.I., 1974b. Föroreningssituationen i Nordsjön och Östersjön. Sver. Natur Årsbok, 1974: 151—162.
Ehlin, U., 1973. Kylvattenutsläpp i sjöar och hav. Sveriges Met. Hydr. Inst. Notiser och Preliminära Rapporter Ser. Hydr., 28: 1—26.
Ehlin, U., 1974. Spridning och avkylning av kylvatten. Kylvatten, effekter på miljön. Statens Naturvårdsverk Publ., 25: 13—36.
Ehlin, U., Mattisson, I. and Zachrisson, G., 1974. Computer based calculations of volumes of the Baltic area. Proc. 9th Conf. Baltic Oceanogr., Kiel, 17—20 April, 1974, pp. 115—128 (mimeogr.).
Engberg, Å., 1976. Undersøgelser vidrørande tungmetaller i hav og kustfisk. Statens Levnedsmiddelinstitut, Copenhagen, Publ., 33: 1—69.
Engström, S. and Fonselius, S.H., 1974. On the oxygen conditions in the southern Baltic during 1973—1974 and on the general development in the Baltic. Proc. 9th Conf. Baltic Oceanogr., Kiel, 17—20 April, 1974, pp. 327—339 (mimeogr.).
Engwall, R., 1972. Swedish municipal and industrial discharges into the Baltic. Ambio, Spec. Rep., 1: 41—49.
Falkenmark, M. and Mikulski, Z., 1975. The Baltic Sea — a semi-enclosed sea as seen by the hydrologist. Nord. Hydrol., 6: 115—136.
Fonselius, S.H., 1969. Hydrography of the Baltic deep basins III. Fish. Board Sweden, Ser. Hydrogr., 23: 1—97.
Fonselius, S.H., 1970. Stagnant sea. Environment, 12(6): 1—46.
Fonselius, S.H., 1976. Havet, all världens soptipp. Forsk. Framsteg, 7: 29—37.
Fonselius, S.H., 1977. An inflow of unusually warm water into the Baltic deep basins. Medd. Havsfiskelab. Göteborg, 229: 1—3.
Fonselius, S.H., 1978. On nutrients and their role as production limiting factors in the Baltic. Acta Hydrochim. Hydrobiol., 6(4): 329—339.
Funck, B., 1978. Kronologisk förteckning över iakttagna oljeutsläpp under 1977 jämte andra anmärkningsvärda händelser. Mimeogr. Rep. till Nordiska Oljeskyddsunionen.
Ganning, V., 1970. Östersjön, oljan, saneringen. Forsk. Framsteg, 8: 25—27.
Grimås, U., 1974. Värmeeffekter på skilda nivåer i ekosystemet. Kylvatten, effekter på miljön. Statens Naturvårdsverk Publ., 25: 37—73.
Hallberg, R., 1978. Östersjön — i går, i dag, i morgon. In: A. Åkerblom (Editor), Diagnos Östersjön. Statens Naturvårdsverk, Vällingby, Rapp., pp. 8—19.
Häsänen, E., 1973. Mercury pollution in Finland. Proc. European Coll. Problems of the Contamination of Man and His Environment by Mercury and Cadmium. EUR, Luxembourg, 5075: 109—119.
Häsänen, E., 1975. Elohopea ympäristöongelmana Suomessa. Ympäristö ja Terveys, 7: 515—528.
Häsänen, E. and Sjöblom, V., 1968. Kalojen elohopeapitoisuus Suomessa vuonna 1967. Suom. Kalatalous, 36: 1—24.
Heino, A., 1972. The Palva oil tanker disaster in the Finnish southwestern archipelago. I. The extent of oil pollution in the sea area between Utö and Kökar in 1969 and 1970.

Aqua Fenn., 1972: 116—121.
Helle, E., Olsson, M. and Jensen, S., 1976. DDT and PCB levels and reproduction in ringed seal from the Bothnian Bay. Ambio, 5(4): 188—189.
Henriksson, R., 1967. Föroreningars inverkan på bottenfaunan i Öresund. Öresundsvattenkomm. Unders. 1959—1964, pp. 205—263.
Henriksson, R., 1971. Kvicksilversituationen i Öresund. Öresundsvattenkomm. Unders. 1965—1970, pp. 173—194.
Höglund, H. and Koczy, F., 1971. Elströmsförsöken vid Västervik. Hydrografiska undersökningar vid elströmsförsöken vid Västervik. Medd. Havsfiskelab. Lysekil, 105: 1—21.
Hult, J., 1968. P.M. angående utstjälpning av avfall från fartyg. Fish. Board Sweden, Mimeogr. Rep., 6 pp.
ICES, 1970. Report of the ICES Working Group on Pollution of the Baltic Sea. ICES Coop. Res. Rep. Ser. A, 15: 1—86.
ICES, 1974. Report of the Working Group for the International Study of the Pollution of the North Sea and Its Effects on Living Resources and Their Exploitation. ICES, Coop. Res. Rep., 39: 1—191.
ICES, 1975. Report of the Working Group on Effects on Fisheries of Marine and Gravel Extraction. ICES, Coop. Res. Rep., 46: 1—57.
ICES/SCOR, 1977. Studies of the pollution of the Baltic Sea. ICES, Coop. Res. Rep., 63: 1—97.
IVL, 1976. Analys av olja i havet runt Sveriges kuster. IVL Nytt, 10(1): 7—9.
Jansson, A.-M., 1978. Bottnarnas fauna och flora. In: A. Åkerblom (Editor), Diagnos Östersjön. Statens Naturvårdsverk, Vällingby, Rapp., pp. 100—109.
Jensen, S., Johnels, A., Olsson, M. and Otterlind, G., 1969. DDT and PCB in marine animals from Swedish waters. Nature, London, 224: 247—250.
Jensen, S., Johnels, A., Olsson, M. and Otterlind, G., 1972a. DDT and PCB in herring and cod from the Baltic, the Kattegat and the Skagerrak. Ambio, Spec. Rep., 1: 71—85.
Jensen, S., Johnels, A., Olsson, M. and Westermark, T., 1972b. The avifauna of Sweden as indicators of environmental contamination with mercury and chlorinated hydrocarbons. Proc. XVth Int. Ornithological Congr., 1972, pp. 455—465.
Järvekülg, A., 1970. Bottom fauna as an indicator of pollution of the marine benthal in the vicinity of Tallinn. Academy of Sciences of the Estonian S.S.R., Estonian Committee for IBP, Tartu 1970, pp. 158—193.
Kallings, L.O., 1961. Adenovirusinfektioner. Möjligheter till vattenburen smitta illustrerade av Nackaepidemien 1959. Vattenhygien, 17: 1—7.
Kändler, R., 1953. Hydrographische Untersuchungen zum Abwasserproblem in den Buchten und Förden der Ostseeküste Schleswig-Holsteins. Kieler Meeresforsch., 9: 176—200.
Karlgren, L., 1978. Närsalternas roll i kustvattnen. In: A. Åkerblom (Editor), Diagnos Östersjön. Statens Naturvårdsverk, Vällingby, Rapp., pp. 56—63.
Karlgren, L. and Ljungström, K., 1969. P.M. angående kylvattenutsläpp i Lilla Värtan från Värtaverkets kärnkraftaggregat. Statens Naturvårdsverk, Mimeogr. Rep., 8 pp.
Krongård Kristensen, K., 1971. Kvalitative og kvantitative Salmonella-undersøgelser af spildevand og Öresundsvand. Öresundsvattenkomm. Unders. 1965—1970, pp. 226—254.
Kühnhold, W.W., 1972. The influence of crude oils on fish fry. In: M. Ruivo (Editor), Marine Pollution and Sea Life. Fishing News Ltd., London, pp. 315—318.
Landner, L., Nilsson, K. and Rosenberg, R., 1977. Assessment of industrial pollution by means of benthic macrofauna surveys along the Swedish Baltic coast. Vatten, 33(3): 324—379.
Lehmusluoto, P.O. and Pesonen, L., 1973. Eutrophication in the Helsinki and Espoo sea areas measured as phytoplankton primary production. Oikos Suppl., 15: 202—208.

Lemmetyinen, R., 1966. Jäteöljyn vesilinnuille aiheuttamista tuhoista Itämeren alueella. Suomen Riista, 19: 63—71.
Leppäkoski, E., 1975. Assessment of degree of pollution on the basis of macrozoobenthos in marine and brackish-water environments. Acta Acad. Åbo. Ser. B, 35: 1—90.
Lindén, O., 1975. Acute effects of oil and oil/dispersant-mixture on larvae of Baltic herring. Ambio, 4(3): 130—133.
Lindgren, L., 1975. Algal zonation on rocky shores outside Helsinki as a basis for pollution monitoring. Merentutkimuslaitoksen Julk. Havsforskningsinst. Skr., 239: 344—347.
Linko, R., Kaitaranta, J., Ratamäki, P. and Eronen, L., 1974. Occurrence of DDT and PCB compounds in Baltic herring and pike from the Turku archipelago. Environ. Pollut., 7: 193—207.
Lithner, G., 1974. Rönnskärsundersökningen 1973. Statens Naturvårdsverk PM, 497: 1—174.
Lithner, G. and Samberg, H., 1976. Tungmetallföroreningar i Skellefteåbukten och angränsande kustavsnitt. Acta Univ. Oulu. A 42. Biol., 3: 17—21.
Luckas, B., Berner, M. and Pscheidl, H., 1977. On the contamination of cod livers with chlorinated hydrocarbons in the Baltic in 1976/77. ICES C.M. 1977/E:53, 4 pp. (mimeogr.).
Lund, E., 1971. Undersøgelser over forekomst af virus i Øresundsvand. Öresundsvattenkomm. Unders. 1965—1970, pp. 292—298.
Mellgren, L., 1973. Clostridium perfringens and coliform bacteria as indicators of faecal pollution. Oikos Suppl., 15: 195—201.
Melvasalo, T. and Viljamaa, H., 1975. Plankton composition in the Helsinki area. Merentutkimuslaitoksen Julk. Havsforskningsinst. Skr., 239: 301—310.
Miettinen, V., 1971. Mercury pollution in the Finnish coastal waters. ICES C.M. 1971/E:23, 3 pp. (mimeogr.).
Mohr, H., 1967. Observations on the atlanto-scandian herring with respect to schooling and reactions to fishing gear. FAO Conf. on Fish Behaviour in Relation to Fishing Techniques and Tactics, Bergen. Paper E/29, 8 pp. (mimeogr.).
Neuman, E., 1974. Temperaturens inverkan på rörelseaktiviteten hos fisk i en Östersjövik. Statens Naturvårdsverk PM, 477: 1—36.
Nordenberg, B., 1971. Studier av sandsugningens effekt på bottnar i Öresund. Öresundsvattenkomm. Unders. 1965—1970, pp. 195—206.
Norin, L. and Waern, M., 1973. The zone of algal low standing crop near Stockholm. The nutrients and their influence on the algae in the Stockholm archipelago during 1970. Oikos Suppl., 15: 179—184.
Nyman, L., 1975. Behaviour of fish influenced by hotwater effluents as observed by ultrasonic tracking. Rep. Inst. Freshwater Res. Drottningholm, 54: 63—74.
Odsjö, T., 1971. Klorerade kolväten och äggskalsförtunning hos fiskgjuse. Fauna Flora, 66(3): 90—100.
Olsson, M., 1978. DDT, PCB och Ostersjöns fauna. In: A. Åkerblom (Editor), Diagnos Östersjön. Statens Naturvårdsverk, Vällingby, Rapp., pp. 126—137.
Olsson, M., Jensen, S. and Renberg, L., 1973. PCB in coastal areas of the Baltic. PCB Conference II. Statens Naturvårdsverk 1973/4E, pp. 59—68.
Otterlind, G., 1978. Fisken och fisket. In: A. Åkerblom (Editor), Diagnos Östersjön. Statens Naturvårdsverk, Vällingby, Rapp., pp. 84—99.
Pautsch, F., 1978. Review of experiments on the chronic toxicity exerted by some pollutants on animal species from the Bay of Gdansk. Kieler Meeresforsch. Sonderh., 4: 335—359.
Pelkonen, K. and Tulkki, P., 1972. The Palva oil tanker disaster in the Finnish SW archipelago. III. The littoral fauna of the oil polluted area. Aqua Fenn., 1972: 129—136.

Rasmussen, E., 1955. Emigranter i København sydhavn. Nat. Verden, 1955: 1—8.

Ravanko, O., 1972. The Palva oil tanker disaster in the Finnish SW archipelago. V. The littoral and aquatic flora of the polluted area. Aqua Fenn., 1972: 142—146.

Rosén, B. (Editor), 1969. Fartygens och fritidsbåtarnas förorening av farvattnen med fast och flytande avfall. Vatten, 25(1): 21—66.

Rönnberg, O., 1975. The effects of ferry traffic on the rocky shore macrofauna in the southern Åland archipelago. Merentutkimuslaitoksen Julk. Havsforskningsinst. Skr., 239: 325—330.

Rosenberg, R., Nilsson, K. and Landner, L., 1975. Effects of a sulphate pulp mill on the benthic macrofauna in a firth of the Bothnian Sea. Merentutkimuslaitoksen Julk. Havsforskningsinst. Skr., 239: 289—300.

Schramm, W., 1972. Untersuchungen über den Einfluss von Ölverschmutzungen auf Meeresalgen. I. Die Wirkung von Rohölfilmen auf den CO_2-Gaswechsel ausserhalb des Wassers. Mar. Biol, 14: 189—198.

Schulz, S., 1968. Rückgang des Benthos in der Lübecker Bucht. Meeresber. Dtsch. Akad. Wiss. Berlin, 10(10): 748—754.

Simola, K., Jaakkola, T., Miettinen, J., Voipio, A. and Niemistö, L., 1977. Plutonium in Baltic sediments. 4th International Congress of the International Radiation Protection Assoc., Paris, April 24—30, 1977. Communic., 205: 1—4.

Soikkeli, M. and Virtanen, J., 1972. The Palva oil tanker disaster in the Finnish southwestern archipelago. II. Effects of oil pollution on the eider (Somateria mollissima) population in the archipelago of Kökar and Föglö, southwestern Finland. Aqua Fenn., 1972: 122—128.

Somer, E., 1977. Heavy metals in the Baltic. ICES C.M. 1977/E:9, 23 pp. (mimeogr.).

Somer, E. and Appelqvist, H., 1974. Changes and differences in mercury level in the Baltic and Kattegat compared to the North Atlantic using Uria sp (Guillemot sp) and Cepphus Grylle (Black Guillemot) as indicators. Proc. IXth Conf. Baltic Oceanogr., Kiel, 17—20 April, 1974, pp. 379—388 (mimeogr.).

S.O.U., 1967. Miljövårdsforskning I. 239 pp. Statens Offentliga Utredningar 1967: 47, Jordbruksdepartementet.

Svansson, A., 1975. Physical and chemical oceanography of the Skagerrak and the Kattegat. I. Open sea conditions. Fish. Board Sweden, Inst. Mar. Res., Rep., 1: 1—88.

Theede, H., Scholz, N. and Fischer, H., 1979. Temperature and salinity effects on the acute toxicity of cadmium to Laomedea loveni (Hydrozoa). Mar. Ecol. Prog. Ser., 1: 13—19.

Ungsgård, Y., 1976. Minsprängning och fiskdöd. Fiskerinytt, 1976(2): 13—14.

Vasseur, E., 1966. Progress in sulfite pulp pollution abatement in Sweden. J. Water Pollut. Control Fed., 38: 27—37.

Viitanen, R., 1971. Helsingin läntisen merialueen pohja- ja rantaeläimistöstä. About the bottom and littoral fauna in the western sea area of Helsinki. Vesiensuojelulaboratorion Tiedonantoja. Reports of the Water Conservation Laboratory, 3(8): 1—90 (in Finnish, with Swedish and English summaries).

Voipio, A. and Salo, A., 1971. On the balances of ^{90}Sr and ^{137}Cs in the Baltic Sea. Nord. Hydrol., 2: 57—63.

v. Wachenfeldt, T., 1971. Alg- och fytoplanktonundersökningar i Öresund. Öresundsvattenkomm. Unders. 1965—1970, pp. 139—172.

Wallentinus, I., 1976. Environmental influences on benthic macrovegetation in the Trosa — Askö area, northern Baltic Proper. I. Hydrographical and chemical parameters, and the macrophytic communities. Contrib. Askö Lab., Univ. Stockh., 15: 1—138.

Weil, J.G., 1974. Verification of heated water jet. Numerical model. Sveriges Met. Hydr. Inst. Rapp. Hydrologi och Oceanografi, RHO, 1: 1—75.

Westerberg, H., 1975. Counter current orientation in the migration of the European eel (Anguilla anguilla L.). Göteborgs Univ., Oceanogr. Inst., Rep., 9: 1—18, figs.
Westöö, G. and Rydälv, M., 1971. Metylkvicksilverhalter i fisk fångad mars 1968—april 1971. Vår Föda, 7—8: 179—318.
Witting, R., 1922. Fjärdarna kring Helsingfors med hänsyn till vattenomsättningen och föroreningarna. Die Meeresbuchten um Helsingfors, ihre Wasserumsetzung und Verunreinigung. Merentutkimuslaitoksen Julk. Havsforskningsinst. Skr., 11: 1—115 (in Swedish, with a German summary).
Żmudziński, L., 1975. The Baltic Sea pollution. Polskie Arch. Hydrobiol., 22: 601—614.
Żmudziński, L., 1976. Eutrophierung der Ostsee und ihrer Randgewässer. Limnologica, 10: 419—424.

Chapter 8

INTERNATIONAL MANAGEMENT AND COOPERATION

VEIKKO SJÖBLOM and AARNO VOIPIO

A. INTERNATIONAL MANAGEMENT OF BALTIC SEA FISHERIES*

To prevent overexploitation of fish stocks, the countries around the Baltic Sea have agreed to international management of the fisheries. For this purpose, the Convention on Fishing and Conservation of the living Resources in the Baltic Sea and the Belts was signed in Gdansk on 13 September 1973. The Convention follows in principle the North-East Atlantic Fisheries Convention.

According to the Baltic Convention, the member states recognize the importance of maximum and stable productivity of the living resources of the Baltic Sea and their joint responsibility for the rational exploitation of these. Bearing this in mind, the contracting parties undertake to co-operate closely with a view to preserving and increasing the living resources, and in particular to expanding and co-ordinating studies toward these ends. The contracting states agree to prepare and put into effect organizational and technical projects, including measures for artificial reproduction of valuable fish species, and/or to contribute financially to such measures on a just and equitable basis.

To enforce the provisions of the Convention, an International Baltic Sea Fishery Commission has been established. It is the duty of the Commission to keep under review the living resources and fisheries in the Convention area, to work out proposals with regard to co-ordination of scientific research, and to prepare and submit recommendations based as far as possible on results of scientific research.

Each contracting state has an opportunity to consider whether it wishes to observe the recommendation of the Commission. The contracting state will, however, be bound by the recommendation if it has not raised any objection within a given time.

The Commission maintains its office in Warsaw.

The area covered by the Convention comprises the whole Baltic Sea and Belts excluding internal waters. The contracting states are thus required to implement the recommendations of the Commission within their territorial sea and the waters under their fisheries jurisdiction.

The most important task of the Commission has been to agree upon fishing regulations, to fix annual catch quotas and to create an enforcement scheme.

---

* By Veikko Sjöblom.

TABLE 8.I

The allowable catch (in 1000 metric tons) of Baltic herring, sprat and cod in 1978

	Baltic herring	sprat	cod
Total allowable catch (TAC)	444.0	184.3	173.8
Management Unit 3 (Bothnian Bay, Bothnian Sea and the northern Baltic Sea proper north of latitude 59°30′ N)			
Finland and Sweden	74.0[a]		
Denmark	20.2[b]	8.2	43.8
Federal Republic of Germany	15.6	2.8	16.6
Finland	15.5	5.5[c]	0.1[d]
German Democratic Republic	60.8	12.0	8.4
Poland	61.0	48.4	49.2
Sweden[e]	81.0	24.0	27.2
USSR	115.9	83.4	28.5

[a] The catch can be increased or reduced against a corresponding reduction or increase of the Finnish quota in the Convention area outside Management Unit 3.
[b] Transfer of 10% of the Danish quota into sprat is possible.
[c] Sprat catches in mixed fishery are excluded.
[d] Bycatches are not included.
[e] It is understood that when fixing allowable catches by species within the extended fishing zone, Sweden will take into account her allocation, as indicated.

TABLE 8.II

Catch of Baltic herring, sprat and cod in 1976 (in metric tons)

	Baltic herring	sprat	cod
Denmark	23,086	13,054	77,429
Federal Republic of Germany	6543	731	23,360
Finland	73,600	5780	196
German Democratic Republic	58,077	7493	8550
Poland	63,850	56,079	70,500
Sweden	46,237	800	20,188
USSR	164,878	73,200	49,097
Total	436,271	157,137	249,320

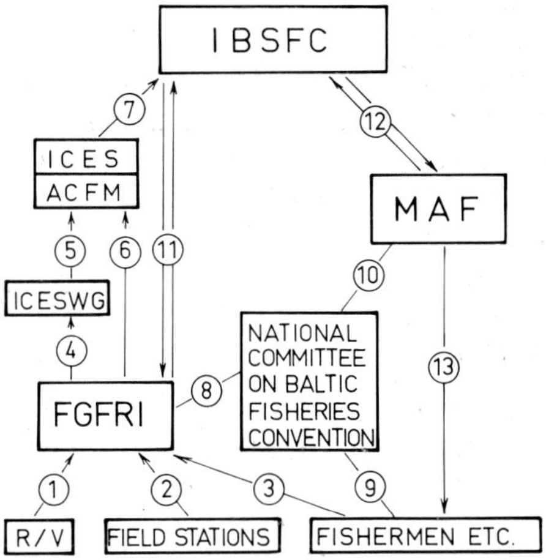

Fig. 8.1. International management of Baltic Sea fishery in Finland.

1—2: Data for assessment of the fish stocks are obtained by the research vessel (R/V) and the field stations of the Finnish Game and Fisheries Research Institute (FGFRI).

3: Information on fishing and catches is supplied by the fishermen and on the use of the catch by the fishmonger, fur farmers etc.

4: After the data have been processed in the FGFRI, stock assessments are made in the working groups of the International Council for the Exploration of the Sea (ICES) in cooperation with the other member states.

5—6: Reports of the working groups are discussed at the Advisory Committee on Fishery Management (ACFM).

7: Report of the ICES is delivered to the International Baltic Sea Fishery Commission (IBSFC).

8—10: The matters taken up in IBSFC are discussed by the National Committee on the Baltic Fisheries Convention formed by the representatives of various spheres of interest in fishing industry.

11—12: The Finnish Delegation to IBSFC is composed of representatives of the Ministry of Agriculture and Forestry (MAF) and the FGFRI, and of other experts, as needed.

13: Dissemination of information on the decisions of the IBSFC and their enforcement in cooperation with the Coast Guard is the duty of MAF.

At its first session in 1974, the Commission approved regulations concerning fishing of the most important fish species, viz., Baltic herring, sprat, cod and flounder. At its second session in 1976, the Commission recommended catch quotas for Baltic herring, sprat and cod, and adopted the regulations for salmon developed by the Baltic Salmon Convention. In addition, the enforcement scheme was approved.

The basis for allocations has been, as in the North East Atlantic Fisheries Commission (NEAFC), the earlier catches. Therefore, the states with undeveloped fishery have been in danger of coming off worse than those with an already efficient fishing industry. In consequence, Sweden felt that she had no option but to extend her fishing zone to the midline from 1 January 1978.

According to the Final Act of the Baltic States Diplomatic Conference at Gdansk, 4—14 September, 1973, the contracting states took note that none of them claimed jurisdiction beyond a limit of twelve nautical miles, and agreed that if any of them claimed a wider jurisdiction, the Convention should be renegotiated. The amendments of the Convention caused by the establishment of national fishery zones have before the fourth session of the Commission in 1978 been ratified by five member states.

Because of the extension of the Swedish fishing zone the allocation of the quotas at the third session of the Commission in 1977 (Table 8.I) differed considerable from the catches (Table 8.II). Quotas were reduced for the other states, especially the USSR (Baltic herring and cod) and Poland (cod).

In Finland the work of the Commission has had a stimulating effect on scientific research into fisheries and has led to the creation of a special organizational procedure (Fig. 8.1).

B. INTERNATIONAL CO-OPERATION AS THE BASIS OF THE PROTECTION OF THE MARINE ENVIRONMENT OF THE BALTIC SEA*

International co-operation in the field of marine sciences dates back many years in the Baltic Sea Area. Since 1902, the International Council for the Exploration of the Sea (ICES) has provided opportunities for contact between the Baltic and the North Atlantic oceanographers. The Intergovernmental Convention regarding ICES, signed in 1964, gave a firm basis for further co-operative activities. However, the political situation after World War II created certain difficulties for the co-operation between the countries round the Baltic Sea. These were in part overcome by the establishment of an informal scientific community composed of the leading oceanographers in the different countries. Their first meeting was held in Helsinki in 1957, i.e., long before the solution of the problem of the German Republics. Since then, the marine scientists working on the Baltic Sea have been able to co-operate effectively through the Conference of the Baltic Oceanographers. Later the marine biologists established their own organization, called the Baltic Marine Biologists.

It was the work of these scientific bodies, and of the various institutions and individual scientists behind them, that yielded the observations confirming the need to protect the Baltic Sea. At present, important work is also being carried out by the Working Group on the Study of the Pollution of the Baltic Sea, which was jointly established by ICES and SCOR (Scientific Committee on Oceanic Research). For instance, this group recently finished a baseline study of heavy metals and chlorinated hydrocarbons in living organisms in the Baltic Sea. All seven countries participated in this exercise.

In addition to the multilateral co-operation, the bilateral work also de-

* By A. Voipio.

serves mention. Within the framework of this kind of activity, a number of local studies are being carried out, which are also very useful for the purpose of monitoring the whole Baltic Sea area.

At the United Nations Conference on the Human Environment, held in Stockholm in June 1972, the Finnish delegation indicated that the Government of Finland was willing to host a conference on the protection of the Baltic Sea. A meeting of governmental experts was convened in Helsinki at the end of May 1973.

On the basis of the discussions held during this meeting it was agreed that the Government of Finland should prepare a draft convention for a Second Meeting of Experts (November 1973).

The final step was taken by the Diplomatic Conference on the Protection of the Marine Environment of the Baltic Sea Area, held in March 1974, again in Helsinki, where the Final Act was signed on 22 March, 1974, by the seven ministers of the Contracting Parties. The whole period from the United Nations conference in Stockholm to the Diplomatic Conference in Helsinki was only some 19 months. This is a remarkably short time, as is clearly seen when it is compared with the time that elapsed between the signing of the Convention and its ratification by the first of the Contracting Parties (Finland). The latter period, needed for purely national measures, was about 15 months. And in October, 1978, only three additional countries have ratified the Convention. These countries are Sweden, the German Democratic Republic and Denmark*.

The Convention consists of 29 articles and six annexes. It is characterized by a broad approach to the different aspects of pollution of a sea area; the ultimate goal is to regulate and reduce all the factors resulting in the pollution of the Baltic Sea. The Convention covers different kinds of land-based pollution and ship-generated wastes. It takes into account the fact that pollution may also be airborne or carried into the sea by rivers. In addition, the Convention includes articles prohibiting dumping of wastes, with a few exceptions, and pollution caused by pleasure craft and by the exploration and exploitation of the sea-bed. The articles also deal with co-operation in combating marine pollution, responsibility for damage, and other items.

The strictest measures in the Convention text are those laid down in Article 5 in respect of hazardous substances. According to this article, the signatories undertake to counteract the introduction into the Baltic Sea of substances listed in Annex 1. In its original form, Annex 1 comprises only two groups of chemical substances, viz., DDT and its derivatives and the polychlorinated biphenyls. The only exception to the prohibition of these substances is trace amounts in wastes or dredge spoils that can be labelled "insignificant amounts".

---

* Note added in proof: The ratification process was completed in March 1980. Consequently, the Convention came into force in May 1980, and the Secretariat of the Commission commenced its activities in July 1980 in Helsinki.

Article 5 gives a means of attacking one of the most serious problems in the pollution of the Baltic Sea, viz., the increased contents of some environmental poisons in the living resources of certain areas of the Baltic Sea. The implementation of this article "can involve the use of appropriate technical means, prohibitions and regulations of the transport, trade, handling, application and final deposition of products containing such substances". The list is short at present, but other substances are readily added, if necessary. Mercury and cadmium have already been discussed, but the fact that they are normally present in nature will give rise to some difficulties: for instance, it will be difficult to define the limit between the terms "natural concentration" and "not insignificant amount", etc.

Article 6 gives the principles and obligations concerning land-based pollution. In addition to the general obligation to take "all appropriate measures to control and minimize the land-based pollution, the Contracting Parties shall take special measures to control and strictly limit pollution by noxious substances and materials in accordance with Annex II".

Here we can find a resemblance to the conventions on oceanic dumping. In these conventions there is a list containing substances the dumping of which is totally banned (often called the "black list" in everyday speech), and another list, called the "grey list", containing substances the release of which is brought under the strict control of the national authorities. As already mentioned, the "black list" in the Helsinki Convention comprises only two groups of substances. However, the "grey list", which is given in Annex II includes noxious substances and materials. It is divided into two sub-groups, the first of which comprises mercury, cadmium, and their compounds. The introduction of these substances into the Baltic Sea is a subject "for urgent consideration".

The list given in Annex II is rather inhomogeneous including, as it does, virtually all the substances and materials that are likely to cause marine pollution. Some of these are noxious even if introduced in small quantities, the others, like acids and alkalis, only if discharged in large quantities.

One of the essential points regarding the introduction of the noxious materials is that, according to Paragraph 5, Article 6, "the Contracting Parties shall endeavour to establish and adopt common criteria for issuing permits for discharges". This kind of work has already started within the framework of the Provisional Scientific-Technological Working Group.

The first difficulty is the definition of the phrase "significant quantities". If the noxious substance is not present "in significant quantities" in the wastes discharged, no special discharge permit will be needed.

Another problem is posed by a difference in the basic "philosophy" of the regulation measures; in the socialistic countries norms are set for the various substances while, for instance, in the Nordic countries certain kinds of, say, industrial enterprises apply for general permits to discharge wastes into the sea. In the latter case the authorities must consider all the possible substances released by the said enterprise that are likely to cause pollution.

In order to find the common criteria needed, it seems to be necessary to reconcile the immission criteria philosophy and the enterprise philosophy with the emission criteria philosophy.

Annex III of the Convention gives "Goals, Criteria and Measures concerning the Prevention of Land-based Pollution". It is said in the Convention text itself that "the Contracting Parties shall take all appropriate measures to control and minimize land-based pollution...", and this annex gives somewhat more concrete definitions of the obligations of the signatories.

"Municipal sewage shall be treated in an appropriate way so that the amount of organic matter does not cause harmful changes in the oxygen content of the Baltic Sea Area and the amount of nutrients does not cause harmful eutrophication of the Baltic Sea Area".

"The polluting load of industrial wastes shall be minimized in an appropriate way in order to reduce the amount of harmful substances, organic matter and nutrients".

Annex III also lists certain technological and other principles to be followed in reducing land-based pollution.

Articles 7—10 deal with the prevention of pollution from ships, pleasure craft, dumping, and exploration and exploitation of the sea-bed and its subsoil. Annex IV gives regulations on the prevention of pollution from ships. It follows the principles outlined by IMCO. It was considered necessary to include the contribution of pleasure craft to the pollution of the Baltic Sea in a special Article, which mentions the need of adequate reception facilities for wastes from pleasure craft. There is a general ban on all dumping, except that of dredged spoils, for which additional regulations are given in Annex V.

Article 11 and Annex VI deal with the obligation of the Contracting Parties to co-operate in combating marine pollution.

Articles 12—15 contain administrative regulations, providing in particular for the establishment of the main operational body, "The Baltic Marine Environment Protection Commission", which shall ensure the implementation of the Convention, give recommendations on measures, review the contents of the Convention, etc.

Besides the scientific and technological co-operation, which deserves special attention, the remaining Articles include regulations on responsibility for damage, settlement of disputes and other legal matters, on the operation of the Convention itself, and on its revision, signing, ratification, entry into force, etc.

Scientific and technological co-operation, which has a central position in the realization of the Helsinki Convention, is dealt with in Article 16:

"The Contracting Parties undertake directly, or when appropriate through competent regional or other international organizations, to co-operate in the fields of science, technology and other research, and to exchange data as well as other scientific information for the purposes of the present Convention.

Without prejudice to Paragraphs 1, 2 and 3 of Article 4 of the present Convention the Contracting Parties undertake directly, or when appropriate through competent regional or other international organizations, to promote studies, undertake, support or contribute to programmes aimed at developing ways and means for the assessment of the nature and extent of pollution, pathways, exposures, risks and remedies in the Baltic Sea Area, and particularly to develop alternative methods of treatment, disposal and elimination of such matter and substances that are likely to cause pollution of the marine environment of the Baltic Sea Area.

The Contracting Parties undertake directly, or when appropriate through competent regional or other international organizations and, on the basis of the information and data acquired pursuant to Paragraphs 1 and 2 of this Article, to co-operate in developing intercomparable observation methods, in performing baseline studies and in establishing complementary or joint programmes for monitoring.

The organization and scope of work connected with the implementation of tasks referred to in the preceding Paragraphs should primarily be outlined by the Commission".

Scientific and technological co-operation is needed for three kinds of activities: first, marine scientific research aimed at describing to the Contracting Parties the actual situation regarding the pollution of the Baltic Sea and the factors which have led to such a situation; secondly, the technological research urgently needed for finding means of limiting or sometimes even putting a complete stop to the introduction into the sea of polluting substances and materials.

Between these two fields of activity lies a third kind of study, which describes how the changes in the waste load affect the state of the Baltic Sea. In planning technical measures, it is naturally very important to know something about their effectiveness and consequences. Unfortunately, we cannot expect that it will be possible in the immediate future to create a quantitative model for the whole Baltic Sea that includes all physical and chemical aspects of the ecosystem in detail to permit realistic prognoses. While we continue our efforts to create the quantitative models needed, we will have to find some provisional course of action for the present-day protection of the Baltic Sea. One way has been labelled the "best practical means" strategy. In spite of the lack of quantitative knowledge of the exact cause—consequence relations, we should carry out all practically possible measures to reduce the pollution input.

The results obtained with mercury and DDT, against which strong measures have been taken in all the Baltic Sea countries, should encourage us to continue our efforts to find appropriate methods to reduce the load of the substances included in the Helsinki Convention or those which will be added to it in the future.

Something has been done, but many things are still waiting for joint action by all the countries round the Baltic Sea. The Helsinki Convention is a suitable instrument for this purpose.

AUTHORS' INDEX

Aarkrog, A., 371, *376*
Aarnisalo, J., see Tuominen, H.V. et al.
Aartolahti, T., 55, 57, *117*
Abshagen, G., *265*
Ackefors, H., 239, 240, 242, 245, 248, 249, 250, *265*
Ackefors, H. and Fonselius, S.H., 375, *376*
Ackefors, H. and Hernroth, L., 239, 243, *265*
Ackefors, H., Hernroth, L., Lindahl, O. and Wulff, F., 243, *265*
Ackefors, H. and Lindahl, O., 221, 242, 247, 251, 252, 253, *265*
Ackefors, H. and Rosén, C.-G., 249, *265*
Ackefors, H., see Dybern, B.I. et al., see also Hällfors, G. et al., see also Hernroth, L. and Ackefors, H.
Agalcova, E.N., see Volkolokov, F.K. et al.
Agnedal, P.O., 371, *376*
Ahl, T., 365, *376*
Ahlnäs, K., 143, *175*
Ahnoff, M. and Johnsson, L., 370, *376*
Alasaarela, E., 219, 221, 227, *265*
Alasareela, E. and Siira, J., 229, *265*
Alexander, G., see Stewart, W.D.P. and Alexander, G.
Alhonen, P., 60, *117*
Alm, G., 293, 299, *341*
Ambjörn, C., see Ehlin, U. and Ambjörn, C.
Andersin, A.-B., Lassig, J., Parkkonen, L. and Sandler, H., 220, 255, 262, *265*, *266*
Andersin, A.-B., Lassig, J. and Sandler, H., 258, 259, 260, 261, *265*
Andrushaitis, G., 360, 361, *376*
Anger, K., 361, *376*
Anheden, H., see Svärdson, G. and Anheden, H.
Ankar, S., 262, 263, 264, *266*
Ankar, S. and Elmgren, R., 262, 264, *266*
Anonymous, *341*; 369, 375, *376*
Antonovicha, L., Kondratovich, E. and Uzars, D., 318, *341*
Anttila, R., 361, *376*
Appelqvist, H., see Somer, E. and Appelqvist, H.
Ardt, G., 361, *376*
Arntz, W.E. and Brundwig, D., *266*
Arrhenius, G., 187, *213*
Aschan, O., 211, *213*
Asplund, C. and Södergren, S., 309, *341*
Aurimaa, K., see Pietikäinen, S. et al.
Axberg, S., see Ignatius, H. et al., see also Thorslund, P. and Axberg, S.

Bågander, L.E., 210, 211, *213*
Bågander, L.E., see Hallberg, R.O. et al.
Bagge, O., 311, 313, 314, 320, 332, *341*; 361, *376*
Bagge, O. and Müller, A., 315, *342*
Bagge, O., Tiews, K., Lamp, F. and Otterlind, G., 313, *342*
Bagge, O., see Müller, A. and Bagge, O.
Bagge, P. and Lehmusluoto, P.O., 221, *266*; 361, *376*
Bagge, P. and Niemi, Å., 221, *266*
Balzer, R., 202, *213*
Banarescu, P., Papadopol., M. and Mikhailova, L., 307, *342*
Bansemir, K. and Rheinheimer, G., 210, *213*
Barkov, L.K., see Logvinenko, N.V. et al.
Beers, J.R. and Stewart, G.L., 251, *266*
Bengtsson, B.-E., 369, *376*
Berg, L.S., 291, 292, *342*
Bergqvist, E., 21, *117*
Berkow, B., see Mörner, N.-A. et al.
Berner, M., 313, *342*
Berner, M., see Luckas, B. et al.
Biriukov, N.P., 313, *342*
Bitjukov, E.P., 285, 286, *342*
Bjerknes, V., 157, *175*
Black, H.J., see Schwenke, H. et al.
Bladh, J.-O., 212, *213*
Blagoboline, N., see Lillienberg, D. et al.
Blažčišin, A.I., 58, 63, 68, 103, *117*
Blažčišin, A.I. Boldyrev, V.L. and Šuiskii, Ju.D., 114, 115, 116, *117*
Blažčišin, A.I., see Varencov, I.M. and Blažčišin, A.I.
Blegvad, H., 322, 324, 333, *342*

Blomquist, Å., Pilo, C. and Thompson, T., 174, *175*
Bock, K.-H., 137, *175*
Boëtius, J., 309, *342*
Böhnecke, G. and Dietrich, G., 150, *175*
Boldyrev, V.L., see Blažčišin, A.I. et al.
Bonde, G., 360, *376*
Bontemps, S., see Erm, V. et al., see also Volskis, R. et al.
Borg, K., Wanntorp, H., Erne, K. and Hanko, E., 367, *376*
Boström, R., see Laurén, L. et al.
Boulanger, Ju., Deumlich, F., Entin, I., Joo, I., Kashin, L., Hristov, V., Lillienberg, D., Setounskaya, L., Vyskochil, P., Wyrzykowsky, T. and Zotin, M., 21, *117*
Brangulis, A., Kala, E., Mardla, A., Mens, K., Pirrus, E., Sabaljauskas, V., Fridrihsone, A. and Jankauskas, T., 43, *117*
Brännström, B., see Flodén, T. and Brännström, B.
Brattberg, G., 226, *266*
Brattberg, G., see Lindahl, G. et al.
Brogmus, W., 126, 127, 128, 129, *133*; 171, *175*
Brogmus, W., see Simojoki, H. and Brogmus, W., see also Wüst, G. and Brogmus, W.
Brosin, H.J., 164, 166, 167, *175*
Brown, M., 212, *213*
Brügmann, L., 204, *213*; 365, *376*
Brunswig, D., see Arntz, W.E. and Brunswig, D.
Brzyski, B., see Kornaś, J. et al.
Buch, K., 194, *213*; 221, 226, 227, *266*
Buczowska, Z. and Nowicka, B., 360, *376*
Bursa, A., Wojtusiak, H. and Wojtusiak, R.J., 236, *266*
Busnel, R.G., 375, *377*
Bylinskaya, L., see Lillienberg, D. et al.

Carlberg, S.R., 212, *213*
Carlin, B., 293, 295, 301, 302, *342*
Carlin, B. and Johansson, N., 295, *342*
Caspers, H., 361, *377*
Christensen, O., 304, *342*, *343*
Christensen, O. and Johansson, N., 294, *342*
Cięglewicz, W., *343*
Cięglewicz, W., Draganik B. and Zukowski, C., 326, *343*
Cooper, L.H.N., 224, *266*
Cox, R.A. and Culkin, F., 184, *213*
Cox, R.A., Culkin, F., Greenhalgh, R. and Riley, J.P., 183, *213*
Csanady, G.T., 156, 158, *175*
Culkin, F., *213*
Culkin, F., see Cox, R.A. and Culkin, F., see also Cox, R.A. et al.

Dadlez, R., 36, *117*
Dahl-Madsen, K.I., see Gargas, E. et al.,
Dahlin, H., 139, 143, *175*; 227, *266*
Dahlström, B., 126, 127, 129, *133*
Defant, F., 163, *175*
Dementieva, T.F., 361, *377*
Demina, L.L., see Morozov, N.P. et al.
Denman, K.L., 165, *175*
Denman, K.L. and Miyake, M., 165, *175*
Dera, J. and Olszewski, J., 170, *175*
Deumlich, F., see Boulanger, Ju. et al.
Deutsches Hydrographisches Institut, 174, *175*
Dickson, R.R., 147, *176*
Dietrich, G., 172, *176*
Dietrich, G., Kall, K., Krauss, W. and Siedler, G., 161, *176*
Dietrich, G., see Böhnecke, G. and Dietrich, G., see also Wooster, W.S., et al.
Donner, J.J., 56, *117*
Doronin, Ju.P., 176, *176*
Draganik, B., see Cięglewicz, W. et al.
Druet, Cz., Hupfer, P. and Shadrin, I., 157, *176*
Du Rietz, G.E., 233, *266*
Dutt, S., see Kändler, R. and Dutt, S.
Dybern, B.I., 354, 356, *377*
Dybern, B.I., Ackefors, H. and Elmgren, R., 221, 239, 265, *266*
Dyrssen, D.W. and Uppström, L.R., 187, *213*

Edler, L., 219, 220, 221, 223, 227, 228, *266*
Ehlers, M., see Schott, F. et al.
Ehlin, U., 372, *377*
Ehlin, U. and Ambjörn, C., 143, *176*
Ehlin, U., Mattisson, I. and Zachrisson, G., 165, 167, *176*; 352, *377*
Ehlin, U. and Zachrisson, G., 126, *133*
Ehrhardt, M., 199, 212, *213*
Ekman, F.L., 146, 151, *176*
Ekman, T., 338, *343*

Ekman, V.W., 157, *176*
Elhammar, A., see Mörner, N.A. et al.
Elmgren, R., 256, 261, 262, 263, 264, *267*
Elmgren, R. and Ganning, B., 236, 237, *267*
Elmgren, R., see Ankar, S. and Elmgren, R., see also Dybern, B.I. et al.
Elson, P.E. and Shearer, W.H., 309
Elwertowski, J., 279, 283, 284, 286, 287, 288, 289, 290, 291, *343*
Elwertowski, J. and Popiel, J., 289, *343*
Emeljanov, E.M., 205, 206, 207, *213*
Enequist, P., 308, 309, *343*
Engberg, Å., 368, *377*
Engström, S. and Fonselius, S.H., 358, *377*
Engvall, A.G., see Hallberg, R.O. et al.
Engwall, R., 365, *377*
Entin, I., see Boulanger, Ju. et al.
Erlenkeuser, H., Suess, E. and Wilkomm, H., 102, *117*
Erlenkeuser, H., see Suess, E. and Erlenkeuser, H.
Erm, V., Bontemps, S., Volskis, R. and Spirina, L.I., 308, *343*
Erm, V., 339, *343*
Erm, V., see Kublickas, A. et al., see also Volskis, R. et al.
Erne, K., see Borg, K. et al.
Eronen, L., see Linke, R. et al.
Eronen, M., 60, *118*
Evtjuchova, B.K., see Ojaveer, E. et al.

Fabricius, E. and Gustafson, K.-J., 306, *343*
Fabricius, E. and Lindroth, A., 307, *343*
Fagerholm, H.P., 256, *267*
Falkenmark, M. and Kikulski, Z., 124, 133, *134*; 352, *377*
Fenchel, T., 255, *267*
Fester, U., 329, *343*
Filkine, V., see Lillienberg, D. et al.
Filuk, J., 338, *343*
Fischer, H., see Theede, H. et al.
Fjeldstad, J.E., 159, *176*
Fleming, R.H. 199, *214*
Flodén, R., 1, *118*
Flodén, T. and Brännström, B., 84, *118*
Flodén, T., see Mörner, N.-A. et al.
Flyg, C., see Wulff, F. et al.
Foberg, M., see Wulff, F. et al.
Fogelin, P., 309, *343*
Fonselius, S.H., 138, 139, 141, 143, 144, 146, 147, 148, 149, 159, *176*; 190, 191, 193, 196, 197, 208, 212, *214*; 238, *267*; 358, 361, 374, *377*
Fonselius, S.H., see Ackefors, H. and Fonselius, S.H., see also Engström, S. and Fonselius, S.H.
Forsman, B., 239, *267*
Francke, E. and Nehring, D., 139, *176*
Francke, E., see Nehring, D. and Francke, E., see also Sturm, M. et al.
Fredén, C., 61, *118*
Fridrihsone, A., see Brangulis, A. et al.
Friers, C.-C., see Rechlin, O. and Friess, C.-C.
Fromm, E., 34, 57, 85, *118*
Funck, B., 370, *377*

Ganning, B., see Elmgren, R. and Ganning, B.
Ganning, V., 370, *377*
Gargas, E., Dahl-Madsen, K.I., Schroeder, H. and Rasmussen, H., 119, *214*
Geer de G., 57, *117*
Gieskes, J.M. and Grasshoff, K., 194, 196, 208, *214*
Gontarev, E.A., see Logvinenko, N.V. et al.
Gonzales, J.G., see Zillioux, E.J. and Gonzales, J.G.
Gorbatschev, R., 25, *118*
Gorelov, S., see Lillienberg, D. et al.
Gorshkova, T.I., 107, *118*
Gorijanskii, V.Ju., see Plissov, A.A. et al.
Gosteeva, M.N., 339, *343*
Gottberg, G., 339, *343*
Grandlund, E., see Magnusson, N. et al.
Granqvist, G., 143, 145, *176*; 190, *214*
Grasshoff, K., 184, *214*
Grasshoff, K., see Gieskes, J.M. and Grasshoff, K.
Grauman, G.N., 283, 290, 313, 314, 315, 316, *343*, *344*
Greenhalgh, R. and Riley, J.P., 187, *214*
Greenhalgh, R., see Cox, R.A. et al.
Grimas, U., 373, *377*
Gripenberg, S., 63, 103, *118*; 184, 185, 187, 200, 201, *214*
Gudelis, V.K., 54, 55, 70, 83, 86, 95, *118*
Gudelis, V.K. and Litvin, V.M., 54, *118*
Gustafson, K.-J., see Fabricius, E. and Gustafson, K.-J.
Gustafsson, O., see Olausson, E. et al.

Gustafsson, T. and Kullenberg, B., 151, 152, *176*
Guterstam, B., 236, 237, *267*
Guterstam, B., see Schramm, W. and Guterstam, B.

Haage, P., 257, *267*
Haahtela, I., 275, *344*
Hagelin, L.-O., 308, *344*
Hagelin, L.-O. and Steffner, N., 308, *344*
Hagström, Å., 251
Halbach, P., 113, *118*
Hallberg, R.O., 206, 207, 209, *214*; 358, *377*
Hallberg, R.O., Bågander, L.E., Engvall, A.-G. and Schippel, F.A., 208, *214*
Hällfors, G., 221, 223, 225, 232, 233, 235, *267*
Hällfors, G., Kangas, P. and Lappalainen, A., 236, 256, 257, *267*
Hällfors, G. and Niemi, Å, 219, 220, 221, 223, *267*
Hällfors, G., see Luther, H. et al., see also Lappalainen, A. et al.
Halme, E., 244, *267*
Halme, E. and Korhonen, M., 338, *344*
Hankimo, J., 171, 172, 173, *176*
Hanko, E., see Borg, K. et al.
Hansen, U.J., 329, *344*
Hansson, R. and Sandström, O., 307, *344*
Hansson, S., see Wulff, F. et al.
Hardisty, M.W. and Potter, I.C., 309, *344*
Hargrave, B., see Parsons, T.R. et al.
Härme, M., 4, *118*
Häsänen, E., 100, *118*; 368, *377*
Häsänen, E. and Sjöblom, V., 368, *377*
Hasle, G.R., 221, *267*
Haxner, H., see Mörner, N.-A. et al.
Heino, A., 370, *377*
Heincke, F.R., see Möbius, K. and Heincke, F.R.
Hela, I., 138, 143, 144, 145, 146, 162, 164, 171, *176*, *177*
Hela, I., see Kullenberg, B. and Hela, I.
Helle, E., Olsson, M. and Jensen, S., 363, 364, *378*
Heller, E., 116
Hempel, G. and Nellen, W., 276, 335, *344*
Henking, H., 338, *344*
Henning, D., 128, 129, *131*
Henriksson, R., 361, 367, *378*
Hensen, V., 220, *267*
Hentse, 361
Hernroth, L., 240, *267*
Hernroth, L. and Ackefors, H. 239, 242, 244, 245, 246, 247, *268*
Hernroth, L., see Ackefors, H. and Hernroth, L., see also Ackefors, H. et al.
Hessle, Chr., 278, 279, 282, 283, 290, 299, 310, *344*
Hickel, W., 223, *268*
Hill, M.N., *217*
Hillebrandt, M., 251, *268*
Hinrichsen, D., 310, *344*
Hobro, R. and Nyqvist, B., 221, *268*
Hobro, R., Larsson, K. and Wulff, F., 224, *268*
Hoffmeister, H., 329, *344*
Hofman-Bang, O., 201, *214*
Høјerslev, H.K., 168, 169, 170, *177*
Höglund, H. and Koczy, F., 375, 378
Hollan, E., 154, 158, 159, 161, 162, *177*; 193, *214*
Hoppe, G., 97, *118*
Horner, R.A., 223, *268*
Horstmann, U., 198, *214*; 226, *268*
Hristov, V., see Boulanger, Ju. et al.
Hübel, H., 221, 229, *268*
Hübel, H. and Hübel, M., 226, *268*
Hübel, M., see Hübel, H. and Hübel, M.
Hubrich, L.M., see Schott, F. et al.
Hult de Geer, E., 55, 57, *118*
Hult, J., 373, *378*
Hupfer, P., 147, 171, 172, *177*
Hupfer, H.A., see Wojewodzki, T. et al., see also Druet, Cz. et al.
Hutley, H.T., see Savidge, G. and Hutley, H.T.
Huttunen, M., see Melvasalo, T. et al.
Hyyppä, E., 57, 58, *118*
Hyvärinen, H., 55, 56, 57, 59, 60, *118*

ICES, 205, *215*; 290, 304, 318, 319, 323, *344*; 362, 369, 371, 374, 375, *378*
ICES/SCIR, 353, 354, 355, 363, 368, *378*
Ignatius, H., 58, 67, 68, 69, *118*
Ignatius, H., Axberg, S., Niemistö, L. and Winterhalter, B., 54
Ignatius, H., Kukkonen, E. and Winterhalter, B., 58, 63, 65, 66, 67, 68, 95, *119*; 209, *215*
Ignatius, H. and Niemistö, L., 55, *119*
Ignatius, H., Niemistö, L. and Voipio, A., 100, 101, *119*

Ignatius, H. and Tynni, R., 62, 118, *119*
Ilus, E., *268*
International Baltic Sea Fishery Commission, 290, *344*
IVL, 370, *378*

Jaakkola, T., see Simola, K. et al.
Jacobs, W.C., 172, *177*
Jacobsen, T.S., 131, *134*
Jakovleva, V.I., see Volkolakov, F.K. et al.
Jamieson, A., 312, *344*
Jamieson, A. and Otterlind, J., 312, *344*
Jankauskas, T., see Brangulis, A. et al.
Jansson, A.-M., 220, 224, 232, 236, 237, 256, 257, *268*; 361, *378*
Jansson, A.-M. and Kautky, N., 236, *268*
Jansson, B.-O., 237, 238, *268*
Jansson, B.-O. and Källander, C., 257, *268*
Jansson, B.-O. and Wulff, F., 236, 237, *268*
Järvekulg, A., 292, *344*; 361, *378*
Järvi, T.H., 295, 298, 299, *345*
Jensen, A.J.C., 146, 147, *177*; 277, 278, 282, 283, 291, *345*
Jensen, S., Johnels, A., Olsson, M. and Otterlind, G., 362, 364, *378*
Jensen, S., Johnels, A., Olsson, M. and Westermark, T., 363, 364, 368, *378*
Jensen, S., see Helle, E. et al., see also Olsson, M. et al.
Jerbo, A., 63, 65, 68, *119*
Jerlov, N.G., 167, 168, 169, 170, 171, 173, *177*; 211, *215*
Jerlov, N.G. and Liljequist, G., 169, *177*
Jerlov, N.G. and Nygard, K., 169, *177*
Jespersen, P., 285, 286, *345*
Johansson, B., 307, *345*
Johansson, N., see Garlin, B. and Johansson, N., see also Christensen, O. and Johansson, N., see also Wulff, F. et al.
Johnels, A.G., 309
Johnels, A.G., see Jensen, S. et al.
Johnsson, L., see Arhnoff, M. and Johnsson, L.
Johnston, W.R., see Thompson, T.G. et al.
Jones, K. and Stewart, W.D.P., 226, *268*
Joo, I., see Boulanger, Ju., et al.
Josefsson, B. 213, *215*
Josefsson, B., Lindroth, P. and Östling, G., 213, *215*
Joseph, J. and Sindner, H., 166, *177*
Jurkovskis, A.K. and Luke, M.P., 212, *215*

Kaiser, W. and Schulz, S., 221, 223, 224, *269*
Kaitaranta, J., see Linko, R. et al.
Kala, E., see Brangulis, A. et al.
Kaleis, M.V., 147, 167, *177*
Källander, C., see Jansson, B.-O. and Källander, C.
Kalle, K., 190, 194, 211, *215*
Kalle, K., see Dietrich, G. et al.
Kalleberg, H., 295, 296, 297, *345*
Kallings, L.O., 360, *378*
Kändler, R., 312, 326, 328, *345*; 361, *378*
Kändler, R. and Dutt, S., 284, *345*
Kändler, R. and Pirwitz, W., 321, 324, 326, 327, *345*
Kändler, R. and Thurow, F., 322, 328, *345*
Kangas, P., see Hällfors, G. et al., see also Lappalainen, A. et al., see also Luther, H. et al.
Karlgren, L., 361, *378*
Karlgren, L. and Ljungström, K., 372, *378*
Karlsson, S. and Larsson, C., 307, *345*
Karlström, Ö., 295, 296, 297, 302, 305, *345*
Kashin, L., see Boulanger, Ju. et al.
Kautsky, H., see Wulff, F. et al.
Kautsky, N., see Jansson, A.-M. and Kautsky, N.
Kay, H., 212, *215*
Kell, V., 228, *269*
Ketchum, B.H., see Redfield, A.C. et al.
Keunecke, K.H., see Kielmann, J. et al.
Kielmann, J., Krauss, W. and Keunecke, K.H., 153, 155, *177*
Kielmann, J., Krauss, W. and Magaard, L., 153, *177*
Kielmann, J., see Krauss, W. et al.
King, R.J. and Schramm, W., 236, *269*
Kinne, O., 248, *269*
Klintberg, T., see Wulff, F. et al.
Knudsen, M., 183, *215*
Koczy, F., see Höglund, H. and Koczy, F.
Kögler, F.-C. and Larsen, B., 58, 66, 100, *119*
Kolp, O., 55, 59, *119*
Kondratovich, E., see Antonovicha, L. et al.

Korhonen, M., see Halme, E. and Korhonen, M.
Kornaś, J., 233, *269*
Kornaś, J. and Medwecka-Kornaś, A., 233, *269*
Kornaś, J., Pancer, E. and Brzyski, B., 236, *269*
Koroleff, F., 199, 204, 205, 208, *215*
Koroleff, F., see Sen Gupta, R. and Koroleff, F.
Koske, P., see Krauss, W. et al.
Kostrichkine, E.M. and Starodub, M.L., 285, 287, *345*
Kouvo, O., see Simonen, A. and Kouvo, O.
Kowalik, Z. and Taranowska, S., 153, *177*
Kowalik, Z., see Sarkisyan, A.S. et al.
Krause, G., 139, *177*
Krauskopf, K.B., 202, *215*
Krauss, W., 153, 154, 155, 156, 158, 159, *177*, *178*
Krauss, W., Koske, P. and Kielmann, J., 161, *178*
Krauss, W. and Magaard, L., 154, *178*
Krauss, W., see Dietrich, G. et al., see also Kielmann, J., et al., see also Magaard, L. and Krauss, W.
Kremling, K., 184, 185, 186, 187, 188, 204, *215*
Kremling, K. and Petersen, H., 205, *215*
Kremling, K., see Weigel, H.-P. and Kremling, K.
Kremser, U. and Matthäus, W., 162, 164, *178*
Kremser, U., see Matthäus, W. and Kremser, U.
Krongård Kristensen, K., 360, *378*
Krzykawska, I., see Zalachowski, W. et al.
Krzykawski, S., see Zalachowski, W. et al.
Kublickas, A., Zheltenkova, M.V., Erm, V. and Murzabekova, N.M., 308, *345*
Kühnel, I., 174, *178*
Kühnhold, W.W., 371, *378*
Kukkamäki, T.J., 21, *119*
Kukkonen, H., see Ignatius, H. et al.
Kullenberg, G., 147, 150, 164, 165, 166, 167, 169, *178*
Kullenberg, B. and Hela, I., 152, 153, 156, 159, *178*
Kullenberg, B. and Sen Gupta, R., 187, *215*
Kullenberg, B., see Denman, K.L. and Kullenberg, B., see also Gustafsson, T. and Kullenberg, B.
Kumpas, M., 36, 51, *119*
Kwiecinski, B., 185, *215*

Laevastu, T., 173, *178*
Lamp, F., see Bagge, O. et al.
Landner, L., Nilsson, K. and Rosenberg, R., 361, 373, *378*
Landner, L., see Rosenberg, R. et al.
Lappalainen, A., Hällfors, G. and Kangas, P. 234, 236, *269*
Lappalainen, A., see Hällfors, G. et al., see also Luther, H. et al.
Larsen, B., see Kögler, F.-C. and Larsen, B.
Larsson, C., see Karlsson, S. and Larsson, C.
Larsson, H.-O. and Larsson, P.-O., 298, *345*
Larsson, P.-O., 301, *345*
Larsson, P.-O., see Larsson, H.-O. and Larsson, P.-O.
Larsson, U., see Hobro, R. et al.
Lassig, J., Leppänen, J.-M., Niemi, Å. and Tamelander, G., 221, 222, 223, 237, *269*
Lassig, J. and Leppäkoski, E., 254
Lassig, J. and Niemi, Å., 227, *269*
Lassig, J., see Andersin, A.-B. et al., see also Hällfors, G. et al.
Launiainen, J. and Danielson, R., 206
Laurén, L., Lehtovaara, J., Boström, R. and Tynni, R., 4, *119*
Lawacz, W., see Pęcherzewski, K. and Lawacz, W.
Lee, A.J., see Wooster, W.S. et al.
Lehmusluoto, P.O., 221, *269*
Lehmusluoto, P.O. and Pesonen, L., 360, *378*
Lehnusluoto, P.O., see Bagge, P. and Lehmusluoto, P.O.
Lehtonen, H., 338, 339, *345*, *346*
Lehtonen, H. and Toivonen, J., 333
Lehtovaara, J., see Laurén, L. et al.
Lemmetijinen, R., 370, *379*
Lemons, F., see Rasmus, J. et al.
Lenz, W., 137, *178*; 251, *269*
Leppäkoski, E., 255, 260, *269*; 361, *379*
Leppäkoski, E., see Hällfors, G. et al., see also Lassig, J. and Leppäkoski, E.
Leppänen, J.-M., see Lassig, J. et al.

Leppäranta, M., see Valli, A. and Leppäranta, M.
Levring, T., 230, 233, *269*
Liljequist, G., 169
Liljequist, G., see Jerlov, N.G. and Liljequist, G.
Lillienberg, D., Setounskaya, L., Blagoboline, N., Bylinskaya, L., Gorelov, S., Nikonov, A., Rozanov, L., Serebryannyi, L. and Filkine, V., 21, *119*
Lillienberg, D., see Boulanger, Ju. et al.
Lindahl, G., Wallström, K. and Brattberg, G., 226, *269*
Lindahl, O., 221, 223, 227, 242, 247, 252, *269*
Lindahl, O., see Ackefors, H. et al., see also Ackefors, H. and Lindahl, O.
Lindahl, P.E.B., see Melen, K.E.R. and Lindahl, P.E.B.
Lindén, O., 371, *379*
Lindgren, L., 360, *379*
Lindner, A., 233, *269*
Lindquist, A., 143, *178*; 243, *269*; 310, *346*
Lindroth, A., 292, 295, 296, 297, 298, 299, 301, 302, 303, 305, 306, 307, 309, 311, *346*
Lindroth, A., see Fabricius, E. and Lindroth, A.
Lindroth, P. see Josefsson, B. et al.
Linko, R., Kaitaranta, J., Ratamäki, P. and Eronen, L., 363, *379*
Lishev, M.N. and Reins, E.Ja., 295, 297, 298, 302, *346*
Lisitzin, E., 146, 155, *178*
Liszkowski, J., 21, *119*
Lithner, G., 267, *379*
Lithner, G. and Samberg, H., 367, 368, *379*
Litvin, V.M., see Gudelis, V.K. and Litvin, V.M.
Ljungström, K., see Karlgren, L. and Ljungström, K.
Logvinenko, N.V., Barkov, L.K. and Gontarev, E.A., *97*, *119*
Lohmann, H., 230, *269*
Lomniewski, K., Mankowski, W. and Zaleski, J., 290, *346*
Lowenstam, H.A. and McConnell, D., 187, *215*
Luckas, B., Berner, M. and Pscheidl, H., 262, *379*
Luke, M.P., see Jurkovskies, A.K. and Luke, M.P.
Lund, E., 360, *379*
Lundgren, O.R., 162, *178*
Lundgren, B., 168, *178*
Lundqvist, G., 57, *119*
Lundqvist, G., see Magnusson, N. et al.
Lundstedt, K., 212, *215*
Luther, H., 230, 232, 233, 236, *270*
Luther, H., Hällfors, G., Lappalainen, A. and Kangas, P., 236, *270*

Määr, A., 310, 311, *346*
Maë, H., 156, 158, *178*
Magaard, L., 154, 160, *178*
Magaard, L. and Krauss, W., 154, 155, *178*
Magaard, L., see Kielmann, J. et al., see also Krauss, W. and Magaard, L.
Magnusson, N., Lundqvist, G. and Grandlund, E. 55, *119*
Mälkki, P., 153, 156, 159, *178*
Mälkki, P., see Voipio, A. and Mälkki, P.
Manheim, F.T., 107, 113, *119*; 207, 209, *216*
Mankowski, W., 285, 286, *346*
Mankowski, W., see Lonniewski, K. et al.
Mardla, A., see Brangulis, A. et al.
Martinsson, A., 30, 31, 84, *119*
Matthäus, W., 142, 143, 144, 146, 162, 163, 164, 171, 173, *178*; 191, 192, 193, *216*
Matthäus, W. and Kremser, U., *179*
Matthäus, W., see Kremser, U. and Matthäus, W., see also Sturm, M. et al.
Mattisson, I., see Ehlin, U. et al.
McCarthy, J.J., 199, *216*
McConnell, D., see Lowenstam, H.A. and McConnell, D.
Medwecka-Kornaś, A. see Kornaś, J. and Medwecka-Kornaś, A.
Melin, K.E.R. and Lindahl, P.E.B., 226, *270*
Melin, T., see Olausson, E. et al.
Mellgren, L., 360, *379*
Mellin, T., see Olausson, E. et al.
Melvasalo, T., 220, 229, *270*
Melvasalo, T. and Viljamaa, H., 361, *379*
Melvasalo, T., Viljamaa, H. and Huttunen, M., 220, *270*
Melvasalo, T., see Rinne, I. et al.
Menge, J.L., 230, *270*
Mens, K., see Brangulis, A. et al.

Miskues, E., 221, 227, *270*
Michanek, G., 221, *270*
Miettinen, J., see Simola, K. et al.
Miettinen, V., 368, *379*
Mikhailova, L., see Banarescu, P. et al.
Mikulski, Z., 123, 125, 126, 129, *134*
Mikulski, Z., see Falkenmark, M. and Mikulski, Z.
Mironov, C.N., see Petipa, T.S. et al.
Miyaki, M., see Denman, K.L. and Miyaki, M.
Möbius, K. and Heincke, F.-R., 311, *346*
Mohr, H., 375, *379*
Molin, G., see Svärdson, G. and Molin, G.
Møller, D., 293, 346
Mörner, N.-A., 21, 55, 56, 58, *120*
Mörner, N.-A., Flodén, T., Beskow, B., Elhammar, A. and Haxner, H., 97, *120*
Moroz, V.N., see Papadopol, M. et al., see also Sukhanova, E.R. et al., see also Volskis, R. et al.
Morozov, N.P., Demina, L.L., Sokolova, L.M. and Prohoryčeva, N.P., 204, 205, *216*
Morris, A.W. and Riley, J.P., 185, *216*
Mulicki, Z., 322, 324, *346*
Müller, A. and Bagge, O., 315, *346*
Müller, A., see Bagge, O. and Müller, A.
Munch-Petersen, S., *346*
Munthe, H., 61, *120*; 275, *347*
Murzabekova, N.M., see Kublickas, A. et al.
Muus, B.J., 255, *270*

Naglis, A.K., see Ojaveer, E. et al.
Nehring, D. and Francke, E., 193, *216*
Nehring, D. and Rohde, K.-H., 184, *216*
Nehring, D., see Francke, E. and Nehring, D.
Nellen, W., see Hempel, G. and Nellen, W.
Netzel, J., 313, *347*
Neuhaus, E., 339, *347*
Neumann, E., 146, 154, 155, 159, 161, *179*; 372, *379*
Nielsen, A., 155, *179*
Nielsen, E. 330, *347*
Niemi, Å., 197, 201, 202, *216*; 221, 223, 224, 225, 226, 229, *270*
Niemi, Å. and Ray, I.-L., 227, *270*
Niemi, Å., see Bagge, P. and Niemi, Å., see also Hällfors, G. and Niemi, Å., see also Hällfors, G. et al., see also Lassig, J. et al., see also Lassig, J. and Niemi, Å., see also Pietikäinen, S. et al., see also Rinne, I. et al.
Niemistö, L., Tervo, V. and Voipio, A., 102, 103, *120*
Niemistö, L. and Tervo, V., 102, *120*; 201, 206, 207, *216*
Niemistö, L., see Ignatius, H. et al., see also Ignatius, H. and Niemistö, L., see also Rinne, I. et al., see also Simola, K. et al., see also Tarkiainen, E. et al., see also Voipio, A. and Niemistö, L.
Nikolaev, I.I., 227, *270*
Nikonov, A., see Lillienberg, D. et al.
Nilsson, E., 59, *120*
Nilsson, H. and Swanson, A., 147, *179*
Nilsson, K., see Landner, L. et al., see also Rosenberg, R. et al.
Nordenberg, B., 375, *379*
Norin, L. and Waern, M., 361, *379*
Nowicka, B., see Buczowska, Z. and Nowicka, B.
Nygård, K., see Jerlov, N.G. and Nygård, K.
Nijman, L., 372, *379*
Nijman, L., see Westin, L. and Nijman, L.
Nyguist, G., 214, *216*
Nygvist, B., see Hobro, R. and Nygvist, B.

Ochocki, S., see Renk, H. et al.
Odén, S., 211, *216*
Odsjö, T., 363, *379*
Odum, H.T., 251, *270*
Oertzen, van, J.A., 257, *270*
Ojaveer, E., 275, 276, 277, 278, 282, 283, 284, 285, 287, 288, 331, *347*
Ojaveer, E., Evtjuchova, B.K. and Naglis, A.K., 289, 291, *347*
Ojaveer, E. and Simm, M., 277, 283, *347*
Okubo, A., 167, *179*
Olausson, E., Gustafsson, O., Mellin, T. and Svensson, R., 102, *120*; 206, 207, *216*
Olsson, B., 236, *270*
Olsson, M., 363, *379*
Olsson, M., Jensen, S. and Renberg, L., 364, *379*
Olsson, M., see Helle, E. et al., see also Jensen, S.
Olszewski, J., see Dera, J. and Olszewski, J.
Omstedt, A., see Udin, I. and Omstedt, A.
Öpik, A.A., 32, *120*
Österdahl, L., 297, 298, *347*

Österdahl, L., see Södergren, S. and Österdahl, L.
Östling, G., see Josefsson, B. et al.
Öström, B., 226, *270*
Otterlind, G., 278, 279, 281, 287, 313, 314, 320, 324, *347*; 361, *379*
Otterlind, G., see Bagge, O. et al., see also Jamieson, A. and Otterlind, G., see also Jensen, S. et al.
Overbeck, J., 233, *271*
Ozolinš, J., see Zarinš, E. and Ozolinš, J.

Palmén, E., 150, 151, 154, 156, 157, 158, 171, 173, *179*
Palmén, E. and Söderman, D., 128, *134*
Palosuo, E., 138, 139, 173, 174, *179*, 219, *271*
Panasenko, J.D., 22, *120*
Pancer, E., see Kornaś, J. et al.
Pankow, H., 221, *271*
Papadopol, M., Volskis, R., Moroz, V.N. and Vladimirov, M.Z., 299, *347*
Papadopol, M., see Banarescu, P. et al.
Papunen, H., 67, *120*; 209, *216*
Parkkonen, L., see Andersin, A.-B. et al.
Parmanne, R., see Sjöblom, V. and Parmanne, R.
Parson, T.R., Takahashi, M. and Hargrave, B., 224, *271*
Pautsch, F., 371, *379*
Pavlova, E.V., see Petipa, T.S. et al.
Pearson, T.H. and Rosenberg, R., 258, *271*
Pęcherzewski, K., 200, *216*
Pęcherzewski, K. and Lawacz, W., 212, *216*
Pedersen, F.B., 160, *179*
Pelkonen, K. and Tulkki, P., 370, *379*
Penttilä, E., 22, *120*
Pesonen, L., see Lehmusluoto, P.O. and Pesonen, L.
Petersen, C.G.J., 264, *271*
Petersen, H., see Kremling, K. and Petersen, H.
Peterson, H., 299, 306, *347*
Petipa, T.S., Pavlova, E.V. and Mironov, C.N. *271*
Petrén, O. and Walin, G., 160, *179*
Petterson, O., 190, *216*
Piechura, J., 162, *179*
Pietikäinen, S., Niemi, Å., Tulkki, P. and Aurimaa, K., 227, *271*
Pilo, C., see Blomquist, Å. et al.
Pirrus, E., see Brangulis, A. et al.
Pirwitz, W., see Kändler, R. and Pirwitz, W.
Plissov, A.A., Gorijanskii, V.Ju., Vanderflit, E.K. and Sapoznikova, P.S., 43, *120*
Polivajko, A.G., see Veldre, I. and Polivajko, A.G.
Polivkov, I.A., see Volkolakov, F.K. et al.
Pomeranec, K.S., 171, 172, *179*
Popiel, J., 277, 278, 282, 285, 286, 287, *347*
Popiel, J. and Strzyżewska, K., 281, 289, *348*
Popiel, J., see Elwertowski, J. and Popiel, J., see also Strzyżewska, K. and Popiel, J.
Potter, I.C., see Hardisty, M.W. and Potter, I.C.
Poulsen, E., 279, 283, 291, *347*
Pratje, O., 86, *120*
Prohoryčeva, N.P., see Morozov, N.P. et al.
Pscheidl, H., see Luckas, B. et al.
Pustelnikov, O.S., 101, *120*; 209, *217*

Quadfased, D., see Schott, F. et al.

Ramah, L.A., 277, 282, 284, 288, 289, *347*
Rankama, K., 1, 33, *120*
Rankama, K. and Sahama, T.G., 202, *217*
Rasmus, J., Lemons, F. and Zimmerman, C., 230, *271*
Rasmussen, E., 373, *380*
Rasmussen, J., see Gargas, E. et al.
Ratamäki, P., see Linko, R. et al.
Rautiainen, H. and Ravanko, O., 237, *271*
Ravanko, O., 232, 236, *271*; 370, *380*
Ravanko, O., see Rautiainen, H. and Ravanko, O.
Ray, I.-L., see Niemi, Å. and Ray, I.-L.
Rechlin, O., 282, 288, 289, *348*
Rechlin, O. and Friess, C.-C., 287, 289, 291, *348*
Redfield, A.C., Ketchum, B.H. and Richard, F.A., 199, *217*
Remane, A., 242, 255, *271*; 275, *348*
Renberg, L., see Olsson, M. et al.
Renk, H., 221, *271*
Renk, H., Torbicki, H. and Ochocki, S., 221, *271*
Rheinheimer, G., see Bansemir, K. and Rheinheimer, G.

Richard, F.A., see Redfield, A.C. et al.
Ricker, W.E., 300, *348*
Riley, J.P. see Cox, R.A. et al., see also Greenhalgh, R. and Riley, J.P., see also Morris, A.W. and Riley, J.P.
Ringer, Z., 227, 228, *271*
Rims, E.Ja., see Lishev, M.N. and Rims, E.Ja.
Rinne, I., 198, *217*
Rinne, I., Melvasalo, T., Niemi, Å. and Niemistö, L., 198, *217*; 226, *271*
Rinne, I. and Tarkiainen, E., 226, *271*
Rinne, I., see Tarkiainen, E. et al.
Rohde, K.-H., 184, *217*
Rohde, K.-H., see Nehring, D. and Rohde, K.-H.
Rönnberg, O., 232, 236, *271*; 375, *380*
Rothe, F., 227, *271*
Rosén, B., 305, *348*; 373, *380*
Rosén, C.-G., see Ackefors, H. and Rosén, C.-G.
Rosenberg, R., Nilsson, K. and Landner, L., 361, *380*
Rosenberg, R., see Landner, L. et al., see also Pearson, T.H. and Rosenberg, R.
Rozanov, L., see Lillienberg, D. et al.
Rubey, W.W., 185, *217*
Ruppin, E., 193, *217*
Rutkowicz, S., 313, *348*
Rydälv, M., see Westöö, G. and Rydälv, M.

Sabaljauskas, V., see Brangulis, A. et al.
Sahama, T.G., see Rankama, K. and Sahama, T.G.
Salo, A. and Saxén, R., 212, *217*
Salo, A., see Voipio, A. and Salo, A.
Samberg, H., see Lithner, G. and Samberg, H., see also Wulff, F. et al.
Sandler, H., see Andersin, A.-B. et al.
Sandman, J.A., 321, *348*
Sandström, O., see Hansson, R. and Sandström, O.
Sapoznikova, P.S., see Plissov, A.A. et al.
Sarkisyan, A.S., Stashkevich, A. and Kowalik, Z., 160, *179*
Sawala, J., 253, *271*
Sauramo, M., 55, 57, 58, 59, 60, 61, 63, 65, *120*
Savidge, G. and Hutley, H.T., 199, *217*
Saxén, R., see Salo, A. and Saxén, R.
Schippel, F.A., see Hallberg, R.O. et al.
Schmidt, D., 204, *217*
Schmidt, J., 312, *348*
Schnack, S., 251, *272*
Schnese, W., 229, *272*
Scholz, N., see Theede, H. et al.
Schopka, S.A., 316, *348*
Schott, F., Ehlers, M., Hubrich, L.M. and Quadfased, D., 166, *179*
Schramm, W., 236, *272*; 370, *380*
Schramm, W. and Guterstam, B., 236, *272*
Schramm, W., see Schwenke, H. et al.
Schroeder, H., see Gargas, E. et al.
Schultz, S., 127, *134*
Schulz, S., 361, *380*
Schulz, S., see Kaiser, W. and Schulz, S.
Schwarz, S., 244, *272*
Schwenke, H., 230, 232, 233, 234, *272*
Schwenke, H., Schramm, W. and Black, H.J., 237, *272*
Segerstrale, S.G., 236, 242, 255, 264, *272*
Seleckaja, A.V., 279, *348*
Sendner, H., see Joseph, J. and Sendner, H.
Sen Gupta, R., 196, 197, 198, 199, 200, 204, 205, *217*; 221, *272*
Sen Gupta, R. and Koroleff, F., 199, 200, *217*
Sen Gupta, R., see Kullenberg, B. and Sen Gupta, R.
Seppänen, H. and Voipio, A., 210, *217*
Serebryannyi, L., see Lillienberg, D. et al.
Setounskaya, L., see Boulanger, Ju. et al., see also Lillienberg, D. et al.
Shadrin, J., see Druet, Cz. et al., see also Wojewodzki, T. et al.
Shaffer, G., 156, 158, 165, *179*
Sick, K., 312, *348*
Siedler, G., see Dietrich, G. et al.
Siira, J., see Alasaarela, E. and Siira, J.
Siivola, J., see Winterhalter, B. and Siivola, J.
Sillén, L.-G., 187, *217*
Simojoki, H., 127, 128, *134*; 162, 164, 171, 172, *180*
Simola, K., Jaakkola, T., Miettinen, J., Voipio, A. and Niemistö, L., 371, *380*
Simm, M., see Ojaveer, E. and Simm, M.
Simonen, A., 1, *120*
Simonen, A. and Kouvo, O., 25, *120*
Sjöberg, K., 308, 309, *348*
Sjöblom, V., 277, 282, 287, *348*, *382*
Sjöblom, V. and Parmanne, R., 237, *272*; 287, 288, 291, 313, 314, *348*

Sjöblom, V., see Häsänen, E. and Sjöblom, V.
Skärlund, K., see Wulff, F. et al.
Skopintsev, B.A., 212, *217*
Skvorcova, T.A., 276, *348*
Smetacek, V., 220, 228, *272*
Smith, M.A.K. and Thorpe, A., 339, *348*
Södergren, S. and Österdahl, L., 298, *349*
Södergren, S., see Asplund, C. and Södergren, S.
Söderholm, B., see Tuominen, H.V. et al.
Söderman, D., see Palmén, E. and Söderman, D.
Soikkeli, M. and Virtanen, J., 370, *380*
Sokolova, L.M., see Morozov, N.P. et al.
Somer, E., 365, 366, *380*
Somer, E. and Appelqvist, H., 368, *380*
Sörlin, T., see Wulff, F. et al.
Soskin, I.M., 130, 131, 132, *134*; 143, 147, 148, *180*; 193, *217*
S.O.U., 357, *380*
Spirina, L.I., see Erm, V. et al.
Starodub, M.L., see Kostrichkine, E.M. and Starodub, M.L.
Stashkevich, A., see Sarkisyan, A.S. et al.
Steemann-Nielsen, E., 132, *134*; 221, *272*
Steffner, N., see Hagelin, L.-O. and Steffner, N.
Stewart, G.L., see Beers, J.R. and Stewardt, G.L.
Stewart, W.D.P. and Alexander, G., 226, *272*
Stewart, W.D.P., see Jones, K. and Stewart, W.D.P.
Stigebrandt, A., 159, *180*
Strickland, J.D.N., 251, *272*
Strzyżewska, K., 284, 287, 291, *349*
Strzyżewska, K. and Popiel J., 287, 288
Strzyżewska, K., see Popiel, J. and Strzyżewska, K.
Sturm, M., 171, 172, *180*
Sturm, M., Francke, E. and Matthäus, W., 173, *180*
Suess, E. and Erlenkeuser, H., 102, *120*; 206, 207, *217*
Suess, E., see Erlenkeuser, H. et al.
Šuiskii, Ju. D., see Blažčišin, A.I. et al.
Sukhanova, E.R., Voskis, R., Morox, V.N. and Erm, V., 308, *348*
Sukhanova, E.R., see Volskis, R. et al.
Suskina, A.P., 277, 285, 286, *349*
Svansson, A., 131, *134*; 155, *180*; 353, *380*
Svansson, A., see Nilsson, H. and Svansson, A.
Svärdson, G., 299, 305, 306, 307, 310, 311, *349*
Svärdson, G. and Anheden, H., 305, *349*
Svärdson, G. and Molin, G., 339, *349*
Svedelius, N., 230, *272*
Svensson, R., see Olausson, E. et al.
Sverdruk, H.U., 224, *272*
Swanson, A., see Nilsson, H. and Swanson, A.
Szekielda, K.-H., 212, *217*
Szypula, J., see Zalachowski, W. et al.

Takahashi, M., see Parson, T.R. et al.
Talbot, G.A., see Talbot, J.W. and Talbot, G.A.
Talbot, J.W. and Talbot, G.A., 163, *180*
Tamelander, G., see Lassig, J. et al.
Taranowska, S., see Kowalik Z. and Taranowska, S.
Tarkiainen, E., Rinne, I. and Niemistö, L., 103, 104, *121*; 197, *217*; 226, *272*
Tarkiainen, E., see Rinne, I. and Tarkiainen, E.
Tervo, V., 204, 212
Tervo, V., see Niemistö, L. and Tervo, V., see also Niemistö, L. et al.
Theede, H., Scholz, N. and Fischer, H., 369, *380*
Thompsen, H.A., 221, *273*
Thompson, T.G., Johnson, W.R. and Wirth, H.E., 185, *217*
Thompson, T., see Blomquist, A. et al.
Thorpe, A., see Smith, M.A.K. and Thorpe, A.
Thorslund, P., 25, 28, 35, 107, *121*
Thorslund, P. and Axberg, S., 25, *121*
Thorson, G., 239, *273*
Thurow, F., 298, 303, 318, 319, *349*
Thurow, F., see Kändler, R. and Thurow, F.
Tiews, K., see Bagge, O. et al.
Toft, R., 295, *349*
Toivonen, J., 298, 304, 306, 335, 340, *349*
Toivonen, J., see Lehtonen, H. and Toivonen, J.
Thorbicki, H., see Renk, H. et al.
Tornquist, A., 4, *121*
Trahms, O.-K., 229, *273*
Trei, T., 232, 236, *273*
Troitskij, S.K., 308, *349*

Trzosińska, A., 185, *218*
Tulkki, P., 69, 83, 86, 100, 101, *121*; 260, *273*
Tulkki, P., see Pelkonen, K. and Tulkki, P., see also Pietikäinen, S. et al.
Tulley, J.P., 144, *180*
Tuominen, H.V., Aarnisalo, J. and Söderholm, B., 4, *121*
Tynni, R., 2, 24, *121*, 224
Tynni, R., see Ignatius, H. and Tynni, R., see also Laurén, L. et al.

Udin, I. and Ornstedt, A., 174, *180*
Udin, I. and Ullerstig, A., 174, *180*
Ullerstig, A., see Udin, I. and Ullerstig, A.
UNESCO, 183, *218*, *273*
Ungsgard, Y., 375, *380*
Uppström, L.R., see Dyrssen, D.W. and Uppström, L.R.
Utermöhl, H., 220, *273*
Uzars, A., 316, 317, *349*
Uzars, D., see Antonovicha, L. et al.

Välikangas, I., 255, *273*
Valle, K.J., 291, 292, *349*
Valle, A. and Leppäranta, M., 175, *180*
Vanderflit, E.K., see Plissov, A.A. et al.
Varencov, I.J., 107, 113, *121*
Varencov, I.M. and Blažčišin, A.I., 107, 108, *121*; 201, 205, *218*
Vasseur, E., 360, *380*
Veldre, I., 279, 283, 285, 288, *349*
Veldre, I. and Polivajko, A.G., 289, 291, *349*
Veltheim, V., 2, 24, 25, *121*
Viitanen, R., 361, *380*
Viljamaa, H., see Melvasalo, T. and Viljamaa, H., see also Melvasalo, T. et al.
Vintish, M., 320, *329*
Vistanen, J., see Soikkeli, M. and Vistanen, J.
Vladimirov, M.Z., see Papadopol, M. et al., see also Volskis, R. et al.
Voipio, A., 196, 199, 200, 201, 203, 208, *218*; 224, 227, *273*; *383*
Voipio, A. and Mälkki, P., 139, *180*
Voipio, A. and Niemistö, L., 101, *121*
Voipio, A. and Salo, A., 371, *380*
Voipio, A., see Ignatius, H. et al., see also Niemistö, L. and Voipio, A., see also Niemistö, L. et al., see also Seppänen, H. and Voipio, A., see also Simola, K. et al.
Volkolakov, F.K., Polivko, I.A., Agalkova, E.N. and Jokovleva, V.I., 49, 105, 106, *121*
Volskis, R., 308, *350*
Volskis, R., Erm, V., Vladimirov, M.Z., Moroz, V.M., Bontemps, S. and Sukhanova, E.R., 308, *349*
Volskis, R., Morox, V.N. and Sukhanova, E.R., *350*
Volskis, R., see Erm, V. et al., see also Papadopol, M. et al., see also Sukhanova, E.R. et al.
Von Bonsdorff, P.A., 183, *218*
Von Brandt, A., 184, *218*
Vorma, A., 1, *121*
Vyskochil, P., see Boulanger, Ju. et al.

Wachenfeldt, v., T., 360, *380*
Waern, M., 230, 233, 234, *273*
Waern, M., see Norin, L. and Waern, L.
Walin, G., 156, 158, 160, 165, *180*
Walin, G., see Petrén, O. and Walin, G.
Wallentinus, I., 232, 233, 236, *273*; 361, *380*
Wallerius, D., 170, *180*
Wallström, K., see Lindahl, G. et al.
Wanntorp, H., see Borg, K. et al.
Weber, W., *350*
Weidemann, H., 163, *180*
Weigel, H.-P. and Kremling, K., 205, *218*
Weil, J.G., 372, *380*
Welander, P., 160, *180*
Westerberg, H., 375, *381*
Westermark, T., see Jensen, S. et al.
Westin, L. and Nijman, L., 310, *350*
Westöö, G. and Rydälv, M., 367, *381*
Whitton, B.A., 226, *272*
Widborn, B., see Wulff, F. et al.
Wikgren, B.J.P., 309, 338, *350*
Wilkomm, H., see Erlenkeuser, H. et al.
Winterhalter, B., 1, 25, 28, 29, 68, 70, 83, 86, 97, 101, 105, 107, 110, 113, *121*; *218*
Winterhalter, B. and Siivola, J., 107, 112, *121*
Winterhalter, B., see Ignatius, H. et al.
Wittig, H., 184, *218*
Witting, R., 150, 151, 157, 170, *180*; 211, *218*; 360, *381*
Wojewódzki, T., Hupfer, H.A. and Shadrin, J., 161, *181*

Wojtusiak, H., see Bursa, A. et al.
Woltusiak, R.J., see Bursa, A. et al.
Wooster, W.D., Lee, A.J. and Dietrich, G., 183, *218*
Wulff, F., 221, *273*
Wulff, F., Flyg, C., Foberg, M., Hansson, S., Johansson, S., Kautsky, H., Klintberg, T., Samberg, H., Skärlund, K., Sörlin, T. and Widborn, B., 236, 243, 244, *273*
Wulff, F., see Ackefors, H. et al., see also Hobro, R. et al., see also Jansson, B.-O. and Wulff, F.
Wüst, G. and Brogmus, W., 167, *181*
Wyrtki, K., 130, 133, *134*; 139, *181*; 191, *218*
Wyrzykowsky, T., see Boulanger, Ju. et al.
Zachrisson, G., see Ehlin, U. et al., see also Ehlin, U. and Zachrisson, G.
Zalachowski, W., Szypula, J., Krzykawski, S. and Krzykawska, I., 317, *350*
Zaleski, J., see Lomniewski, K. et al.
Zarnecke, S., 305, *350*
Zarinš, E. and Ozolinš, J., 185, *218*
Zemskaya, K.A., 322, *350*
Zenkevitsch, L., 219, 238, 255, 256, *274*
Zheltenkova, M.V., see Kublickas, A. et al.
Zillioux, E.J. and Gonzales, J.G., 245, *274*
Zimmerman, C., see Rasmus, J. et al.
Żmudziński, L., 260, *274*; 361, *381*
Zotin, M., see Boulanger, Ju. et al.
Zsolnay, A., 212, *218*
Zukowski, C., see Cięglewicz, W. et al.

SUBJECT INDEX

Abiotic factors, 249
Abramis ballerus, *see* Bream, silver
Abramis brama, *see* Bream
Acartia, 247, 254, 285
Acartia bifilosa, 240, 242, 243, 245
Acartia longiremis, 240, 242, 243
Accumulation
—, coastal sediments, 86
—, pollutants, 352—353
Achnanthes taeniata, 223, 227
Acoustic pollution, 375
Acrosiphonia centralis, 231
Actiniaria, 256
Actinocyclus ehrenbergii, 62
Actinocyclus octonarius, 226, 228
Adenovirus, 7, 360
Advection, 141, 163
Aerological method, 128
Age of fishes, 287—288, 312—313, 317—318, 322
Ahnfeltia plicata, 230, 231, 234
Åland Deep, 29, 85, 97
Åland Islands, 2,4, 31, 85
Åland Sea, 1, 4, 28—31, 69, 84, 85, 97
Alburnus alburnus, see Bleak
Algae, 231
—, blue-green, 225—228
Alisma gramineum, 235
Alkalinity, 186
Amber, 53, 112, 114, 116
Ammocoetes, 309
Ammodytes tobianus, *see* Sandeel
Ammonia, 197, 224
—, ammonia-nitrogen, 197, 198, 200
Amphipoda, 257, 260, 262, 285, 329
Amphora robusta, 62
Anabaena lemmermannii, 226
Anadromous fishes, 292—311
Anchovy, 292
Ancylus
— Lake, 60, 61—63, 254
— stage, 67
Ancylus fluviatilis, 61
Anguilla anguilla, *see* Eel, European
Anguilla rostrata, *see* Eel, American
Anhydrite, 48
Annelida, 255
Anomoeoneis sphaerophora, 62
Anoxic, 103, 111
— conditions, 191
— water, 208
Anthropochorous, 255
Aphanizomenon flos-aquae, 223, 226, 227, 228
Apparent oxygen utilization (AOU), 196, 200
Archipelago Sea, 2, 31, 84, 85
Arctic char, 333
Arenaceous, 43, 44, 45, 47, 50
Arenicola marina, 324
Arenite, 48, 50
Argillaceous, 43, 44, 45, 46, 48, 51, 52
Argillite, 48, 50, 52
Arkona Basin, 84
Arbose, 1
Aricidea suecica, 261
Artificial stocking of fishes, 301—302
Ascophyllum nodosum, 231
Asp, 333
Aspius aspius, *see* Asp
Assimilation, 221
Astacus astacus, *see* Crayfish
Astarte borealis, 225, 261
Atlantic Ocean, 123
Atmospheric circulation, 148
Attenuation, 168
Aurelia aurita, *see* Medusa

Bacteria, 251, 253
Balanus improvisus, *see* Barnacle
Baltic Ice Lake, 57, 59, 62
Baltic Ice Sea, 58
"Baltic limestone", 28
Baltic Marine Biologists (BMB), 265, 385
Baltic Proper, 2, 33—54, 123
Baltic Shield, 1, 3, 33
Baltic Syneclise, 33, 43, 44, 106
Barium, 206
Barnacle, 257
Baroclinity, 160
Bathymetry, 71—76, 77—82
Bathyporeia, 322
Belone belone, *see* Garpike
Belt Sea, 123

Benthic
— communities, 262
— fauna, 254—265, 361
— —, macrofauna, 258—261
— —, meiofauna, 257, 261, 262
— —, recolonization, 260
— — research, 264—265
— heterotrophic subsystem, 257
— secondary production, 262
— system, 220
— vegetation, 229—238
Von Bertalanffy growth parameters, 318
Billingen, 59
Biochemical oxygen demand (BOD), 354, 355
Biogeochemical processes, 205
Bioherm, 106
Biological oceanography, 219—265
Biotic factors, 249
Bituminous, 46, 48, 105, 107
Bivalves, 254, 255, 256, 324, 332
Bjerknes' circulation theorem, 157
Black goby, 331
Blackfly, 307
Bladderwrack, 338
Bleak, 333, 335
Blicca bjoerkna, *see* Bream, white
Blidingia minima, 231
Blue clay, 2, 31, 32
Böda Hamn drilling, 44
"Boddengewässer", 229, 233
Boric acid, 186
Bornholm, 50, 52, 83
— Basin, 84, 100, 102
Boron, 113, 187
Bosmina, 285
Bosmina coregoni maritima, 240, 242, 243, 245, 247, 248, 251, 254
Bothnian Bay, 2, 4, 23—25, 84, 86, 100, 101
Bothnian Sea, 2, 3, 4, 25—28, 45, 84, 86, 97, 101, 102
Bottoms
—, hard, 233, 234, 236
—, soft, 234, 235, 255, 261
Box models, 159
Bream, 333, 335, 337, 338, 340, 341
—, silver, 333
—, white, 333, 335
Brill, 327
Brittle star, 328
Bromide, 186
Bryopsis plumosa, 231
Bryozoa, 238, 257
Bulk aerodynamic method, 127, 128
Buntsandstein, 50
Burbot, 295, 298, 334, 338, 339, 341
Butterfish, 333

Cadmium, 102, 204, 206, 365, 369, 388
Caesium-137, 371
Cainozoic rocks, 53
Calanus finmarchicus, 240
Calcarous, 46, 53
Calcium, 183, 184, 185, 186
Callithamnion roseum, 231
Callitriche autumnalis, 235
Caloneis amphisbaena, 62
Caloneis latiuscula, 62
Cambrian, 3, 4, 24, 25, 28, 31, 32, 36, 37—45, 105, 106
Campylodiscus echeneis, 59
Cannibalism, 318
Cap rock, 105
Capitella capitata, 262
Caprella, 329
Carassius carassius, *see* Crusian carp
Carbon
—, authigenic, 103
—, C/N ratio, 103
—, C:N:P ratio, 199, 200
— dioxide, 188, 208, 209, 210, 221
—, dissolved organic (DOC), 211, 212, 213
—, "labile", 212
—, total organic (TOC), 211, 212
—, *see also* Organic matter
Carbon-14, 221
Carboniferous, 36, 48
Cardium, 256, 322
—, *see also* Cockle
Cardium edule, 322, 324
Cardium glaucum, 240
Cardium hauniense, 240
Catadromous fishes, 292—311
Catch of fishes, 290—291, 304—305, 319, 323, 341
Cation and anion concentrations, 187
Cell,
— counts, 220
— volume, 220
Cellular type waves, 161
Central Basin (Baltic Proper), 84
Central Swedish end moraine, 56
Centropages, 285
Centropages hamatus, 240, 242, 243,

245, 247, 248, 250, 254
Ceramium rubrum, 231
Ceramium tenuicorne, 231, 232
Ceratium, 228
Ceratium fusus, 227
Ceratium tripos, 227
Chaetoceros, 228
Chaetoceros danicus, 223, 225, 226, 227, 228
Chaetoceros holsaticus, 223, 227
Chaetoceros mitra, 62
Chaetoceros subsecundus, 62
Chaetoceros wighami, 223, 225, 227, 228
Chaetognath, 251
Chalk, 53
Chara aspera, 235
Charophytes, 234, 256
Chekhon, 334
Chemical composition of sea water
— — — — —, anomalies, 183—188
— — — — —, heavy metals, 204
Chemical oceanography, 183—213
Chemical oxygen demand (COD), 212
Chironomida, 256, 257, 307
Chlorinity, 183
Chlorophyceae, 225, 231
Chlorophyll, 221
Chorda filum, 231
Chorda tomentosa, 231
Chromium, 206
Chrysochromulina birgeri, 223
Chrysophyceae, 223
Chub, 334
Ciliates, 253
Circulation, 150, 157, 159
—, transverse, 158
—, vertical, 161
Cladocera, 242, 243, 245, 248, 250, 251, 253, 254, 285
Cladophora belt, 257
Cladophora aegagropila, 230, 234, 235
Cladophora glomerata, 230, 232, 235, 237, 256, 360, 361
Cladophora rupestris, 231
Clam, 262
Clastic
— deposits, 86
— dykes, 31
Clay
—, glacial, 65, 66, 67, 109
—, gyttja-banded, 65
—, homogenous, 59
— sulphide, 59, 66, 67, 69
— transition, 65, 66, 67, 68
Clostridium perfringens, 360
Clupea harengus, *see* Herring
Cnidarian, 248
Coal, 50, 52
Coastal
— boundary layer, 155—157, 158, 161
— processes, 83
— sand, 95
— zone, 172
Coasts
—, morphogenetic types, 70
Cobalt, 113, 114, 205, 206
Cobitidae, 334
Cobitis taenia, *see* Spined loach
Cocconeis disculus, 62
Cocconeis scutellum, 62
Cockle, 262
Cod, 312—319, 329, 330, 331, 332, 384
—, Atlantic, 312
—, Baltic, 312
— liver oil, 363
Codium fragile, 231
Coelosphaerium kuetzingianum, 228
Colour index, 170
Concentration factors, 205
Concretion
—, aggregates, 110
—, composition of, 110
—, concretic layering, 114
—, crusts, 107, 112
—, discoidal, 107, 108, 112, 114
—, ferromanganese, 107—114
—, nucleus, 108
—, pyrite and marcasite, 68
—, spheroidal, 107, 108, 111, 112
—, structure of, 110, 114
—, trace elements, 114
Conference of the Baltic Oceanographers, 385
Conductivity, 183
Convection, 166
Convention, 383, 386
Copepoda, 239, 242, 245, 248, 251, 253, 254, 285, 307
Copper, 102, 113, 114, 204, 205, 206, 209
Corallina officinalis, 231
Coregonus albula, *see* Vendace
Coregonus lavaretus, *see* Whitefish
Coregonus nasus, 333
Coregonus peled, *see* Whitefish, peled
Coriolis effect, 150, 157, 158

Corophium, 325, 329
Coscinodiscus, 227
Coscinodiscus granii, 226, 228
Coscinodiscus lacustris, 62
Cottidae, 334
Cottus gobio, *see* Miller's thumb
Cottus poecilopus, *see* Mottlefoot sculpin
Crangon crangon, *see* Shrimp
Crayfish, 310
Cretaceous, 36, 50, 52—53, 115
Crinoidea, 256
Crusian carp, 334
Crustaceae, 255, 256, 257, 261, 262, 285, 291, 308, 322, 325, 333
Crustal
— downwarping, 58
— uplift, 4, 21, 86
— —, *see also* Land uplift
Cryptophyceae, 223, 225
Crystalline
— basement, 1, 5—8, 24, 33
— complex, 23
Ctenolabrus rupestris, *see* Gold sinny
Current, 151, 152, 154, 159, 160
—, baroclinic, 153
—, barotropic, 153
— measurements, 161
—, rip, 162
— spectra, 153
— velocities, 150
—, wind-induced, 157
Cyanea capillata, 240
Cyanide, 374
Cyclops, 243
Cyclopterus lumpus, *see* Lumpsucker
Cyclostome, 308
Cymatopleura elliptica, 62
Cymbella prostata, 62
Cyprina islandica, 328
Cyprinidae, 307, 333, 334, 335, 340
Cyprinus carpio, 334

Dab, 327—328
Dace, 334
Danian, 53
Danish Sounds, 59, 61, 69, 83, 96, 123, 129, 352
Danish Straits, *see* Danish Sounds
Danish-Polish Depression, 50, 51, 53, 106
Daphnia, 243
Daphnia cristata, 243
DDT, 102, 362, 364, 373, 387
Deglaciation, 55, 57
Delesseria sanguinea, 231
Demersal fishes, 311—333
Denitrification, 199
Density
—, maximum, 141
—, stratification, 149—150
Depression, 69
Derbesia marina, 231
Desmarestia aculeata, 231
Detonula confervacea, 228
Detritus, 230, 251
Devonian, 36, 48, 106
Diabase, 28
Diaptomus, 243
Diastylis rathkei, 261, 322, 329
Diatactic, 64, 65, 66
Diatom, 223, 225, 227, 237, 252, 285
—, frustules, 196, 202
—, stratigraphy, 58
Diatoma elongatum, 223, 227
Diatoma vulgare, 227
Dictyosiphon chordaria, 231
Dictyosiphon foeniculaceus, 231, 232
Dieldrine, 363
Diffusion, 164, 166, 167
Dinobryon balticum, 225, 227, 228
Dinobryon petiolatum, 225
Dinoflagellates, 223, 225, 227
Dinophysis, 228
Dinophysis acuminata, 225
Dinophysis acuta, 227
Dinophysis norvegica, 225
Diploneis didyma, 62
Diploneis domblittensis, 62
Diploneis interrupta, 62
Diploneis mauleri, 62
Diploneis smithii, 62
Dipteran, 257
Distribution of fishes, 320, 323, 325
Diversity, 238, 254, 261
Dolomites, 48
Downtonian, 47—48
Drainage basin, 123
Drumlin, 54
Dumping of solids, 373
Dykes, 1

Earthquakes, 4, 22
East European platform, 1, 3, 4, 33, 46, 48, 53
East European sedimentary complex, 1
Echeneis Sea, 59
Echinodermata, 238

Echiuroidea, 256
Ecosystem, 219
—, model, 236
—, sensitivity of, 353
Ectocarpus siliculosus, 231, 232
Eel, 332, 334
—, American, 309
—, European, 292, 309—311
—, silver, 310
—, yellow, 310
Eelpout, 331
Eemian, 54, 60
—, microfossils, 58
Eggs of fishes, 284, 295, 306—307, 308, 313, 315—316, 320, 321, 322, 324, 326, 327, 328, 330, 331, 332, 339
Ekman theory, 157
Elatine hydropiper, 235
Electra crusculenta, 257
Eleocharias acicularis, 234, 235
Energy,
— balance method, 128
— circuit language, 251
— exchange, 173
— flow, 251, 252
— flux, 173
Engraulis encrasicholus, *see* Anchovy
Eningi-Lampi Lake, 113
Enteromorpha, 232
Enteromorpha intestinalis, 231
Enteromorpha linza, 231
Enteroviruses, 360
Eocene, 53, 115
Epiphytes, 237
Epithemia hyndmanni, 62
Epithemia turgida, 62
Epizootics, 291
Eriocheir sinensis, 255
Erosion, 86
Escarpment, 43, 45, 84
Esker, 97, 99, 100
Esocidae, 333
Esox lucius, *see* Pike
Estuarine, 145, 160
Eudesme virescens, 231
Euglenophyceae, 225
Eunotia clevei, 62
Euphotic layer, 224, 226
Euryhaline species, 247, 248
Eurytemora, 240, 242, 243, 245, 247, 254, 285
Eurytherm, 245
Eustatic, 59, 69, 86
Eutreptia, 225
Eutreptiella, 225
Eutrophication, 238, 353—361
Evadne, 251, 285
Evadne nordmanni, 240, 242, 243, 245, 247, 248, 251, 253
Evaporation, 126—129, 172, 173
Exchange,
— coefficient for oxygen, 164
— coefficient for salt, 165
— of heat, 172
— of water, 147
Exploitation
— of fishes, 302, 311, 318, 323, 325, 327
—, sand and gravel, 374—375
—, *see also* Natural resources

Fårö, 43
Father lasher, 330
Faults, 4, 13—20, 52, 83
Fecal bacteria, 360
Fecundity of fishes, 284, 295, 306, 316, 321, 324, 326, 327, 331, 332
Feeding of fishes, 285—287, 306, 307, 308, 316, 322, 324, 326, 327, 329
Fennoscandian Border Zone, 33, 50, 52, 53
Fifteen-spined stickleback, 291, 298
Fish, 275—341
— fauna, 275—276
— taste, 361, 371
Fishery, 275—341, 383
— zones, 386
Fishing gear, 290, 302—303, 305, 307, 311, 331, 340—341
Flagellates, 224, 285
Flatfish, 323
Flint, 53
Flounder, 320—323, 360, 386
—, bank flounder, 326
Fluorescence, 168, 169
Fluoride, 186, 187
Flux,
— of energy, 173
— of sensible heat, 172
— of various substances, 159
Fontinalis, 234, 235
Food, *see* Feeding of fishes
Four-bearded rockling, 329—330
Four-horned cottus, 330—331
Fractures, 4, 36, 83, 86
Freezing point, 141
Fresh-water fishes, 333-341

Frictional force, 157
Fritillaria borealis, 240, 242, 243, 245, 247, 248, 252
Fry, 301, 307
Fucus, 338
Fucus serratus, 230, 231, 233
Fucus spiralis, 21
Fucus vesiculosus, 230, 231, 232, 233, 234, 237, 257
Furcellaria fastigiata, 231, 232

Gadidae, 334
Gadus morhua, *see* Cod
Gammarus, 257, 306, 307, 322
Garnet, 117
Garpike, 291
Gas
— containers, 374
— vacuoles, 226
Gasterosteidae, 334
Gasterosteus aculeatus, *see* Three-spined stickleback
Gastropoda, 240, 254, 255, 256, 257, 292, 306, 307
Gävle Bay, 2, 25, 97, 99
Gävle sandstone, 25
Gdansk Depression, 69, 83, 84
Gelbstoff, *see* Yellow substance
Geolittoral, 233
Geology, 1—117
—, pre-quaternary, 1—54
—, quaternary, 54—104
Geomorphology, 69
GESAMP, 351
Glacial,
— deposition, 69, 86
— drift, 54, 83, 97, 98
— erosion, 69
— gouging, 54
— isostasy, 58
— scouring, 83
Glaciation
—, Pleistocene, 54
—, Weichselian, 55, 58
Glaciofluvial, 95
Glasiogenic sediments, 230
Glauconite, 46, 53, 112, 114, 116
Gneiss, 23
Gobio gobio, *see* Gudgeon
Gobius minutus, *see* Sand goby
Gobius niger, *see* Black goby
Gold sinny, 333
Gomposphaeria lacustris, 225
Gonyaulax catenata, 223, 227
Gonyaulax triacantha, 225
Goosander, 295, 308
Gothian, 1
Gotland, 2, 44, 46, 47, 97, 105
— Deep, 55, 83, 84, 100, 102, 103, 111
— Sea, 84
Gotska Sandön, 2, 34, 44, 84, 97
Grammatophora oceanica, 62
Granite, 23
Grayling, 292, 299, 306, 333, 335, 337, 339
Grazing, 254
Greater sandeel, 328—329
Growth of fishes, 287—288, 317, 318, 322, 325, 330, 339
Gudgeon, 334
Gulf of Bothnia, 2, 62, 84, 85, 113, 190
Gulf of Finland, 2, 31—33, 62, 84, 113, 123, 190
Gulf of Gdansk, 48
Gulf of Riga, 2, 84, 97, 113, 123
Gull, 308
Gypsum, 48
Gymnocephalus cernua, *see* Ruff
Gyrosigma attenuatum, 62

Haemoglobin, 312
"Haffs", 229
Hailuoto, 24
Halicryptus spinulosus, 255
Halocline, 95, 143, 144, 145
—, permanent, 145
—, primary, 138
—, secondary, 139, 145
Hanö Bay, 2, 44, 50, 51, 52, 53
Harmothoë, 317
Harmothoë sarsi, 240, 242, 248, 262, 316, 322, 324, 329
Härnösand Deep, 70, 86, 97
Harpactocoida, 262
Hatching, 301, 307
Heat
— balance, budget, 170—173
— content, 171
— exchange, 163, 171, 172
— storage, 173
— transfer, 162, 164
Heavy metals, 102, 368
— content in sediments, 206
— —, *see also* Trace metals
Herring, 277, 279, 281, 285, 287, 291, 298, 316

—, Atlantic, 276
—, autumn, 276, 277, 278, 282, 283, 284, 288
—, Baltic, 277, 386
—, catches, 290
—, distribution, 280
—, gulf, 277, 278, 279, 286
—, maximum age, 288
—, mortality, 290
—, sea, 277, 278, 279
—, spawn, 283
—, spring, 276, 277, 278, 282, 283, 284, 285
Heterocapsa triquetra, 225
Heterocope, 243
Heterotrophic benthic subsystem, 257
Hiiumaa, 2, 44, 46, 84
Hoburg Bank, 84
Hogland, 2, 33
Holeurysaline species, 248
Holocene, 4
Holoplankton, 238
Holsteinian Interglacial, 54
Homeoosmotic species, 248
Homing of fishes, 279, 295, 306, 338
Humic substances, 211, 212
Humus, 103, 213
Hyalodiscus scotius, 62
Hydrobia, 256, 334
Hydrobia neglecta, 255
Hydrobia ulvae, 255
Hydrobia ventrosa, 255
Hydrobiidae, 255
Hydrocarbons, 105, 107
Hydrogen sulphide, 102, 185, 210, 357, 358, 372
Hydroid, 369
Hydrolittoral, 231, 233, 235
Hydrology, 123—133
Hydromorphology, 123—124
Hydroxylamine, 199
Hyperoplus lanceolatus, *see* sandeel

Iaera albifrons, 257
Ice
— conditions, 174—175
— cover, 219
—, marginal positions, 55, 56, 57
ICES, 386
Ictaluridae, 334
Ictalurus nebulosus, 334
Ide, 334, 335, 340, 341
Idotea, 257
Idotea granulosa, 329
Igneous rocks, 1
Ilmenite, 117
Inertial motion, 152, 153, 156, 161
Inflow, 133
—, mean, 130
—, river water, 123, 125—126
—, salt water, 136, 143
Insects, 291
Insolation, 251
—, *see also* Heat balance
Interglacial deposits, 54
Internal waves, 154, 158, 160, 161
International Baltic Sea Fishery Commission, 319, 325, 383
International Baltic Year, 193
International Hydrological Program, 128
Interstitial water, 210
Iron, 102, 104, 107, 111, 113, 204
—, Fe/P ratio, 103
—, Mn/Fe ratio, 111, 114
— oxyhydrate, 107, 109
— phosphate, 196, 207
— in seston, 205
— sulphide, 59, 103, 208, 209
Irradiance, 169, 170
Isoëtes lacustris, 236
Isopoda, 257, 260, 262, 329
Isopycnals, 158
Isostatic rebound, 58

Jotnian, 1, 2, 25, 28, 29, 30, 36, 43
Jurassic, 36, 52, 53

Kalmar Strait, 44
Karelian Ice Sea, 58
Kattegatt, 123
Keratella, 243, 248
Keratella cochlearis recurvispina, 240, 242
Keratella cruciformis eichwaldi, 240
Keratella quadrata platei, 240, 242
Keratella quadrata quadrata, 240
Keuper, 50
Kinorhyncha, 262
Klint, 44, 84
Knudsen relations, 130, 131, 132, 139

Lamellibranch, 260, 261
Laminaria digitata, 231
Laminaria saccharina, 231
Lampetra fluviatilis, *see* Lamprey

Lampetra fluviatilis planeri, *see* Lamprey, brook
Lamprey, 292, 300, 308—309, 334
—, brook, 308
Land uplift, 61, 86, 95
Landsort Deep, 1, 34, 36, 43, 69, 84
Laomedea loveni, 369
Larus argentatus, *see* Gull
Larus fuscus, *see* Gull
Larvae of fishes, 281—283, 285, 288, 292, 295, 312, 315, 332
Late Cenozoic (Pleistocene) Glaciation, 293
Lawicka Slupska, 84
Lead, 102, 113, 204—206
Lead-210 dating, 100, 101
Length of fishes, *see* Growth of fishes
Leucoxene, 117
Leuciscus cephalus, *see* Chub
Leuciscus idus, *see* Ide
Leuciscus leuciscus, *see* Dace
Lias, 52
"Light blue earth", 114, 115
Light intensity, 223
Lignin
—, input of, 355
—, sulphonates, 213
Limanda limanda, *see* Dab
Limestone, 25, 31, 43, 45, 50, 52, 53, 105
Limiting factor, 361
Limnaea Sea, 69
Limnetic plants, 256
Limnocalanus, 285
Limnocalanus marcurus, 240, 242, 243, 245, 248
Limosella aquatica, 235
Lindane, 362, 363
Lineaments, 4, 13—20, 23
Liparis liparis, *see* Sea snail
Litorina Sea, 61, 62, 63, 100, 254, 275
Littoral
— system, 219
— zones, 233, 256—257
Littorina littorea, 63, 257
Littorina saxatilis, 257
Llandovery, 46
Lontova
— beds, 4, 32
— stage, 43
Lota lota, *see* Burbot
Lota vulgaris, 298
Ludlow, 46, 47
Lükati
— beds, 4, 32
— stage, 43
Lumparn, 31, 45
Lumpenus lampretaeformis, *see* Snake blenny
Lumpsucker, 332
Lymnaea, 257, 307

Mackerel, 292, 360
Macoma baltica, 240, 261, 262, 322, 324, 326, 328
Macrofauna, *see* Bentic
Madreporaria, 256
Magnesium, 184, 186
Magnetite, 117
Major constituents, 183
Manganese, 107, 109, 111, 113, 114, 204, 209, 210
Marcasite, 209
Marine pollution, definition of, 351
Mass-balance, 101, 366
Mastogloia elliptica, 62
Mastogloia Sea, 63, 67
Mastogloia smithii, 62
Maturity of fishes, 283, 286
Medusa, 242, 253
Meiofauna, *see* Benthic
Melosira arctica, 223, 227
Melosira arenaria, 62
Melosira islandica, 62
Melosira juergensi, 62
Melosira moniliformis, 62
Membranoptera alata, 231
Mercierella enigmatica, 373
Mercury, 206, 367, 388
Merganser, 295
Mergus merganser, *see* Goosander
Mesidotea entomon, 255, 260, 262, 316, 317
Mesoplankton, 238
Mesozoic, 4, 43, 50—53, 106
Metachronous, 63
Metals, 365—369
—, heavy, 102, 368
—, mass-balance for, 366
—, salts, 367
Metamorphic rocks, 1
Methane formation, 210
Microdeutopus gryllotalpa, 257
Midsjö Bank, 84
Migration of fishes, 281, 293, 307, 309, 310, 312—313, 320, 324, 325, 338

Military waste, 373—374
Miller's thumb, 334
Mines, 369, 374
Minnow, 334, 369
Mixing, 159, 161, 167
— conditions, 160—162
—, horizontal, 162—167
— rates, 163
—, vertical, 149, 162—167
Mollusca, 308, 328
Molybdenum, 113, 206, 209
Monostroma grevillei, 231
Moraine
—, end-, 54
—, ground-, 54
Mortality, 290—291, 295, 300
Mottlefoot sculpin, 334
Mud
—, laminated, 68
—, post-glacial, 65, 68
—, recent, 102, 103
Muhos-formation, 2, 23, 24
Muschelkalk, 50
Mustard gas, 374
Mya, 262
Mya arenaria, 240, 322, 324
Mycella bidentata, 329
Myoxocephalus quadricornis, 275
Myoxocephalus scorpius, *see* Father lasher
Myriophyllum spicatum, 234
Mysidacea, 257, 285, 292, 317
Mysis, 306, 307
Mysis mixta, 245, 255, 316, 322, 329
Mysis relicta, 245, 255
Mytilus, 262, 324
Mytilus edulis, 63, 232, 240, 242, 256, 257, 322

Natural resources, 105—117
Nauplii, 253
Navicula digitoradiata, 62
Navicula elegans, 62
Navicula peregrina, 62
Nemacheilus barbatulus, *see* Stoneloach
Nematoda, 257, 262
Neogene, 53, 115
Neotectonic, 4, 23
Nephthys ciliata, 322
Nerophis ophidion, 292
Niche diversification, 255
Nickel, 113, 114, 206
Nitrification, 198
Nitrogen, 361
—, circulation of, 197
— compounds, 198
— cycle, 226
— fixation, 226
—, molecular, 198, 199, 226
—, N:P ratio, 227
—, nitrate-nitrogen, 197, 200, 223, 224, 226
—, nitrite-nitrogen, 198, 200, 224
—, organically bound, 197
—, total, 199, 200, 354, 355
Nitzschia circumsuta, 62
Nitzschia (fragilariopsis) cylindrus, 223
Nitzschia frigida, 223
Nitzschia navicularis, 62
Nitzschia punctata, 62
Nitzschia tryblionella, 62
Nodularia spumigena, 226, 228
Non-deposition, 86, 95
Nutrients, 194—203, 219, 251, 357, 358, 359

Ocdogoniaceae, 230
Octocorallia, 256
Odonata, 256
Oikopleura dioica, 240
Oil
— accidents, 370
— breakdown, 369
— deposits, 106
— and gas, perspectives, 105
— pollution, 369—371
Oithona similis, 240, 248, 251
Öland, 2, 23, 44, 46, 97, 105
Oligochaeta, 257
Oncocottus quadricornis, *see* Four-horned cottus
Onos cimbrius, 225
Oocytes, 225
Optical properties, 167—170
Ordovician, 25, 28, 36, 43, 45—46
— Klint, 45, 46
— Sea, 45
Organic matter, 357
— — content, 68
— —, dissolved, 211—213
— —, particulate (POM), 251
— —, production of, 372
— —, *see also* Carbon
Organochlorines, 362—364
—, *see also* DDT, Polychlorinated biphenyls

Oscillatoria agardhii, 229
Osmeridae, 333
Osmerus eperlanus, *see* Smelt
Ostracoda, 262, 329
Outflow, 123, 133
Overfertilization, 360
Oxidation of sulphides, 211
Oxidizing conditions, 358
Oxygen
—, chemical oxygen demand (COD), 212
— content, 188—194, 358
— —, medium fluctuations, 193
— —, short-term fluctuations, 193, 194
— deficiency, 103, 188
— depletion, 361
— distribution, 190
— utilization, 354—359

Paleocene, 53
Paleogene, 53, 114, 115
Paleozoic, 3, 25, 28, 31, 43, 84, 85
Parasites, 291, 372
Parr, 293, 295, 296, 297, 299, 301, 302, 305, 306
Particle distribution, 168
Particulate organic matter (POM), 251
PCB, *see* Polychlorinated biphenyl
Pelagial system, 219
Pelagic fishes, 276—292
Pelecus cultratus, 334
Pelvetia canaliculata, 231
Perca fluviatilis, *see* Perch
Perch, 334, 335, 341, 372
Percidae, 334
Periphyton, 237
Permian, 36, 48—50, 106
Petroperspective, 105, 106
Phaeophyceae, 231
Phanerozoic, 3
Pholis gunellus, *see* Butterfish
Phosphorus, 102, 103, 104, 113, 114, 200, 227, 361
— content in sediments, 209
—, Fe/P ratio, 103
—, phosphate, 195, 196, 221, 223, 224, 226, 358
—, total, 224, 354, 355
Phosphorite, 112, 114, 116
Photosynthetic sulphur bacteria, 211
Phoxinus phoxinus, *see* Minnow
Phragmites australis, 256
Phycodrys rubens, 231
Phyllophora, 231, 234
Physical oceanography, 135—175
Physical pollution, 372—375
Phytoplankton, 220—229, 251
—, bloom, 224, 227
—, dynamics of phytoplankton community, 251—254
—, *see also* Primary production
Pike, 295, 333, 335, 338, 340, 341
Pike-perch, 334, 335, 336, 338, 339, 340, 341
Pilayella littoralis, 231, 232
Placer deposits, 116
Plaice, 323—325, 360
Plankton
—, larvae, 239
—, nets, 239
—, *see also* Phytoplankton, Zooplankton
Platichthys flesus, *see* Flounder
Pleistocene, 4, 24
Pleopsis polyphemoides, 240, 242, 243, 245, 248, 249, 250, 251, 253
Pleurobrachia pileus (larvae), 240, 245, 248
Pleuro platessa, *see* Plaice
Plutonium, 371
Podon, 243, 251, 285
—, *see also Pleopsis polyphemoides*
Podon intermedius, 240, 245, 248
Podon leuckarti, 240, 253, 304
Poikilo-osmotic species, 248
Pollen zonation, 58
Pollution, 102, 351—375
—, accumulation of, 352
—, coastal waters, 361
—, effects on ecosystems, 359
—, effects on fish, 361
—, global aspects of, 351
—, mercury, 367, 368
—, metals, 365
—, oil, 369—371
—, physical, 372
—, radioactive, 371
—, spread of, 352
Polychaeta, 254, 261, 316, 322, 324, 328, 329, 372
Polychlorinated biphenyl (PCB), 102, 362, 363, 364, 373, 387
Polyides rotundus, 231
Polyps, 257
Polysiphonia nigrescens, 231, 232
Pontoporeia affinis, 255, 260, 261, 262, 316, 322
Pontoporeia femorata, 255, 261, 262, 316

Porifera, 238, 256
Porpoise, 300
Postsmolts, 297—298
Potamogeton, 256
Potamogeton filiformis, 234, 235
Potamogeton gramineus, 236
Potamogeton perfoliatus, 235
Potamogeton pusillus, 235
Potamogeton vaginatus, 234, 235
Potamopyrgus, 255
Potamopyrgus jenkinsi, 255
Potassium, 184, 186
Prasinophyceae, 223, 225
Precipitation, 126—129
Pressure gradient force, 157
Prey-predator relationship, 251, 253
Priapulid, 255
Primary production, 220—238, 251, 253, 254
— —, littoral, 236
Prorocentrum balticum, 225
Prorocentrum micans, 227
Proterozoic, 1, 3, 5—8, 23, 24, 25, 29, 33, 36, 43
Protoperidinium brevipes, 225
Protoperidinium pellucidum, 228
Proximal varves, 64, 66
Prymnesiophyceae, 223
Pseudocalanus, 285
Pseudocalanus m. elongatus, 240, 242, 243, 245, 246, 247, 248, 250, 251, 252, 254
Pungitius pungitius, *see* Ten-spined stickleback
Pygospio elegans, 240
Pyramimonas, 225
Pyrite, 209

Quark, 86
Quartzite, 43
Quaternary deposits, 87—94

Radiance distribution, 169
Radiation, 173
Radiactive pollution, 371
Radionuclides, 212
Rapakivi-granite, 1, 5—8, 29
Redbreasted merganser, 295
Redox potential, 101, 209
Reducing environment, 107, 358
Reef
—, algal, 45
—, bank, 47
— barrier, 47
—, coral, 46
—, formation of, 46, 47
—, limestone, 106
— structure, 45
— —, oil bearing, 106
Regression, 63, 69, 86, 95
Relict, 330
Reproduction of fishes, 313, 320, 324, 326
Residence time, 139, 160, 352
Rhabdonema arcuatum, 62
Rhaetian, 50, 51
Rhinonemus cimbrius, *see* Four-bearded rockling
Rhizophytes, 234, 235
Rhizosolenia, 228
Rhizosolenia calvar avis, 62
Rhodochorton purpureum, 231
Rhodochrosite, 209
Rhodomela confervoides, 231, 232
Rhodophyceae, 231
Rhopaloidia musculus, 62
Riphean, 1, 2
Rithropanopeus harrisi, 311
River life, 295
Rivularia atra, 235
Roach, 334, 335, 340, 372
Rotatoria, *see* Rotifera
Rotifera, 242, 243, 245, 251, 253, 257
Rudd, 334, 335
Ruff, 334, 335
Run-off, 123, 129
— regulations, 126
Ruppia maritima, 234
Rutile, 117
Rutilus rutilus, *see* Roach

Saaremaa, 2, 46, 48, 84
Sagitta elegans baltica, 240, 248, 251
Sagitta setosa, 240
Salinity, 135—139
— distribution, 135, 137, 150
—, mean, 146
—, minimum, 142
—, S/Cl relationship, 183
—, stratification, 135, 143
—, variations, 143—149
Salmo gairdneri, *see* Trout, rainbow
Salmo salar, *see* Salmon, Atlantic
Salmo trutta, *see* Trout
Salmon, 293—303
—, Atlantic, 292, 297, 334

Salmonella, 360
Salmonidae, 333, 334, 335
Salpausselkä end moraine, 56
Salt deposits, 48
Salvelinus alpinus, *see* Arctic char
Salvelinus namaycush, *see* Trout, lake
Sambian Peninsula, 2, 53, 114, 115
Sand goby, 331—332
Sand and gravel, submarine, 116, 374—375
Sandeel, 328, 332
Sandstone, 1, 24, 25, 30, 31, 48, 52, 105
Sarsia tubalosa, 240
Satakunta sandstone, 1, 25
Scania, 2, 50, 52, 53, 105
Scaphopoda, 256
Scardinius erythrophthalmus, *see* Rudd
Scattering of light, 168—169
Scenedesmus, 225
Schist, 23
Scientific Committee on Oceanic Research (SCOR), 387
Scoloplos armiger, 261, 262, 324
Scomber scombrus, *see* Mackerel
Scophthalmus maximus, *see* Turbot
Scophthalmus rhombus, *see* Brill
Scytosiphon lomentaria, 231
Sea cucumber, 329
Sea level, *see* Water level
Sea scorpion, 330
Sea snail, 332—333
Seal, 301, 363
Secondary production, 253, 254
Sediment-water interactions, 205—211
Sedimentation
—, active 95, 101
—, non-, 109
—, post-glacial, 67
— rate, 67, 100, 101
—, re-, 100
—, retarded, 86
Sediments
—, Ancylus, 64
—, content of heavy metals, 206, 209, 210
—, C/N ratio, 200
—, lag, 97
—, late-glacial, 63, 85, 86
—, Litorina, 64
—, methane formation, 210
—, pH of, 209, 210
—, phosphorus content, 209
—, post-glacial, 63, 85, 86
—, Quaternary, 86, 97, 99
—, redeposited, 97
—, varved, 55, 59, 67
Seiches, 154, 155, 159
Seismicity, 22
Seston, 205
Sewage discharge, 354
Shale, 105
—, alum, 45, 105
—, bituminous, 107
Sheatfish, 334
Shoreline displacement, 58
Shores,
—, exposed, 235
—, rocky, 232, 235
—, *see also* Coastal
Shrimp, 262, 318, 322, 326, 329, 371
Silicate, 201, 202, 203, 223, 224
Sill depth, 369
Siltstone, 1, 31, 105
Silurian, 36, 46—48
—, Klint, 46
Siluridae, 334
Silurus glanis, *see* Sheatfish
Skeletonema costatum, 223, 227, 228
Smelt, 333, 335, 340, 341
Smolt, 293, 297—298, 300, 301, 306
Snake blenny, 317, 329
Sodium, 184, 186
Solid waste, 373
Sound, *see* Danish Sounds
Sound scattering, 161
Spawning, 281—283, 286, 295, 305, 306, 308, 309, 313—315, 320—321, 327, 328, 331, 332, 333, 338
Spawning migration, 293, 303
Spectral measurements, 169
Sphacelaria arctica, 231, 232
Spinachia spinachia, *see* Fifteen-spined stickleback
Spined loach, 334
Spongomorpha pallida, 231, 232
Sprat, 276, 277, 279, 280, 281, 283, 284, 285, 287, 288, 290, 291, 298, 316, 386
Sprattus sprattus, *see* Sprat
Spring flood, 126
Stability, 149, 150
Stagnation, 101, 102, 143, 190, 191
Stenotherm, 245
Stephanodiscus astraea, 62
Stictyosiphon tortilis, 231
Stizostedion lucioperca, *see* Pike-perch

Stoneloach, 334
Stratification of water, 189
Strontium, 186
Strontium-90, 371
b-Cambrian peneplane, 1, 27, 31, 33—43, 45
b-Jotnian peneplane, 30
ublittoral, 231, 233
zone, 257—262
ıbularia aquatica, 235
ulphate, 185, 186
Sulphate-reducing bacteria, 210
Sulphide, mono-, black-, 69
Surf zone, 162
ırface layer, 138
ecofennian, 23
vecokarelian, 1
Symmict, 65
ynchaeta, 240, 242, 243, 247, 251, 253
ynchaeta baltica, 243
Synchronous deposition, 63
Synedra crystallina, 62
Synedra tabulata, 62
Syngnathus typhte, 292

Tagging of fishes, 301, 309, 311, 312, 314, 320, 324
Tarulus bubalis, see Sea scorpion
Tectonic
- activity, 53
— depressions, 1
—, *see also* Neotectonic
Temora longicornis, 240, 242, 243, 245, 247, 248, 250, 252, 254, 285
Temperature, 139—143
—, distribution, 137
—, long-term variations, 143—149
Ten-spined stickleback, 334, 335
Tench, 334
Terbellides strömi, 322
Tertiary, 4
Thalassionema nitzschioides, 62
Thalassiosira baltica, 62, 223, 227
Thalasiosira lacustris, 223
Theodoxus, 307
Theodoxus fluviatilis, 255, 257
Thermal regime, 173
Thermocline, 141, 142, 143, 145, 224
Three-spined stickleback, 334, 335
Thymallidae, 333
Thymallus thymallus, see Grayling
Tides, 69, 151, 154
Time-dependent motion, 151—155, 158
Tin, 206
Tinca tinca, see Tench
Titanium, 113
Tornqvist-line, 3, 4, 33, 36
Toxic matter, 362—369
Trace metals, 203—205
— —, *see also* Heavy metals
Transgression, 69, 86
Transition zone, 123
Transparency, 168
Triassic, 36, 50—51, 52
Trichoptera, 256, 257
Tritium, 371
Trophic
— relations, 263
— status, 237—238
Trophogenic layers, 219
Trough, 69
Trout, 300, 303, 305
—, brown (sea), 292, 303, 334
—, lake, 334
—, rainbow, 334
—, sea-running brown, 303—306
—, sea, 299, 303, 305, 306, 334
Tunicata, 238
Turbellaria, 257, 262
Turbot, 325—327, 331, 332
Tyndall measurements, 168

Ulothrix subflaccida, 231
Ulva lactuca, 231
Uppsala Esker, 97, 99
Upwelling, 157, 158, 163, 220
Urea, 199
Uroglena americana, 225
Urospora penicilliformis, 231, 232

Valvata, 307
Vanadium, 113, 206
Varved clay chronology, 57, 59
Vegetation, 220—238
Vendace, 333, 335, 336, 339, 340, 341
Vendian, 1, 31, 32, 36, 43, 105
Vimba, 292, 299, 307—308, 334, 339
Vimba vimba, see Vimba

Warm-water effects, 372—373
Waste discharge, 220, 353—354
Water
— balance, 123—133
— colour, 170, 211
—, exchange of, 133
— level, 152, 155

—, storage of, 133
Waves, 154, 375
— action, 232
—, "base" of, 69, 95
— surface, 160
—, transport of, 162
Weichselian glaciation, 55, 58
White-tailed eagle, 363
Whitebait, 281, 285, 288
Whitefish, 292, 299, 306—307, 333, 335, 338, 339, 340, 341
—, migratory, 333
—, peled, 334
"Wild earth", 112, 115
Wind
— fetch, 160, 163
— force, 163
— stress, 154
Winter dormancy, 245

Year-classes of fishes, 288—290
Yellow substance, 168, 169, 170, 211, 212
Yoldia arctica, 59
Yoldia Sea, 59, 60, 62
— —, late-glacial, 58, 59
— —, preboreal, 57, 59
Younger Diyas, 56

Zannichellia palustris, 233, 234
Zechstein, 48
Zinc, 102, 113, 204, 206, 209, 365, 369
Zirkon, 117, 206
Zoarces viviparus, *see* Eelpout
Zonation of vegetation, 232
Zoobenthos, 256
—, production of, 262—264
—, utilization of, 262—264
—, see also Benthic fauna
Zooplankton, 238—251, 280, 361
—, biomass, 246, 254
—, carnivorous, 251
—, composition of, 242
—, diel migration, 249
—, distribution of, 240, 249, 250
—, effect of salinity, 247
—, effect of temperature, 245
—, herbivorous, 251
—, macrozooplankton, 239
—, microzooplankton, 239, 244, 251
—, omnivorous, 251
—, oxygen consumption, 248
—, production, 247
— sampling, 239
Zostera marina, 233, 234
Zygnemaceae, 230